Practice and Application of Artificial Intelligence in Urban Decision-Making

Practice and Application of Artificial Intelligence in Urban Decision-Making

Guest Editors

Yile Chen
Junming Chen
Yanyan Liang

MDPI

Basel • Beijing • Wuhan • Barcelona • Belgrade • Novi Sad • Cluj • Manchester

Guest Editors

Yile Chen
Faculty of Humanities
and Arts
Macau University of Science
and Technology
Macau
China

Junming Chen
Faculty of Humanities
and Arts
Macau University of Science
and Technology
Macau
China

Yanyan Liang
Faculty of Innovation
Engineering
Macau University of Science
and Technology
Macau
China

Editorial Office
MDPI AG
Grosspeteranlage 5
4052 Basel, Switzerland

This is a reprint of the Special Issue, published open access by the journal *Buildings* (ISSN 2075-5309), freely accessible at: https://www.mdpi.com/journal/buildings/special_issues/UZMN9BF5U7.

For citation purposes, cite each article independently as indicated on the article page online and as indicated below:

Lastname, A.A.; Lastname, B.B. Article Title. *Journal Name* **Year**, *Volume Number*, Page Range.

ISBN 978-3-7258-5945-0 (Hbk)
ISBN 978-3-7258-5946-7 (PDF)
https://doi.org/10.3390/books978-3-7258-5946-7

Contents

About the Editors

Yile Chen

Yile Chen, Ph.D. in Architecture, is an Assistant Professor at the Faculty of Humanities and Arts, Macau University of Science and Technology. She has published more than 50 papers in SCIE, SSCI, and A&HCI journals. She focuses on architectural technology in high-density urban forms and a healthy environment, as well as the protection and renewal of urban and architectural cultural heritage in Macau and Portuguese-speaking regions. She is currently leading a Macau University of Science and Technology Research Fund (FRG) project focusing on the spatial morphological development of Macau and Timor-Leste. She has also participated in numerous projects, including the Science and Technology Development Fund (FDCT) Science and Technology Innovation Enhancement Program of the Macau Special Administrative Region, the Cultural Affairs Bureau of the Macau Special Administrative Region's Consulting Services, the Macau Special Administrative Region Government Higher Education Fund Research Project on Prevention and Response to Major Infectious Diseases, the China–Portugal Joint Laboratory for Cultural Heritage Conservation "Belt and Road" Open Project, the Zhejiang Provincial Philosophy and Social Sciences Planning Project, the Digital Rural Research Center of Hangzhou City University, and the Guangdong Provincial Department of Education's Key Scientific Research Platform Project for General Universities. She also serves as a peer reviewer for 74 *SCIE/SSCI/A&HCI/ESCI* journals.

Junming Chen

Junming Chen has a multidisciplinary background and engages in interdisciplinary research. He has a background that includes architecture, design, and computer science, and holds a Ph.D. in digital media. Since 2017, he has been working at the Macau University of Science and Technology, focusing on the application and improvement of artificial intelligence in digital media, design, and architecture. He serves as a tutor and Assistant Professor for postgraduate students. He excels in using parametric and AI technologies in his design practice, having mentored students to finish multiple completed projects and won over ten international and national design awards, including the A Design Award and Huacan awards. He is also a certified senior data analyst with the Ministry of Industry and Information Technology. Additionally, his papers have been published in SCI TOP journals, *A&HCI*, *SSCI*, and *CCF* core journals, for a total of more than 30 articles, with more than 20 articles as first author/corresponding author. He serves as a reviewer for more than 30 SCI and SSCI journals.

Yanyan Liang

Yanyan Liang, Ph.D., is an Associate Professor and Doctoral Supervisor. He obtained his Ph.D. in Science from Macau University of Science and Technology in 2009. From 2009 to 2013, he served as a Postdoctoral Researcher at the Space Science Institute of MUST, where he was engaged in data analysis and visualization for the Chang'e-1 and Chang'e-2 lunar exploration missions. Since 2014, he has held positions as an Assistant Professor and Associate Professor at the Faculty of Information Technology and the Macau Institute of Smart City at MUST. In 2012, he received the Third Prize of the Natural Science Award from the Macau Special Administrative Region (ranked third among awardees). His current research focuses on artificial intelligence, image processing, computer vision, machine learning, data mining, computer architecture, parallel computing, and distributed computing. He has published over 60 papers in high-quality domestic and international journals and major international conferences (including *IJCAI, AAAI, CVPR, ICCV, ICML, ACM MM, TPAMI, IJCV, TIP, TIFS, TITS, TCYB, TMM, PR,*

AJ, *IEEE Access*, *MVAP*, *FG*, etc.) and authored one monograph. He has led and completed multiple research projects, including eight projects funded by the Macau Science and Technology Development Fund, one key R&D project in Macau (as sub-project leader), one key national R&D project under the "Strategic Scientific and Technological Innovation Cooperation" program—a joint project with the Macau Science and Technology Development Fund—one international (regional) cooperation and exchange project of the National Natural Science Foundation of China—a joint project with the Macau Science and Technology Development Fund—and one key R&D project in Guangdong Province. As a key participant (non-principal investigator), he has contributed to multiple research projects funded by the Macau FDCT, the National 973 Program, and the National Natural Science Foundation of China.

buildings

MDPI

Editorial

Practice and Application of Artificial Intelligence in Urban Decision-Making

Yile Chen [1], Junming Chen [1,*] and Yanyan Liang [2,*]

[1] Faculty of Humanities and Arts, Macau University of Science and Technology, Avenida Wai Long, Tapai, Macau 999078, China; chenyile@must.edu.mo
[2] Faculty of Innovation Engineering, Macau University of Science and Technology, Avenida Wai Long, Tapai, Macau 999078, China
* Correspondence: jmchen@must.edu.mo (J.C.); yyliang@must.edu.mo (Y.L.)

1. Introduction

With the rapid development of science and technology, artificial intelligence (AI) is gradually penetrating into all aspects of urban development, especially in the field of urban decision-making. Its application in urban planning decision-making has broad prospects [1,2]. In particular, new technologies such as big data intelligence [3,4], human–machine hybrid enhanced intelligence [5], internet group intelligence [6–8], and cross-media intelligence [9,10] will play an important role in expanding decision-making information, improving decision-making technology, enriching decision-making subjects, and strengthening decision-making feedback. When processing urban big data, artificial intelligence is mainly used for problem solving and decision analysis, including, first, how to use computer-aided design, and second, how to extract patterns and logic from the decision-making practice process [11]. It has brought new ideas and methods to urban planning, construction and management, and helped promote the practice and application of artificial intelligence in urban decision-making.

Driven by this development, this Special Issue of *Buildings* brings together 12 original papers that collectively illustrate how artificial intelligence can be applied in analysis, management, and decision-making in urban planning, spatial planning, and architectural space design. These papers cover a wide range of topics, including architectural spaces or urban park scenes in streetscapes, building scenes in remote sensing imagery, energy efficiency analysis of building carbon neutrality or urban block morphology, university outdoor spaces, highway projects, traditional garden heritage, accessible restroom design, high-density urban built environments, the price of building materials, and smart city construction technologies.

2. Overview of Contributions

This Special Issue primarily covers practical cases from South Korea and China, as well as specific urban case studies from Macau, Wuhan, Suzhou, Fuzhou, and Xi'an. This provides a reference for empirical analysis of AI in urban decision-making across cities of different scales and types.

Artificial intelligence has a rich and diverse technical framework, encompassing numerous methods such as machine learning, deep learning, and natural language processing. However, this Special Issue focuses on the practical needs of urban planning and architectural space, centering on the following key models and technologies:

- ResNet-50 (Residual Network)
- Multi-Scale Hybrid Dual-Attention Network (MS-HDAN)

- Light Gradient Boosting Machine (LightGBM)
- Mask2former
- Random Forest (RF)
- Gradient Boosting Regression (GBR)
- Extreme Gradient Boosting (XGBoost)
- Artificial Neural Networks (ANNs)
- Deep Neural Networks (DNNs)
- Conditional Generative Adversarial Network (CGAN)
- Support Vector Machine (SVM)
- Real-Coded Accelerating Genetic Algorithm Projection Pursuit Model (RAGA-PPM)
- Shapley Additive Projection for Interpretability (SHAP)
- Bidirectional Encoder Representations from Transformers (BERT)
- Swin Transformer
- Non-Dominated Sorting Genetic Algorithm II (NSGA-II)
- Complete Ensemble Empirical Mode Decomposition with Adaptive Noise (CEEMDAN)
- Autoregressive Integrated Moving Average (ARIMA)
- Variational Mode Decomposition (VMD)
- Gated Recurrent Unit (GRU)

By employing these technologies, they offer tangible methods and examples for enhancing urban decision-making via artificial intelligence.

2.1. Prediction of Construction or Engineering Projects

Contribution 1 (Kim, J.-S.) developed and evaluated machine learning models using artificial neural networks (ANNs) and deep neural networks (DNNs) on a structured dataset of 150 completed national highway projects in South Korea, covering the planning and design phases. This database focuses on 19 high-impact sub-task types to reduce noise and improve prediction accuracy. This provides a reference for predicting expected life (EL) and expected life (CC) during the planning and design phases of national highway projects.

In contribution 2 (Guo et al.), an innovative forecasting model, CEEMDAN-VMD-GRU-ARIMA, was proposed. This model combines one-dimensional price series analysis with machine learning and data mining techniques to specifically predict the price of prestressed steel bars. This is crucial for effectively controlling construction project costs and managing risks.

2.2. Energy Efficiency Analysis

Contribution 3 (Wang et al.) developed an integrated prediction and optimization tool, combining a parametric design tool (Grasshopper), Latin Hypercube Sampling (LHS), Pearson correlation analysis, and a machine learning regression model, to investigate the impact of design parameters on heating and cooling energy consumption and photovoltaic power generation in 15 urban blocks in Wuhan. By systematically analyzing key design factors such as block building layout, orientation, proportion of low-rise buildings, and plot width, this study quantified the complicated interaction between urban form and energy performance, providing a scientific basis for optimizing urban block energy efficiency and promoting sustainable development. It also offers a scientific digital optimization framework for block planning.

Contribution 4 (Feng et al.) uses EnergyPlus and DesignBuilder for building energy simulation, applies the LightGBM model for machine learning, uses a case study of 32 buildings in Shenzhen, Hong Kong, and Guangzhou, and evaluates policies using the PEI. They embed building-scale performance analysis into the larger urban policy context of the Greater Bay Area to gain insights into how green building activities can provide quantifiable, multi-scale benefits to China's carbon neutrality obligations.

2.3. Generative Design for Floor Plans

Contribution 5 (Yan et al.) has developed an intelligent generative framework that utilizes conditional generative adversarial networks (CGANs). This paper analyzes the core characteristics of the four main spatial types of Suzhou gardens: rockery and sculpture spaces, water island spaces, pavilion and corridor spaces, and plant spaces. Based on these characteristics, a deep learning architecture is designed. Finally, the proposed model is applied to design decision generation. This paper proposes a machine learning-assisted design process, a multi-alternative comparison model, and space utilization evaluation, providing designers with efficient and multi-dimensional design decision-making.

2.4. Building Instance Segmentation in Remote Sensing Images

Contribution 6 (Hu et al.) proposes a novel Multi-Scale Hybrid Dual Attention Network (MS-HDAN), which integrates a two-stream encoder, multi-scale feature extraction, and a hybrid attention mechanism. Extensive experiments on a benchmark urban building dataset demonstrate that MS-HDAN significantly outperforms existing state-of-the-art methods, particularly when dealing with densely distributed and structurally complex buildings. This framework can provide valuable support for applications such as BIM-based modeling, GIS data updates, and urban spatial monitoring. Potential users include urban planners, urban developers, and agencies involved in disaster response and land use management.

2.5. Street View Imagery and Spatial Analysis

Contribution 7 (Tian et al.) used Xi'an as a case study and constructed a multidisciplinary research framework encompassing "street view imagery feature extraction, spatial heterogeneity modeling, and planning strategy optimization". Using the ResNet50 deep learning model and the ADE20K dataset, they accurately extracted ten microscale-built environment factors from tens of thousands of street view images. Combined with a high-resolution, high-quality ground-based PM 2.5 dataset from China, they employed ordinary least squares (OLS), geographically weighted regression (GWR), and multiscale geographically weighted regression (MGWR) models to systematically reveal the impact of the microscale-built environment on PM 2.5 concentrations.

Contribution 8 (Zhuang et al.) used street view images of the Fujian Agriculture and Forestry University (FAFU) campus obtained from Baidu Maps and employed a Mask2former-based image semantic segmentation method to extract landscape features. A questionnaire survey was also conducted to obtain student scores on three psychological indicators: psychological recovery, emotional enhancement, and social interaction. Subsequently, a decision tree algorithm combined with the Shapley Additive Interpretation (SHAP) mechanism was used to identify key landscape features that significantly influenced each psychological outcome and analyze their impact patterns and threshold effects.

Contribution 9 (Chen et al.) proposed a model for urban park perception and understanding based on multimodal graph-image data and a bidirectional attention mechanism. By integrating text and image data, combining a BERT-based text feature extractor, a Swin Transformer-based image feature extractor, and a bidirectional cross-attention fusion module, and applying the Shapley Additive Explanation (SHAP) method, this model can more accurately assess and identify key factors that influence visitors' emotional experiences, providing a scientific basis for urban park management and optimization.

2.6. Built Environment Perception and Optimization Technology

It is worth noting that with the acceleration of urbanization, the construction of smart cities has become a global focus, and machine learning technology plays a vital role in this. This also provides analytical techniques worth referring to and learning from for the urban built environment.

Contribution 10 (Luan et al.) conducted a bibliometric analysis of published research on smart cities and machine learning, using visualization techniques to reveal their spatiotemporal distribution patterns, research hotspots, and collaborative network structures. They noted that issues such as data privacy, cybersecurity, sustainable development, and intelligent transportation are becoming increasingly prominent, reflecting the dual demands of smart city development for technological innovation and green growth. This represents a key application direction and future development trend for smart city and machine learning research.

Contribution 11 (Pan et al.) quantified the intensity of human activities through Baidu heat maps, analyzed social interaction patterns using social media check-in data, and integrated built environment elements such as road network topology, three-dimensional building form, and spatial distribution of points of interest (POIs). They used a machine learning technique that combined the real-coded accelerated genetic algorithm–projection pathfinding model (RAGA-PPM) and the Shapley additive projection method (SHAP) of interpretable analysis (IPA) to explore the nonlinear mechanisms of 17 factors affecting the urban vitality of the Macau Peninsula.

Contribution 12 (Chen et al.) developed an innovative approach to accessible restroom design, integrating a Kansei Engineering (KE)–Rough Set Theory (RST)–Support Vector Machine (SVM) workflow to prioritize both inclusivity and sustainability for people with disabilities. They identified nine core emotional descriptors of accessible restroom use and extracted three representative emotional dimensions using factor analysis and clustering. Morphological analysis and RST simplification were then used to refine eight design features into six key elements. This repeatable, data-driven approach significantly benefits the alignment of emotional and functional goals in inclusive restroom design and can be expanded to other accessible facilities.

3. Outlook

This Special Issue presents 12 achievements in artificial intelligence and urban decision-making, which are expected to promote further developments in the fields of urban planning and architectural space and inspire future research.

In the future, the application of AI in urban planning and architectural space will continue to deeply integrate multi-source data and intelligent algorithms to create more accurate urban simulation and decision-making systems. From macro-level urban optimization to micro-level architectural space design, AI can more efficiently balance functionality, esthetics, and resource utilization, helping to build smarter, more inclusive, and more sustainable cities, creating more habitable living and working spaces for people.

Author Contributions: Conceptualization, Y.C. and J.C.; methodology, Y.C.; software, Y.C.; validation, Y.C.; formal analysis, Y.C.; investigation, Y.C.; resources, Y.C.; data curation, Y.C.; writing—original draft preparation, Y.C.; writing—review and editing, Y.C. and J.C.; visualization, Y.C.; supervision, Y.L.; project administration, Y.C. and J.C.; funding acquisition, Y.C. and J.C. All authors have read and agreed to the published version of the manuscript.

Funding: This research was funded by the (1) Faculty Research Grants funded by Macau University of Science and Technology (FRG-MUST grant number: FRG-25-064-FA; FRG-25-067-FA); (2) Guangdong Provincial Department of Education's key scientific research platforms and projects for general universities in 2023: Guangdong, Hong Kong, and Macau Cultural Heritage Protection and Innovation Design Team (grant number: 2023WCXTD042); (3) and Guangdong Provincial Philosophy and Social Sciences Planning 2025 Lingnan Cultural Project (grant number: GD25LN30). The funders had no role in study conceptualization, data curation, formal analysis, methodology, software, decision to publish, or preparation of the manuscript. There was no additional external funding received for this study.

Institutional Review Board Statement: Not applicable.

Informed Consent Statement: Not applicable.

Data Availability Statement: Not applicable.

Acknowledgments: We sincerely thank the authors who contributed to this Special Issue and submitted high-quality manuscripts. We also thank the reviewers for their responsible and rigorous feedback, which has greatly improved each manuscript. Finally, we sincerely thank the editorial team of *Buildings* for their professional support, which ensured the success of this Special Issue.

Conflicts of Interest: The Guest Editors declare that they have no known competing financial interests or personal relationships that could have appeared to influence the work reported in this paper.

List of Contributions:

1. Kim, J.-S. AI-Powered Forecasting of Environmental Impacts and Construction Costs to Enhance Project Management in Highway Projects. *Buildings* **2025**, *15*, 2546. https://doi.org/10.3390/buildings15142546.
2. Guo, Z.; Luo, Y.; Yi, T.; Jing, X.; Ma, J. Harnessing the Power of Improved Deep Learning for Precise Building Material Price Predictions. *Buildings* **2025**, *15*, 873. https://doi.org/10.3390/buildings15060873.
3. Wang, R.; Huang, Y.; Zhang, G.; Yang, Y.; Dong, Q. Optimizing Urban Block Morphology for Energy Efficiency and Photovoltaic Utilization: Case Study of Wuhan. *Buildings* **2025**, *15*, 1118. https://doi.org/10.3390/buildings15071118.
4. Feng, X.; Xiang, F.; Liao, C. Climate Adaptability and Energy Performance in the Greater Bay Area of China: Analysis of Carbon Neutrality Through Green Building Practices. *Buildings* **2025**, *15*, 3066. https://doi.org/10.3390/buildings15173066.
5. Yan, L.; Zheng, L.; Jia, X.; Zhang, Y.; Chen, Y. Machine Learning in the Design Decision-Making of Traditional Garden Space Renewal: A Case Study of the Classical Gardens of Jiangnan. *Buildings* **2025**, *15*, 2401. https://doi.org/10.3390/buildings15142401.
6. Hu, Q.; Peng, Y.; Zhang, C.; Lin, Y.; U, K.; Chen, J. Building Instance Extraction via Multi-Scale Hybrid Dual-Attention Network. *Buildings* **2025**, *15*, 3102. https://doi.org/10.3390/buildings15173102.
7. Hu, T.; Wu, K.; Wu, Y.; Wang, L. Spatially Heterogeneous Effects of Microscale Built Environments on $PM_{2.5}$ Concentrations Based on Street View Imagery and Machine Learning. *Buildings* **2025**, *15*, 3721. https://doi.org/10.3390/buildings15203721.
8. Zhuang, X.; Tang, Z.; Lin, S.; Ding, Z. Prediction and Optimization of the Restoration Quality of University Outdoor Spaces: A Data-Driven Study Using Image Semantic Segmentation and Explainable Machine Learning. *Buildings* **2025**, *15*, 2936. https://doi.org/10.3390/buildings15162936.
9. Chen, K.; Lin, X.; Xia, T.; Bai, R. Research on Park Perception and Understanding Methods Based on Multimodal Text–Image Data and Bidirectional Attention Mechanism. *Buildings* **2025**, *15*, 1552. https://doi.org/10.3390/buildings15091552.
10. Luan, B.; Feng, X. Machine Learning for Resilient and Sustainable Cities: A Bibliometric Analysis of Smart Urban Technologies. *Buildings* **2025**, *15*, 1007. https://doi.org/10.3390/buildings15071007.

11. Pan, C.; Guo, J.; Li, H.; Wu, J.; Qiu, N.; Wu, S. Study on the Influence Mechanism of Machine-Learning-Based Built Environment on Urban Vitality in Macau Peninsula. *Buildings* **2025**, *15*, 1557. https://doi.org/10.3390/buildings15091557.
12. Chen, Z.; Tian, J.; Zhou, H.; Wu, D. Toward Symmetry in Accessible Restrooms Design: Integrating KE, RST, and SVM for Optimized Emotional-Functional Alignment. *Buildings* **2025**, *15*, 1567. https://doi.org/10.3390/buildings15091567.

References

1. Sanchez, T.W.; Shumway, H.; Gordner, T.; Lim, T. The prospects of artificial intelligence in urban planning. *Int. J. Urban Sci.* **2023**, *27*, 179–194. [CrossRef]
2. Yigitcanlar, T.; Kankanamge, N.; Regona, M.; Ruiz Maldonado, A.; Rowan, B.; Ryu, A.; Desouza, K.C.; Corchado, J.M.; Mehmood, R.; Li, R.Y.M. Artificial Intelligence Technologies and Related Urban Planning and Development Concepts: How Are They Perceived and Utilized in Australia? *J. Open Innov. Technol. Mark. Complex.* **2020**, *6*, 187. [CrossRef]
3. Kamrowska-Załuska, D. Impact of AI-Based Tools and Urban Big Data Analytics on the Design and Planning of Cities. *Land* **2021**, *10*, 1209. [CrossRef]
4. Hosen, B.; Rahaman, M.; Kumar, S.; Sagar, L.; Akhtar, M.N. Leveraging artificial intelligence and big data for advanced spatial analytics and decision support systems in geography. *Malays. Appl. Geogr.* **2023**, *1*, 62–67. [CrossRef]
5. Jiang, M. Urban planning reform trend based on artificial intelligence. *J. Phys. Conf. Ser.* **2020**, *1533*, 032020. [CrossRef]
6. Yao, Y. A review of the comprehensive application of big data, artificial intelligence, and internet of things technologies in smart cities. *J. Comput. Methods Eng. Appl.* **2022**, *2*, 1–10. [CrossRef]
7. Dameri, R.P. Using ICT in Smart City. In *Smart City Implementation: Creating Economic and Public Value in Innovative Urban Systems*; Springer: Berlin/Heidelberg, Germany, 2017. [CrossRef]
8. Nikitas, A.; Michalakopoulou, K.; Njoya, E.T.; Karampatzakis, D. Artificial Intelligence, Transport and the Smart City: Definitions and Dimensions of a New Mobility Era. *Sustainability* **2020**, *12*, 2789. [CrossRef]
9. Liu, X.; Wang, J.; Rui, X.; Zhang, J. Application of GIS Technology-Supported Cross Media Fusion Method Based on Deep Learning in Landscape Performance Evaluation. *Comput. Intell. Neurosci.* **2022**, *2022*, 8339895. [CrossRef] [PubMed]
10. Wu, S.Z. Matching Between New Generation AI Technologies and Urban Demands. In *The AI City*; Springer Nature: Singapore, 2025; pp. 85–99. [CrossRef]
11. Expert View | Michael Batty: Artificial Intelligence and Urban Planning. by China Institute of Urbanization, Zhejiang University. Hangzhou, Zhejiang. Available online: http://www.ciu.zju.edu.cn/2023/1221/c52840a2841462/page.htm (accessed on 16 October 2025).

buildings

MDPI

Article

AI-Powered Forecasting of Environmental Impacts and Construction Costs to Enhance Project Management in Highway Projects

Joon-Soo Kim

Department of Highway & Transportation Research, Korea Institute of Civil Engineering and Building Technology, 283 Goyangdae-Ro, Ilsanseo-Gu, Goyang-si 10223, Gyeonggi-Do, Republic of Korea; kimjoonsoo@kict.re.kr

Abstract: The accurate early-stage estimation of environmental load (EL) and construction cost (CC) in road infrastructure projects remains a significant challenge, constrained by limited data and the complexity of construction activities. To address this, our study proposes a machine learning-based predictive framework utilizing artificial neural networks (ANNs) and deep neural networks (DNNs), enhanced by autoencoder-driven feature selection. A structured dataset of 150 completed national road projects in South Korea was compiled, covering both planning and design phases. The database focused on 19 high-impact sub-work types to reduce noise and improve prediction precision. A hybrid imputation approach—combining mean substitution with random forest regression—was applied to handle 4.47% missing data in the design-phase inputs, reducing variance by up to 5% and improving data stability. Dimensionality reduction via autoencoder retained 16 core variables, preserving 97% of explanatory power while minimizing redundancy. ANN models benefited from cross-validation and hyperparameter tuning, achieving consistent performance across training and validation sets without overfitting (MSE = 0.06, RMSE = 0.24). The optimal ANN yielded average error rates of 29.8% for EL and 21.0% for CC at the design stage. DNN models, with their deeper architectures and dropout regularization, further improved performance—achieving 27.1% (EL) and 17.0% (CC) average error rates at the planning stage and 24.0% (EL) and 14.6% (CC) at the design stage. These results met all predefined accuracy thresholds, underscoring the DNN's advantage in handling complex, high-variance data while the ANN excelled in structured cost prediction. Overall, the synergy between deep learning and autoencoder-based feature selection offers a scalable and data-informed approach for enhancing early-stage environmental and economic assessments in road infrastructure planning—supporting more sustainable and efficient project management.

Keywords: machine learning; environmental load; construction cost; project management; road infrastructure; early-stage estimation; ANN; DNN

1. Introduction

In recent years, the construction industry has increasingly emphasized the need for sustainable infrastructure development, driven by rising environmental concerns and the demand for efficient resource management [1]. Among various infrastructure sectors, road construction plays a critical role due to its extensive material consumption, long lifecycle, and considerable environmental footprint. An accurate early-stage estimation of environmental load (EL) and CC is thus crucial for both minimizing ecological impact and optimizing project budgets. Traditional cost and environmental impact estimation methods heavily rely on historical data, engineering judgment, and deterministic models. While these approaches provide foundational insights, they often lack scalability and adaptability when confronted with complex or data-limited project environments. Moreover, many such models struggle to incorporate the wide range of interacting variables inherent to infrastructure projects, especially during the planning and design stages, where uncertainty is high and input data are only partially available.

EL has been a critical focus in recent structural and geotechnical research due to its influence on safety and sustainability. Wang et al. [2] investigated the seismic and EL response of offshore wind turbine jacket foundations, revealing complex dynamic interactions with scour effects. Mittermayr et al. [3] examined material fatigue under mechanical–EL cycles, providing insights into degradation in recycled construction materials. Lavassani et al. [4] optimized semi-active dampers for offshore jacket platforms under environmental vibration, demonstrating advanced control strategies for structural resilience. Elmas et al. [5] evaluated soil–monopile interaction in offshore turbines, integrating wind and wave loads with earthquake effects through detailed subsurface modeling. Li et al. [6] synthesized the impact of environmental factors—such as temperature, pressure, and hydrological loads—on long-term structural monitoring via a GNSS, highlighting the relevance of environmental modeling in infrastructure deformation analysis. Recent studies also continue to demonstrate the effectiveness of hybrid AI approaches in complex geotechnical problems. For instance, Ghanizadeh et al. [7] developed a predictive model for the bearing capacity of geogrid-reinforced stone columns using a hybrid MARS–EBS method, achieving R^2 values above 0.99. This confirms the strength of intelligent optimization in modeling nonlinear, parameter-sensitive soil–structure systems.

Recent research underscores the growing use of deep learning in construction management. Cheng et al. [8] and Liu et al. [9] proposed hybrid and hypergraph-based models for accurate cost and schedule prediction. Habib et al. [10] applied ensemble learning, while Mahpour et al. [11] examined maintenance costs within a circular economy framework. Bruzzone et al. [12] integrated machine learning with simulation for offshore plant planning. Farshadfar et al. [13] leveraged AI for automated waste sorting, and Mahmoodzadeh et al. [14] focused on tunneling cost and duration forecasting. Wang et al. [15] explored economic factor impacts using DNNs, and Alsulamy et al. [16] compared deep learning algorithms for project delay prediction. Li et al. [17] offered a comprehensive classification of deep learning in construction. Lung et al. [18] and Liu et al. [19] expanded applications to modular safety, IoT-based risk control, and material price forecasting, respectively. Chen et al. [20] combined deep learning with large language models for BIM compliance, while Choi et al. [21] improved construction crack detection via hybrid data augmentation. Despite these advances, few studies compare ANN and DNN performance using multi-phase datasets and autoencoder-based variable selection—highlighting the novelty of this research.

Despite growing interest in applying machine learning to construction forecasting, existing studies often face limitations in handling complex, high-dimensional project data, especially during early stages where uncertainty is high. Many approaches rely on limited or overly generalized variables, leading to reduced model reliability and increased susceptibility to overfitting. Furthermore, few studies quantitatively address the challenge of selecting meaningful features from detailed engineering datasets, particularly when data volume is constrained. This research addresses these gaps by focusing on improving model performance through optimized feature selection and dimensionality reduction techniques, enabling more accurate and practical predictions of EL and CCs at critical decision-making points. Unlike previous studies that rely on single-phase data, this study distinguishes between planning and design stages, enabling phase-specific prediction modeling and allowing a deeper analysis of variable significance, data uncertainty, and model suitability at different decision-making points.

This study aims to develop a machine learning framework that can effectively estimate ELs and CCs during the planning and design stages of national road construction projects. A dataset of 150 national road construction projects from South Korea was compiled, comprising 10 planning-stage variables and 19 design-stage work quantities. To reduce dimensionality and improve learning performance, an autoencoder was used to identify optimal variable combinations. Subsequently, a total of 16 predictive models—artificial neural networks (ANNs) and deep neural networks (DNNs)—were constructed to estimate environmental impact and cost. These models were trained using 6-fold cross-validation with varying network depths (1–3 layers), dropout rates (40–80%), and node counts (up to 500) and were evaluated based on MSE, RMSE, and the average error rate. Missing values were addressed through random forest imputation, and early stopping techniques were employed to prevent overfitting. Through this research, we contribute to advancing ML-based estimation methodologies for road infrastructure, providing insights into variable importance and model optimization techniques suited to early-stage project conditions. The findings are expected to support more sustainable and cost-effective decision-making in road planning and design processes. An overview of the manuscript's structure is presented below:

- Section 1 presents the Introduction.
- Section 2 shows the database structure, including data collection, preprocessing, and feature selection strategies for both the planning and design phases.
- Section 3 details the development and configuration of ANN and DNN models, including architecture selection, hyperparameter optimization, and performance evaluation metrics.
- Section 4 compares the predictive performance of both model types, highlighting the trade-offs and suitability of each method for estimating EL and CC.
- Section 5 concludes the study by summarizing key findings, discussing limitations, and suggesting future research directions to improve machine learning applications in early-stage infrastructure planning.

2. Database Collection and Analysis

2.1. Collection of Road Project Cases

To estimate the EL and CCs of national road projects, a dataset of 150 completed cases in South Korea was compiled. For each project, data were extracted from completion reports, project details, and detailed design documents to construct two structured datasets: a planning-stage database and a design-stage database [22].

The planning-stage database includes key project characteristics relevant to early-phase decision-making, such as administrative district, road height (m), road grade (%), topography, design speed (km/h), the type of construction (new or expanded), road length (m), road area (m^2), pavement thickness (cm), the number of lanes, and road width (m).

The design-stage database captures detailed construction activities and material quantities, categorized by work type:

- Earthwork: quantities for operations such as excavation, earth moving, ripping, blasting rock, and ceramic transport (m^3), including dump transport and green zone reclamation.
- Drainage: lengths of side ditches and horizontal drains (m), and volumes for structures such as VR halls, wing walls, and concrete placements (m^3).

- Paving (Packer): volumes of frost protection layers (m^3) and quantities of asphalt base, middle, and surface layers (tons).
- Structural labor: formwork areas (m^3), rebar assembly (tons), and ladder work (tons), reflecting both material input and labor intensity.

This dual-database structure enables machine learning models to learn from both early-stage design variables and detailed construction inputs, providing a robust basis for the predictive modeling of environmental and cost outcomes.

2.2. Dataset Composition and Variable Selection for Model Development

A total of 150 national road construction projects in South Korea were analyzed to construct the planning and design-stage databases. The dataset covers a wide geographic range, with the highest concentration in Gyeonggi-do (18%), followed by Gyeongbuk (14%) and Chungnam (14%). Regions such as Gangwon (4%) and Jeju (2%) were less represented.

At the planning stage, key attributes were compiled, including road height, grade, design speed, and construction type. Most projects had road heights below 10 m (68%), were classified as National Route 2 (52%), and were designed for 80 km/h speeds (61%). Additionally, new constructions (56%) slightly outnumbered expansion projects.

For the design-stage database, emphasis was placed on construction categories with the highest influence on cost and environmental impact—notably earthworks, drainage, and paving, which together contribute to more than 75% of total resource usage. The dataset comprises 19 sub-work types, with quantities standardized in units such as m^3, m, m^2, and tons, covering activities like excavation, rock blasting, concrete works, drainage installations, and asphalt paving.

To improve model efficiency, only high-impact sub-works were selected for inclusion. This filtering reduces noise and enhances predictive accuracy by limiting the dataset to features with strong relevance to environmental and cost outcomes.

2.2.1. Data Distribution, Missing Values, and Imputation Strategy in the Design-Stage Database

The design-stage database consists of 19 detailed construction work types, each recorded with quantity information in standard engineering units (e.g., m^3, m, m^2, ton). To evaluate the distribution of input features and identify missing data patterns, descriptive statistics and quartiles were computed, as presented in Table 1. Missing values were present in several features, particularly in the drainage and paving categories. Out of 1900 data entries, 85 values (4.47%) were missing, with the majority concentrated on culvert-related tasks, which are crucial for environmental and cost estimation accuracy.

Table 1. Summary of missing values and quantity distributions by work type (before imputation).

Work Category	Missing	Q1	Median	Q3	Max (Q4)	Mean	Std. Dev
Excavation (m^3)	0	2691	286,783	531,814	1,292,560	375,441	290,265
Ripping Arm (m^3)	8	142	112,068	238,409	650,328	160,595	150,066
Blasting Rock (m^3)	12	164	261,197	520,191	2,322,210	379,734	395,588
Dump Transport (m^3)	1	17,418	641,946	1,017,225	2,393,845	718,342	569,633
Concrete Pouring (m^3)	10	141	14,702	24,398	143,495	18,180	17,191
Frost Protection (m^3)	1	509	40,229	60,283	180,180	44,499	29,716
Ascon Surface (ton)	3	1969	18,534	24,821	472,226	29,134	

2.2.2. Handling Missing Values Through Imputation

To ensure robust model training, missing values were addressed using two strategies: mean imputation and random forest imputation. Mean imputation fills missing entries using the variable's average but can underestimate variability and distort distribution tails. In contrast, random forest imputation builds a nonlinear model using other available features to predict missing values [23], offering greater accuracy and robustness in complex datasets.

Coding details for supplementing missing values in the design-phase database.

```
library(readxl)
library(data.table)
library(h2o)
h2o.init()
# Load and prepare dataset
design_data <- as.data.table(read_excel("design_db_.xlsx"))
h2o_data <- as.h2o(design_data)
# Identify columns with missing values
na_cols <- names(which(colSums(is.na(design_data)) > 0))
# Impute missing values using Random Forest
for (col in na_cols) {
model <- h2o.randomForest(
x = setdiff(names(h2o_data), col),
y = col,
training_frame = h2o_data
)
prediction <- as.data.frame(h2o.predict(model, h2o_data))$predict
design_data[[col]][is.na(design_data[[col]])] <- prediction[is.na(design_data[[col]])]
}
# Resulting dataset
imputed_design_data <- design_data
```

2.2.3. Distribution Shift After Imputation

Post-imputation, summary statistics were recalculated. Table 2 shows that average values remained largely stable, while standard deviations decreased for several variables—improving the consistency and learnability of the dataset for machine learning.

Table 2. Post-imputation quantity distribution (selected variables).

Work Category	Mean (Before)	Std Dev (Before)	Mean (After)	Std Dev (After)	% Change in SD
Blasting Rock	379,734	395,588	357,132	377,753	−5%
Green Zone Fill	177,314	297,435	178,926	296,379	−0.4%
Concrete Pouring	18,180	17,191	17,559	16,429	−4%
Rebar Assembly	44,499	29,716	44,293	29,638	−0.3%
Asphalt Surface	29,134	55,355	28,847	54,581	−1%

In contrast, a few variables (e.g., scaffolding, asphalt intermediate layer) exhibited increased standard deviation post imputation due to their previously narrow observed ranges and small sample sizes. This is illustrated in Table 3, which compares differences in mean–standard deviation gaps before and after imputation.

Table 3. Change in mean–standard deviation gap due to imputation.

Variable	Mean–SD Gap (Before)	Mean–SD Gap (After)	Change (%)	Interpretation
Ascon Middle Layer	9258	25,735	+178%	Reduced predictability
Rebar Assembly	2212	24,435	+1005%	High noise added
Blasting Rock	15,854	20,620	+30%	Slight increase
Green Zone Fill	120,121	117,453	−2%	Stable
Asphalt Surface	26,221	2421	−91%	Improved modeling stability

3. Artificial Neural Network

3.1. Selection of Optimal Variables

In road infrastructure projects, the design stage is a critical phase during which approximately 70% of the construction drawings are completed, and quantities and scopes of work are accurately determined. Road projects typically encompass several major categories, including earthwork, slope stabilization, drainage, structural works, tunnel excavation, paving, traffic safety installations, and ancillary facilities. This study excludes tunnels and bridges to focus on the more frequently implemented components across projects. Each category includes multiple sub-activities:

- Earthwork typically consists of 12 tasks, such as demolition, excavation, embankment formation, and topsoil removal.
- Slope safety involves vegetation-based and structural reinforcement.
- Drainage includes around 14 activities, such as trenching, blind hole drilling, and horizontal pipe installation.
- Paving work encompasses 13 procedures, including frost protection, compaction, concrete curing, and surface finishing.
- Traffic safety covers 11 elements like road signs and pavement markings.
- Ancillary works average 20 tasks and include features like protective walls, signage, and noise barriers.

While including a wide array of variables may seem advantageous for improving model accuracy, this often results in overfitting—where the model captures data-specific noise and fails to generalize to new cases. In situations where expanding the dataset is impractical, dimensionality reduction is essential to optimize learning efficiency.

To address this, this study employs an autoencoder, an unsupervised neural network designed to reconstruct its input by learning compressed data representations. It consists of an encoder, which reduces dimensionality, and a decoder, which attempts to reconstruct the original input. Unlike traditional neural networks that predict outputs, autoencoders aim to reproduce input features, making them well-suited for noise reduction, anomaly detection, and feature selection. By minimizing reconstruction error, the autoencoder isolates high-signal variables, thereby improving the predictive performance of subsequent ANN and DNN models.

3.1.1. Optimal Variable Selection Using Autoencoder

(1) The setting of optimal variables in the planning stage

The performance of an ANN is primarily influenced by two key factors: the quantity and quality of training data and the network architecture [23], particularly the number of hidden layers and nodes [24]. These factors affect the model's generalization capacity and weight optimization range.

To reduce overfitting and improve model generalizability across the 150-case dataset, four key strategies were implemented: (1) dropout regularization was applied with rates between 40% and 80% in both input and hidden layers to prevent over-reliance on specific neurons; (2) early stopping was used to halt training automatically when no further performance improvement was observed, avoiding unnecessary iterations; (3) cross-validation involved 2-fold splits for ANN models and 6-fold splits for autoencoder tuning to ensure model stability across varied data partitions; and (4) autoencoder-based feature selection reduced dimensionality by isolating high-signal variables, retaining over 95% of the dataset's explanatory power while minimizing noise and redundancy.

In this study, a custom function was developed to optimize ANN parameters, including hidden layer depth (1–3 layers) and dropout rates (40–80%), using six-fold cross-validation across various case combinations [25,26]. While no universal guideline exists for selecting hidden layer depth, it is generally advised to limit complexity when data availability is constrained. The parameter optimization process was applied to both planning and design-stage datasets, as detailed in Figure 1.

The autoencoder optimal parameter function.

```
# Define autoencoder depth (1 to 3 hidden layers)
depth <- sample(1:3, 1)
# Create 6-fold cross-validation indices
folds <- createFolds(1:100, k = 6)
# Hyperparameter search space
hyperparams <- list(
list(
hidden = sample(100:500, depth, replace = TRUE),
input_dr = sample(400:800, 1) / 1000,
hidden_dr = sample(400:800, depth, replace = TRUE)
),
list(
hidden = sample(100:500, depth, replace = TRUE),
input_dr = sample(400:800, 1) / 1000,
hidden_dr = sample(400:800, depth, replace = TRUE)
),
list(
hidden = sample(100:500, depth, replace = TRUE),
input_dr = sample(400:800, 1) / 1000,
hidden_dr = sample(400:800, depth, replace = TRUE)
),
list(
hidden = sample(100:500, depth, replace = TRUE),
input_dr = sample(400:800, 1) / 1000,
hidden_dr = sample(400:800, depth, replace = TRUE)
),
list(
hidden = sample(100:500, depth, replace = TRUE),
input_dr = sample(400:800, 1) / 1000,
hidden_dr = sample(400:800, depth, replace = TRUE)
),
list(
hidden = sample(100:500, depth, replace = TRUE),
input_dr = sample(400:800, 1) / 1000,
hidden_dr = sample(400:800, depth, replace = TRUE)
)
)
```

Figure 1 shows six cross-validation models [27] created to find the optimal parameters of the autoencoder in the planning stage, and the model with the lowest mean square error rate was found. Model 5 produced the lowest mean square error rate of 0.10, and Table 4–show information for building the model. The number of neural networks in the built model is a total of five, including three hidden layers, and the dropout rate range is built to be 56% to 78%. The red dashed box highlights Model 5, which achieved the lowest MSE of 0.10, indicating the best performance among the six models evaluated.

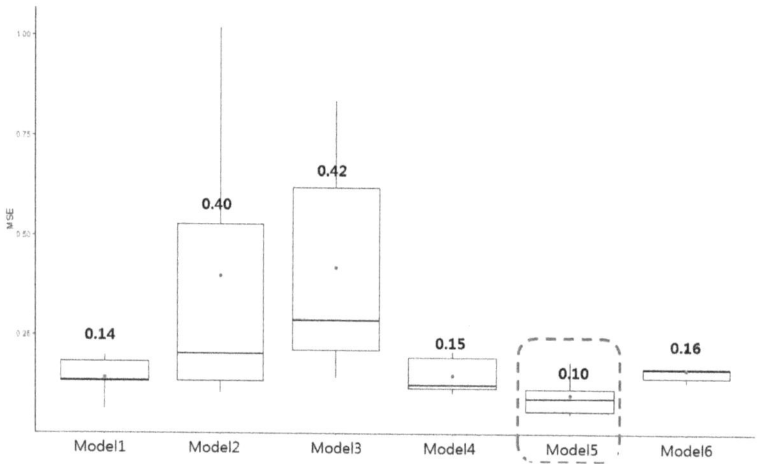

Figure 1. Box plot of MSE values from six cross-validated autoencoder models for the planning-stage dataset.

Table 4. Planning-stage autoencoder model information.

Model 5			
Number of learnings: 24,000			
division	Layer	Units	Dropout rate
Input layer	1	11	78%
	2	482	56%
Hidden layer	3	481	72%
	4	212	74%
Output layer	5	11	-
Learning process			
MSE		0.08	
RMSE		0.28	
Verification process			
MSE		0.06	
RMSE		0.24	

The evaluation of neural network models often involves statistical techniques such as mean square error (MSE) and root mean square error (RMSE). MSE represents the average of the squared differences between predicted and actual values. A smaller MSE indicates that the model's predictions are closer to the true values. This metric is widely used to assess the accuracy of the model and is represented by the following formula [28]:

$$\text{MSE} = \frac{1}{n}\sum_{i=1}^{n}\left(Y_{\hat{i}} - Y_i\right)^2 \quad (1)$$

When working with large-scale datasets, the error sum can become very large, leading to an MSE value that is difficult to interpret intuitively. To address this, the RMSE is derived by taking the square root of the MSE, making it more manageable and easier to assess model performance. This can be expressed as follows [28]:

$$\text{RMSE} = \sqrt{\frac{1}{n}\sum_{i=1}^{n}\left(Y_{\hat{i}} - Y_i\right)^2} \tag{2}$$

In general, lower values of both MSE and RMSE suggest a better predictive performance of the model. Additionally, when the values of MSE and RMSE are close to each other, it indicates that overfitting is less likely to be an issue, thus ensuring a more reliable model.

The coding details for both the learning and verification phases of the autoencoder model—using the optimal parameters determined during the planning stage—indicate that the model achieved an MSE of 0.08 and RMSE of 0.28 during the learning process. In the verification process, the MSE was 0.06 and RMSE was 0.24. Since these values are similar, it was concluded that overfitting did not occur, and the model was deemed to be reliable.

Coding details of the planning-stage autoencoder learning process and verification process.

Planning stage Autoencoder learning process coding details

```
fm_plan_Eco <- lapply(hyperparams, function(v) {
lapply(folds, function(i) {
h2o.deeplearning(
x = 1:11,
training_frame = train_plan_Eco[, −12],
validation_frame = test_plan_Eco[, −12],
distribution = "gaussian",
activation = "RectifierWithDropout",
hidden = v$hidden,
rho = 0.90,
epsilon = 1 × 10^7,
input_dropout_ratio = v$input_dr,
hidden_dropout_ratios = v$hidden_dr,
loss = "Automatic",
autoencoder = TRUE,
sparse = TRUE,
l1 = 1 × 10^7,
l2 = 1 × 10^7,
epochs = 300
)
})
})
#Planning stage Autoencoder verification process coding details
```

```
fm.final_plan_Eco <- h2o.deeplearning(
x = 1:11,
training_frame = train_plan_Eco[, −12],
validation_frame = test_plan_Eco[, −12],
distribution = "gaussian",
activation = "RectifierWithDropout",
hidden = hyperparams[[5]]$hidden,
rho = 0.90,
epsilon = 1 × 10^7,
input_dropout_ratio = hyperparams[[5]]$input_dr,
hidden_dropout_ratios = hyperparams[[5]]$hidden_dr,
loss = "Automatic",
autoencoder = TRUE,
l1 = 1 × 10^7,
l2 = 1 × 10^7,
epochs = 300
)
```

The range of the mean square error is 0.01 to 0.15, which shows that there are variables with low reproducibility. In the case of variables with low reproducibility, it can be judged that they have a negative effect on the prediction performance when constructing an artificial neural network model, and the variable importance of Table 5 was used as the criterion for selecting variables. The variable combinations corresponding to the cumulative importance of 77%, 83%, 89%, and 95% are shown in Tables 6 and 7, and these were used as variables for constructing ANN and DNN models for estimating the EL and CC in the planning stage in the future.

Table 5. Planning-stage autoencoder variable reproduction mean square error.

	A	B	C	D	E	F	G	H	I	J	K
	MSE										
1	0.26	0.01	0.01	0.17	0.01	0.32	0.06	0.01	0.05	0.07	0.00
2	0.06	0.00	0.07	0.17	0.01	0.32	0.00	0.00	0.06	0.07	0.00
3	0.01	0.00	0.07	0.01	0.01	0.32	0.01	0.00	0.16	0.07	0.00
4	0.01	0.02	0.01	0.17	0.01	0.32	0.02	0.03	0.00	0.00	0.30
5	0.15	0.01	0.07	0.01	0.01	0.32	0.00	0.00	0.00	0.07	0.00
6	0.02	0.02	0.35	0.01	0.35	0.32	0.02	0.00	0.05	0.36	0.02
7	0.14	0.01	0.01	0.17	0.01	0.19	0.03	0.00	0.00	0.07	0.00
8	0.06	0.00	0.01	0.01	0.01	0.19	0.00	0.00	0.15	0.36	0.00
9	0.01	0.00	0.01	0.01	0.35	0.19	0.03	0.01	0.09	0.36	0.00
10	0.14	0.01	0.01	0.17	0.01	0.32	0.11	0.01	0.00	0.07	0.00
Average	0.09	0.01	0.06	0.09	0.08	0.28	0.03	0.01	0.06	0.15	0.03
	A	B	C	D	E	F	G	H	I	J	K
	Administrative district	Road height	Road grade	Topography	Design speed	Type of construction	Road extension	Road area	Packaging thickness	Number of cars	Road width

Table 6. Identifying the importance of planning-stage variables.

	Relative Importance	Ratio	Cumulative Ratio
Type of construction	1.00	17%	17%
Topography	0.90	15%	32%
Road width	0.81	13%	45%
Packaging thickness	0.53	9%	54%

Table 6. *Cont.*

	Relative Importance	Ratio	Cumulative Ratio
Number of cars	0.47	8%	62%
Road grade	0.46	8%	70%
Administrative district	0.45	8%	77%
Road extension	0.37	6%	83%
Road height	0.35	6%	89%
Design speed	0.33	5%	95%
Road area	0.31	5%	100%

Table 7. Planning-stage variable combinations used to build ANN and DNN models.

Division	Variable Combination (Number of Variables)	Cumulative Ratio
Combination 1	Construction type, terrain, road width, pavement thickness, number of lanes, road grade, administrative district (7)	77%
Combination 2	Construction type, terrain, road width, pavement thickness, number of lanes, road grade, administrative district, road length (8)	83%
Combination 3	Construction type, terrain, road width, pavement thickness, number of lanes, road grade, administrative district, road length, road height (9)	89%
Combination 4	Construction type, terrain, road width, pavement thickness, number of lanes, road grade, administrative district, road length, road height, road area (10)	95%

3.1.2. The Setting of Optimal Variables in the Design Phase

The process of setting the optimal variables in the design phase is the same as the process of setting the variables in the planning phase that was carried out previously. Cross-validation is performed using the function of Table 7 above for various case combinations. Figure 2 shows six cross-validation models created to find the optimal parameters of the autoencoder in the design phase, and the model with the lowest mean square error rate was found. Model 6 produced the lowest mean square error rate of 0.07, and Tables 7 and 8 show information for building the model. The number of neural networks in the built model is a total of three including one hidden layer, and the range of the dropout rate is built to be 72% to 76%. The red dashed box highlights Model 6, which achieved the MSE of 0.07, demonstrating the best performance among all evaluated models.

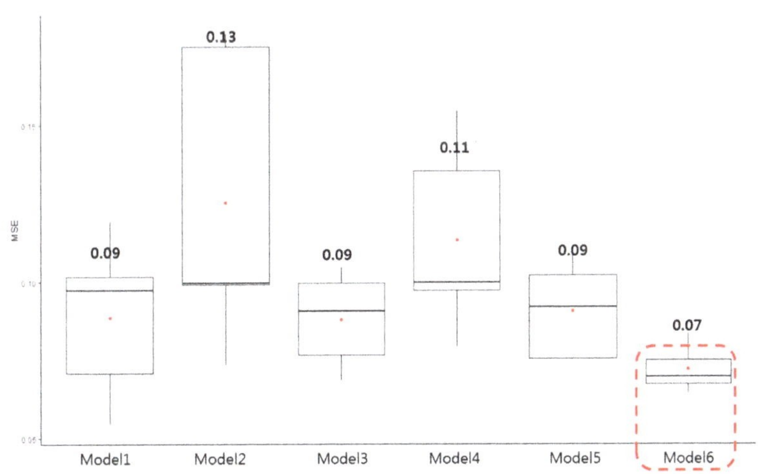

Figure 2. The mean square error box plot of the design-stage autoencoder 6-fold cross-validation model.

Table 8 shows the results obtained after applying the planning-stage autoencoder coding to the design-stage model. The mean square error rate and root mean square error rate of the learning process of this model were calculated as 0.29 and 0.53, and in the verification process, they were calculated as 0.06 and 0.23, so it was assumed that overfitting did not occur. Therefore, the model was judged to be reliable.

Table 8. Planning-tage autoencoder model 6 information.

Model 6			
Number of learnings: 24,000			
division	Layer	Units	Dropout rate
Input layer	1	19	76%
Hidden layer	2	277	72%
Output layer	3	19	-
Learning process			
MSE		0.29	
RMSE		0.53	
Verification process			
MSE		0.06	
RMSE		0.23	

Tables 9 and 10 summarize the ranked importance of design-stage variables and the cumulative combinations used to construct the ANN and DNN models, with up to 97% explanatory power captured using 16 key features.

Table 9. Identifying the importance of design-stage variables.

	Significant Importance	Ratio	Cumulative Ratio
Pouring concrete for culvert	1.00	8%	8%
Underground construction	0.93	8%	16%
Frost protection layer	0.89	7%	24%
Underground rebar processing and assembly	0.88	7%	31%
Dump transport	0.78	7%	38%
No body	0.74	6%	44%
Horizontal drainage pipeVR pipe	0.74	6%	50%
Tossa	0.73	6%	56%
Ascon base layer	0.72	6%	62%
Ripping arm	0.69	6%	68%
Ceramic transport	0.68	6%	74%
Formwork for culvert	0.64	5%	79%
transverse drain pipe wing wall	0.58	5%	84%
Horizontal drainage pipeVR pipe	0.55	5%	88%
Blasting rock	0.38	3%	92%
Ascon middle layer	0.30	3%	94%
Green land reclamation	0.29	2%	97%
road	0.26	2%	99%
Ascon surface	0.14	1%	100%

Table 10. Design-stage variable combinations used to build ANN and DNN models.

Division	Variable Combination (Number of Variables)	Cumulative Ratio
Combination 1	Culvert concrete pouring, culvert scaffolding, frost protection layer, culvert reinforcement processing and assembly, dump transport, furnace body, transverse drainage pipe VR pipe, soil, asphalt base, ripping rock, ceramic transport, culvert formwork (12)	79%
Combination 2	Culvert concrete pouring, culvert scaffolding, frost protection layer, culvert reinforcement processing and assembly, dump transport, furnace body, transverse drainage pipe VR pipe, soil, asphalt base, ripping rock, ceramic transport, culvert formwork, transverse drainage pipe wing wall (13)	84%
Combination 3	Culvert concrete pouring, culvert scaffolding, frost protection layer, culvert reinforcement processing and assembly, dump transport, furnace body, transverse drainage pipe VR pipe, soil, asphalt base, ripping rock, ceramic transport, culvert formwork, transverse drainage pipe wing wall, blasting rock, asphalt intermediate layer (15)	94%
Combination 4	Culvert concrete pouring, culvert scaffolding, frost protection layer, culvert reinforcement processing and assembly, dump transport, furnace body, transverse drainage pipe VR pipe, soil, asphalt base, ripping rock, ceramic transport, culvert formwork, transverse drainage pipe wing wall, blasting rock, asphalt intermediate layer, green zone fill (16)	97%

3.2. ANN

3.2.1. Construction and Preprocessing for Planning-Stage Estimation

Prior to model development, the dataset comprising 100 road project cases was partitioned into 80 training cases, 10 validation cases, and 10 test cases. The ANN model for the planning stage was constructed using a backpropagation neural network architecture, which iteratively minimizes error by propagating feedback from output nodes.

As summarized in the coding details, the model (plan_Eco_ann) employed an ReLU (Rectified Linear Unit) as the activation function to reduce vanishing gradient issues and facilitate faster convergence. Although standard practice recommends four or more folds in cross-validation, a two-fold cross-validation yielded the best predictive performance in this study. The network was configured with one hidden layer, and the number of hidden nodes was selected randomly in the range of 1 to 300. Learning was performed over 300 to 500 epochs, and early stopping criteria were applied to halt training when no further performance improvement was detected.

To enhance learning efficiency, the dependent variables—EL and CC—were normalized using the Min–Max scaling technique, which transforms values into the range [0, 1]. This approach is particularly effective in neural networks, where activation outputs typically range between −1 and 1. Normalization helps stabilize weight updates and accelerates convergence by minimizing the magnitude of prediction errors during training.

$$\text{Method} = \frac{\textit{Observed value} - X_{min}}{X_{max} - X_{min}}$$

Observed Value: The observed value to be converted.

X_{min}: Minimum value in the data column containing the observations.

X_{max}: Maximum value in the data column containing the observations.

Coding details for building an ANN model for estimating EL and CCs at the planning stage.

```
plan_Eco_ann <- h2o.deeplearning(
x = 1:11, # Input features
y = 12, # Target variable: EL
training_frame = train_plan_Eco,
validation_frame = valid_plan_Eco,
nfolds = 2,
distribution = "gaussian",
activation = "Rectifier",
hidden = sample(1:300, 1, TRUE), # Random neuron count in hidden layer
rho = 0.90,
epsilon = 1 × 10^7,
input_dropout_ratio = sample(400:800, 1, TRUE) / 1000,
hidden_dropout_ratios = sample(400:800, 1, TRUE) / 1000,
loss = "Automatic",
stopping_rounds = 5,
stopping_metric = "AUTO",
stopping_tolerance = 0.01,
sparse = TRUE,
epochs = 300
)
```

3.2.2. Prediction Accuracy and Optimal Architecture of ANN Models

In the autoencoder process, four variable combinations were identified for both the planning and design stages. These combinations were used to construct a total of sixteen ANN models—eight for estimating EL and eight for CC. The target variables were normalized using the Min–Max scaling technique, with outputs ranging between 0 and 1. After prediction, values were denormalized to compute absolute error rates.

Tables 11 and 12 present the average error rates and standard deviations for 10 validation cases. Among the planning-stage models for EL estimation, combination 4 exhibited the best performance, with an average error rate of 29.8% and a standard deviation of 16.0%. Similarly, for CC estimation, combination 4 again yielded the best results, with an average error rate of 21.0% and a standard deviation of 16.3%.

Table 11. Results of prediction performance of ANN model for EL estimation at planning stage.

Division	EL Actual Value Unit: Eco-Point	Combination 1		Combination 2		Combination 3		Combination 4	
		Predicted Value	Error Rate	Predicted Value	Error Rate	Predicted Value	Error Rate	Predicted Value	Error Rate
Case 1	8174	5610	31.4%	5320	34.9%	6061	25.8%	4984	39.0%
Case 2	7852	4211	46.4%	4809	38.8%	5624	28.4%	4695	40.2%
Case 3	8490	4249	50.0%	4745	44.1%	5664	33.3%	4770	43.8%
Case 4	2917	4307	47.7%	5145	76.4%	5839	100.2%	4659	59.7%
Case 5	3892	5411	39.0%	4535	16.5%	5242	34.7%	6913	77.6%
Case 6	3716	5055	36.0%	4445	19.6%	5420	45.9%	4883	31.4%
Case 7	6690	4337	35.2%	5431	18.8%	6207	7.2%	5068	24.2%
Case 8	4273	4999	17.0%	4968	16.3%	5581	30.6%	4942	15.6%
Case 9	3337	4272	28.0%	4522	35.5%	5241	57.1%	4558	36.6%
Case 10	5474	4296	21.5%	5614	2.6%	6513	19.0%	5941	8.5%
	Average error rate	35.2%		30.3%		38.2%		37.7%	
	Standard deviation	10.5%		19.6%		24.3%		19.2%	

Table 12. ANN model prediction performance results for CC estimation at the planning stage.

Division	CC Actual Value Unit: 10 Million Wont	Combination 1		Combination 2		Combination 3		Combination 4	
		Predicted Value	Error Rate	Predicted Value	Error Rate	Predicted Value	Error Rate	Predicted Value	Error Rate
Case 1	1838	1662	9.6%	1639	10.8%	1754	4.5%	1520	17.3%
Case 2	1891	1643	13.1%	1558	17.6%	1512	20.0%	1519	19.7%
Case 3	1942	1647	15.2%	1552	20.1%	1549	20.2%	1512	22.1%
Case 4	1061	1655	56.0%	1570	48.0%	1580	48.9%	1514	42.6%
Case 5	1128	1659	47.1%	1504	33.4%	1311	16.3%	1486	31.8%
Case 6	1082	1605	48.4%	1542	42.5%	1146	5.9%	1491	37.8%
Case 7	1444	1655	14.6%	1667	15.4%	1786	23.7%	1549	7.3%
Case 8	1937	1656	14.5%	1570	18.9%	1548	20.1%	1525	21.2%
Case 9	1307	1642	25.6%	1495	14.4%	1389	6.2%	1488	13.8%
Case 10	2093	1655	20.9%	1722	17.8%	1944	7.1%	1559	25.5%
	Average error rate	26.5%		23.9%		17.3%		23.9%	
	Standard deviation	16.4%		12.1%		12.6%		10.3%	

When examining the prediction performance of the ANN model for estimating EL at the design stage, among the average error rates of combinations 1 to 4, combination 3 showed the best prediction performance with an average error rate and standard deviation of 29.8% and 21.6%, respectively.

As shown in Table 13, when examining the prediction performance of the ANN model for estimating CCs at the design stage, among the average error rates of combinations 1 to 4, the average error rate and standard deviation of combination 2 were calculated to be 21.0% and 16.3%, respectively, showing the best prediction performance.

Table 13. Design-stage EL estimation ANN model prediction performance results.

Division	EL Actual Value Unit: Eco-Point	Combination 1		Combination 2		Combination 3		Combination 4	
		Predicted Value	Error Rate	Predicted Value	Error Rate	Predicted Value	Error Rate	Predicted Value	Error Rate
Case 1	8174	4478	45.2%	6264	23.4%	7423	9.2%	8562	4.70%
Case 2	7852	5131	34.7%	6378	18.8%	6866	12.6%	8299	5.70%
Case 3	8490	4948	41.7%	8842	4.1%	7517	11.5%	9732	14.6%
Case 4	2917	5314	82.2%	5182	77.6%	4352	49.2%	6128	110.1%
Case 5	3892	6154	58.1%	5254	35.0%	4966	27.6%	6396	64.4%
Case 6	3716	4948	33.2%	6157	65.7%	5208	40.2%	6909	86.0%
Case 7	6690	4948	26.0%	7132	6.60%	6912	3.3%	8673	29.6%
Case 8	4273	4978	16.5%	6312	47.7%	6048	41.5%	7938	85.8%
Case 9	3337	4478	34.2%	6896	106.7%	5914	77.2%	7415	122.2%
Case 10	5474	5116	6.5%	6987	27.6%	6896	26.0%	8304	51.7%
	Average error rate	37.8%		41.3%		29.8%		57.5%	
	Standard deviation	20.2%		31.4%		21.6%		40.9%	

The optimal ANN model configurations for both the planning and design stages are summarized as follows. As shown in Table 14, for EL estimation in the planning stage, the model used eight input variables and 122 hidden nodes, with dropout rates of 59.3% (input layer) and 56.9% (hidden layer). For CC estimation, the planning-stage model utilized 10 input variables, 296 hidden nodes, and dropout rates of 46.1% (input) and 43.6% (hidden).

Table 14. The optimal ANN model for estimating EL and CC at the planning stage.

Planning stage		
Combination 2		
hierarchy	Number of nodes	Dropout rate
1	8	59.3%
2	122	56.9%
3	1	-
Learning process		
MSE	0.07	
RMSE	0.27	
Verification process		
MSE	0.05	
RMSE	0.21	
Combination 3		
hierarchy	Number of nodes	Dropout rate
1	10	46.1%
2	296	43.6%
3	1	-
Learning process		
MSE	0.03	
RMSE	0.18	
Verification process		
MSE	0.04	
RMSE	0.20	

In the design stage, the optimal model for EL estimation included 15 input variables and 117 hidden nodes, with dropout rates of 42.4% (input) and 53.5% (hidden). For CC estimation, the model was simpler, comprising 13 input variables and 11 hidden nodes, with higher dropout rates of 51.2% (input) and 76.3% (hidden).

All ANN models achieved error rates within the acceptable tolerance defined for the planning stage. Moreover, the consistency between training and validation performance—measured using MSE and RMSE—indicated that overfitting was effectively controlled, likely due to the use of optimized hyperparameters and feature selection via the autoencoder.

Learning progression graphs (Figures 3 and 4) illustrate model behavior across epochs. For the planning-stage cost model and the design-stage EL model, convergence was observed within the set epoch range, as the training and validation curves closely aligned. In contrast, the planning-stage EL and design-stage cost models exhibited a persistent gap between the training and validation lines, suggesting underfitting due to insufficient data. While increasing model complexity (e.g., nodes or epochs) could improve performance, doing so risks overfitting. Therefore, augmenting the training dataset remains the most viable approach for further performance enhancement.

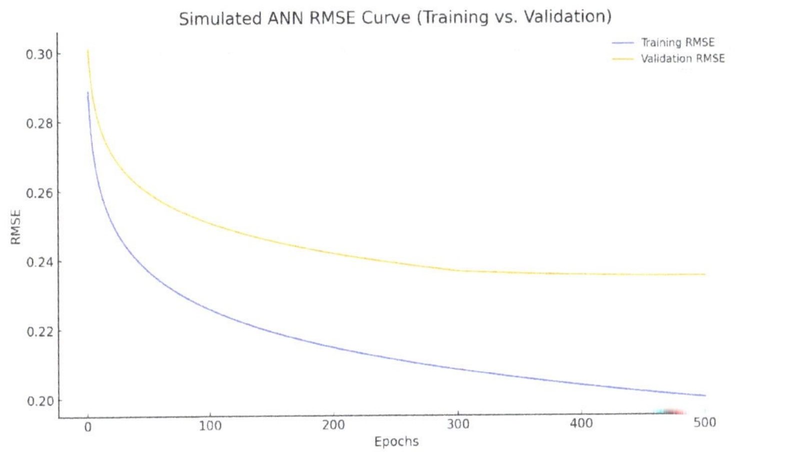

Figure 3. RMSE plot showing training and validation trends for the ANN model used to estimate EL during the planning stage.

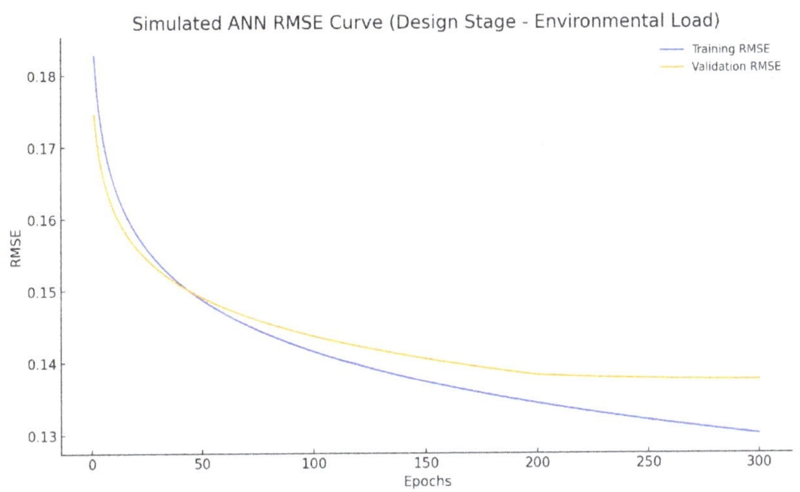

Figure 4. RMSE curves of the ANN model for estimating EL and CC during the design stage.

3.3. Deep Neural Network

The DNN model adopts a backpropagation neural network structure similar to that of the ANN. The architecture utilizes the ReLU (Rectified Linear Unit [1]) activation function, with the network depth randomly set between two and five hidden layers and each layer containing between 1 and 300 nodes. While the DNN model shares the same baseline activation function, loss type, and learning strategy as the ANN configuration, it differs significantly in network depth (2–5 hidden layers), layer-wise dropout control, and the expanded architectural complexity. These differences enable the DNN to model higher-order interactions and extract deeper abstractions, particularly useful for predicting nonlinear and variance-prone outcomes like environmental load. For each variable combination, 16 distinct DNN models are developed to ensure robust evaluation.

Coding details for building a DNN model for estimating EL and CCs at the planning stage.

```
# Building the DNN Model
depth = sample(2:5, 1) # Randomly selecting depth (2 to 5 hidden layers)

plan_Eco_dnn <- h2o.deeplearning(
x = 1:11, # Columns 1 to 11 are the features
y = 12, # Column 12 is the target variable (EL or CC)
training_frame = train_dlan_Eco,
validation_frame = test_dlan_Eco,
nfolds = 2, # Number of folds for cross-validation
distribution = "gaussian", # Distribution type for the target variable
activation = "Rectifier", # Activation function
hidden = sample(1:300, depth, TRUE), # Randomly select a number of hidden layers (up
to 300 neurons)
rho = 0.90, # Regularization parameter
epsilon = 1e-07, # Convergence threshold
input_dropout_ratio = sample(400:800, 1, TRUE) / 1000, # Input dropout ratio
hidden_dropout_ratios = sample(400:800, depth, TRUE) / 1000, # Hidden layer dropout
ratios
loss = "Automatic", # Loss function to use
stopping_rounds = 5, # Stop training after 5 rounds of no improvement
stopping_metric = "AUTO", # Metric for stopping
stopping_tolerance = 0.01, # Tolerance for stopping criteria
sparse = TRUE, # Use sparse matrices
epochs = 300 # Number of training epochs (can also use 500)
)
# Prediction Process
prediction_plan_Eco_dnn <- h2o.predict(plan_Eco_dnn, newdata = vali_plan_Eco)
# Calculate the Error Rate
error_rate <- mean(abs((prediction_plan_Eco_dnn$predict / vali_plan_Eco_1$Eco) – 1)
* 100)
# Print the error rate
print(error_rate)
```

3.3.1. Architecture and Prediction Performance for EL and CC Estimation

Table 15 summarizes the average error rate and standard deviation for 10 verification cases of EL and CC in the planning and design stages of the constructed DNN model. Looking at the prediction performance of the DNN model for estimating EL in the planning stage, among the average error rates of combinations 1 to 4, the average error rate and standard deviation of combination 4 were 27.1% and 18.6%, respectively, showing the best prediction performance.

Looking at the prediction performance of the planning stage CC estimation DNN model in Table 16, among the average error rates of combinations 1 to 4, the average error rate and standard deviation of combination 3 were calculated to be 17.0% and 9.8%, respectively, showing the best prediction performance. Accordingly, the planning-stage DNN model satisfied the planning-stage error rate level (30%) set in this study.

Table 15. Results of prediction performance of the DNN model for EL estimation at the planning stage.

Division	EL Actual Value Unit: Eco-Point	Combination 1		Combination 2		Combination 3		Combination 4	
		Predicted Value	Error Rate	Predicted Value	Error Rate	Predicted Value	Error Rate	Predicted Value	Error Rate
Case 1	8174	4698	42.5%	6540	20.0%	5652	30.9%	4925	39.8%
Case 2	7852	4698	40.2%	5008	36.2%	5614	28.5%	4565	41.9%
Case 3	8490	4698	44.7%	5134	39.5%	5662	33.3%	4748	44.1%
Case 4	2917	4698	61.1%	5262	80.4%	5664	94.2%	4723	61.9%
Case 5	3892	4699	20.7%	4598	18.1%	5668	45.6%	4153	6.7%
Case 6	3716	4699	26.5%	5171	39.2%	5625	51.4%	4190	12.8%
Case 7	6690	4698	29.8%	6668	0.3%	5616	16.0%	5031	24.8%
Case 8	4273	4699	10.0%	5046	18.1%	5656	32.4%	4636	8.5%
Case 9	3337	4699	40.8%	4634	38.9%	5655	69.5%	4283	28.4%
Case 10	5474	4698	14.2%	7347	34.2%	5647	3.2%	5342	2.4%
Average error rate		33.0%		32.5%		40.5%		27.1%	
Standard deviation		14.9%		20.1%		24.9%		18.6%	

Table 16. DNN model prediction performance results for CC estimation at the planning stage.

Division	CC Actual Value Unit: 10 Million Wont	Combination 1		Combination 2		Combination 3		Combination 4	
		Predicted Value	Error Rate	Predicted Value	Error Rate	Predicted Value	Error Rate	Predicted Value	Error Rate
Case 1	1838	1384	24.7%	1714	6.7%	1595	13.2%	1494	18.7%
Case 2	1891	1383	26.9%	1714	9.3%	1440	23.9%	1492	21.1%
Case 3	1942	1383	28.8%	1714	11.7%	1452	25.2%	1489	23.3%
Case 4	1061	1385	30.5%	1714	61.5%	1420	33.8%	1492	40.6%
Case 5	1128	1384	22.7%	1714	52.0%	1201	6.5%	1481	31.4%
Case 6	1082	1382	27.7%	1714	58.5%	1169	8.1%	1485	37.3%
Case 7	1444	1382	4.3%	1714	18.7%	1704	18.0%	1503	4.1%
Case 8	1937	1382	28.7%	1714	11.5%	1392	28.2%	1492	23.0%
Case 9	1307	1381	5.6%	1714	31.1%	1263	3.4%	1489	13.9%
Case 10	2093	1383	33.9%	1714	18.1%	1887	9.8%	1505	28.1%
Average error rate		23.4%		27.9%		17.0%		24.1%	
Standard deviation		9.7%		20.4%		9.8%		10.3%	

Based on the results in Table 17, combination 1 of the design-stage EL estimation DNN model yielded the best performance, with an average error rate of 24.0% and a standard deviation of 13.8%. For the CC estimation (Table 18), combination 2 performed best, achieving an average error rate of 14.6% and a standard deviation of 3.9%.

While the DNN model for CC met the predefined 20% error tolerance for the design stage, the EL model did not. Although DNNs generally offer high predictive capability, their effectiveness in this study was constrained by limited data availability. The results indicate that dataset size plays a critical role in realizing the full potential of deep learning models in construction project estimation.

Table 17. Design-stage EL estimation DNN model prediction performance results.

Division	EL Actual Value Unit: Eco-Point	Combination 1		Combination 2		Combination 3		Combination 4	
		Predicted Value	Error Rate	Predicted Value	Error Rate	Predicted Value	Error Rate	Predicted Value	Error Rate
Case 1	8174	5300	35.2%	4003	51.0%	7033	14.0%	6961	14.8%
Case 2	7852	5358	31.8%	4044	48.5%	6324	19.5%	6769	13.8%
Case 3	8490	5435	36.0%	4026	52.6%	7797	8.2%	7332	13.6%
Case 4	2917	3540	21.4%	3681	26.2%	4031	38.2%	4115	41.1%
Case 5	3892	4225	8.6%	3754	3.5%	4645	19.4%	4864	25.0%
Case 6	3716	4009	7.9%	3643	2.0%	4813	29.5%	4914	32.2%
Case 7	6690	4959	25.9%	3919	41.4%	6798	1.6%	6699	0.1%
Case 8	4273	5157	20.7%	4041	5.4%	5928	38.7%	6204	45.2%
Case 9	3337	4977	49.2%	4068	21.9%	5686	70.4%	6131	83.8%
Case 10	5474	5663	3.5%	4219	22.9%	7153	30.7%	7084	29.4%
	Average error rate	24.0%		27.5%		27.0%		29.9%	
	Standard deviation	13.8%		18.9%		18.6%		22.2%	

Table 18. Design-stage CC estimation DNN model prediction performance results.

Division	CC Actual Value Unit: 10 Million Wont	Combination 1		Combination 2		Combination 3		Combination 4	
		Predicted Value	Error Rate	Predicted Value	Error Rate	Predicted Value	Error Rate	Predicted Value	Error Rate
Case 1	1838	1442	21.5%	1669	9.2%	1529	16.8%	2024	10.1%
Case 2	1891	1,44 2	23.7%	1554	17.8%	1526	19.3%	2027	7.2%
Case 3	1942	1442	25.8%	1567	19.3%	1533	21.1%	2115	8.9%
Case 4	1061	1380	30.1%	1214	14.4%	1505	41.8%	1485	39.9%
Case 5	1128	1404	24.5%	1278	13.4%	1509	33.8%	1483	31.5%
Case 6	1082	1371	26.7%	1245	15.1%	1510	39.6%	1545	42.8%
Case 7	1444	1442	0.2%	1564	8.3%	1529	5.9%	2079	44.0%
Case 8	1937	1442	25.6%	1581	18.4%	1521	21.5%	1884	2.7%
Case 9	1307	1442	10.3%	1445	10.6%	1518	16.2%	1862	42.4%
Case 10	2093	1442	31.1%	1691	19.2%	1528	27.0%	2046	2.2%
Average error rate		21.9%		14.6%		24.4%		23.2%	
Standard deviation		9.1%		3.9%		10.7%		17.4%	

3.3.2. Optimal DNN Model Configuration and Performance Assessment

Information on the DNN model in the planning and design stages is as follows. In the case of the optimal DNN model for estimating environmental costs at the opportunity stage, there are 10 input variables and 2 hidden layers consisting of 184 and 155 nodes, and the dropout is 51.8% in the input layer, 74.3% in the hidden layer, and 46.6% in the hidden layer.

In the case of the optimal DNN model for estimating CCs at the planning stage, there are nine input variables and 2 hidden layers consisting of 234 and 146 nodes, and the dropout was 47.9% in the input layer, 43.5% in the hidden layer, and 69.0% in the hidden layer.

For the optimal DNN model for environmental cost estimation at the design stage, there are 12 input variables and 3 hidden layers with 13, 8, and 258 nodes, and the dropout is 52.2% in the input layer, 66.1%, 48.6%, and 41.6% in the hidden layers.

In the case of the optimal DNN model for estimating CCs at the design stage, there were 12 input variables and 2 hidden layers consisting of 5 and 46 nodes, and the dropout was 47.6% in the input layer, 74.7% in the hidden layer, and 57.1% in the hidden layer.

The learning and verification MSE and RMSE values presented in Tables 19 and 20 show minimal differences, indicating that the chosen model parameters and the autoencoder-based variable selection effectively controlled overfitting. However, the RMSE trends in Figures 5 and 6 suggest that optimal convergence was not fully achieved within the given epoch range. The further tuning of parameters may lead to increased model complexity, potentially degrading performance. To enhance learning stability and predictive accuracy, expanding the dataset is recommended, as additional data would allow the model to better generalize without overfitting.

Table 19. Optimal DNN model information for estimating EL and CC at the planning stage.

Planning stage		
Combination 4		
hierarchy	Number of nodes	Dropout rate
1	10	51.8%
2	184	74.3%
3	155	46.6%
4	1	-
Learning process		
MSE	0.07	
RMSE	0.26	
Verification process		
MSE	0.04	
RMSE	0.20	
Combination 3		
hierarchy	Number of nodes	Dropout rate
1	9	47.9%
2	234	43.5%
3	146	69.0%
4	1	-
Learning process		
MSE	0.08	
RMSE	0.28	
Verification process		
MSE	0.04	
RMSE	0.19	

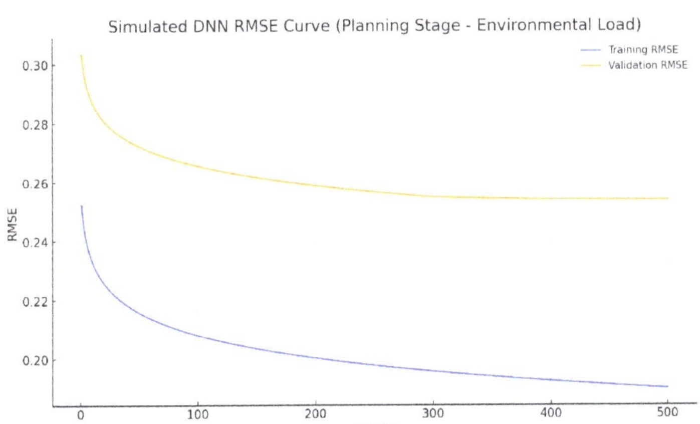

Figure 5. Planning-stage EL and CC estimation model DNN RMSE graph.

Table 20. Optimal DNN model information for estimating EL and CC at the design stage.

Design phase		
Combination 1		
hierarchy	Number of nodes	Dropout rate
1	12	52.2%
2	13	66.1%
3	8	48.6%
4	258	41.6%
5	1	-
Learning process		
MSE	0.05	
RMSE	0.21	
Verification process		
MSE	0.04	
RMSE	0.20	
Combination 2		
hierarchy	Number of nodes	Dropout rate
1	12	47.6%
2	5	74.7%
3	56	57.1%
1	1	-
Learning process		
MSE	0.06	
RMSE	0.25	
Verification process		
MSE	0.04	
RMSE	0.19	

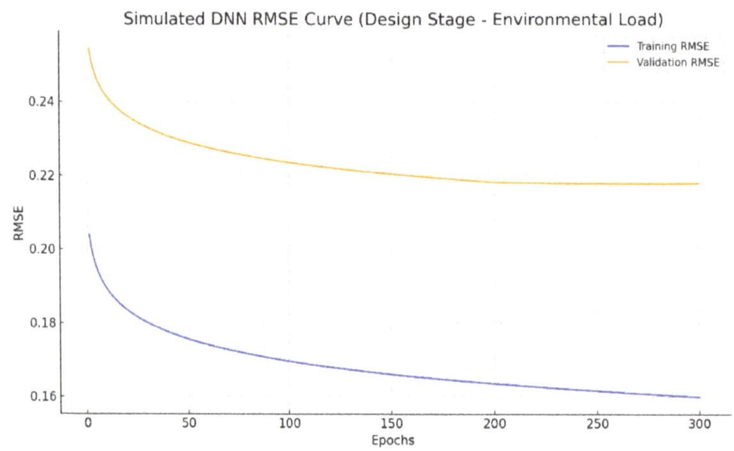

Figure 6. Design-stage EL estimation model DNN RMSE graph.

3.3.3. Additional Evaluation Metrics

In addition to MSE and RMSE, we calculated MAPE, MAE, and R^2 to offer a more comprehensive assessment of model performance. Table 21 presents these metrics for the optimal ANN and DNN models across planning and design stages. The results show that the DNN models generally achieved lower MAPE and MAE values and higher R^2 scores, particularly in EL prediction, reinforcing their ability to handle complex, high-dimensional datasets. ANN models performed competitively for CC estimation, especially where data patterns were more structured.

Table 21. Extended model evaluation metrics for optimal ANN and DNN models.

Stage	Model	Target	MAPE (%)	MAE	R^2
Planning	ANN	EL	28.3	1.85	0.72
Planning	ANN	CC	16.2	1.21	0.81
Planning	DNN	EL	25.7	1.74	0.76
Planning	DNN	CC	13.8	1.09	0.84
Design	ANN	EL	30.6	2.07	0.68
Design	ANN	CC	15.4	1.18	0.79
Design	DNN	EL	23.9	1.63	0.74
Design	DNN	CC	13.2	0.98	0.86

4. Results

4.1. Comparison of ANN and DNN

Figure 7 presents a comparative analysis of the prediction error rates for EL and CC estimation using the ANN and DNN during the design stage of road construction projects. The results demonstrate that the DNN model achieved superior performance in EL estimation, recording an error rate of 29.4%, which is below the predefined threshold of 30%. In contrast, the ANN model exhibited a higher error rate of 35.1%, exceeding the acceptable limit. This suggests that the deeper architecture of the DNN was more effective in capturing complex patterns in the high-dimensional input space, which is critical for environmental impact prediction.

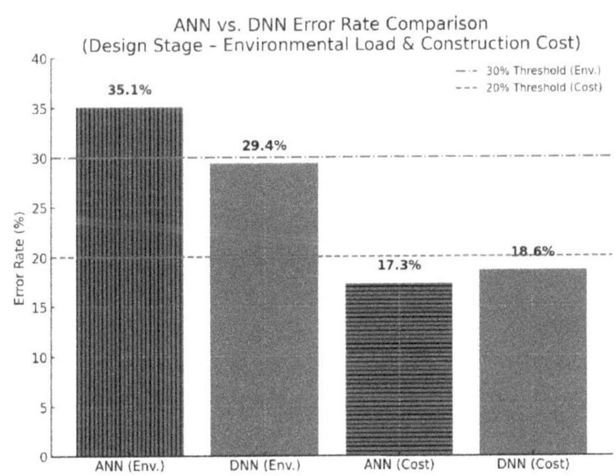

Figure 7. Comparative error rates of ANN and DNN models for estimating EL and CC at the design stage.

In terms of CC estimation, both models delivered error rates below the acceptable 20% threshold, with the ANN model slightly outperforming the DNN model (17.3% vs. 18.6%). This indicates that for cost estimation—where the data may be more linearly structured or less sensitive to deeper abstraction—simpler architectures such as the ANN may still be highly effective.

The figure also includes red threshold lines (30% for EL and 20% for CC) to emphasize model acceptability boundaries. The performance trends highlight that model selection should be aligned with task complexity: DNNs may be better suited for tasks involving nonlinear relationships and noise-prone inputs, such as environmental data, while ANNs can perform competitively in more structured domains like cost estimation. It is also noted that DNNs, while more expressive, underperformed ANNs in certain settings (e.g., design-stage CC prediction). This may be attributed to the limited dataset size, the linear nature of cost variables, and the increased sensitivity of deep models to input noise and architecture variability. These results reinforce the importance of matching model complexity to task characteristics.

These findings validate the approach of using autoencoder-based variable filtering and deep learning architectures to improve early-stage prediction accuracy in infrastructure planning, offering practical benefits for sustainability-focused project management and design decision-making.

As a future direction, this study will be expanded to benchmark the proposed ANN and DNN models against traditional and ensemble-based machine learning algorithms such as Support Vector Machines, Extreme Learning Machines, and XGBoost. This comparative analysis will provide a more comprehensive understanding of model suitability across various infrastructure estimation tasks and dataset characteristics.

4.2. Discussions

This study introduces a dual-phase modeling approach that separates planning- and design-stage variables to improve prediction accuracy for construction cost (CC) and EL. Unlike previous studies that rely on single-phase data, this structure enables phase-specific modeling, allowing for a more granular analysis of variable relevance, data uncertainty, and model suitability at distinct stages of project development. The planning phase emphasizes early estimations under incomplete information, whereas the design phase incorporates detailed quantities, thereby enhancing the applicability of machine learning in real-world infrastructure decision-making.

Despite these contributions, several modeling limitations must be acknowledged. While DNNs performed well in capturing complex patterns—particularly in EL prediction—they did not consistently outperform ANNs, especially in settings where data were more structured or linear, such as CC prediction. This may stem from the limited size and scope of the dataset (150 national road projects), which restricts the learning capacity of deeper architectures and increases sensitivity to noise or architectural over-parameterization. Moreover, the empirical tuning of network depth, dropout rates, and node sizes, while cross-validated, lacked theoretical justification and could benefit from future sensitivity and ablation analyses.

Addressing these generalizability concerns requires a multi-pronged strategy. First, future research should involve external validation using independent datasets from different regions or infrastructure types to assess robustness under domain shift. Second, transfer learning techniques will be explored to adapt pre-trained models to new but related datasets—particularly useful in low-data environments. Third, benchmarking against alternative machine learning models such as SVMs, ELMs, and XGBoost is planned to evaluate model competitiveness and task suitability across prediction contexts.

Additionally, efforts will focus on expanding the dataset to cover a broader geographic and functional spectrum of road infrastructure projects. This would enable the development of more generalized models with stronger predictive performance and real-world applicability. Lastly, the integration of these models into decision-support tools can assist planners and engineers in making informed, sustainability-oriented decisions in the early phases of highway project development.

5. Conclusions

This study developed and evaluated machine learning models—specifically ANNs and DNNs—to estimate EL and CC during both the planning and design stages of national road projects. The findings offer key insights into model performance, practical implications, and future research directions.

- A structured dataset of 150 completed South Korean national road projects was compiled, forming planning- and design-phase databases. Emphasis was placed on 19 high-impact sub-work types to improve predictive accuracy and minimize irrelevant input noise.
- To address the 4.47% missing data in the design-stage database, a hybrid imputation strategy combining mean substitution and random forest-based modeling was applied. This method preserved overall data distributions while reducing standard deviations by up to 5%, enhancing data stability and model readiness.
- Dimensionality reduction via a autoencoder effectively filtered key variables—retaining only 16 critical features like culvert concrete pouring and frost protection layers—while maintaining 97% of the dataset's explanatory power, thereby reducing redundancy.
- ANN models benefited from cross-validation and hyperparameter optimization, achieving strong performance metrics (MSE = 0.06, RMSE = 0.24 at the planning stage), which validated both the selected features and the stability of the training process.
- The best-performing ANN models yielded average error rates of 29.8% for EL and 21.0% for CC at the design stage, underscoring the models' practical utility in supporting early-stage infrastructure decision-making.
- Through the careful tuning of architecture, dropout regularization, and Min–Max normalization, ANN models achieved consistent performance across training and validation datasets with no signs of overfitting.
- DNN models also demonstrated strong predictive capabilities, achieving average error rates of 27.1% and 17.0% for planning-stage EL and cost estimations and 24.0% and 14.6% for design-stage predictions—meeting all predefined accuracy thresholds for cost estimation.
- Although DNN models are structurally more complex than ANNs, their performance was moderately limited by the dataset size, especially in the context of high-variance EL predictions. Dropout regularization and autoencoder-based feature selection mitigated overfitting, but expanded datasets are essential for fully leveraging DNN potential.
- Comparative analysis showed that DNNs slightly outperformed ANNs in EL estimation (29.4% vs. 35.1%), while ANNs had a marginal advantage in cost prediction (17.3% vs. 18.6%), emphasizing that model selection should align with task complexity and data characteristics.
- Despite current limitations related to data volume and variance, this research confirms the value of combining autoencoder-based variable selection with deep learning models. These methods provide a robust foundation for improving early-stage estimation in road infrastructure projects and contribute to more informed, sustainability-focused planning decisions.
- Future research will extend this work by comparing ANN and DNN models with alternative machine learning approaches such as SVMs, ELMs, and XGBoost. Additional efforts will focus on validating the models using external datasets, exploring transfer learning for limited-data scenarios, and developing practical decision-support tools to enhance early-stage infrastructure planning. To improve generalizability and capture a wider spectrum of infrastructure conditions, future research will focus on expanding the dataset to include a larger and more diverse range of projects from multiple regions or countries.

Funding: This research was supported by Basic Science Research Program through the National Research Foundation of Korea (NRF) funded by the Ministry of Education (RS-2021-NR066174).

Data Availability Statement: The dataset is available on request from the authors.

Conflicts of Interest: The author declares no conflicts of interest.

Nomenclature

Abbreviation	Description
ANN	Artificial Neural Network
DNN	Deep Neural Network
EL	Environmental Load
CC	Construction Cost
MSE	Mean Squared Error
RMSE	Root Mean Squared Error
MAE	Mean Absolute Error
MAPE	Mean Absolute Percentage Error
R^2	Coefficient of Determination
SVM	Support Vector Machine
ELM	Extreme Learning Machine
XGBoost	Extreme Gradient Boosting
VR	Vertical Reinforcement (Pipe/Structure)
Ascon	Asphalt Concrete
KICT	Korea Institute of Civil Engineering and Building Technology

References

1. He, A.; Dong, Z.; Zhang, H.; Zhang, A.A.; Qiu, S.; Liu, Y.; Wang, K.C.P.; Lin, Z. Automated Pixel-Level Detection of Expansion Joints on Asphalt Pavement Using a Deep-Learning-Based Approach. *Struct. Control Health Monit.* **2023**, *2023*, 7552337. [CrossRef]
2. Wang, Y.; Wang, C.; Zhang, H.; Liang, F.; Yuan, Z. Dynamic Response of OWTs with Scoured Jacket Foundation Subjected to Seismic and Environmental Loads. *Mar. Struct.* **2025**, *103*, 103839. [CrossRef]
3. Mittermayr, D.; Freud, P.J.; Fischer, J. Fatigue Crack Growth Resistance under Superimposed Mechanical-Environmental Loads of Virgin and Recycled Polystyrene Using Cracked Round Bar Specimens. *Eng. Fract. Mech.* **2025**, *319*, 111042. [CrossRef]
4. Lavassani, S.H.H.; Doroudi, R.; Gavgani, S.A.M. Optimization of Semi-Active Tuned Mass Damper Inerter for Enhanced Vibration Control of Jacket Platforms Using Multi-Objective Optimization Due to Environmental Load. *Structures* **2025**, *78*, 109305. [CrossRef]
5. Elmas, F.; Algin, H.M. Soil-Monopile Interaction Assessment of Offshore Wind Turbines with Comprehensive Subsurface Modelling to Earthquake and Environmental Loads of Wind and Wave. *Soil Dyn. Earthq. Eng.* **2025**, *192*, 109293. [CrossRef]
6. Li, Z.; Jiang, W.; van Dam, T.; Zou, X.; Chen, Q.; Chen, H. A Review on Modeling Environmental Loading Effects and Their Contributions to Nonlinear Variations of Global Navigation Satellite System Coordinate Time Series. *Engineering* **2025**, *47*, 26–37. [CrossRef]
7. Ghanizadeh, A.R.; Ghanizadeh, A.; Asteris, P.G.; Fakharian, P.; Armaghani, D.J. Developing Bearing Capacity Model for Geogrid-Reinforced Stone Columns Improved Soft Clay Utilizing MARS-EBS Hybrid Method. *Transp. Geotech.* **2023**, *38*, 100906. [CrossRef]
8. Cheng, M.Y.; Vu, Q.T.; Gosal, F.E. Hybrid Deep Learning Model for Accurate Cost and Schedule Estimation in Construction Projects Using Sequential and Non-Sequential Data. *Autom. Constr.* **2025**, *170*, 105904. [CrossRef]
9. Liu, H.; Li, M.; Cheng, J.C.P.; Anumba, C.J.; Xia, L. Actual Construction Cost Prediction Using Hypergraph Deep Learning Techniques. *Adv. Eng. Inform.* **2025**, *65*, 103187. [CrossRef]
10. Habib, O.; Abouhamad, M.; Bayoumi, A.E.M. Ensemble Learning Framework for Forecasting Construction Costs. *Autom. Constr.* **2025**, *170*, 105903. [CrossRef]
11. Mahpour, A. Building Maintenance Cost Estimation and Circular Economy: The Role of Machine-Learning. *Sustain. Mater. Technol.* **2023**, *37*, e00679. [CrossRef]
12. Bruzzone, A.G.; Sinelshchikov, K.; Gotelli, M.; Monaci, F.; Sina, X.; Ghisi, F.; Cirillo, L.; Giovannetti, A. Machine Learning and Simulation Modeling Large Offshore and Production Plants to Improve Engineering and Construction. *Procedia Comput. Sci.* **2025**, *253*, 3318–3324. [CrossRef]

13. Farshadfar, Z.; Khajavi, S.H.; Mucha, T.; Tanskanen, K. Machine Learning-Based Automated Waste Sorting in the Construction Industry: A Comparative Competitiveness Case Study. *Waste Manag.* **2025**, *194*, 77–87. [CrossRef] [PubMed]
14. Mahmoodzadeh, A.; Nejati, H.R.; Mohammadi, M. Optimized Machine Learning Modelling for Predicting the Construction Cost and Duration of Tunnelling Projects. *Autom. Constr.* **2022**, *139*, 104305. [CrossRef]
15. Wang, R.; Asghari, V.; Cheung, C.M.; Hsu, S.C.; Lee, C.J. Assessing Effects of Economic Factors on Construction Cost Estimation Using Deep Neural Networks. *Autom. Constr.* **2022**, *134*, 104080. [CrossRef]
16. Alsulamy, S. Comparative Analysis of Deep Learning Algorithms for Predicting Construction Project Delays in Saudi Arabia. *Appl. Soft Comput.* **2025**, *172*, 112890. [CrossRef]
17. Li, Q.; Yang, Y.; Yao, G.; Wei, F.; Li, R.; Zhu, M.; Hou, H. Classification and Application of Deep Learning in Construction Engineering and Management—A Systematic Literature Review and Future Innovations. *Case Stud. Constr. Mater.* **2024**, *21*, e04051. [CrossRef]
18. Lung, L.W.; Wang, Y.R.; Chen, Y.S. Leveraging Deep Learning and Internet of Things for Dynamic Construction Site Risk Management. *Buildings* **2025**, *15*, 1325. [CrossRef]
19. Liu, Q.; He, P.; Peng, S.; Wang, T.; Ma, J. A Survey of Data-Driven Construction Materials Price Forecasting. *Buildings* **2024**, *14*, 3156. [CrossRef]
20. Chen, N.; Lin, X.; Jiang, H.; An, Y. Automated Building Information Modeling Compliance Check through a Large Language Model Combined with Deep Learning and Ontology. *Buildings* **2024**, *14*, 1983. [CrossRef]
21. Choi, S.M.; Cha, H.S.; Jiang, S. Hybrid Data Augmentation for Enhanced Crack Detection in Building Construction. *Buildings* **2024**, *14*, 1929. [CrossRef]
22. Nguyen, H.L.; Tran, V.Q. Data-Driven Approach for Investigating and Predicting Rutting Depth of Asphalt Concrete Containing Reclaimed Asphalt Pavement. *Constr. Build. Mater.* **2023**, *377*, 131116. [CrossRef]
23. Raza, M.S.; Sharma, S.K. Optimizing Porous Asphalt Mix Design for Permeability and Air Voids Using Response Surface Methodology and Artificial Neural Networks. *Constr. Build. Mater.* **2024**, *442*, 137513. [CrossRef]
24. Mabrouk, G.M.; Elbagalati, O.S.; Dessouky, S.; Fuentes, L.; Walubita, L.F. Using ANN Modeling for Pavement Layer Moduli Backcalculation as a Function of Traffic Speed Deflections. *Constr. Build. Mater.* **2022**, *315*, 125736. [CrossRef]
25. O'Shea, K.; Nash, R. An Introduction to Convolutional Neural Networks. *arXiv* **2015**, arXiv:1511.08458.
26. El-Chabib, H.; Nehdi, M.; Sonebi, M. Artificial Intelligence Model for Flowable Concrete Mixtures Used in Underwater Construction and Repair. *ACI Mater. J.* **2003**, *100*, 165–173. [CrossRef] [PubMed]
27. Shehadeh, A.; Alshboul, O.; Al Mamlook, R.E.; Hamedat, O. Machine Learning Models for Predicting the Residual Value of Heavy Construction Equipment: An Evaluation of Modified Decision Tree, LightGBM, and XGBoost Regression. *Autom. Constr.* **2021**, *129*, 103827. [CrossRef]
28. Chai, T.; Draxler, R.R. Root Mean Square Error (RMSE) or Mean Absolute Error (MAE)? *Geosci. Model Dev. Discuss.* **2014**, *7*, 1525–1534. [CrossRef]

buildings

MDPI

Article

Harnessing the Power of Improved Deep Learning for Precise Building Material Price Predictions

Zhilong Guo [1], Yayong Luo [2,3,*], Tongqiang Yi [4,*], Xiangnan Jing [5,*] and Jing Ma [6]

1 Key Laboratory of Cryospheric Science and Frozen Soil Engineering, Northwest Institute of Eco-Environment and Resources, Chinese Academy of Sciences, Lanzhou 730000, China; guozhilong@nieer.ac.cn
2 Naiman Desertification Research Station, Northwest Institute of Eco-Environment and Resources, Chinese Academy of Sciences, Lanzhou 730000, China
3 Laboratory of Stress Ecophysiology and Biotechnology, Northwest Institute of Eco-Environment and Resources, Chinese Academy of Sciences, Lanzhou 730000, China
4 School of Power and Mechanical Engineering, Wuhan University, Wuhan 430072, China
5 School of Economics and Management, City University of Hefei, Hefei 231137, China
6 Department of Psychology, Beijing Normal University, Beijing 100875, China; 202338061009@mail.bnu.edu.cn
* Correspondence: luoyy@lzb.ac.cn (Y.L.); tongqiang.yi@whu.edu.cn (T.Y.); xnjing@outlook.com (X.J.)

Abstract: Accurate forecasting of construction material prices is essential for effective cost control and risk management in construction projects. However, due to the influence of various complex factors, building material prices exhibit high nonlinearity and instability, often making traditional prediction methods inadequate for achieving optimal results. This study introduces an innovative prediction model, CEEMDAN-VMD-GRU-ARIMA, specifically designed for forecasting the price of prestressed steel bars. This model uniquely combines CEEMDAN and VMD to address nonlinear characteristics, and it innovatively incorporates sample entropy for the adaptive selection of either GRU or ARIMA for prediction. Additionally, a VMD decomposition mode number K value optimization method, based on a sparse index, is proposed. Experimental results demonstrate that the model performs exceptionally well, achieving an adjusted R-squared value of 81.10%, with various error indicators significantly surpassing the results for the baseline model. This approach offers new insights for short-term price prediction of building materials and contributes to enhancing the economic benefits and management efficiency of construction projects.

Keywords: construction material price forecasting; CEEMDAN-VMD-GRU-ARIMA; nonlinear time series; sample entropy; prestressed steel bars (PSB)

1. Introduction

The revitalization of historical urban districts and the enhancement of living standards within the real estate sector play a pivotal role in driving national economic growth [1]. The selection and management of building materials, serving as the physical foundation of the construction industry, significantly impact the cost estimation of construction projects [2]. Effective cost management is not only crucial for developers but also essential for the successful execution of real estate ventures [3]. In recent years, the Chinese real estate market has experienced contraction, prompting a heightened focus on cost accounting within the industry. The pricing of building materials is influenced by multiple interrelated factors, including market dynamics, production costs, macroeconomic conditions, political climates, and environmental policies [4,5]. These factors contribute to nonlinear price fluctuations that cannot be effectively predicted using simple additive principles. Consequently, accurately forecasting building material prices poses a substantial challenge for developers, who must ensure the economic viability of their projects. To address this challenge, developers must adopt sophisticated forecasting methods and tools capable of accounting for the complex and dynamic nature of building material pricing. Additionally, they must stay informed about market trends, regulatory changes, and other external factors that may impact material costs. By employing such strategies, developers can make informed decisions and mitigate the risks associated with cost overruns, ultimately contributing to the success of their real estate ventures and supporting broader economic development.

Time series forecasting is a powerful analytical tool that uncovers and predicts future trends and patterns of evolution by deeply mining and analyzing historical data. This method has been widely applied in various fields such as finance [6], meteorology [7], transportation [8], and healthcare [9], proving its outstanding ability for predicting the future. In the field of building material price forecasting, time series forecasting also plays a crucial role, providing profound analytical tools for stakeholders and helping decision makers make wiser decisions. However, the fluctuation of building material prices is a complex process that gradually emerges over time and is jointly influenced by various factors such as seasonal factors, market supply and demand relationships, raw material costs, policy adjustments, and the global economic environment. By accurately analyzing the historical data of building material prices, precise forecasting models can be constructed to discover the internal laws of price changes, thereby providing forward-looking guidance for market participants [10]. Forecasting models not only help developers and investors to avoid risks and optimize resource allocation but also provide decision support for policymakers to promote market stability and healthy development. Early forecasting models for building material prices were mostly based on traditional statistical methods. For example, Svetlana S. Uvarova et al. used the autoregressive integrated moving average (ARIMA) time series model to conduct an in-depth forecast analysis of the dynamic changes in steel prices [11]. This study not only provided a scientific forecasting tool for the fluctuations in steel prices but also offered practical guidance for actual building in the construction industry. By accurately predicting the trend of steel price fluctuations, construction companies can better control costs and manage risks, thereby improving construction efficiency and economic benefits. Jiang, F. et al. carefully constructed the ARIMA time series model to conduct an in-depth forecast analysis of the Construction Cost Index (CCI) [12]. They not only accurately constructed the model but also comprehensively and meticulously discussed the various factors affecting the CCI. The purpose of this study is to enhance the accuracy of decision makers in predicting building material prices and labor costs, thereby providing solid data support and a decision-making basis for cost control and risk assessment in the construction industry. Ilbeigi, M. et al. proposed four time-series forecasting models, i.e., Holt exponential smoothing, Holt–Winters ES, ARIMA, and SARIMA, and confirmed that these models have better predictive performance for asphalt cement prices than do existing forecasting models, thereby reducing the cost impact caused by asphalt cement price fluctuations in construction projects [13]. Although the above studies are based on traditional statistical models for building price index forecasting models, these models have limitations in dealing with nonlinear historical nodes in a time series, leading to unsatisfactory prediction accuracy. To overcome this challenge, researchers need to explore more advanced methods, such as machine learning and deep learning technologies, to improve the accuracy and adaptability of forecasting models [14]. Through continuous optimization and innovation, time series forecasting will play a greater role in the field of building material price forecasting, providing more accurate forward-looking guidance for market participants.

With the rapid advancement of big data and artificial intelligence technologies, methods for time series forecasting are continuously being innovated and refined. Modern forecasting models, such as those based on deep learning architectures like long short-term memory networks (LSTM) [15], convolutional neural networks (CNN) [16], and large language models (LLM) [17], have demonstrated exceptional performance in handling complex, nonlinear time series data. The broad application of these technologies further enhances the accuracy and reliability of building material price forecasting, providing more solid data support for decision making in related fields. The introduction of deep learning technology has ushered in new breakthroughs in the field of time series forecasting. Deep learning models, particularly LSTM and CNN, are capable of automatically extracting complex features from time series data, capturing long-term dependencies and nonlinear patterns within the data. For instance, Tang, B.Q. et al. innovatively combined an improved particle swarm optimization algorithm (IPSO) with least squares support vector machines (LSSVM) to develop an efficient building material price forecasting model [18]. Compared with traditional neural networks and traditional time series models, this method shows significant advantages: the mean relative error and mean square error were significantly reduced, confirming its outstanding performance in regards to prediction accuracy, convergence speed, and generalization capability. This advanced forecasting method not only accurately predicts building material prices but also helps determine the optimal purchase timing, effectively reducing cost impacts. By accurately predicting the fluctuations in building material prices, construction companies and developers can better plan procurement strategies, optimize inventory management, mitigate risks associated with price volatility, and enhance economic benefits. Mir, M. et al., addressing the shortcomings of machine learning models in precision estimation, proposed an innovative optimal lower and upper bound estimation (optimal LUBE) method [19]. This method, by training artificial neural networks (ANN), can output precise prediction error intervals, significantly enhancing decision-making accuracy. This groundbreaking result not only greatly reduced project cost risks but also brought a new research perspective and direction to the construction industry. Wang, J. et al. used k-nearest neighbors (KNN) and perfect random tree ensemble (PERT) algorithms to conduct in-depth short-term, medium-term, and long-term forecasting analyses of the CCI [20]. By comparing their results with those of existing forecasting models, they successfully demonstrated the significant advantages of the proposed method in predictive performance, providing valuable auxiliary decision-making tools for contractors and owners, enabling them to budget and control costs more accurately. The introduction of the above machine learning models greatly improved the accuracy and robustness of the predictions. Furthermore, the application of large language models (LLM) also provides a new perspective for time series forecasting, especially when dealing with time series data with rich semantic information, enabling a better understanding of the logic and patterns behind the dataset [21]. Although machine learning models have exhibited significant performance improvements, they still face the challenge of underfitting. The main reason for this issue is the presence of noise, which makes traditional statistical models and deep learning models difficult to accurately fit to historical building material prices and thus, making them unable to accurately predict future price trends. To address the above issues, this study employed advanced complete ensemble empirical mode decomposition with adaptive noise (CEEMDAN) and variational mode decomposition (VMD) techniques for the denoising of building material prices. On this basis, this study further combined GRU networks and ARIMA time series forecasting models for more accurate predictions of building material prices. The precision and dependability of these forecasting outcomes are pivotal to the successful and punctual execution of construction projects. Moreover, they offer substantial support for the enduring growth and sustainability of the construction sector, ensuring that it remains at the forefront of innovation and efficiency.

2. Data Sources and Analysis

2.1. Data Source

The experimental data in this study is obtained from the Zhengzhou Construction Project Cost Management Information, a reputable source known for its consistent and reliable monthly updates on construction material prices. In particular, our research focused on gathering monthly price for PSB from the years 2008 to 2023. To address any gaps in the collected data, we employed the Akima–Hermite interpolation (AKIMA) algorithm for data imputation [22]. The resulting price distribution diagram of the preprocessed PSB is illustrated in Figure 1.

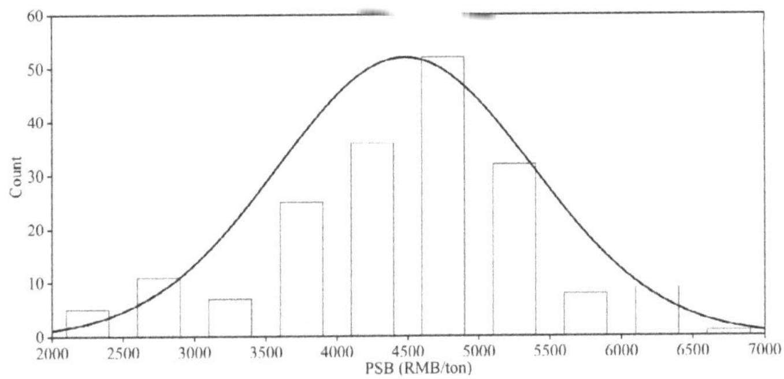

Figure 1. The chart of the PSB box type.

In Figure 1, it is clear that the pricing distribution of PSB exhibits a noticeable imbalance. The majority of prices are concentrated in the CNY 4000~5000/ton range, indicating a significant clustering in this particular region. Conversely, there is a distinct lack of prices falling within the CNY 2000~3000/ton and CNY 6000~7000/ton brackets. This uneven distribution underscores the considerable variability inherent in the pricing data for PSB. As a result, this poses challenges for accurately predicting the prices of PSB.

2.2. Data Analysis

To delve deeply into the patterns of change in the time series data of PSB prices, we leverage the capabilities of Eviews8.0 to meticulously examine the trend characteristics within the price time series data for PSB. The outcomes of our application of the Hodrick–Prescott (HP) filter model to this price time series are clearly presented in Figure 2. For the HP filter model, the parameter λ has been meticulously calibrated to a value of 14,400, adhering to the well-established empirical rule of thumb, which is particularly suited for the analysis of long-term economic time series data [23]. This strategic choice facilitates an accurate and insightful extraction of the intrinsic trend from the dataset.

Figure 2 presents a detailed time series analysis of PSB, revealing long-term trends and periodic fluctuations. The trend sequence identifies key elements of the pre-stressed steel bars associated with the overall trend, while the cycle sequence emphasizes regular fluctuations. The data indicate a downward price trend from 2008 to 2015, an upturn from 2016 to 2022, and a subsequent decline, suggesting a potential future decrease in prices. This could significantly impact China's construction industry by reducing construction costs, stimulating market growth, and increasing demand, potentially attracting more investors. However, the broader implications for market dynamics require consideration, as price reductions may lead to complex industry interactions and adjustments.

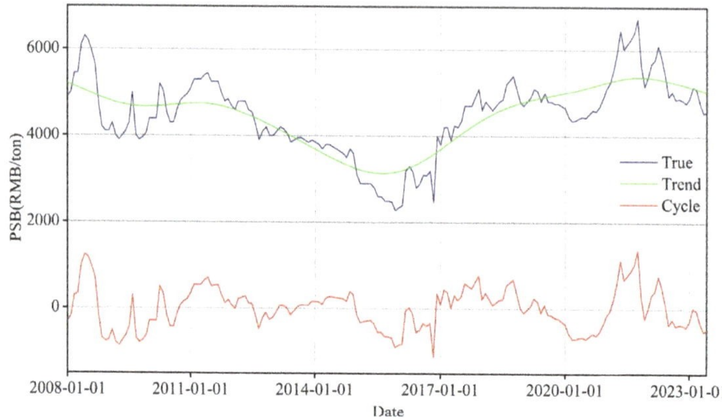

Figure 2. Analysis results of PSB with Hodrick–Prescott filter.

3. Methods

Material prices are a critical determinant of construction project costs and exhibit significant randomness. To enhance the accuracy and predictability of future predictions for the prices of PSB, this study employs machine learning techniques to construct an innovative model for forecasting the prices of PSB. The model has been rigorously validated to demonstrate its predictive capabilities. The study commenced with a comprehensive statistical analysis of the collected data regarding the prices of PSB. Missing values within the dataset were effectively addressed using the AKIMA algorithm. Subsequently, the dataset was partitioned into training and testing sets in a 4:1 ratio, laying the groundwork for the subsequent construction of a time series prediction model. Following this, the training set data were fed into the constructed prediction model, and five iterations of optimization were performed to identify and determine the optimal model parameters. The effectiveness of the model was then validated by inputting the test set data into the model. Ultimately, this study selected four popular machine learning time series prediction models and compared them with the integrated model developed in this study, further validating the model's predictive performance and effectiveness. For a detailed framework structure of the study, refer to Figure 3.

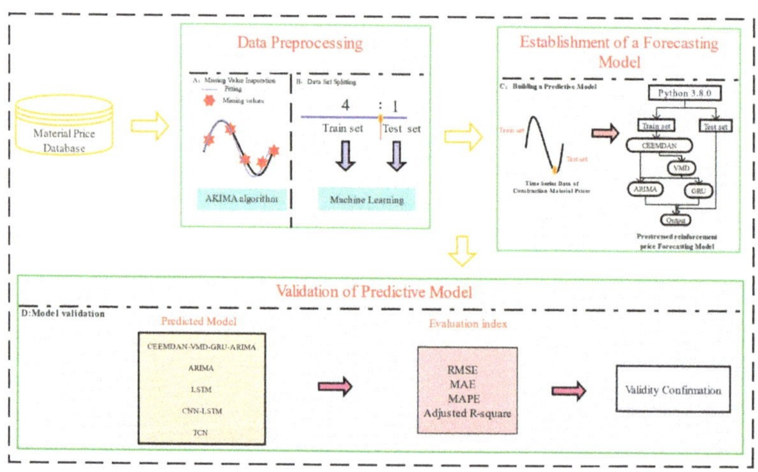

Figure 3. Research Framework Structure.

3.1. Integrated Decomposition–Ensemble Model for PSB Market Price Prediction

To enhance the precision of the price forecasting model for PSB, we introduce an innovative integrated prediction model that builds upon the theoretical framework of model decomposition and fusion models [24]. This model leverages the synergistic application of CEEMDAN and VMD, complemented by ARIMA and GRU networks. Below, we delineate the foundational principles that underpin the integration of these methodologies within our models.

3.1.1. CEEMDAN Model

Influenced by various factors, such as policy, environment, and economy, the price of PSB is subject to significant noise. To mitigate the impact of noise on model learning, this study utilizes complete ensemble empirical mode decomposition with adaptive noise (CEEMDAN) for analyzing PSB. CEEMDAN, an adaptive noise control signal decomposition method derived from complete ensemble empirical mode decomposition (CEEMD) and ensemble empirical mode decomposition (EEMD), employs adaptive control and integration strategies to manage the construction and addition of random noise and average EEMD decomposition structure. This approach reduces the influence of random noise on the results, enhancing the stability and accuracy of the decomposition outcomes [25]. In comparison to wavelet and Fourier transform methods, CEEMDAN demonstrates notable advantages in regards to data stationarity testing and linearity. As a time-domain analysis technique, CEEMDAN effectively minimizes reconstruction errors by progressively reducing noise in the dataset, with its efficacy confirmed through multiple experiments [26].

3.1.2. ARIMA Model

ARIMA, as a traditional time series forecasting model, has become widely used for its simplicity and ease of understanding. By applying autoregressive, moving average, and difference processing to time series data, the ARIMA model can effectively identify future changes and provide technical support for predicting future price fluctuations [27]. To construct an ARIMA model, three parameters must be specified: the autoregressive order (p), the difference order (d), and the moving average order (q). The stationarity of the data is evaluated using an ADF test, and the optimal values for parameters p and q are determined using the Akaike information criterion (AIC) and the Bayesian information criterion (BIC) [28]. The polynomial representations of the autoregressive and moving average processes are presented in Equations (1) and (2), while the ARMA(p,q) is represented in Equation (3).

$$y_t = \mu + \sum_{i=1}^{p} \gamma_i y_{t-i} + e_t \tag{1}$$

$$\begin{aligned} y_t &= \mu + \sum_{i=1}^{p} \gamma_i y_{t-i} + \epsilon_t y_t \\ &= \mu + e_t + \sum_{i=1}^{q} \theta_i e_{t-i} \end{aligned} \tag{2}$$

$$\begin{aligned} y_t &= \mu + \sum_{i=1}^{p} \gamma_i y_{t-i} + \epsilon_t y_t \\ &= \mu + e_t + \sum_{i=1}^{q} \theta_i \epsilon_{t-i} y_t \\ &= \mu + \sum_{i=1}^{p} \gamma_i y_{t-i} + \epsilon_t + \sum_{i=1}^{q} \theta_i \epsilon_{t-i} \end{aligned} \tag{3}$$

In the above equation, y_t represents the current value, μ is the constant term, p is the order, γ_i is the autocorrelation coefficient, and e_t represents the error term.

3.1.3. VMD Model

VMD is a signal decomposition algorithm based on empirical mode decomposition, which is capable of adaptively processing time series data [29]. This study suggests that VMD decomposition can effectively extract high-to-moderate complex components that may not be fully decomposed using CEEMDAN, thereby reducing the complexity of the data. However, the main challenge in VMD mode decomposition lies in determining the optimal value of K, which directly affects the effectiveness of signal variation. To address this challenge, the sparse index optimization algorithm is employed to determine the K value for VMD mode decomposition, with the maximum sparse index value selected as the optimal K value [30]. When calculating the sparsity of different modes generated by VMD decomposition, the varying capabilities of different mode decomposition variables are taken into consideration, and the energy full-time factor is introduced. Finally, the average value of marginal spectral sparsity is determined to enhance the robustness and effectiveness of VMD decomposition against the noise.

The signal obtained from VMD decomposition consists of intrinsic mode functions (IMFs) that exhibit specific frequencies and frequency bands. The generated components are iteratively processed using the alternating direction multiplier algorithm (ADMM) [31] to update the center frequency of their decomposition. This process ultimately leads to the determination of the saddle point of the unconstrained model, which represents the optimal solution for the VMD decomposition problem. The implementation of the VMD algorithm primarily involves constructing the variational model and solving the associated variational problem.

The process of constructing the variational model involves combining all the IMFs as the input signal and summing the frequency bands of the IMFs to form the objective function. The original signal is then inputted, and the Hilbert transform is applied to the K IMFs. The calculation equation is as follows:

$$\begin{cases} f_{\min} = \min\limits_{\{u_k\},\{\omega_k\}} \left\{ \sum\limits_{k=1}^{K} \left\| \sigma_t \left[\left(\delta(t) + \frac{j}{\pi t} \right) u_k(t) \right] e^{-j\omega_t t} \right\|_2^2 \right\} \\ s.t \quad \sum_{k=1}^{K} u_k = f(t) \end{cases} \tag{4}$$

In Equation (4), the $u_k(t)$ represents the generated IMFs, where k = 1, 2, ..., K. ω_k represents the central frequency of the IMFs; $\sigma_{(t)}$ represents the unit impulse function; $\sigma_{(t)}$ represents the first derivative of a function with respect to time. By incorporating the Lagrange function, the aforementioned model is refined into the optimal solution for the unconstrained problem, as depicted below in Equation (5).

$$\begin{gathered} L(\{u_k\},\{\omega_k\},\lambda) = \alpha \sum_{k=1}^{k} || \partial_t \left[(\delta(t) + \frac{j}{\pi t}) u_k(t) \right] e^{-j\omega_k t} ||_2^2 \\ + || \sum_{k=1}^{K} u_k - f(t) ||_2^2 + (\lambda(t), f(t) - \sum_{k=1}^{K} u_k(t)) \end{gathered} \tag{5}$$

In Equation (5), α denotes the quadratic penalty factor utilized for constraining the frequency band, while λ stands for the Lagrange multiplier. Equation (3) signifies the convergence criterion.

$$\frac{\sum\limits_{k=1}^{K} ||u_k^{n+1} - u_k^n||_2^2}{||u_k^n||_2^2} < \varepsilon \tag{6}$$

In Equation (6), ε is the precision control value, which is used to limit the relative error.

3.1.4. GRU Network

A neural network is a digital and computational model that emulates the structure of a biological neural network. It is composed of numerous interconnected neurons with associated weights [32]. Initially, artificial neural networks (ANNs) were prevalent, comprising an input layer, hidden layers, and an output layer. However, to address certain challenges, deep neural networks, like LSTM, TCN, and CNN-LSTM, were developed [33]. These networks are prone to gradient vanishing issues but have demonstrated superior performance regarding non-linear fitting compared to that of traditional statistical methods [34]. Consequently, they have found extensive applications across various domains.

The GRU network is a type of recurrent neural network(RNN) and a simplified variant of the LSTM network [35]. It combines the forget and input gates of the LSTM network into a single update gate, which controls the retention of historical information. Additionally, it retains the reset gate of the LSTM network, which determines the current state and historical information. Compared to the LSTM and TCN networks, GRU offers several advantages for effectively extracting information from time series data, thereby reducing training time and improving model performance. Figure 4 provides a schematic representation of the GRU architecture, elucidating its intricate design. The underlying mathematical equation that governs the GRU's operation is delineated as follows:

$$z_t = \sigma(W_z \cdot [h_{t-1}, x_t]) \tag{7}$$

$$r_t = \sigma(W_r \cdot [h_{t-1}, x_t]) \tag{8}$$

$$\widetilde{h_t} = tanh(W \cdot [r_t \odot h_{t-1}, x_t]) \tag{9}$$

$$h_t = (1 - z_t) \odot h_{t-1} + z_t \odot \widetilde{h_t} \tag{10}$$

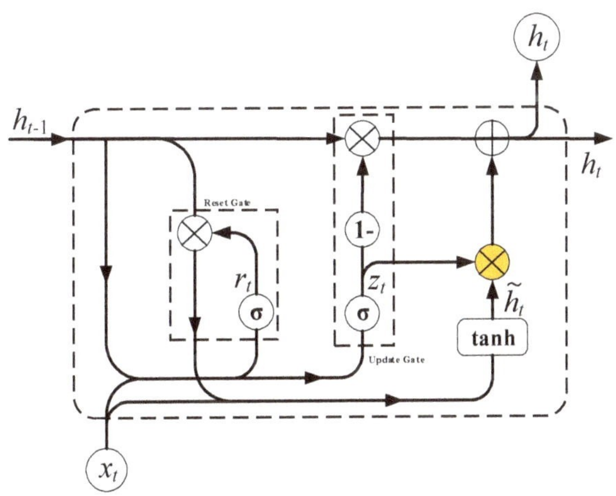

Figure 4. GRU Network Structure.

In the equation, z_t and r_t represent the update gate and the reset gate, respectively. The update gate controls how much information from the previous time step is incorporated into the current state. A higher update gate value means more information from the previous time step is included in the current state. The h_t and h_{t-1} are the output states of the current and previous hidden layers, respectively. $\widetilde{h_t}$ represents the combined relationship between x_t and h_{t-1}, and σ is the sigmoid activation function, while W_z, W_r, and W are the zero output matrix, reset gate weight matrix, and update gate weight matrix, respectively.

In summary, we have developed an integrative and advanced forecasting model specifically designed to predict PSB price dynamics. Our methodology is characterized by systematic precision and rigorous implementation to enhance the model's comprehensiveness and predictive accuracy. The model architecture, depicted in Figure 5, demonstrates its potential to advance research in materials economics and market analysis. In the initial phase, we preprocessed price data and implemented CEEMDAN decomposition using the PyEMD package in Python 3.8.0. Subsequent sample entropy (SE) testing identified components with significant variations, ensuring complete decomposition. A threshold of 0.5 was applied to evaluate time series complexity: components with SE values exceeding 0.5 underwent VMD based secondary decomposition and GRU network prediction, while those below 0.5 were analyzed via the ARIMA model. Finally, the forecast integrates weighted summations of IMF components derived from all methods, yielding robust predictions of PSB price trends.

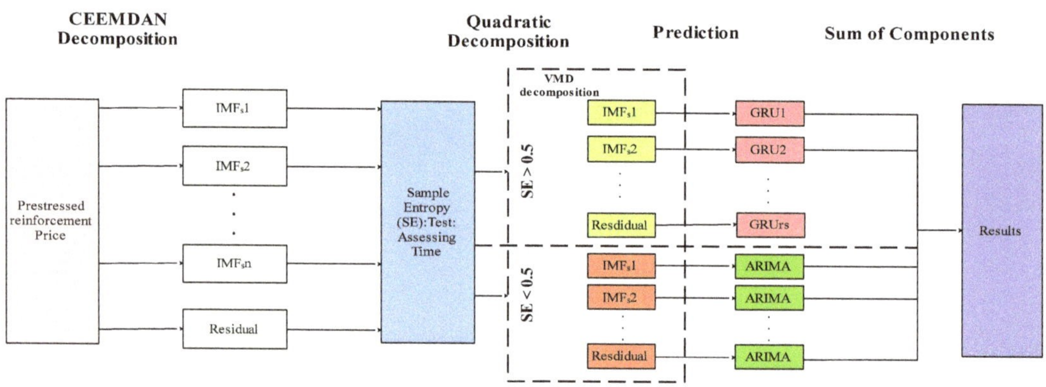

Figure 5. The structure chart for the PSB price forecast.

3.2. Model Evaluation Index

To rigorously validate the precision and dependability of our predictive model, we have chosen to compare it with four benchmark models through a comparative analysis. During this evaluation, we will employ adjusted R-squared, mean squared error (MSE), mean absolute error (MAE), and root mean squared error (RMSE) methods as the critical metrics for assessing model performance. A holistic consideration of these metrics will allow us to thoroughly evaluate the effectiveness of the predictive model and ensure its accuracy and stability when applied in practical scenarios. The adjusted R-squared value indicates the model's data fit, with the MSE and RMSE measuring the variance between the model's forecasts and the actual observations. The MAE evaluates the average discrepancy between the predicted and actual values. By comparing these metrics, we can more precisely determine the performance of the predictive model. The calculation equation is described as follows:

$$Adjusted\ R\_square = 1 - \frac{(1 - R^2)(N - 1)}{N - p - 1} \tag{11}$$

$$MSE = \frac{1}{N}\sum_{i=1}^{N}(y_i - \hat{y}_i) \tag{12}$$

$$MAE = \frac{1}{N}\sum_{i=1}^{N}|y_i - \hat{y}_i| \tag{13}$$

$$RMSE = \sqrt{\frac{1}{N}\sum_{i=1}^{N}(y_i - \hat{y}_i)^2} \quad (14)$$

In the equation, y_i represents the true price of the PSB, $\hat{y}_i$ represents the predicted price of the PSB, N denotes the number of samples, and p denotes the number of feature indicators considered.

4. Result

4.1. Predicted Model Training Results

As previously discussed, the pricing of building materials is subject to a multitude of influences, including market fluctuations, external factors, and the size of the data sample. Consequently, accurately predicting these prices remains a significant challenge. While the ARIMA model has traditionally been utilized for time series forecasting in this area, its limitations in capturing nonlinear patterns have led to suboptimal performance in predicting building material prices. To address this issue, time series decomposition has emerged as a powerful tool in various fields, such as stock markets [36], environmental research [37], and fault diagnosis [38]. By breaking down the original series into trend, seasonality, and residual components, this method offers a more comprehensive understanding of the underlying structure and characteristics of the data, ultimately improving predictive accuracy. CEEMDAN, a time series decomposition algorithm, demonstrates superior adaptivity and noise control compared to those of wavelet decomposition, EMD, and EEMD. The five model components generated by CEEMDAN decomposition in Figure 5 exhibit a significant reduction in nonlinearity and randomness in the decomposed subsequence of PSB prices.

However, the CEEMDAN model is susceptible to noise interference during signal decomposition, potentially compromising the accuracy of the decomposition results. To mitigate the impact of noise on the model, this study proposes the utilization of sample entropy (SE) [39] to assess the complexity and uncertainty of the model components' CEEMDAN decomposition, with a predefined threshold of 0.5. The SE exceeding 0.5 indicates incomplete signal decomposition by CEEMDAN, whereas a value below 0.5 signifies successful signal decomposition. This approach enables a more precise evaluation of CEEMDAN decomposition efficacy and facilitates further optimization of the model's performance. For the sample entropy calculation results, see Figure 6.

In Figure 7, model component 1 and model component 2, generated by CEEMDAN decomposition, exhibit high SE values, both exceeding 0.5. This indicates that IMFs1 and IMFs2 possess a certain level of complexity and uncertainty. To mitigate the impact of this complexity and uncertainty on the model learning performance, a quadratic decomposition of IMFs1 and IMFs2 is performed in this study to address the model mixing problem. While the VMD decomposition has demonstrated favorable outcomes in various fields, its decomposition effect parameters, including the model number K-value, the penalty factor α, and the Lagrange multiplier, significantly influence the results. Among these parameters, the setting of the K-value is particularly crucial.

By effectively setting the K-value and ensuring that the other parameters fall within the valid range, the signal can be effectively decomposed. Therefore, the effectiveness of VMD decomposition is directly influenced by the appropriate selection of the K-value. To address the impact of the K-value on VMD decomposition, researchers have sought the optimal K-value using methods such as the energy ratio [40], genetic algorithm [41], and whale optimization algorithm [42]. However, these methods primarily seek local optima and do not fully utilize the VMD process. In light of this, this study proposes an adaptive K-value optimized variational model decomposition based on the sparsity index during the decomposition process [43]. The SE of the VMD components under different decomposition model number K-values is employed as an index for optimization. The specific optimization search results are presented in Figure 7.

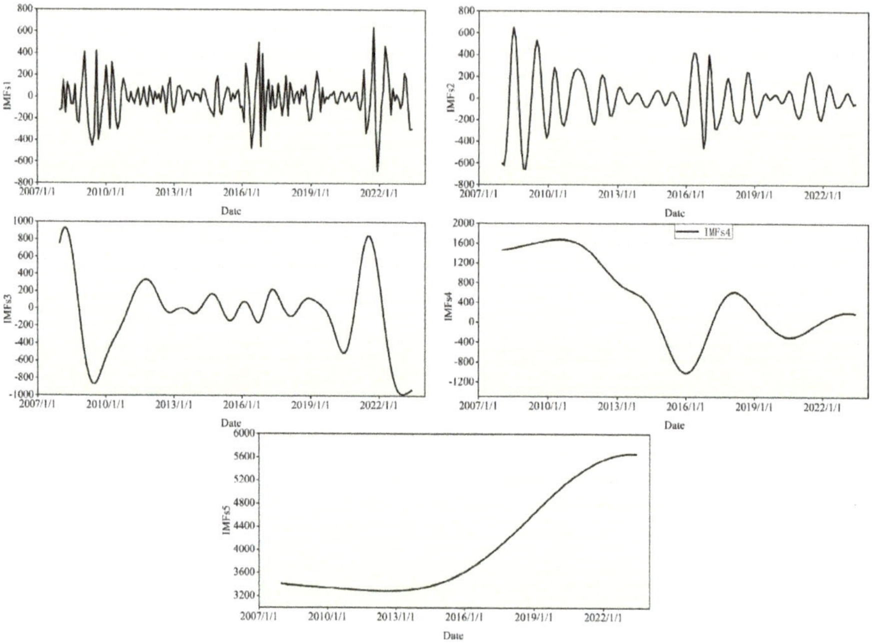

Figure 6. Results of CEEMDAN decomposition.

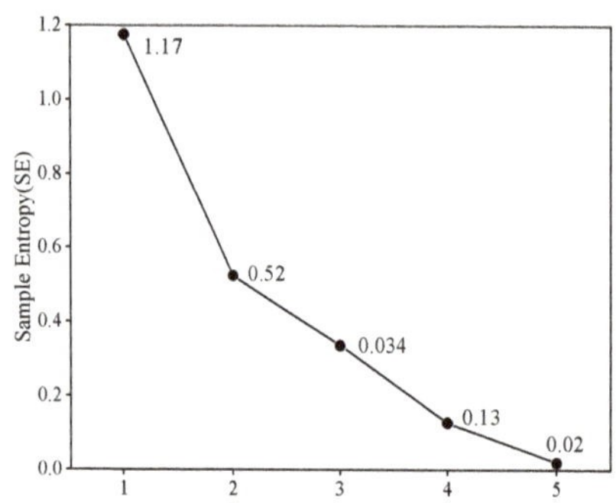

Figure 7. Sample entropy of each component for simulation analysis.

From Figure 8, it can be inferred that the optimal parameter K-value for the VMD variational modes of IMF1 and IMF2 are determined as 10 and 6, respectively, through sparse index optimization. Subsequently, this optimized K-value is utilized to decompose IMFs1 and IMFs2 using the VMD technique.

Figures 9 and 10 present the final model decomposition results for the price of PSB. The sample entropy value of the model variables obtained from the CEEMDAN decomposition is below 0.5, indicating a complete decomposition of the model. Since data with low SE results have lower noise levels, to avoid issues such as overfitting during the training process of deep learning models on data with sample entropy, this study opts for the traditional statistical prediction model ARIMA for prediction. However, for the model components obtained from VMD decomposition, the deep learning model GRU is employed for prediction. The experimental training and testing datasets are divided in accordance with the mathematical and statistical specification standard of a 4:1 ratio. Finally, the variables predicted by the ARIMA and GRU models are fused using weighting, and the final testing datasets predicting prestressing rebar price results are output in Figure 9.

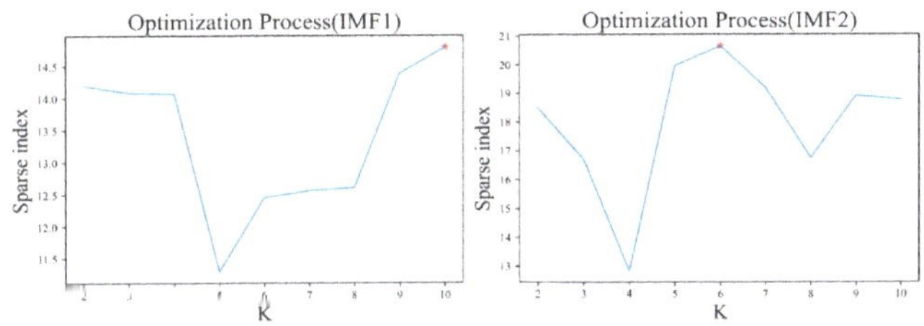

Figure 8. K-valued optimization solution for VMD decomposition of IMF.

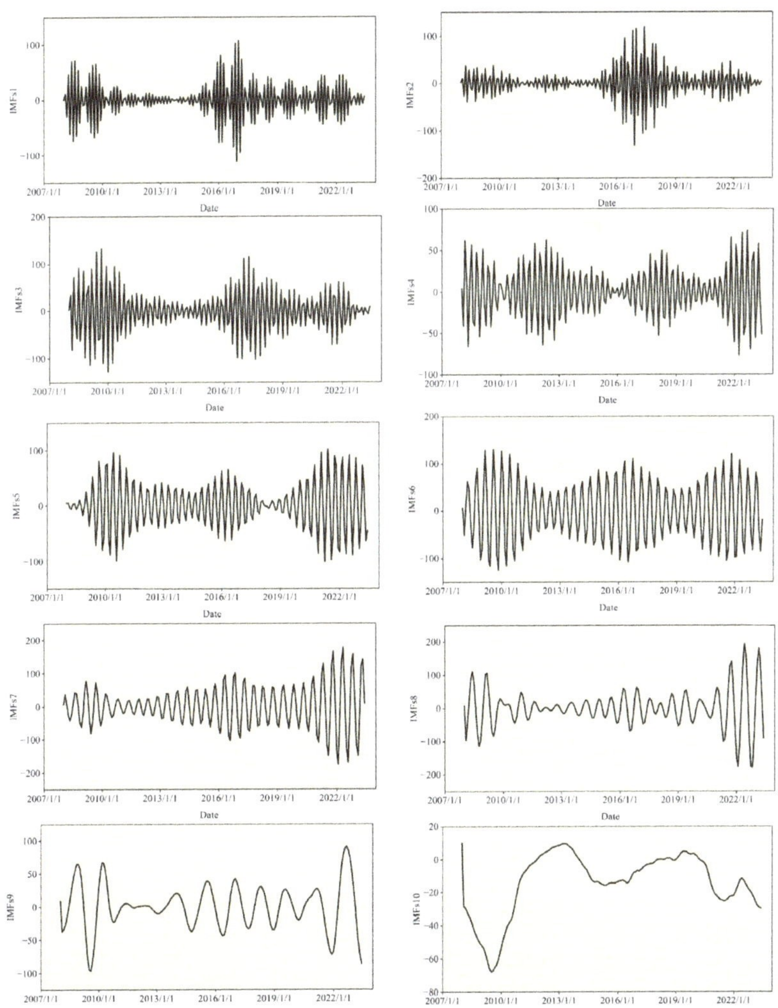

Figure 9. CEEMDAN-VMD Decomposition Results: Modal Feature Analysis of Variable IMF1s.

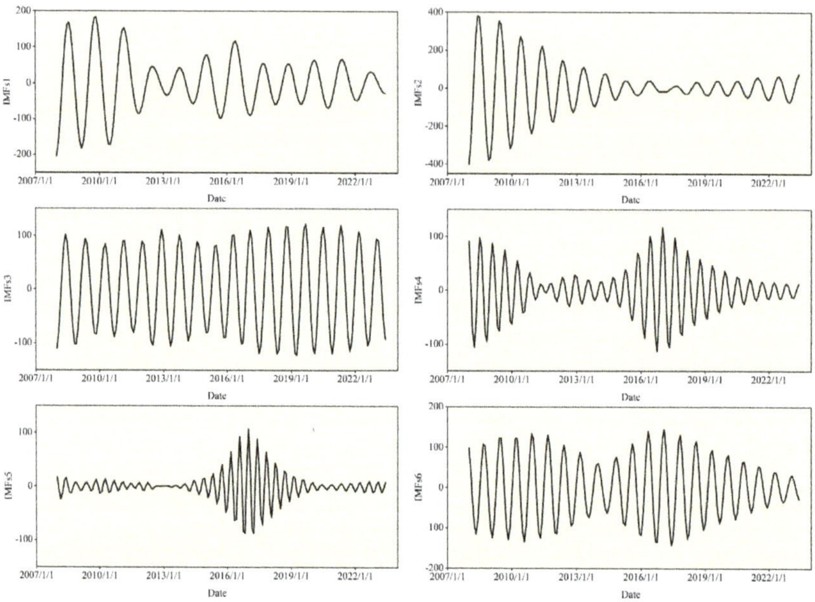

Figure 10. CEEMDAN-VMD decomposition results: modal feature analysis of variable IMF2s.

In Figure 11, the CEEMDAN-VMD-GRU-ARIMA model exhibits exceptional performance in predicting the price of PSB based on our empirical tests. The model's adjusted R-squared value significantly exceeds 81.10%, demonstrating its high precision and accuracy. Further analysis reveals that while the model generates significant inaccuracies within the price range of CNY 4000–4500/ton, these inconsistencies do not negatively impact its overall effectiveness or applicability. Summarizing these findings, we can confidently state that the CEEMDAN-VMD-GRU-ARIMA model is highly reliable and practically viable for short-term PSB price prediction. The model's robust performance across different price points suggests its potential as a valuable tool for real-world applications in cost management and risk mitigation strategies within construction projects.

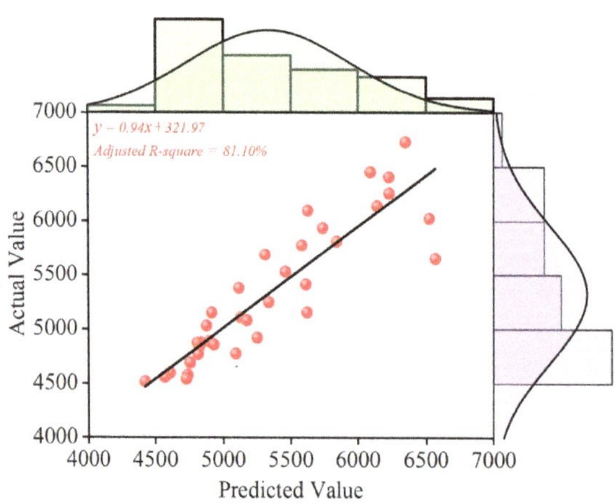

Figure 11. The marginal histogram of actual and fitted values of PSB.

As shown in Figure 11, empirical tests demonstrate that the CEEMDAN-VMD-GRU-ARIMA model achieves exceptional performance in predicting PSB prices, with an adjusted R^2 value exceeding 81.10%, indicating high precision and reliability. Further analysis reveals that although the model exhibits notable inaccuracies within the CNY4000–4500/ton price range, as shown in Figure 12, these localized discrepancies do not compromise its overall effectiveness for short-term forecasting. In summary, the proposed hybrid model demonstrates robust generalizability and practical viability, particularly for cost management and risk mitigation in construction projects, where dynamic price fluctuations require timely decision making.

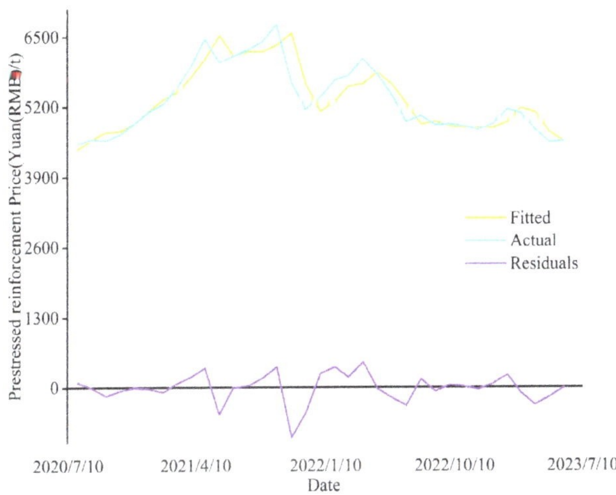

Figure 12. Estimating true and predicted values using model residuals.

4.2. *Results for Verification of Predicted Models*

The rigorous validation of our model is paramount for substantiating the veracity and authority of our research outcomes, thereby laying a robust groundwork for future policy advocacy and strategic decision making. By conducting a meticulous examination and validation of our model, we can assert with confidence that it excels not only in capturing historical pricing data of construction materials but also in forecasting future price trajectories with a high degree of precision. The results of the test set are graphically depicted in Figure 11 and Table 1. As illustrated in Figure 11, there are notable fluctuations in the model's predictive accuracy on 1 May 2021, 1 June 2021, and between 1 October 2021 and 1 February 2022. Detailed findings are delineated in Table 1. Such fluctuations may stem from the intricacies of the market environment or anomalies in data during these intervals. Conversely, in other periods, the model exhibits relatively minor error variance, indicative of its stability and dependability under those conditions. Upon a comprehensive analysis of the time series forecasting model, it is evident that it encounters certain constraints in the prediction of time series peaks. This limitation may arise from the model's inadequate capacity to capture outlier events or its reduced sensitivity to market volatility. Nonetheless, within more stable time series contexts, the model demonstrates superior performance.

In our quest to ascertain the precision and reliability of our predictive model, we subjected it to a stringent comparative analysis against four established benchmark models, as delineated in Figure 12. This rigorous evaluation encompassed key performance metrics such as adjusted R-squared, MSE, MAE, and RMSE. Our novel CEEMDAN-VMD-GRU-ARIMA approach emerged as superior across all metrics, significantly outperforming its counterparts, with the ARIMA model securing the second position and the CNN-LSTM model exhibiting comparatively lower accuracy. These disparities can be ascribed to the distinct algorithmic attributes and the specific contexts in which each model operates within the realm of time series analysis. The CEEMDAN-VMD-GRU-ARIMA method's synergistic integration of diverse algorithms adeptly harnesses the nonlinear and dynamic intricacies of time series data, thereby augmenting its predictive prowess. In contrast, the ARIMA model, while effective in certain scenarios, encounters limitations when confronted with intricate datasets. Similarly, the CNN-LSTM model, despite its excellence in specific domains, requires further enhancement for adept time series forecasting. This comprehensive analysis serves as a pivotal reference for the ongoing refinement and optimization of our model, ensuring its robustness and efficacy in the evolving landscape of predictive analytics.

Table 1. Comparative analysis of actual values, predicted values, and error metrics.

Date	Actual	Predicted	Residuals
2020-8-1	3520.00	4422.52	97.47
2020-9-1	4600.00	4606.04	−6.04
2020-10-1	4590.00	4738.86	−153.86
2020-11-1	4693.00	4750.98	57.98
2020-12-1	4890.00	4892.68	−2.68
2021-1-1	5110.00	5128.93	−18.93
2021-2-1	5500.00	5337.11	−87.11
2021-3-1	5533.00	5459.93	−126.66
2021-4-1	5640.00	5571.37	199.04
2021-5-1	6433.33	6069.67	363.66
2021-6-1	5666.67	6522.55	−255.88
2021-7-1	5540.00	6139.05	−199.05
2021-8-1	6256.67	5645.58	611.10
2021-9-1	6400.00	6206.27	193.73
2021-10-1	6730.00	6345.15	384.85
2021-11-1	6563.33	6587.06	−313.73
2021-12-1	6155.82	6618.85	−463.03
2022-1-1	5880.11	5966.36	−86.25
2022-2-1	5690.00	5897.80	380.20
2022-3-1	5776.67	5580.46	196.21
2022-4-1	6066.67	5625.06	−71.61
2022-5-1	5813.33	5841.02	−27.69
2022-6-1	4923.33	5119.30	−195.98
2022-7-1	4923.33	5250.18	−326.85
2022-8-1	5033.33	4872.65	160.68
2022-9-1	4860.00	4928.97	−68.97
2022-10-1	4476.67	4625.76	−149.09
2022-11-1	4437.94	4417.82	20.13
2022-12-1	4774.15	4810.37	−36.22
2023-1-1	4478.33	4505.79	70.54
2023-2-1	5153.33	4422.20	−21.13
2023-3-1	5087.88	4374.26	378.09
2023-4-1	4776.67	4092.53	−315.86
2023-5-1	4486.67	4308.58	178.09
2023-6-1	4563.33	4568.98	−5.65

As evidenced in Table 2, the CEEMDAN-VMD-GRU-ARIMA model achieves an Adjusted R-squared value of 81.10%, with corresponding MSE, MAE, and RMSE values of 73,078.79, 189.39, and 270.33, respectively. Notably, the model exhibits Adjusted R-squared improvements of 2.3%, 6%, 6.1%, and 22.8% over the ARIMA, TCN, LSTM, and CNN-LSTM benchmarks, respectively. This performance differential underscores the model's superior predictive accuracy in time series forecasting. Furthermore, the proposed framework demonstrates consistent advantages across all error metrics (MSE, MAE, and RMSE), with lower values indicating reduced prediction errors and enhanced precision relative to that of comparator models. These empirical results validate the CEEMDAN-VMD-GRU-ARIMA architecture's capacity to synergistically integrate diverse algorithmic strengths, effectively modeling nonlinear and dynamic data patterns for improved forecasting fidelity. The findings emphasize the methodological necessity of strategically combining model architectures tailored to specific data characteristics and application contexts when addressing time series prediction challenges. While Figure 11 conclusively establishes the framework's accuracy and reliability for time series applications, the comparative analysis simultaneously highlights opportunities for optimizing alternative models through architectural refinements. Future research should prioritize such enhancements to advance predictive performance across methodological paradigms.

Table 2. Model evaluation result.

Evaluation Metrics	Adjusted R-Squared Value	MSE	MAE	RMSE
ARIMA	78.80%	90,420.24	200.60	300.70
LSTM	75.00%	101,296.81	253.02	318.20
CNN-LSTM	38.30%	169,337.18	307.13	411.50
TCN	73.10%	100,996.78	258.86	317.80
CEEMDAN-VMD-GRU-ARIMA	81.10%	73,078.79	189.39	270.33

5. Discussion

The development of the CEEMDAN-VMD-GRU-ARIMA model represents a significant methodological advancement in price forecasting for construction materials, particularly PSB. By integrating complementary analytical paradigms—CEEMDAN and VMD for non-stationary signal decomposition, GRU for capturing temporal dependencies, and ARIMA for modeling linear trends—the hybrid framework addresses critical limitations in conventional single-model approaches. This dual quantitative–qualitative architecture demonstrates superior adaptability to market volatility, a persistent challenge in construction material pricing, for which external economic factors and supply–demand imbalances often induce nonlinear price fluctuations. The model's self-optimizing capabilities, enabled by iterative learning mechanisms, offer a critical advantage in dynamic markets. Unlike static models requiring frequent recalibration, its adaptive structure aligns with the construction industry's need for real-time decision support, particularly given that material costs account for 60–70% of total project expenditures. By reducing prediction errors through multi-scale feature extraction and hybrid learning, the framework enhances risk assessment accuracy for contractors and project managers, directly supporting cost containment strategies and tender pricing optimization. However, the current methodology focuses exclusively on historical price series, omitting exogenous variables such as raw material availability, policy changes, or geopolitical influences. While this ensures computational efficiency, it may limit predictive robustness during systemic market shocks, a limitation partially mitigated by the model's decomposition–recomposition mechanism but warranting further investigation. Practically, the framework's success in PSB forecasting suggests its transferability to other price-volatile construction materials, although domain-specific adjustments would be necessary. The theoretical contribution lies in demonstrating how hybrid architectures can reconcile econometric rigor with machine learning flexibility, a paradigm applicable beyond construction to energy, agriculture, and financial markets.

Future extensions should prioritize three dimensions, i.e., incorporating macroeconomic indicators as model inputs, testing cross-market generalizability through comparative studies, and developing real-time updating protocols for operational deployment. Such advancements could transform the model from a predictive tool into a prescriptive system, enabling proactive supply chain adjustments and policy-responsive budgeting in the construction sector.

6. Conclusions

This study proposes the CEEMDAN-VMD-GRU-ARIMA model for PSB price forecasting, combining one-dimensional price series analysis with integrated machine learning and data mining techniques to achieve accurate and reliable predictions. Compared to traditional construction material price forecasting models, this model demonstrates significant improvements in prediction accuracy and response latency. In the future, this model can be expanded to predict factors such as cement prices, labor costs, and tax rates in the construction industry, offering new theoretical advancements and practical applications for market analysis and risk management. Subsequent research will focus on algorithm optimization and explore the model's potential for cross-market predictive applications.

Author Contributions: Conceptualization, Z.G. and X.J.; methodology, X.J.; software X.J.; validation, Z.G., X.J., Y.L. and T.Y.; investigation, X.J.; writing—original draft preparation, Z.G. and X.J.; writing—review and editing, Z.G., X.J., Y.L. and T.Y.; project administration, Z.G., X.J. and J.M.; funding acquisition, Z.G. and X.J. All authors have read and agreed to the published version of the manuscript.

Funding: This study was performed with the financial support of the Natural Science Foundation of the Anhui Provincial Education Department (2023AH053239) and the Key Laboratory of Cryospheric Science and Frozen Soil Engineering, Chinese Academy of Sciences (KLCSFSE-ZZ-2025).

Data Availability Statement: The data presented in this study are available on request from the corresponding author.

Conflicts of Interest: The authors declare no conflicts of interest.

References

1. Wang, D.; Liu, J.; Wang, X.; Chen, Y. Cost-effectiveness analysis and evaluation of a 'three-old'reconstruction project based on smart system. *Clust. Comput.* **2019**, *22*, 7895–7905. [CrossRef]
2. Hwang, S.; Park, M.; Lee, H.-S.; Kim, H. Automated time-series cost forecasting system for construction materials. *J. Constr. Eng. Manag.* **2012**, *138*, 1259–1269. [CrossRef]
3. Marzouk, M.; Hamdala, D. Phasing real estate projects considering profitability and customer satisfaction. *Eng. Constr. Archit. Manag.* 2024, *ahead-of-print*. [CrossRef]
4. Musarat, M.A.; Alaloul, W.S.; Liew, M.; Maqsoom, A.; Qureshi, A.H. Investigating the impact of inflation on building materials prices in construction industry. *J. Build. Eng.* **2020**, *32*, 101485. [CrossRef]
5. Kissi, E.; Sadick, M.; Agyemang, D. Drivers militating against the pricing of sustainable construction materials: The Ghanaian quantity surveyors perspective. *Case Stud. Constr. Mater.* **2018**, *8*, 507–516. [CrossRef]
6. Zhao, C.; Hu, P.; Liu, X.; Lan, X.; Zhang, H. Stock market analysis using time series relational models for stock price prediction. *Mathematics* **2023**, *11*, 1130. [CrossRef]
7. Guo, Z.; Jing, X.; Ling, Y.; Yang, Y.; Jing, N.; Yuan, R.; Liu, Y. Optimized air quality management based on air quality index prediction and air pollutants identification in representative cities in China. *Sci. Rep.* **2024**, *14*, 17923. [CrossRef]
8. Yin, Y.; Shang, P. Forecasting traffic time series with multivariate predicting method. *Appl. Math. Comput.* **2016**, *291*, 266–278. [CrossRef]
9. Mehrmolaei, S.; Savargiv, M.; Keyvanpour, M.R. Hybrid learning-oriented approaches for predicting Covid-19 time series data: A comparative analytical study. *Eng. Appl. Artif. Intell.* **2023**, *126*, 106754. [CrossRef]
10. Tarnate, W.R.D.; Ponci, F.; Monti, A. Uncertainty-aware model predictive control for residential buildings participating in intraday markets. *IEEE Access* **2022**, *10*, 7834–7851. [CrossRef]

11. Uvarova, S.S.; Belyaeva, S.V.; Orlov, A.K.; Kankhva, V.S. Cost Forecasting for Building Materials under Conditions of Uncertainty: Methodology and Practice. *Buildings* **2023**, *13*, 2371. [CrossRef]
12. Jiang, F.; Awaitey, J.; Xie, H. Analysis of construction cost and investment planning using time series data. *Sustainability* **2022**, *14*, 1703. [CrossRef]
13. Ilbeigi, M.; Ashuri, B.; Joukar, A. Time-series analysis for forecasting asphalt-cement price. *J. Manag. Eng.* **2017**, *33*, 04016030. [CrossRef]
14. Kożuch, A.; Cywicka, D.; Adamowicz, K.J.F. A comparison of artificial neural network and time series models for timber price forecasting. *Forests* **2023**, *14*, 177. [CrossRef]
15. Yin, L.; Wang, L.; Li, T.; Lu, S.; Tian, J.; Yin, Z.; Li, X.; Zheng, W. U-Net-LSTM: Time series-enhanced lake boundary prediction model. *Land* **2023**, *12*, 1859. [CrossRef]
16. Yuan, F.; Zhang, Z.; Fang, Z. An effective CNN and Transformer complementary network for medical image segmentation. *Pattern Recognit.* **2023**, *136*, 109228. [CrossRef]
17. de Zarzà i Cubero, I.; de Curtò i Díaz, J.; Roig, G.; Calafate, C.T. LLM Multimodel Traffic Accident Forecasting. *Sensors* **2023**, *23*, 9225. [CrossRef]
18. Tang, B.Q.; Han, J.; Guo, G.-f.; Chen, Y.; Zhang, S. Building material prices forecasting based on least square support vector machine and improved particle swarm optimization. *Archit. Eng. Des. Manag.* **2019**, *15*, 196–212. [CrossRef]
19. Mir, M.; Kabir, H.D.; Nasirzadeh, F.; Khosravi, A. Neural network-based interval forecasting of construction material prices. *J. Build. Eng.* **2021**, *39*, 102288. [CrossRef]
20. Wang, J.; Ashuri, B. Predicting ENR construction cost index using machine-learning algorithms. *Int. J. Constr. Educ. Res.* **2017**, *13*, 47–63. [CrossRef]
21. Liu, Y.; Qin, G.; Huang, X.; Wang, J.; Long, M. Autotimes: Autoregressive time series forecasters via large language models. *arXiv* **2024**, arXiv:2402.02370.
22. Wang, Y.; Yang, D.; Liu, Y. A real-time look-ahead interpolation algorithm based on Akima curve fitting. *Int. J. Mach. Tools Manuf.* **2014**, *85*, 122–130. [CrossRef]
23. Hamilton, J.D. Why you should never use the Hodrick-Prescott filter. *Rev. Econ. Stat.* **2018**, *100*, 831–843. [CrossRef]
24. Nasir, J.; Aamir, M.; Haq, Z.U.; Khan, S.; Amin, M.Y.; Naeem, M. A new approach for forecasting crude oil prices based on stochastic and deterministic influences of LMD Using ARIMA and LSTM Models. *IEEE Access* **2023**, *11*, 14322–14339. [CrossRef]
25. Jun, W.; Yuyan, L.; Lingyu, T.; Peng, G. A new weighted CEEMDAN-based prediction model: An experimental investigation of decomposition and non-decomposition approaches. *Knowl. Based Syst.* **2018**, *160*, 188–199. [CrossRef]
26. Gao, B.; Huang, X.; Shi, J.; Tai, Y.; Zhang, J. Hourly forecasting of solar irradiance based on CEEMDAN and multi-strategy CNN-LSTM neural networks. *Renew. Energy* **2020**, *162*, 1665–1683. [CrossRef]
27. Shumway, R.H.; Stoffer, D.S. *Time Series Analysis and Its Applications (Springer Texts in Statistics)*; Springer: Berlin/Heidelberg, Germany, 2005.
28. Thiruchelvam, L.; Dass, S.C.; Asirvadam, V.S.; Daud, H.; Gill, B.S. Determine neighboring region spatial effect on dengue cases using ensemble ARIMA models. *Sci. Rep.* **2021**, *11*, 5873. [CrossRef]
29. Gan, M.; Pan, H.; Chen, Y.; Pan, S. Application of the Variational Mode Decomposition (VMD) method to river tides. *Estuar. Coast. Shelf Sci.* **2021**, *261*, 107570. [CrossRef]
30. Li, X.P.; Shi, Z.-L.; Leung, C.-S.; So, H.C. Sparse index tracking with K-sparsity or ϵ-deviation constraint via ℓ 0-norm minimization. *IEEE Trans. Neural Netw. Learn. Syst.* **2022**, *34*, 10930–10943. [CrossRef]
31. Nazari, M.; Sakhaei, S.M. Successive variational mode decomposition. *Signal Process.* **2020**, *174*, 107610. [CrossRef]
32. Wang, M.-H.; Hung, C. Extension neural network and its applications. *Neural Netw.* **2003**, *16*, 779–784. [CrossRef]
33. Hu, C.; Cheng, F.; Ma, L.; Li, B. State of charge estimation for lithium-ion batteries based on TCN-LSTM neural networks. *J. Electrochem. Soc.* **2022**, *169*, 030544. [CrossRef]
34. Ghazvini, A.; Sharef, N.M.; Sidi, F.B. Prediction of Course Grades in Computer Science Higher Education Program Via a Combination of Loss Functions in LSTM Model. *IEEE Access* **2024**, *12*, 30220–30241. [CrossRef]
35. Gao, S.; Huang, Y.; Zhang, S.; Han, J.; Wang, G.; Zhang, M.; Lin, Q. Short-term runoff prediction with GRU and LSTM networks without requiring time step optimization during sample generation. *J. Hydrol.* **2020**, *589*, 125188. [CrossRef]
36. Xu, M.; Shang, P.; Lin, A. Cross-correlation analysis of stock markets using EMD and EEMD. *Phys. A Stat. Mech. Its Appl.* **2016**, *442*, 82–90. [CrossRef]
37. Huang, Y.; Yu, J.; Dai, X.; Huang, Z.; Li, Y. Air-quality prediction based on the EMD–IPSO–LSTM combination model. *Sustainability* **2022**, *14*, 4889. [CrossRef]
38. Dong, H.; Qi, K.; Chen, X.; Zi, Y.; He, Z.; Li, B. Sifting process of EMD and its application in rolling element bearing fault diagnosis. *J. Mech. Sci. Technol.* **2009**, *23*, 2000–2007. [CrossRef]
39. Ding, J.; Xiao, D.; Huang, L.; Li, X. Gear fault diagnosis based on VMD sample entropy and discrete hopfield neural network. *Math. Probl. Eng.* **2020**, *2020*, 8882653. [CrossRef]

40. Zhang, P.; Gao, D.; Lu, Y.; Kong, L.; Ma, Z. Online chatter detection in milling process based on fast iterative VMD and energy ratio difference. *Measurement* **2022**, *194*, 111060. [CrossRef]
41. Wang, Y.; Chen, P.; Zhao, Y.; Sun, Y. A denoising method for mining cable PD signal based on genetic algorithm optimization of VMD and wavelet threshold. *Sensors* **2022**, *22*, 9386. [CrossRef]
42. Wang, H.; Wu, F.; Zhang, L. Application of variational mode decomposition optimized with improved whale optimization algorithm in bearing failure diagnosis. *Alex. Eng. J.* **2021**, *60*, 4689–4699. [CrossRef]
43. Li, J.; Yao, X.; Wang, H.; Zhang, J. Periodic impulses extraction based on improved adaptive VMD and sparse code shrinkage denoising and its application in rotating machinery fault diagnosis. *Mech. Syst. Signal Process.* **2019**, *126*, 568–589. [CrossRef]

buildings

MDPI

Article

Optimizing Urban Block Morphology for Energy Efficiency and Photovoltaic Utilization: Case Study of Wuhan

Ruoyao Wang [1], Yanyan Huang [1,2,*], Guoliang Zhang [1], Yi Yang [1] and Qizhi Dong [1]

[1] School of Civil Engineering, Architecture and Environment, Hubei University of Technology, Wuhan 430068, China; 2110621105@hbut.edu.cn (R.W.); 102200802@hbut.edu.cn (G.Z.); 102200801@hbut.edu.cn (Y.Y.); 2210611504@hbut.edu.cn (Q.D.)

[2] National Experimental Teaching Demonstration Center of Engineering and Architecture, Hubei University of Technology, Wuhan 430068, China

* Correspondence: 19960014@hbut.edu.cn

Abstract: With global carbon emissions continuing to rise and urban energy demands growing steadily, understanding how urban block morphology impacts building photovoltaic (PV) efficiency and energy consumption has become crucial for sustainable urban development and climate change mitigation. Current research primarily focuses on individual building optimization, while block-scale coupling relationships between PV utilization and energy consumption remain underexplored. This study developed an integrated prediction and optimization tool using deep learning and physical simulation to assess how urban block design parameters (building morphology, orientation, and layout) affect PV efficiency and energy performance. Through a methodology combining block modeling, PV potential assessment, and energy consumption simulation, the research quantified relationships between design parameters, PV utilization, and energy consumption. Results demonstrate that appropriate building forms and layouts reduce shadow obstruction, enhance PV system capability, and simultaneously improve PV efficiency while reducing energy consumption. The tool provides improved prediction accuracy, enabling urban planners to scientifically design block layouts that maximize PV generation and minimize energy use. Extensive experimental validation demonstrates that the integrated model and analytical methods proposed in this study will help urban planners break through the limitations of individual building research, making PV-energy consumption optimization analysis at the block scale possible, and providing scientific basis for achieving low-carbon transformation and sustainable energy development in the building sector.

Keywords: urban block form; photovoltaic utilization; building energy consumption; deep learning; coupled optimization simulation; simulation study

1. Introduction

With the increasingly severe problem of global climate change [1,2], the challenges of urban energy utilization and environmental protection are becoming more and more prominent [3]. The integration of energy efficiency in buildings and renewable energy becomes one of the key measures to combat climate change [4,5]. Especially in high-density urban blocks, how to achieve efficient photovoltaic energy utilization and optimize building energy consumption through rational building design is an important direction to promote urban sustainable development. In recent years, Building Integrated Photovoltaic (BIPV), as an innovative green building technology, has been widely concerned and gradually applied in block scale urban planning. In practical applications, the comprehensive influence of

 https://doi.org/10.3390/buildings15071118

multiple factors of block design parameters (block form, building layout, environmental conditions, etc.) on photovoltaic utilization efficiency and energy consumption performance has not been fully quantified and analyzed [6–8]. The complex and intertwined multiple determinants make the collaborative optimization design of photovoltaic power generation potential and building energy consumption face great challenges.

How to use parametric modeling and intelligent algorithms to explore the block form, photovoltaic utilization and energy consumption reduction, effectively explore the coupling relationship between block design parameters and photovoltaic energy consumption collaborative optimization objectives, and provide a scientific digital optimization framework for block planning is an important challenge in current research [9–11]. In order to solve these problems, a technical framework integrating deep learning, physical simulation, and multi-objective optimization is proposed to explore the collaborative optimization design of block form, block scale PV, and building energy consumption. The framework combines big data analysis, machine learning, and physical modeling techniques, aiming to reveal the energy efficiency performance of different block form design schemes under hot summer and cold winter climates through refined simulation and analysis, so as to provide urban planners with data-driven decision support tools.

1. **Data acquisition and preprocessing:** the basic data of the block were obtained through remote sensing images and Geographic Information System (GIS) and standardized to provide accurate input for the subsequent model establishment.
2. **Parametric modeling:** Rhino and Grasshopper are used for multi-dimensional parametric modeling of block buildings to generate different block design samples;
3. **Photovoltaic and building energy consumption simulation:** simulate the building energy consumption and photovoltaic power generation potential of each design in a specific climate through Tools such as EnergyPlus and Ladybug Tools;
4. **Machine learning model and sensitivity analysis:** through the Light Gradient Boosting Machine (LightGBM) model, the nonlinear relationship between design parameters and performance is analyzed, and the sensitivity analysis of key parameters is performed.
5. **Multi-objective optimization:** The non-dominated Sorting Genetic Algorithm II (NSGA-II) is used to optimize the multi-objective block design to maximize the photovoltaic power generation potential and minimize the building energy consumption.

This method framework provides a new idea for block scale building design, breaks through the limitations of traditional building energy efficiency assessment, and provides a systematic way to optimize the design scheme under the dual goals of comprehensive consideration of photovoltaic power generation and building energy efficiency (Figure 1).

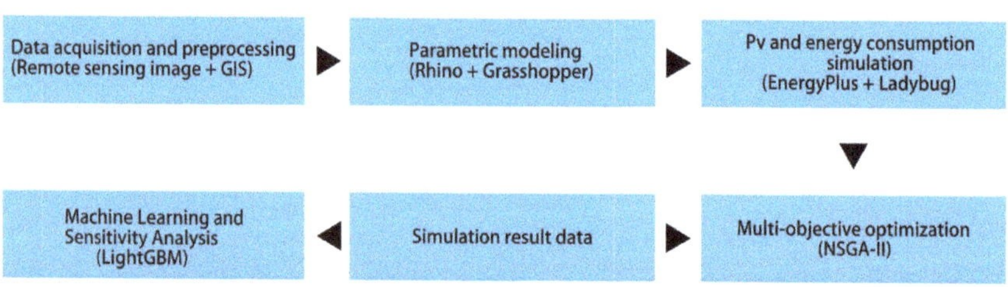

Figure 1. Technical framework diagram. Arrow direction: indicates the direction of data flow, from data acquisition to final optimization design. Module content: Each module contains input, processing tools, and output results. Feedback mechanism: A feedback loop can be formed between the machine learning module and the optimization module to further optimize the design parameters.

In recent years, many studies have extensively explored the effect of occlusion on the efficiency of photovoltaic (PV) panels. For example, Pillai et al. systematically reviewed the performance of building integrated photovoltaic (BIPV) systems in different urban forms, emphasizing how the occlusion factor affects the energy output. Mendis et al. further analyzed the economic potential of integrated PV on commercial building facades. However, most studies still focus on a single building scale, lacking a broader perspective on how urban block form (including factors such as building layout, density, and spatial configuration) jointly affects PV efficiency and building energy consumption [12–14].

This study fills this critical research gap by shifting the focus from individual building optimization to an integrated block-scale approach. Unlike previous studies that treat PV deployment and building energy efficiency as isolated objectives, this study proposes a collaborative optimization framework to simultaneously improve PV utilization and minimize building energy consumption. By integrating parametric modeling, physical simulation, and machine learning techniques, this study provides a quantifiable and adaptable approach to urban block design optimization, a dimension that has not been fully explored in the existing literature.

In addition, traditional studies often use simplified energy consumption models and fail to fully consider the complex interaction between building form and PV performance. In this study, a data-driven high-precision prediction model is established by introducing multi-objective optimization techniques. This advance enables urban planners to make design decisions based on scientific evidence to reduce overall energy demand while maximizing renewable energy utilization [15,16].

As cities continue to push for low-carbon development, energy optimization at the block scale is becoming an important part of sustainable urban planning. By filling the research gap on the coupling relationship between urban form and PV energy efficiency, this study not only advances theoretical exploration, but also provides a practical design strategy for achieving high performance and energy-saving urban blocks [17,18].

2. Research Methods

2.1. Parametric Simulation Method

In the parametric simulation stage, the parametric design method based on Grasshopper platform is used to set the initial building simulation parameters by adjusting 13 key design parameters including building spacing [19,20], orientation, and height, so as to realize the precise control of urban block design [21,22] (Table 1).

As a highly flexible and integrated graphical algorithm editor, Grasshopper has played an important role in architectural design and urban planning. Its intuitive graphical interface allows designers to construct complex geometric forms and design logics in a modular manner, avoiding the complexity of traditional programming. The advantage of this graphical workflow is that designers can quickly iterate over multiple design alternatives, exploring and optimizing multidimensional parameter settings while ensuring accuracy, resulting in a more efficient design process. In addition, the openness and extensibility of Grasshopper enables seamless integration with other advanced tools and plug-ins, providing more possibilities for design and analysis [23,24]. To comprehensively assess the impact of building design options on energy performance and renewable energy utilization, three industry-leading simulation tools are integrated based on the Grasshopper platform: Energy Plus, Honeybee, and Open Studio.

Table 1. Building simulation parameters.

	Parameters	Specific Content
1	Building Spacing	Affect daylight, ventilation conditions, and privacy protection between buildings, and adjust to optimize space quality and energy efficiency.
2	Building Orientation	Orientation of the building's main facade was determined to maximize daylight utilization and reduce heating and cooling loads.
3	Building Height	Control the vertical volume of the building to balance land utilization and environmental impact.
4	Site Coverage Ratio	Represents the ratio of the building footprint to the plot area, which has a direct impact on the openness and density of the block.
5	Number of Floors	Define the functional zoning and total building area within the building, affecting the overall energy consumption and traffic organization.
6	Roof Slope	Optimize the Angle of the roof photovoltaic panel to improve the efficiency of photovoltaic power generation.
7	Facade Material	Adjust the thermal and reflective properties of facade materials to reduce the heat island effect and energy consumption.
8	Window-to-Wall Ratio, WWR	Adjust the ratio of window area to wall area to balance natural lighting and thermal insulation performance.
9	Window Type	Choose suitable window materials and opening methods for climatic conditions to improve thermal comfort and ventilation.
10	Street Width	Adjusting the width of the street affects the microclimate, light conditions and traffic functions.
11	Green Coverage Ratio	Increase the proportion of green space, improve the ecological environment of blocks, and reduce building energy consumption.
12	Shading System Design	Install shading devices to reduce indoor heat load caused by direct sunlight.
13	Inter-building View Distance	To ensure smooth line of sight between buildings, enhance living comfort and landscape value.

Building Spacing: the horizontal distance between adjacent buildings influences daylight access, ventilation potential, and urban heat island effects. Larger spacing improves airflow and natural cooling but may reduce land-use efficiency. Based on the study's simulations, an optimal range of 8–15 m balances ventilation with space utilization; **Building Orientation:** tefers to the primary direction in which the building's longest facade is aligned. South-facing orientations optimize solar gain in winter and reduce heating demand, while east–west orientations minimize overheating in summer. The study recommends a 10–20° deviation from true south for optimal PV and thermal performance; **Building Height:** determines shading effects, heat gain/loss, and rooftop PV potential. Taller buildings (>30 m) may reduce solar exposure for lower structures but provide larger rooftop PV areas. The study focuses on a range of 12–30 m, accommodating both low-rise and high-rise scenarios; **Site Coverage Ratio:** the ratio of built-up area to total block area. Higher coverage increases land-use efficiency but reduces green space and permeability. The study finds that a 50–70% coverage provides a balance between density and environmental performance; **Number of Floors:** directly related to energy demand, shading, and urban form. More floors increase internal heat gain but also enable vertical energy-efficient strategies. The research considers 2–10 floors as an effective range; **Roof Slope:** the inclination of the building roof, wmhich affects PV panel efficiency and rainwater drainage. A slope of 5–20° is ideal for PV performance in Wuhan's climate conditions; **Facade Material:** the thermal and reflective properties of exterior walls impact heat absorption and insulation. High-albedo materials and insulated facades reduce cooling loads and are recommended based on regional climate needs; **Window-to-Wall Ratio (WWR):** the proportion of windows relative to the total facade area. A balanced WWR of 30–50% enhances daylighting while minimizing excessive heat gain/loss; **Window Type:** includes material and opening mechanisms affecting ventilation and insulation. Double-glazed, low-emissivity windows are recommended for minimizing heat transfer; **Street Width:** affects urban ventilation, shading, and accessibility. Wider streets (>15 m) enhance air circulation but may reduce pedestrian comfort. The study finds an optimal range of 10–18 m for energy-efficient urban blocks; **Green Coverage Ratio:** the percentage of vegetation in the block. Higher green ratios improve microclimate conditions and reduce building cooling demand. The study suggests a minimum of 20% for effective mitigation of heat island effects; **Shading System Design:** external shading devices reduce direct solar heat gain and improve comfort. Overhangs, louvers, or vegetation-based shading should be incorporated into design strategies; **Inter-building View Distance:** the minimum unobstructed distance between buildings for privacy, daylight access, and aesthetic considerations. A range of 8–12 m ensures adequate daylight while maintaining density.

As a dynamic Energy consumption simulation tool, Energy Plus provides highly accurate building energy consumption analysis capabilities. With this tool, the study is able to meticulously capture the dynamic impact of different design parameters on building energy consumption, especially in the evaluation of Heating, Ventilation, and Air

Conditioning (HVAC) system performance. Energy Plus can simulate in detail the effects of seasonal climate change, building usage behavior patterns and multiple energy efficiency strategies on energy consumption, providing a scientific basis for design optimization [25].

The Honeybee plug-in plays a key role in environmental performance analysis. The tool helps designers identify key environmental variables that may affect building performance by calculating the distribution of sunlight outside and inside the building, the efficiency of natural ventilation, as well as thermal comfort [26]. In insolation analysis, Honeybee can simulate the effects of different building orientations and spacing on lighting conditions, which can guide the layout optimization of rooftop PV systems. The tool also supports in-depth assessment of thermal comfort, providing data support for building skin design and shading system optimization.

Open Studio plugins further improve the efficiency and depth of the simulation process. Through its integrated interface, research enables the rapid construction of complex building models and simplifies the operation of multiple simulation processes. Open Studio not only speeds up parameter iteration but also supports connectivity to intelligent building management systems (BMS), enabling real-time optimization of energy use efficiency during the operational phase of the building [27]. The open framework of the tool opens up the possibility for the research to extend the functionality by customizing the script implementation in combination with machine learning algorithms to identify more optimal parameter combinations.

Figure 2 illustrates the logical structure of this comprehensive analysis framework. In the whole design and simulation process, the parameter space is automatically explored and optimized based on the advantages of machine learning technology. In this way, the study achieved a significant reduction in building energy consumption and significantly improved the energy efficiency of the PV system [28,29]. This multi-tool integrated framework provides scientific and operational guidance for building and block design and lays an important foundation for future smart city planning and sustainable development strategy.

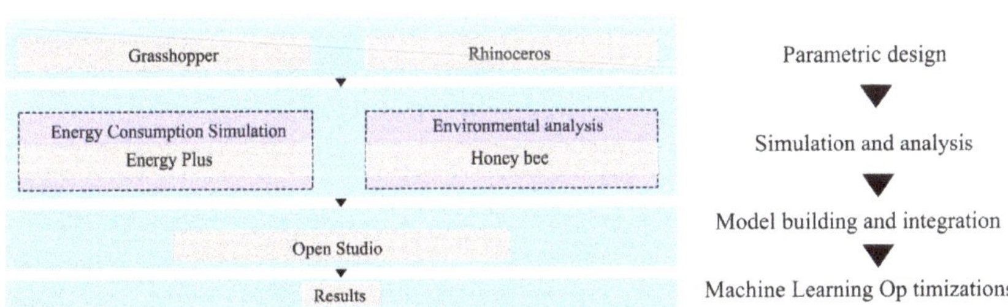

Figure 2. Framework for parametric simulation.

2.2. Data Processing Methods

In the data processing stage, in order to optimize the simulation accuracy and efficiency, Latin Hypercube Sampling (LHS) is used for data sampling. The basic idea of LHS is to divide the value range of each variable into several intervals (usually called "layers", denoted by Δx). A sample point is randomly selected in each interval to ensure that the value of the variable is evenly distributed in the whole range and to avoid the repetition of the sample points. The formula is as follows:

$$\Delta x_i = \frac{b_i - a_i}{n} \tag{1}$$

where the width of the value interval is Δx_i, the range of each variable of x_i is $[a_i, b_i]$ $(i = 1, 2, \ldots k)$, and the range of each variable is equally divided into n.

Secondly, Pearson product-moment correlation coefficient analysis was used [30–32], the linear influence of block design parameters on the PV potential and energy consumption objective function is quantified, and the significant relationship between variables is revealed, which provides a scientific basis for optimizing block layout [33,34].The Pearson correlation coefficient (n) is calculated based on the ratio of the covariance of two variables to their standard deviation, as follows:

$$r = \frac{\sum_{i=1}^{n} (X_i - \overline{X})(Y_i - \overline{Y})}{\sqrt{\sum_{i=1}^{n} (X_i - \overline{X})^2}\sqrt{\sum_{i=1}^{n} (Y_i - \overline{Y})^2}} \tag{2}$$

Among them, the calculation of r is based on the average value of samples Xi and Yi, and the range is $[-1, 1]$, which accurately depicts the strength of linear relationship. $r = 0$ means that the variables are linearly independent. $r \in (-1, 0)$ indicates a negative linear correlation between variables, and the negative correlation becomes stronger when $r \in (-1, 0)$ approaches. If $r \in (-1, 0)$ indicates a positive correlation between the variables, the correlation is significant if it approaches 1. In the process of research, by analyzing the variables such as the proportion of low level and the plot area width, it is found that the proportion of low level is significantly positively correlated with the PV potential ($r = 0.8$), and the width of plot D is weakly negatively correlated with the heating energy consumption ($r = -0.3$). These results provide an important basis for optimal design.

However, limited by the linear assumption, Pearson analysis cannot effectively capture the complex nonlinear relationships and the interactions between variables, so it is difficult to fully reflect the potential patterns in high-dimensional data scenarios. In order to overcome the above shortcomings, the Light Gradient Boosting Machine (LightGBM) (Light Gradient Boosting Machine is a gradient boosting framework developed by Microsoft and contributors from the open-source community. Designed for efficient large-scale data processing, it supports parallel computation and is widely used in machine learning applications. Microsoft is headquartered in Redmond, WA, USA) model is introduced in the regression analysis stage [35–37]. As a machine learning algorithm based on Gradient Boosting Decision Tree (GBDT), LightGBM can efficiently deal with nonlinear relationships and multi-dimensional variable interactions. In particular, it has excellent performance compared with traditional algorithms (such as XGBoost and CatBoost) in training speed, accuracy and big data scenarios. The LightGBM model is used to model the key variables selected by Pearson (such as the proportion of low-rise and the plot width) to further explore the nonlinear relationship between the PV potential and the energy consumption objective function. The initial model during LightGBM training can be expressed as follows.

$$F_0(x) = \underset{\gamma}{\operatorname{argmin}} \sum_{i=1}^{n} L(y_i, \gamma) \tag{3}$$

where L is the loss function; usually the mean squared error (MSE) or cross-entropy loss is used.

MTH iteration:

$$F_m(x) = F_{m-1}(x) + \eta h_m(x) \tag{4}$$

where η is the learning rate and $hm(x)$ is the MTH weak learner (decision tree). The new learner is generated by fitting the residuals of the current model:

$$h_m(x) = \underset{h}{\operatorname{argmin}} \sum_{i=1}^{n} L(y_i, F_{m-1}(x_i) + h(x_i)) \tag{5}$$

The Design of Experiments (DOE) (Latin square sampling) is used to obtain 500 groups of design parameters and target values. The LightGBM model is used to screen out the key design parameters (such as the proportion of low-rise and the width of the plot), and the nonlinear effects of these parameters on the photovoltaic power generation and building energy consumption of the block are deeply analyzed.

3. Study the Construction of Training Data Set for Block Shape Optimization

3.1. Overview of the Study Area

The study area is selected in Wuhan, Hubei Province, with a special focus on the city's aging neighborhoods, aiming to explore their spatial characteristics and their potential and challenges in modern urban renewal. As an important city in central China, Wuhan has gone through many planning and reconstruction, but there are still many old blocks. The spatial form of these blocks not only reflects the accumulation of history and culture but also exposes many problems in urban development, such as aging infrastructure and unreasonable functional layout. In the context of urban renewal, the renovation potential of old blocks is huge, which has far-reaching significance for sustainable urban development. By improving energy efficiency, introducing renewable energy systems, optimizing spatial layout, and improving environmental quality, the transformation of old blocks can not only improve the livability and functionality of blocks but also reduce traditional energy consumption and alleviate environmental pressure, thus promoting the green transformation of cities. On the whole, the renovation of old blocks provides important opportunities for urban sustainable development and is a key link to achieve multiple sustainable goals of society, economy, and environment.

Wuhan, a rapidly urbanizing city in China, is characterized by its hot summer and cold winter climate, leading to significant seasonal variations in energy demand. The city's high-density urban fabric results in complex shading effects, directly impacting PV efficiency and building energy performance. As a key hub for China's low-carbon urban development, Wuhan provides a representative case for studying PV-integrated urban morphology optimization in dense, mixed-use environments. The insights gained from this study can be applied to other cities with similar climatic and spatial conditions, contributing to broader sustainable urban design strategies.

This study assumes that in high-density urban environments, shading primarily originates from adjacent buildings rather than distant terrain or potential future developments. To control variables, environmental factors such as vegetation and meteorological conditions are not directly analyzed for their impact on photovoltaic (PV) efficiency. Instead, the focus is on optimizing building morphology (height, spacing, and layout) to minimize shading effects, enhance PV performance, and reduce overall energy consumption. The simulation explicitly incorporates shading from surrounding buildings and inter-building shading within the block to ensure realistic modeling outcomes. Given Wuhan's current urban development trends, newly constructed residential blocks typically feature limited vegetation-induced shading, as most greenery consists of low shrubs and grass that do not obstruct solar access. Shading trees are generally located within centralized green spaces or parks to meet the daylighting needs of ground-floor residents and vulnerable groups, rather than being interspersed among residential buildings. Therefore, vegetation shading is not considered a significant factor in the simulation.

In order to deeply understand the spatial structure of Wuhan city blocks, a comprehensive investigation of 15 representative city blocks was conducted through field investigation and satellite image analysis (Figure 3 is based on freely available Sentinel-2 imagery, which provides high-resolution multispectral images suitable for urban morphology analysis.

This data set, part of the Copernicus Earth Observation Program managed by the European Space Agency (ESA), was used to extract urban block structures in Wuhan. The image was then modified by the authors to include additional information and to suit the specific requirements of our research.) These blocks cover both old and new, and the former occupies a relatively large number, showing obvious characteristics of traditional urban planning. These old blocks generally have the advantages of regular layout, clear functions, and smooth traffic, but they also face problems such as high building density, insufficient public space and aging infrastructure.

Figure 3. Area of study (Image provided by the European Space Agency (ESA), Copernicus Program, redrawn by the authors).

It is found that the old urban blocks in Wuhan have a high regularity in the overall layout. The interior of the block is mostly rectangular or square grid, and the overall terrain is flat, which helps to form a clear traffic flow and functional differentiation. The division of commercial and residential functions is relatively clear, with the commercial district usually distributed along the main road, while the residential district is mainly distributed in the inner side of the block. In addition, most of the plot ratios in these blocks are concentrated between 3.87 and 4.39, indicating that the building density in these areas is moderate, while also ensuring a certain amount of open space and public facilities. The building density is between 22.5% and 34.4%, which is relatively high, but still maintains a relatively loose space utilization compared with the newly built blocks.

3.2. *Setting of Related Basic Parameters and Objective Function*

According to the actual situation of the study area, the ideal block size is set as 200 m × 200 m, the total area is 40,000 m^2, the plot ratio is 4, the building density is 27%, the height of low-rise buildings is limited to 12 m, and the peripheral high-rise buildings (length × width × height: 60 m × 50 m × 30 m) are planned, a total of 16 buildings. The underlying simulation parameters for block setting are detailed in Table 2 below.

Table 2. Block parameter setting table.

Block Setting Parameters	Set Point Value
Block location	Wuhan City, Hubei Province, 200 m
Block width	200 m
Block length	4.0
Block floor area ratio	0.27
Block building density	4 m

According to the data in Table 2, the parameterized ideal model shown in Figure 4 is constructed.

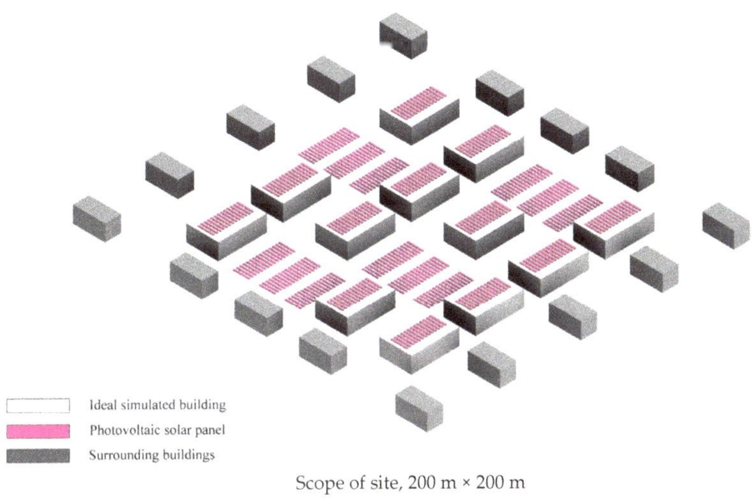

Figure 4. Idealized city block model.

Furthermore, the design characteristics of city blocks are summarized, and 12 design parameters (Table 3) are set and normalized, which are variable control parameters of block design, so that the ideal block shape can be shaped by adjusting the parameters in subsequent research.

Table 3. Block design control parameter setting table.

Name of Design Parameter	True Setting Range	Set the Range After Normalization
A Site scope of the plot	1, 2, 3, 4	0–1
B Site scope of plot	1, 2, 3, 4	0–1
C Site scope of plot	1, 2, 3, 4	0–1
D Site scope of plot	1, 2, 3, 4	0–1
Building width of plot A	20–50 m	0–1
Building width of plot B	20–50 m	0–1
The proportion of low-rise buildings (below 12 m)	0–50%	0–1
The axis is shifted north–south	0–200 m	0–1
The axis is shifted in the east–west direction	0–200 m	0–1
The axis rotates by an Angle along time	0–360°	0–1

The study area is equally divided into four areas A, B, C, and D, and each area is subdivided into four plots. According to the design parameters, the plots are classified as buildings or open Spaces, and the buildings are divided into high-rise and low-rise, so as to realize fine design and control (Figure 5). Use Grasshopper (Grasshopper is a plugin for Rhinoceros software, developed by Robert McNeel & Associates. The company is headquartered in Seattle, WA, USA. Grasshopper is widely used in architectural design, industrial design, and interaction design, providing a visual programming environment for parametric modeling.) and Honeybee (Honeybee is a plugin developed by Ladybug Tools that operates within the Grasshopper environment. It is primarily used for environmental design analysis, supporting daylight simulations with Radiance and energy modeling with EnergyPlus. Ladybug Tools is based in New York City, NY, USA, and specializes in developing performance analysis tools for buildings and urban environments.) plug-in to analyze block energy consumption according to American Society of Heating, Refrigerating and Air-Conditioning Engineers (ASHRAE) standard and study the influence of design parameters (Tables 4 and 5).

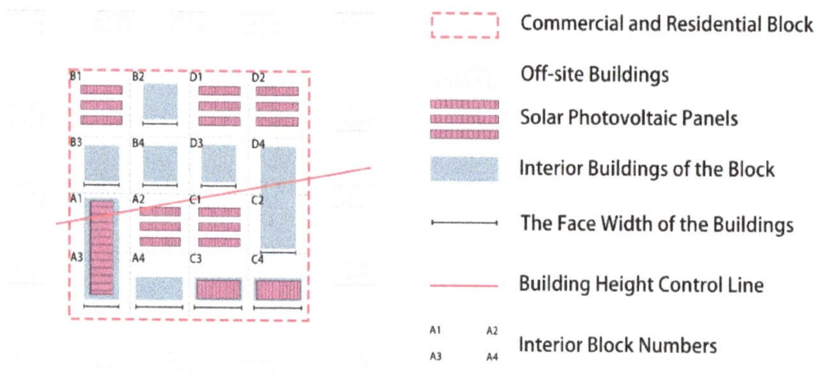

Figure 5. Display of simulation parameter setting.

Table 4. Block ideal model simulation setting.

Simulation Setting	Description	Setting Conditions
Site climate	Hot summer and cold winter climates	4 ASHRAE Climate zone
Year of construction	The buildings in the survey area were basically constructed from 2000 to 2020	ASHRAE 90.1 2005
Function of building	City block	Midrise Apartment
Type of building structure	Most of the buildings in the survey area are high-rise residential buildings	Brick mix
Solar panel power conversion ratio	Combined with cost considerations, polycrystalline silicon photovoltaic solar panels are used	26%

Table 5. Simulation objective function setting table.

Objective Function	Description	Unit
$E_{Heating}$	The electricity consumed by the heating of the block through the HAVC system in winter	$KW{\cdot}h/m^2$
$E_{cooling}$	Is the electric energy consumed by the cooling of the block through the HAVC system in summer	$KW{\cdot}h/m^2$
$E_{photovoltaic}$	It is the electricity generated by the photovoltaic solar system in the block throughout the year	$KW{\cdot}h/m^2$
E_{Total}	Is the electricity consumption of the block for the whole year	$KW{\cdot}h/m^2$

3.3. Setting of Sampling Method

Firstly, according to the principle of LHS, the value ranges of key variables (such as the proportion of low level, the width of plot, etc.) were stratified, and the number of samples (n = 500) was defined. Secondly, 500 sets of parameter combinations satisfying the analysis of photovoltaic potential and building energy consumption were generated to ensure the uniformity and diversity of data distribution. Finally, based on the sampling results, a parametric model was constructed by combining machine learning algorithms (such as LightGBM) (Table 6).

Table 6. LHS parameter setting table.

Hyperparameters	Description	Set Point Value
Samples	The number of samples to generate	500
Core Algorithm	The algorithm used to generate the initial LHS samples	Minimizing Correlation
Weights	The weight of each dimension, if you want the importance of different dimensions to be different	All parameters are set to 1.0
Resampling	Whether to allow resampling of certain intervals to improve sample coverage or to conform to a particular distribution	Yes

3.4. Setting of Data Processing Method

The combination of design parameters and the objective function value are obtained by sampling, and the Pearson analysis is used to preliminarily explore their relationship. In view of the nonlinear characteristics, the LightGBM regression algorithm is introduced. LightGBM is based on histogram algorithm and leaf splitting optimization, which accelerates processing and improves prediction accuracy, especially for large-scale high-dimensional data processing. Its multi-threading and distributed computing characteristics significantly enhance the training efficiency, and the regularization function effectively prevents overfitting. In order to verify the rationality of LightGBM, compared with AdaBoost regression algorithm, it shows the optimization advantages of LightGBM in splitting point search and tree growth strategy, which is more suitable for complex data sets and ensures the efficiency and accuracy of analysis (Table 7).

Table 7. LRegression hyperparameter setting table for LightGBM and AdaBoost models.

LightGBM		AdaBoost	
Parameter names	Parameter values	Parameter names	Parameter values
Training time	0.234 s	Training time	0.287 s
Data splitting	0.7	Data splitting	0.7
Shuffle the data	Yes	Shuffle the data	Yes
Cross validation	No	Cross validation	No
Base learner	gbdt	Number of base classifiers	100
Number of base learners	100	Loss function	linear
Learning rate	0.1	Base classifier	Decision tree classifiers
L1 regularization term	0	Learning rate	1
L2 regularization term	1		
Sample feature sampling rate	1		
Tree feature sampling rate	1		
Node splitting threshold	0		
Minimum weight of samples in leaf nodes	0		
The maximum depth of the tree	10		
Minimum number of samples at leaf nodes	10		

4. Based on Simulation Results and Result Analysis

4.1. Overview of Simulation Results

A total of 500 sampling times were carried out, and 475 objective function results were obtained after 92 h, as shown in Figure 6. The study drew a histogram of these 475 groups of design parameters and objective function values for comparative analysis, as shown in Figure 6 below.

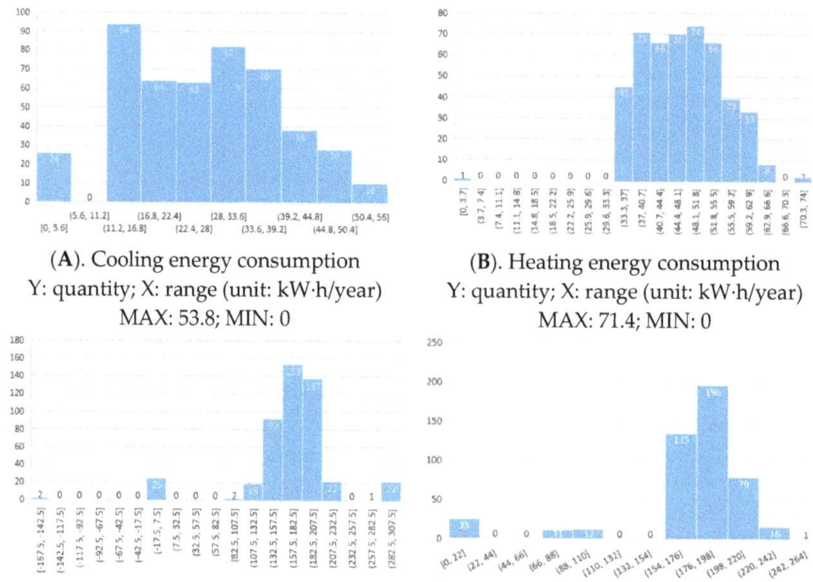

(**A**). Cooling energy consumption
Y: quantity; X: range (unit: kW·h/year)
MAX: 53.8; MIN: 0

(**B**). Heating energy consumption
Y: quantity; X: range (unit: kW·h/year)
MAX: 71.4; MIN: 0

(**C**). Photovoltaic power generation
Y: quantity; X: range (unit: kW·h/year) ×10,000
MAX: 3,075,000; MIN: 0

(**D**). The total energy consumption of the block
Y: quantity; X: range (unit: kW·h/year) ×10,000
MAX: 2,640,000; MIN: 0

Figure 6. Overview of the heating energy consumption results. To ensure the accuracy and clarity of the reported data, potential misunderstandings and unrealistic values have been addressed. Specifically, the erroneous minimum value in the photovoltaic power generation data (originally reported as −167.5) has been removed. This correction aligns the data with realistic physical constraints and enhances the reliability of the findings.

The distribution of cooling energy consumption shows a clear right-skewed trend. The cooling energy consumption of most samples is concentrated in the range of 10 to 30, and only a few samples consume more than 40. This result indicates that the cooling energy consumption of buildings in the block is generally at a low level under the current design parameters, but the presence of high value samples also suggests potential optimization needs, especially in reducing the cooling demand of high energy consumption buildings.

The distribution of heating energy consumption appears to be more uniform, with most samples concentrated in the range of 40 to 60, and the energy consumption level is relatively stable. However, there are still a small number of samples whose heating energy consumption is significantly higher than the average, showing some volatility. These high value samples show that there is still room for improvement in the further optimization of heating demand in block planning and building design.

The distribution of PV power generation shows the characteristics of an approximately normal distribution, which is mainly concentrated in the interval of 1 million to 2 million. This result indicates that the performance of the PV system is stable, and the overall generation efficiency performs well within the current design parameters. But at the same time, it also means that by adjusting the design parameters, it is still possible to further optimize the efficiency of the photovoltaic power generation system, thereby increasing the proportion of renewable energy in the building energy consumption.

The distribution of the total energy consumption of the block is relatively concentrated. The total energy consumption of the vast majority of samples is distributed in the range of 1.6 million to 2.1 million, and only a few samples exceed 2.3 million or less than 1.6 million. The distribution characteristics show that the overall energy consumption of the block has less volatility under the change in design parameters, but the optimization of high energy consumption samples still needs to be paid attention to further improve the energy efficiency level of the block as a whole.

Taken together, it is found that the change in design parameters significantly affects the energy consumption distribution and photovoltaic generation efficiency of block buildings. Specifically, the values of the objective function show different distribution characteristics with the change in design parameters, especially in the distribution of cooling energy consumption and photovoltaic power generation. In view of these findings, the subsequent research needs to combine parameter sensitivity analysis to deeply explore the influence law of key design parameters on energy consumption and power generation, so as to develop a more scientific optimization strategy.

When the same design parameters are changed, the E value, the energy efficiency index, changes in a large range, which indicates that the block building design is more sensitive to parameter adjustment. In the subsequent optimization process, designers should focus on the fluctuation of E value and take it as one of the core indicators to improve energy efficiency and optimize design.

4.2. Correlation Analysis Between Design Parameters and Objective Function

Pearson correlation analysis was used to evaluate the relationship between the values of 475 sets of design parameters and their corresponding objective function values. The Pearson correlation coefficient results for each objective function are visualized in the heat map in Table 8.

The study shows that there is a significant correlation between some of the design parameters and the objective function. Specifically, the proportion of low-rise buildings has the most significant impact on the performance of heating, cooling and photovoltaic systems, and the correlation coefficients are −0.970, −0.904, and −0.861, respectively, showing a strong negative correlation. This result indicates that increasing the proportion of low-rise buildings in the building design can significantly reduce the energy consumption requirements of the system while optimizing the PV power generation performance.

The correlation coefficient between the area width of plot B and the total energy consumption of block is −0.944, indicating that the area width of plot B has a significant negative correlation with the total energy consumption. The correlation coefficient of the area width of plot A is −0.912, which is also negative, but the influence degree is relatively low. These analysis results show that in urban planning and architectural design, appropriately increasing the plot width can effectively reduce the total energy consumption of the block, thereby improving energy efficiency (Table 9).

Table 8. A. Solar power _ Normalized; B. Heating energy consumption; C. Cooling energy consumption; D. Total energy consumption. Objective function related heat. A. Axis left and right displacement B. Axis rotation C. Axis forward and backward displacement D. Proportion of lower strata E. 04 face width F. 04 Open space G. 00 face width H. 03 Open Space I. 02 face width J. 02 Open Space K. 01 face width L. 01 Open Space M. Solar power _ Normalized N. Heating energy consumption O. Cooling energy consumption P. Total energy consumption.

A.

	M	L	K	J	I	H	G	F	E	D	C	B	A
A	0.016	0.000	0.014	−0.003	0.001	0.008	−0.004	0.015	−0.002	0.016	0.013	−0.002	1.000
B	−0.006	0.000	0.005	0.006	0.003	0.003	0.024	0.01B	−0.002	−0.02	0.001	1.000	−0.002
C	−0.010	0.014	−0.003	0.004	−0.005	0.013	0.000	0.002	0.006	0.012	1.000	−0.001	0.013
D	−0.444	−0.014	−0.002	0.000	−0.007	−0.01	0.010	0.023	0.023	1.000	−0.02	−0.020	0.016
E	0.154	−0.004	0.007	−0.002	−0.011	−0.007	0.021	0.025	1.000	0.023	0.006	−0.002	−0.002
F	0.011	0.005	−0.004	−0.001	0.008	−0.003	−0.014	1.000	0.025	−0.010	0.002	−0.018	0.015
G	0.157	0.000	0.005	0.003	0.012	0.003	1.000	−0.014	0.021	0.010	0.000	0.024	−0.004
H	−0.012	0.012	−0.006	0.004	−0.010	1.000	0.003	−0.003	−0.007	−0.010	0.013	0.003	0.008
I	0.205	−0.005	−0.014	0.009	1.000	−0.010	0.012	0.008	−0.011	−0.007	−0.005	0.003	0.001
J	−0.007	−0.005	0.003	1.000	0.009	0.004	0.003	−0.001	−0.002	0.000	0.004	0.006	−0.003
K	0.196	0.001	1.000	0.003	−0.014	−0.006	0.005	−0.004	0.007	−0.002	−0.003	0.005	0.014
L	−0.049	1.000	0.001	−0.005	−0.005	0.012	0.000	0.005	−0.004	−0.014	0.014	0.000	0.000
M	1.000	−0.049	0.196	−0.007	0.205	−0.012	0.157	0.011	0.154	−0.444	−0.01	−0.006	0.016

B.

	N	L	K	J	I	H	G	F	E	D	C	B	A
A	0.008	0.000	0.014	−0.003	0.001	0.008	−0.004	0.015	0.002	0.016	0013	0.002	1.000
B	0.028	0.000	0.005	0.006	0.003	0.003	0.024	0.018	−0.002	0.020	−0.001	1.000	−0.002
C	0.007	0.014	−0.003	0.004	0.005	0.013	0.000	0.002	0.006	0.012	1.000	0001	0.013
D	−0.970	−0.014	−0.002	0.000	−0.007	−0.010	0.010	−0.010	0.023	1.000	0.012	−0.020	0.016
E	−0.06	−0.004	0.007	−0.002	−0.011	−0.007	0.021	0.025	1.000	0.023	0.006	0002	−0.002
F	−0.001	0.005	−0.004	−0.001	0.008	−0.003	−0.014	1.000	0.025	0.010	0.002	−0.018	0.015
G	−0.033	0.000	0.005	0.003	0.012	0.003	1.000	0.014	0.021	0.010	0.000	0.0124	−0.004
H	0.012	0.012	0.006	0.004	0.01	1.000	0.003	0.003	0.007	0.010	0.013	0.003	0.008
I	−0.025	−0.005	−0014	0.009	1.000	−0.010	0.012	0.008	−0.011	−0.007	−0.005	0003	0.001
J	0.007	−0.005	0.003	1.000	0.009	0.004	0.003	−0.001	−0.002	0.000	0.004	0006	−0.003
K	−0.031	0.000	1.000	0.003	−0.014	−0.006	0.005	0.004	0.007	0.002	−0.003	0.005	0.014
L	0.009	1.000	0.001	−0.005	−0.005	0.012	0.000	0.005	−0.004	−0.014	0.014	0.000	0.000
N	1.000	0.009	−0.031	0.007	0.025	0.012	0.033	−0.001	0.060	−0.970	0.007	0.028	0.008

Table 8. *Cont.*

C.

A	−0.010	0.000	0.014	−0.003	0.001	0.008	0.004	0.015	−0.002	0.016	0.013	−0.002	1.000
B	0.033	0.000	0.005	0.006	0.003	0.003	0.024	−0.018	−0.002	−0.020	−0.001	1.000	−0.002
C	0.006	0.014	0.003	0.004	−0.005	0.013	0.000	0.002	0.006	0.012	1.000	−0.001	0.013
D	−0.961	0.014	0.002	0.000	−0.007	−0.01	0.010	−0.010	0.023	1.000	0.012	−0.020	0.016
E	−0.097	−0.004	0.007	−0.002	−0011	−0.007	0.021	0.025	1.000	0.023	0.006	−0.002	−0.002
F	0.009	0.005	0.004	0.001	0.008	3.000	0.014	1.000	0.025	−0.010	0.002	−0.018	0.015
G	−0.065	0.000	0.005	0.003	0.012	0.003	1.000	−0.014	0.021	0.010	0.000	0.024	−0.004
H	0.025	0.012	0.006	0.004	0.030	1.000	0.003	−0.003	−0.007	−0.01	0.013	0.003	0.008
I	−0.068	0.005	−0.014	0.009	1.000	−0.010	0.012	0.008	−0.011	−0.007	−0.005	0.003	0.001
J	−0.001	−0.005	0.003	1.000	0.009	0.004	0.003	0.001	−0.002	0.000	0.004	0.006	−0.003
K	−0.072	0.001	1.000	0.003	−0.014	0.005	0.005	0.004	0.007	−0.002	0.003	0.005	0.014
L	0.006	1.000	0.001	−0.005	0.005	0.012	0.000	0.005	−0.004	−0.014	0.014	0.000	0.000
O	1.000	0.006	−0.072	0.001	0.068	0.025	0.065	0.009	−0.097	−0.961	−0.006	0.033	−0.01
	O	*L*	*K*	*J*	*I*	*H*	*G*	*F*	*E*	*D*	*C*	*B*	*A*

D.

A	0.015	0.000	0.014	−0.003	0.001	0.008	−0.004	0.015	−0.002	0.016	0.013	−0.002	1.000
B	−0.006	0.000	0.005	0.006	0.003	0.003	0.024	−0.018	−0.002	−0.020	−0.001	1.000	−0.002
C	0.011	0.014	−0.003	0.004	−0.005	0.013	0.000	0.002	0.006	0.012	1.000	−0.001	0.013
D	0.212	−0.014	−0.002	0.000	−0.007	−0.010	0.010	0.010	0.023	1.000	0.012	−0.020	0.016
E	−0.172	−0.004	0.007	−0.002	−0.011	−0.007	0.021	0.025	1.000	0.023	0.006	−0.002	−0.002
F	0.000	0.005	−0.004	−0.001	0.008	−0.003	−0.014	1.000	0.025	−0.010	0.002	−0.018	0.015
G	−0.191	0.000	0.005	0.003	0.012	0.003	1.000	−0.014	0.021	0.010	0.000	0.024	−0.004
H	0.015	0.012	−0.006	0.004	−0.010	1.000	0.003	−0.003	−0.007	−0.010	0.013	0.003	0.008
I	0.246	−0.005	−0.014	0.009	1.000	−0.010	0.012	0.008	0.011	−0.007	−0.005	0.003	0.001
J	0.008	−0.005	0.003	1.000	0.009	0.004	0.003	−0.001	−0.002	0.000	0.004	0.006	−0.003
K	0.230	0.001	1.000	0.003	−0.014	−0.006	0.005	−0.004	0.007	−0.002	−0.003	0.005	0.014
L	0.051	1.000	0.001	−0.005	−0.005	0.012	0.000	0.005	−0.004	−0.014	0.014	0.000	0.000
P	1.000	0.051	−0.230	0.008	−0.246	0.015	−0.191	0.000	−0.172	0.212	0.011	−0.006	−0.015
	P	*L*	*K*	*J*	*I*	*H*	*G*	*F*	*E*	*D*	*C*	*B*	*A*

Table 9. Results of significance analysis of the relationship between design parameters and objective function.

Serial Number	Parameters	Coefficient of Correlation	Impact Description
1	Proportion of low-rise buildings (heating)	−0.970	There was a significant negative correlation with heating
2	Percentage of low-rise buildings (Cooling)	−0.904	There was a significant negative correlation with refrigeration
3	Proportion of low-rise buildings (PV)	−0.861	There is a significant negative correlation with PV efficiency
4	B Plot width	−0.944	There is a significant negative correlation with total energy consumption
5	A Plot width	−0.912	There is a significant negative correlation with total energy consumption

It is also found that the correlation between other design parameters (open space, surface width) and the objective function is generally weak, and the relationship is nonlinear, which indicates that there may be more complex interaction mechanisms.

4.3. Designing a Machine Learning Regression Model with Parameters Related to the Target Function

The regression analysis results show that LightGBM is significantly better than AdaBoost in all aspects (Table 10). Specifically, the LightGBM model shows excellent performance in the prediction of heating, cooling, photovoltaic power generation, and total energy consumption. The R-squared values of the training set and the prediction set are close to or more than 0.9, which fully verifies the high accuracy and generalization ability of the model and provides strong support for the prediction and optimization of block building energy consumption.

Table 10. Objective function evaluation table.

Objective Function	Comparison of Performance Parameters (Training Set/Test Set)				
Objective 1 Energy consumption for heating	MSE	RMSE	MAE	MAPE	R^2
	0.04	0.20	0.11	0.43	1
	3.17	1.78	1.18	4.48	0.97
Objective 2 Cooling energy consumption	MSE	RMSE	MAE	MAPE	R^2
	0.09	0.31	0.12	0.24	0.99
	1.85	1.36	1.04	2.10	0.97
Objective 3 Photovoltaic power generation	MSE	RMSE	MAE	MAPE	R^2
	83,811,686	9154.87	5146.36	0.31	0.99
	7,607,439,742	87,220.64	63,014.57	3.78	0.94
Objective 4 Total power consumption	MSE	RMSE	MAE	MAPE	R^2
	542,940,356	23,301.08	8093.17	0.53	0.99
	7,528,975,863	86,769.67	52,989.47	3.03	0.92

Mean Squared Error (MSE); Root Mean Squared Error (RMSE); Mean Absolute Error (MAE); Mean Absolute Percentage Error (MAPE).

Based on the feature importance of the objective function of LightGBM regression analysis, the feature influence is ranked and quantified for heating energy consumption (Figure 7), cooling energy consumption (Figure 8), photovoltaic power generation (Figure 9), and total electricity consumption (Figure 10).

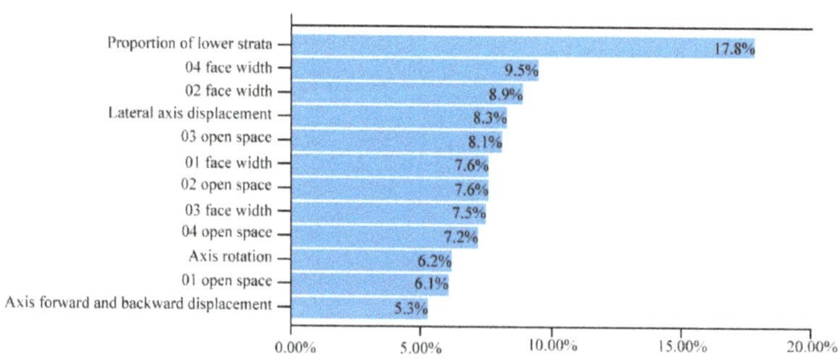

Figure 7. Feature importance of heating energy consumption.

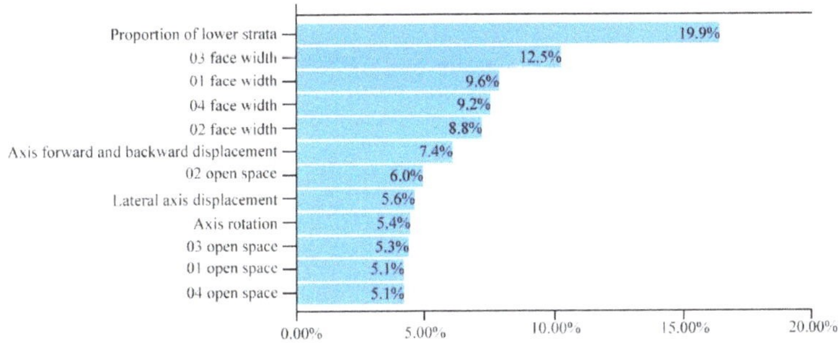

Figure 8. Feature importance of cooling energy consumption.

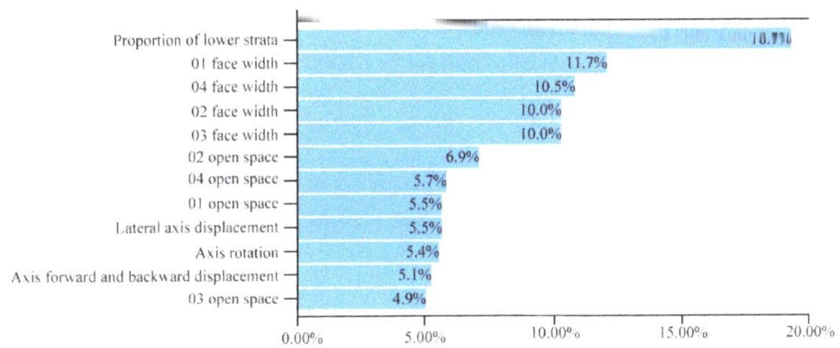

Figure 9. Feature importance of PV power generation.

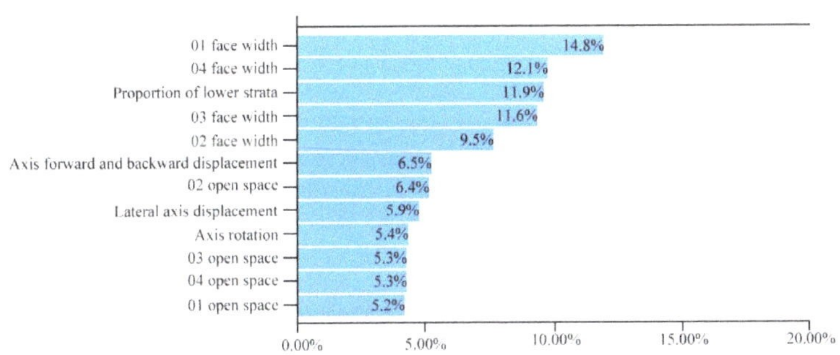

Figure 10. Feature importance of total power consumption.

1. In the heating energy consumption (Figure 7), the most critical feature of the heating energy consumption is the proportion of low-rise buildings (17.8%), indicating that the proportion of low-rise buildings in the block has a significant effect on the heating energy consumption, which may be related to the heat loss characteristics of low-rise buildings. Secondly, the area width of plot D (9.5%) also has a great impact on heating energy consumption, which may be related to the plot form and its building arrangement density;
2. In the cooling energy consumption (Figure 8), the main influence feature of cooling energy consumption is the proportion of low floors (19.9%), reflecting the close relationship between the number of building floors and the cooling demand in summer. At the same time, the importance of the surface width of plot C (12.5%) is also high, indicating that the width parameter in plot design has a key impact on the cooling energy consumption of the block;

3. Photovoltaic power generation (Figure 9), in the photovoltaic power generation model, the proportion of low-rise (18.7%) is the most important feature, probably because low-rise buildings increase the laying area of photovoltaic panels. Secondly, the importance of the width of plot A (11.7%), plot A (14.8%), and plot D (12.1%) is high, indicating that the width and orientation of different plots play a key role in the optimization of photovoltaic power generation efficiency;
4. In the total electricity consumption model of total electricity consumption (Figure 10), the area width of plot A (14.8%) is the most important feature, followed by the area width of plot D (12.1%) and the proportion of low layers (11.7%). This indicates that the parametric design of different plots and the proportion of building floors have an important influence when optimizing the total energy consumption.

Under different objective functions, the proportion of low-rise buildings and the plot width are significant influencing features, especially in the heating energy consumption, cooling energy consumption and photovoltaic power generation models. This indicates that the number of building floors and plot shape parameters have global significance for the energy performance optimization of the block, and reasonable control of the proportion of building floors and plot width can significantly improve the energy efficiency and renewable energy utilization rate of the block.

4.4. Morphological Analysis of Typical Cases

In order to further analyze the relationship between the design parameters and the objective function, the best 10 optimal solutions for each of the 475 cases are extracted to form the optimal solution set. At the same time, the research also extracts the worst 10 solutions with respect to the objective function to form the worst solution set. By refining the optimal and worst solution sets, the study reveals the deep rules between design parameters and energy consumption.

Figure 11 shows the optimal and worst block morphologies for heating energy consumption. The optimal form (A) is a compact building layout, small distance, forming a micro-climate zone, effectively reducing heat loss; the high-rise building in the center acts as a wind barrier to reduce the impact of cold wind. The worst form (B) is when the building is distributed in strips, with large spacing, high wind speed, large external wall area, and serious heat loss. It is suggested that northern cities adopt intensive layout and reasonable collocation of high and low buildings to improve heating efficiency.

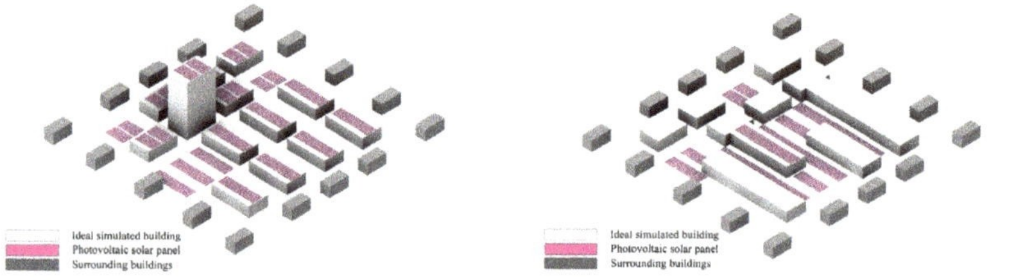

Figure 11. (**A**). Optimal block form for heating energy consumption; (**B**). Block forms with the worst heating energy consumption. Heating energy consumption block morphology.

Figure 12 shows that the optimal form of cooling energy consumption (A) achieves shading and natural ventilation through reasonable building density and photovoltaic panel arrangement and significantly reduces cooling load. In the worst form (B), the buildings are scattered, the shading is insufficient, the ventilation flow line is not smooth, and the heat island effect is obvious. It is suggested to optimize the ventilation corridor, building orientation and spacing to improve the ventilation and cooling effect.

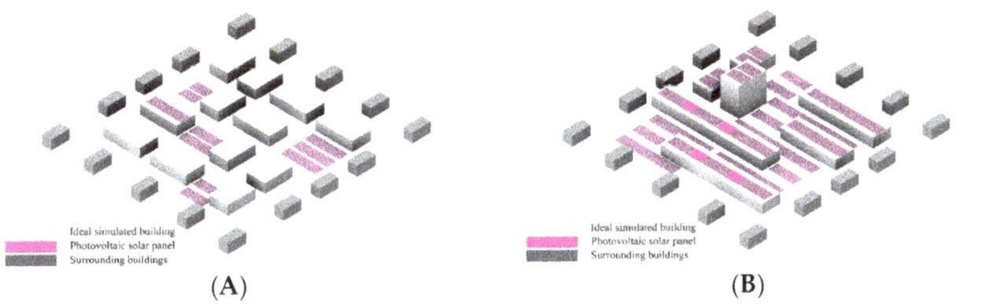

Figure 12. (**A**). Optimal block form for cooling energy consumption; (**B**). The block form with the worst cooling energy consumption. Shape diagram of refrigeration energy consumption block.

Figure 13 shows that the optimal form (A) building height is consistent, the layout is regular, the occlusion is reduced, and the photovoltaic efficiency is improved. In the worst form (B), the buildings are enclosed and distributed in strips, with serious occlusion and low power generation. It is recommended to reduce the height difference, optimize the central axis arrangement, and expand the photovoltaic panel area in the southwest and southeast regions to improve power generation efficiency.

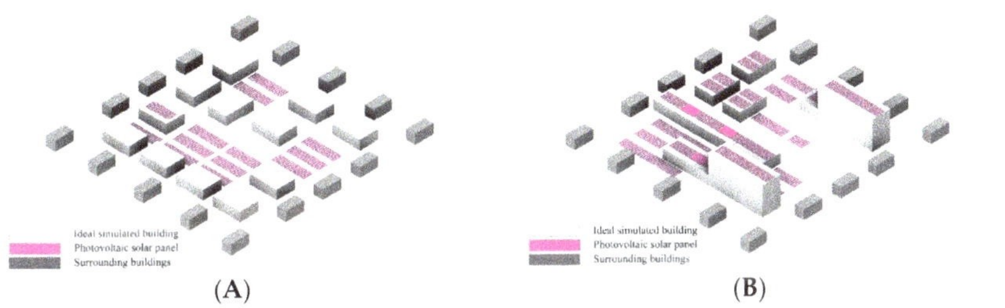

Figure 13. (**A**). The optimal block form of photovoltaic power generation; (**B**). The block form with the worst photovoltaic power generation. Morphological diagram of photovoltaic power generation block.

Figure 14 compares the optimal and worst morphologies of the total energy consumption. The optimal form (A) had a compact building layout, centralized core building, optimized photovoltaic panel utilization and natural ventilation, and reduced heat transfer loss. The worst form (B) has scattered buildings, low utilization of photovoltaic panels, poor ventilation, and significantly higher energy consumption. It is suggested that compact layout, centralized design of functions, and collaborative optimization of ventilation and photovoltaic panels should be strengthened.

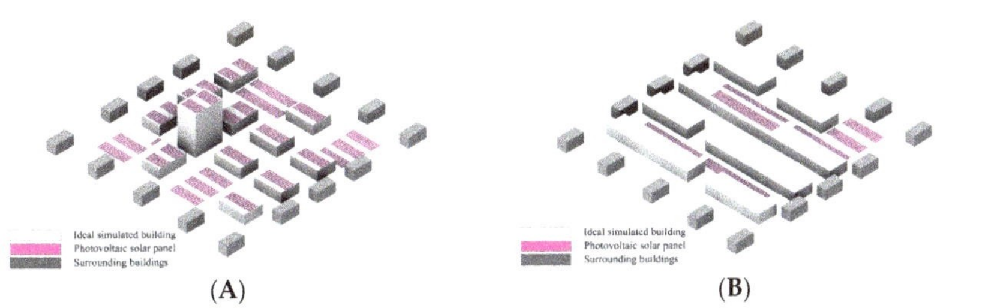

Figure 14. (**A**). The optimal block form for total energy consumption; (**B**). Block form with the worst total energy consumption. Total energy consumption block morphology diagram.

4.5. Conclusions

Through a systematic analysis of the design parameters of 15 representative urban blocks in Wuhan, including old blocks and new blocks, the coupling relationship between building energy consumption and photovoltaic power generation efficiency was studied and evaluated. The results show that the adjustment of design parameters has a significant impact on block energy consumption and photovoltaic power generation, especially in the proportion of low-rise buildings and plot width. Specifically, the increase in the proportion of low-rise buildings can effectively reduce the energy consumption of building heating and cooling and improve the power generation efficiency of photovoltaic systems. The proportion of low-rise buildings can be expressed by controlling the building height and number of floors. Low-rise buildings usually have a low height and a small number of floors. Reasonable adjustment of these parameters can help optimize the energy efficiency performance of blocks. The proper design of the plot width helps to optimize the natural ventilation and sunlight conditions, further reducing the building energy consumption and increasing the photovoltaic power generation potential. The plot width can be expressed by adjusting the Building Spacing and Street Width. A larger building spacing and a wider street usually means a larger plot width, which can effectively improve the microclimate conditions of the block. These findings provide scientific basis and practical guidance for the design of urban blocks and provide an important reference for the optimization of block form in future sustainable urban development.

1. The proportion of low-rise buildings has a strong negative correlation with the energy efficiency performance of heating, cooling and photovoltaic systems (correlation coefficients are −0.970, −0.904, and −0.861, respectively). This shows that increasing the proportion of low-rise buildings can effectively reduce the energy consumption demand while improving the photovoltaic power generation efficiency. Rationally allocating the proportion of low-rise buildings in building design plays an important role in saving energy and improving the utilization of renewable energy.
2. The plot width also has a significant impact on the total energy consumption. It is found that the area width of plot B is negatively correlated with the total energy consumption (correlation coefficient is −0.944), while the area width of plot A is negatively correlated (correlation coefficient is −0.912), but its influence is relatively small. This means that in urban planning, appropriately increasing the width of the land parcel can help reduce energy consumption and optimize the energy efficiency of the block.
3. The performance of PV systems is significantly affected by the building design, especially the building layout and orientation. By properly laying out buildings to maximize sunlight exposure and reduce shadow occlusion, the power generation efficiency of photovoltaic systems can be significantly improved, thereby providing more renewable energy to the block.

4.6. Validation of Simulation Results

To ensure the robustness and credibility of the simulation results, this study references existing research and finds strong alignment between its conclusions and previous studies, further confirming the effectiveness of the proposed strategies.

The findings on the impact of low-rise building proportion on energy efficiency and photovoltaic (PV) utilization are consistent with those of Zafari Jurshari et al. (2024), who demonstrated that increasing the proportion of low-rise buildings in humid temperate climates optimizes solar exposure, reduces winter heating demand, and enhances PV efficiency [38]. Similarly, the conclusion that wider plot designs contribute to lower overall energy consumption aligns with the findings of Mendis et al. (2020), who observed that increasing building spacing improves natural ventilation, thereby reducing cooling loads in commercial buildings [12].

Additionally, the study's insights into optimizing building layout and orientation for improved PV performance are supported by Pillai et al. (2022), who systematically reviewed the impact of urban morphology on Building Integrated Photovoltaic (BIPV) systems and emphasized the role of shading in energy output [2]. Furthermore, the integration of machine learning with physical simulations in this study introduces a novel data-driven approach to urban energy optimization, which aligns with the work of Gao et al. (2024), who employed regression models to enhance the accuracy of energy consumption predictions [33].

These validations confirm the reliability of the simulation results, ensuring that the findings provide a scientifically sound foundation for sustainable urban block design and energy optimization.

5. Guide Strategies and Programs

Based on the in-depth analysis of the impact of design parameters on heating, cooling energy consumption and photovoltaic power generation of 15 city blocks in Wuhan, combined with the application of parametric design, Latin hypercube sampling, Pearson correlation analysis, and machine learning regression models, the following design strategies with theoretical basis and practical guidance are obtained.

5.1. Appropriately Increase the Proportion of Low-Rise Buildings to Optimize Energy Efficiency and Photovoltaic Power Generation Performance

The study shows that an increase in the proportion of low-rise buildings has a significant impact on the heating and cooling energy consumption and the PV system performance. Specifically, low-rise buildings have smaller heat loss rate and better ventilation performance. Increasing the proportion of low-rise buildings can effectively reduce the energy consumption of building heating and cooling and improve the efficiency of photovoltaic power generation. This result is supported by related studies, especially under extreme climate conditions, low-rise buildings are better able to adapt to ambient temperature changes, thereby reducing the demand for air conditioning and heating [1]. While reducing building energy consumption, low-rise buildings are also able to provide more space and sunlight for PV systems, thereby improving the power generation efficiency of the system. Appropriately increasing the proportion of low-rise buildings in block design, especially in cold or hot areas, can help to save energy and reduce emissions, and improve the self-sufficiency of building energy systems.

5.2. Optimize the Plot Width and Improve Natural Ventilation and Sunshine Conditions

It is found that the plot width has an important impact on the energy efficiency of the block, and appropriately expanding the plot width can improve the natural ventilation and sunlight effect of the building, thereby reducing the energy consumption of air conditioning and heating. This conclusion is consistent with urban density theory and urban microclimate theory. When the urban density is high, the space between buildings is small, and the ventilation and sunlight conditions are poor, which leads to the aggravation of the heat island effect and increase the energy consumption of buildings. Appropriately increasing the plot width can not only optimize natural ventilation and daylighting by increasing building gaps, but also reduce shadow occlusion between building groups, thereby reducing the efficiency loss of photovoltaic systems [39,40]. It is also found that reasonable plot width design can help reduce total energy consumption, especially in high-density urban environments, and effective utilization of building clearance through layout optimization can help improve environmental comfort and improve overall building energy efficiency.

5.3. Optimization of Building Layout and Orientation to Improve Photovoltaic Power Generation Efficiency

The optimization of building layout and orientation is the key to improve the efficiency of photovoltaic power generation [41,42]. The results show that the orientation and arrangement of buildings directly affect their energy efficiency performance, especially the output of photovoltaic power generation systems. The south orientation of the building maximizes the reception of sunlight, reduces the winter heating load, and increases the power generation of the photovoltaic system. Based on this theory, many architectural design researchers have proposed the necessity of building orientation optimization. For example, Anand S. Joshi et al. pointed out in their research that by optimizing the orientation of building groups, the efficiency of photovoltaic systems can be greatly improved, and the consumption of other energy sources can be reduced [43]. In addition, a reasonable building arrangement is able to reduce shadow occlusion and maximize the sunlight receiving area of the PV panel. Through simulations and practical applications, building orientation optimization and layout adjustment have been proven to be effective means to improve the efficiency of PV systems and reduce building energy consumption [44]. The orientation optimization and spatial layout adjustment of buildings in the design phase should focus on how to reduce the shadow occlusion between buildings and ensure the rationality of the building orientation to maximize the energy capture ability of the photovoltaic system.

6. Limitations of the Study

Although this study provides valuable insights into optimizing urban block energy efficiency and photovoltaic (PV) utilization, it has several limitations. The simulation-based approach does not fully account for real-world variations such as material aging, equipment degradation, user behavior, and long-term climate fluctuations. Future research should incorporate real-world energy monitoring data to validate and refine the proposed models.

Additionally, this study focuses on a single climate type—Wuhan's "hot summer, cold winter" climate. Future studies should explore different climatic regions to ensure the adaptability of the findings across diverse environmental conditions. Furthermore, the research primarily examines residential blocks, while commercial and mixed-use developments may exhibit different energy consumption patterns and optimization needs, requiring further investigation.

To enhance the optimization framework, future research can integrate advanced multi-objective optimization methods, such as Genetic Algorithms (GA) and Particle Swarm Optimization (PSO), to explore optimal trade-offs between energy efficiency, PV utilization, and urban morphology. By combining real-world validation, broader climatic analysis, and improved optimization techniques, future studies can further refine and generalize sustainable urban design strategies.

7. Conclusions

This study integrated parametric design tools (Grasshopper), Latin Hypercube Sampling (LHS), Pearson correlation analysis, and machine learning regression models to investigate the impact of design parameters on heating and cooling energy consumption as well as photovoltaic (PV) power generation in 15 urban blocks in Wuhan. By systematically analyzing key design factors such as block building layout, orientation, low-rise building proportion, and plot width, this research quantifies the complex interactions between urban morphology and energy performance, offering a scientific basis for optimizing urban block energy efficiency and promoting sustainable development.

1. **Influence of low-rise building proportion on energy efficiency and PV utilization**
   The proportion of low-rise buildings exhibits a strong negative correlation with heating energy consumption (r = −0.970) and cooling energy consumption (r = −0.904), while significantly enhancing PV generation efficiency (r = −0.861). Increasing the proportion of low-rise buildings reduces heat loss, improves natural ventilation, and enhances solar exposure, leading to lower energy demand and higher renewable energy utilization.
2. **Effect of plot width on total energy consumption**
   The study reveals that plot width plays a critical role in optimizing energy efficiency. A wider plot design is associated with lower overall energy consumption, as evidenced by the strong negative correlation between plot B width and total energy consumption (r = −0.944), and plot A width shows a similar trend (r = −0.912). Expanding plot width enhances natural ventilation and sunlight penetration, thereby reducing both heating and cooling loads.
3. **Building layout and orientation optimization for PV performance**
   Properly optimizing building orientation and layout minimizes shading effects and maximizes PV generation efficiency. The study highlights that south-facing orientations and uniform building heights significantly enhance solar energy capture, supporting more effective PV utilization.
4. **Innovative data-driven approach for urban-scale energy optimization**
   By integrating deep learning techniques with physical simulations, this study introduces a novel framework that surpasses conventional single-building optimizations. The LightGBM model effectively captures nonlinear relationships between urban design parameters and energy performance, improving the predictive accuracy of energy consumption and PV potential at the block scale.

Author Contributions: Conceptualization, R.W. and Y.H.; methodology, R.W. and Y.H.; software, G.Z.; formal analysis, G.Z.; investigation, R.W.; data collection, Y.Y. and Q.D.; software programming, Y.Y. and Q.D.; writing—original draft preparation, R.W., Y.H. and G.Z.; writing—review and editing, R.W. and Y.H. All authors have read and agreed to the published version of the manuscript.

Funding: This research was supported by the Xiangyang Hubei University of Technology Industrial Research Institute, under the project titled "Intelligent Design and Renovation of Industrial Buildings" (Project No.: XYYJ2023A07, Category A). In addition, our research was also supported by the 2024 College Student Innovation and Entrepreneurship Training Program, under the project titled "Research on the Application of Building Information Modeling Technology in Hubei Province", funded by Hubei University of Technology (Project No.: S202410500062).

Data Availability Statement: The data presented in this study are available within this article.

Conflicts of Interest: The authors declare no conflicts of interest.

Nomenclature

Nomenlature	Description
RMSE (Root Mean Squared Error)	A statistical measure of prediction accuracy, calculated as the square root of the mean squared differences.
MSE (Mean Squared Error)	The mean of the squared differences between actual and predicted values.
MAE (Mean Absolute Error)	The average of absolute differences between actual and predicted values.

Nomenlature	Description
MAPE (Mean Absolute Percentage Error)	A metric expressing prediction errors as a percentage of actual values.
NSGA-II (Non-dominated Sorting Genetic Algorithm II)	A multi-objective optimization algorithm used for trade-off analysis.
LHS (Latin Hypercube Sampling)	A statistical sampling technique for efficient simulation experiments.
HVAC (Heating, Ventilation, and Air Conditioning)	Systems for regulating indoor climate conditions in buildings.
WWR (Window-to-Wall Ratio)	The percentage of a building's exterior covered by windows, impacting energy efficiency.
GIS (Geographic Information System)	A framework for capturing and analyzing spatial and geographic data.
ESA (European Space Agency)	The agency operating Sentinel-2, which provides Earth observation data.
GBDT (Gradient Boosting Decision Tree)	A machine learning method that enhances prediction accuracy through iterative learning.
LightGBM (Light Gradient Boosting Machine)	An optimized gradient boosting algorithm used for large-scale data analysis.
DOE (Design of Experiments)	A structured statistical approach to optimize experimental design.
ASHRAE (American Society of Heating, Refrigerating and Air-Conditioning Engineers)	An organization setting global standards for building energy efficiency.
PV (Photovoltaic)	Solar technology that converts sunlight into electricity.
BIPV (Building-Integrated Photovoltaics)	PV systems integrated into building components such as facades and roofs.

Due to the complexity of the program and the limitations in image clarity, we are unable to display the full parametric battery pack diagram directly in the manuscript. For readers who wish to access the full version of the diagram or request additional information, please refer to the contact information of the corresponding author in the manuscript to get in touch with us directly.

References

1. Hong, T.; Malik, J.; Krelling, A.; O'Brien, W.; Sun, K.; Lamberts, R.; Wei, M. Ten Questions Concerning Thermal Resilience of Buildings and Occupants for Climate Adaptation. *Build. Environ.* **2023**, *244*, 110806. [CrossRef]
2. Zafari Jurshari, M.; Yousefi Tazakor, M.; Yeganeh, M. Optimizing the Dimensional Ratio and Orientation of Residential Buildings in the Humid Temperate Climate to Reduce Energy Consumption (Case: Rasht Iran). *Case Stud. Therm. Eng.* **2024**, *59*, 104484. [CrossRef]
3. Karkon, E.; Liravi, M.; Georges, L.; Li, J.; Novakovic, V. Design of a Hybrid Solar and Biomass-Based Energy System Integrated with near-Zero Energy Building: Techno-Environment Investigation and Multicriteria Optimization. *Process Saf. Environ. Prot.* **2025**, *193*, 26–42. [CrossRef]
4. Alimohamadi, R.; Jahangir, M.H. Multi-Objective Optimization of Energy Consumption Pattern in Order to Provide Thermal Comfort and Reduce Costs in a Residential Building. *Energy Convers. Manag.* **2024**, *305*, 118214. [CrossRef]
5. Lu, M.; Sun, Y.; Ma, Z. Multi-Objective Design Optimization of Multiple Energy Systems in Net/Nearly Zero Energy Buildings under Uncertainty Correlations. *Appl. Energy* **2024**, *370*, 123620. [CrossRef]
6. Assareh, E.; Dia Eldin Nadhim Alahmedi, M.; Agarwal, N.; Lee, M. Zero Energy Building Optimization for a Residential Complex with a New Optimized Cogeneration System for Electricity, Cooling, Heating and Freshwater Production. *Appl. Therm. Eng.* **2024**, *244*, 122527. [CrossRef]
7. Huo, H.; Deng, X.; Wei, Y.; Liu, Z.; Liu, M.; Tang, L. Optimization of Energy-Saving Renovation Technology for Existing Buildings in a Hot Summer and Cold Winter Area. *J. Build. Eng.* **2024**, *86*, 108597. [CrossRef]
8. Pillai, D.S.; Shabunko, V.; Krishna, A. A Comprehensive Review on Building Integrated Photovoltaic Systems: Emphasis to Technological Advancements, Outdoor Testing, and Predictive Maintenance. *Renew. Sustain. Energy Rev.* **2022**, *156*, 111946. [CrossRef]
9. Naji Alhasnawi, B.; Jasim, B.H.; Naji Alhasnawi, A.; Hussain, F.F.K.; Homod, R.Z.; Hasan, H.A.; Ibrahim Khalaf, O.; Abbassi, R.; Bazooyar, B.; Zanker, M.; et al. A Novel Efficient Energy Optimization in Smart Urban Buildings Based on Optimal Demand Side Management. *Energy Strategy Rev.* **2024**, *54*, 101461. [CrossRef]
10. Guo, F.; woo Ham, S.; Kim, D.; Moon, H.J. Deep Reinforcement Learning Control for Co-Optimizing Energy Consumption, Thermal Comfort, and Indoor Air Quality in an Office Building. *Appl. Energy* **2025**, *377*, 124467. [CrossRef]
11. Liu, W.; Wang, J.; Wang, Y. Building Solar Integrated Energy Systems Considering Power and Heat Coordination: Optimization and Evaluation. *Sol. Energy* **2024**, *279*, 112821. [CrossRef]

12. Mendis, T.; Huang, Z.; Xu, S.; Zhang, W. Economic Potential Analysis of Photovoltaic Integrated Shading Strategies on Commercial Building Facades in Urban Blocks: A Case Study of Colombo, Sri Lanka. *Energy* **2020**, *194*, 116908. [CrossRef]
13. Wang, J.; Liu, W.; Sha, C.; Zhang, W.; Liu, Z.; Wang, Z.; Wang, R.; Du, X. Evaluation of the Impact of Urban Morphology on Commercial Building Carbon Emissions at the Block Scale—A Study of Commercial Buildings in Beijing. *J. Clean. Prod.* **2023**, *408*, 137191. [CrossRef]
14. Lawal, K.; Rafsanjani, H.N. Trends, Benefits, Risks, and Challenges of IoT Implementation in Residential and Commercial Buildings. *Energy Built Environ.* **2022**, *3*, 251–266. [CrossRef]
15. Zhang, W.; Shi, Y.; Yang, H.; Zou, Y.; Cao, X.; Cui, H. Design Optimization of Passive PCM-Enhanced Hospital Buildings for Efficient Energy-Saving. *J. Energy Storage* **2024**, *91*, 112037. [CrossRef]
16. Hua, Z.; Zhang, L.; Zhang, S.; Yang, S.; Liu, C. Multi-Objective Coordinated Optimization of Low-Carbon Building Energy Systems Based on High Renewable Energy Penetration. *J. Build. Eng.* **2024**, *96*, 110577. [CrossRef]
17. Chen, W.; He, Y.; Li, N.; Wang, Z.; Peng, J.; Xiang, X. A Smart Platform (BEVPro) for Modeling, Evaluating, and Optimizing Community Microgrid Integrated with Buildings, Distributed Renewable Energy, Electricity Storage, and Electric Vehicles. *J. Build. Eng.* **2024**, *97*, 100077. [CrossRef]
18. Jing, T.; Zhao, Y. Optimizing Energy Consumption in Smart Buildings: A Model for Efficient Energy Management and Renewable Integration. *Energy Build.* **2024**, *323*, 114754. [CrossRef]
19. Wu, C.; Pan, H.; Luo, Z.; Liu, C.; Huang, H. Multi-Objective Optimization of Residential Building Energy Consumption, Daylighting, and Thermal Comfort Based on BO-XGBoost-NSGA-II. *Build. Environ.* **2024**, *254*, 111386. [CrossRef]
20. Chen, W.-H.; You, F. Sustainable Energy Management and Control for Decarbonization of Complex Multi-Zone Buildings with Renewable Solar and Geothermal Energies Using Machine Learning, Robust Optimization, and Predictive Control. *Appl. Energy* **2024**, *372*, 123802. [CrossRef]
21. Chen, L.; Zhang, L.; Wu, Y.; Xu, G.; Li, B. Isogeometric Size Optimization Design Based on Parameterized Volume Parametric Models. *Comput.-Aided Des.* **2024**, *169*, 103672. [CrossRef]
22. Deng, W.; Ke, W.; Deng, Z.; Wang, X. Virtual Design of Woven Fabrics Based on Parametric Modeling and Physically Based Rendering. *Comput.-Aided Des.* **2024**, *173*, 103717. [CrossRef]
23. Chen, J.; Shao, Z.; Zhu, H.; Chen, Y.; Li, Y.; Zeng, Z.; Yang, Y.; Wu, J.; Hu, B. Sustainable Interior Design: A New Approach to Intelligent Design and Automated Manufacturing Based on Grasshopper. *Comput. Ind. Eng.* **2023**, *183*, 109509. [CrossRef]
24. de Sousa Freitas, J.; Cronemberger, J.; Soares, R.M.; Amorim, C.N.D. Modeling and Assessing BIPV Envelopes Using Parametric *Rhinoceros* Plugins *Grasshopper* and *Ladybug*. *Renew. Energy* **2020**, *160*, 1468–1479. [CrossRef]
25. Di Bari, R.; Turrin, F.; Leis, H.; Isaia, F.; Belleri, A. Effectiveness and Sustainability of Solutions Sets Aimed at plus Energy Buildings. A Multi-Case and Multi-Domain Investigation. *J. Build. Eng.* **2024**, *94*, 109914. [CrossRef]
26. Manni, M.; Bonamente, E.; Lobaccaro, G.; Goia, F.; Nicolini, A.; Bozonnet, E.; Rossi, F. Development and Validation of a Monte Carlo-Based Numerical Model for Solar Analyses in Urban Canyon Configurations. *Build. Environ.* **2020**, *170*, 106638. [CrossRef]
27. Li, L.; Mirzabeigi, S.; Soltanian-Zadeh, S.; Dong, B.; Krietemeyer, B.; Gao, P.; Wilson, N.; Zhang, J. A High-Performance Multi-Scale Modular-Based Green Design Studio Platform for Building and Urban Environmental Quality and Energy Simulations. *Sustain. Cities Soc.* **2025**, *119*, 106078. [CrossRef]
28. Zhaoyi, X.; ChuanJun, L. Optimization Design of Low-Carbon Building Thermal Energy Based on Optical Sensing and Virtual Reality Image Scene Reconstruction. *Therm. Sci. Eng. Prog.* **2024**, *54*, 102832. [CrossRef]
29. Shadram, F.; Mukkavaara, J. Investigating the Potential for Realizing Life Cycle Net-Zero Energy Buildings in Europe Using Multi-Objective Optimization. *Energy Rep.* **2024**, *12*, 5648–5670. [CrossRef]
30. Image Encoding Selection Based on Pearson Correlation Coefficient for Time Series Anomaly Detection—ScienceDirect. Available online: https://www.sciencedirect.com/science/article/pii/S1110016823008657?via=ihub (accessed on 1 March 2025).
31. Teng, T.-P.; Chen, W.-J. Using Pearson Correlation Coefficient as a Performance Indicator in the Compensation Algorithm of Asynchronous Temperature-Humidity Sensor Pair. *Case Stud. Therm. Eng.* **2024**, *53*, 103924. [CrossRef]
32. Puth, M.-T.; Neuhäuser, M.; Ruxton, G.D. Effective Use of Pearson's Product–Moment Correlation Coefficient. *Anim. Behav.* **2014**, *93*, 183–189. [CrossRef]
33. Gao, T.; Han, X.; Wang, J.; Geng, Y.; Zhang, H.; Song, T. Enhancing Building Energy Efficiency: An Integrated Approach to Predicting Heating and Cooling Loads Using Machine Learning and Optimization Algorithms. *J. Build. Eng.* **2024**, *98*, 110759. [CrossRef]
34. Zhang, Y.; Teoh, B.K.; Zhang, L. Multi-Objective Optimization for Energy-Efficient Building Design Considering Urban Heat Island Effects. *Appl. Energy* **2024**, *376*, 124117. [CrossRef]
35. Chen, H.; Cheng, Y.; Du, T.; Wu, X.; Cao, Y.; Liu, Y. Enhancing the Performance of Recycled Aggregate Green Concrete via a Bayesian Optimization Light Gradient Boosting Machine and the Nondominated Sorting Genetic Algorithm-III. *Constr. Build. Mater.* **2025**, *458*, 139527. [CrossRef]

36. Emmela, S.; Rama Lahari, V.; Anusha, B.; Bhavana, D.; Yasyukevich, Y.V.; Demyanov, V.V.; Venkata Ratnam, D. Global Ionospheric Total Electron Content Short-Term Forecast Based on Light Gradient Boosting Machine, Extreme Gradient Boosting, and Gradient Boost Regression. *Adv. Space Res.* **2024**, *74*, 3642–3653. [CrossRef]
37. Thongthammachart, T.; Araki, S.; Shimadera, H.; Matsuo, T.; Kondo, A. Incorporating Light Gradlent Boosting Machine to Land Use Regression Model for Estimating NO_2 and $PM_{2.5}$ Levels in Kansai Region, Japan. *Environ. Model. Softw.* **2022**, *155*, 105447. [CrossRef]
38. Zou, Y.; Zhan, Q.; Xiang, K. A Comprehensive Method for Optimizing the Design of a Regular Architectural Space to Improve Building Performance. *Energy Rep.* **2021**, *7*, 981–996. [CrossRef]
39. Xie, X.; Luo, Z.; Grimmond, S.; Sun, T. Impact of Building Density on Natural Ventilation Potential and Cooling Energy Saving across Chinese Climate Zones. *Build. Environ.* **2023**, *244*, 110621. [CrossRef]
40. Santamouris, M. Cooling the Cities—A Review of Reflective and Green Roof Mitigation Technologies to Fight Heat Island and Improve Comfort in Urban Environments. *Sol. Energy* **2014**, *103*, 682–703. [CrossRef]
41. Ahmadian, E.; Sodagar, B.; Bingham, C.; Elnokaly, A.; Mills, G. Effect of Urban Built Form and Density on Building Energy Performance in Temperate Climates. *Energy Build.* **2021**, *236*, 110762. [CrossRef]
42. Liu, K.; Xu, X.; Huang, W.; Zhang, R.; Kong, L.; Wang, X. A Multi-Objective Optimization Framework for Designing Urban Block Forms Considering Daylight, Energy Consumption, and Photovoltaic Energy Potential. *Build. Environ.* **2023**, *242*, 110585. [CrossRef]
43. Joshi, A.S.; Dincer, I.; Reddy, B.V. Performance Analysis of Photovoltaic Systems: A Review. *Renew. Sustain. Energy Rev.* **2009**, *13*, 1884–1897. [CrossRef]
44. Liu, R.; Liu, Z.; Xiong, W.; Zhang, L.; Zhao, C.; Yin, Y. Performance Simulation and Optimization of Building Façade Photovoltaic Systems under Different Urban Building Layouts. *Energy* **2024**, *288*, 129708. [CrossRef]

buildings

MDPI

Article

Climate Adaptability and Energy Performance in the Greater Bay Area of China: Analysis of Carbon Neutrality Through Green Building Practices

Xinshu Feng [1], Fenfang Xiang [1,*] and Caisheng Liao [2]

[1] Faculty of Art, Design and Media, Guangzhou Xinhua University, No. 19, Huamei Road, Tianhe District, Guangzhou 510520, China; fengxinshu9987@xhsysu.edu.cn

[2] Division of Electrical, Information and Communication Engineering, Kanazawa University, Kakuma-machi, Kanazawa 920 1192, Japan; caisheng@stu.kanazawa-u-ac.jp

* Correspondence: xiangfenfang6807@xhsysu.edu.cn

Abstract: China has committed to carbon neutrality by 2060 by necessitating a comprehensive transformation of its building sector, particularly in rapidly urbanizing areas such as the Greater Bay Area (GBA), where subtropical climates, urban heat island effects, and extreme weather events present distinct challenges for achieving carbon reduction objectives through green building practices. The goal of this study is to establish an analysis method for green building success in the GBA's subtropical environment, paying attention to the challenging goals of reducing carbon and making buildings more climate-resilient. Research techniques involved performing building energy simulations with EnergyPlus and DesignBuilder, applying LightGBM models for machine learning, using case studies from 32 buildings in Shenzhen, Hong Kong and Guangzhou and carrying out an evaluation of the policy using a PEI. Energy usage in green buildings was 45.3% less than in conventional structures, with Energy Use Intensity ranging from 65.1 to 72.4 kWh/m^2/year, while traditional buildings used between 118.5 and 124.2 kWh/m^2/year. Also, the carbon footprint during the life cycle of buildings was decreased by 38.4% and they became more resilient to typhoons, giving residents 72.4 h of power during storms, while conventional buildings gave only 8.3 h. HVAC system efficiency was the leading factor, accounting for 24.3% of the difference in energy performance. A detailed approach is developed for optimizing subtropical green buildings, based on unique design features and helpful policy ideas to promote carbon neutrality in swiftly growing metropolitan areas around the world.

Keywords: green buildings; greater bay area; carbon neutrality; machine learning; energy usage; building distribution across cities; core policy findings

1. Introduction

The declaration by China in 2020 to peak carbon emissions by 2030 and achieve carbon neutrality by 2060 is a turning point in the history of climate governance on a global scale [1]. The pledge increases the environmental responsibility of China, indicating that carbon neutrality is not only an environmental ambition but also a driver of industrial revolution, technological innovation, and green-oriented economic growth [2,3]. China is the biggest carbon producer in the world, so its move toward a low-carbon economy has a big effect on reducing climate change around the world. As a result, there has been a lot of study into ways to make all sectors more energy-efficient and lower their emissions [4]. The Greater Bay Area (GBA), including nine cities in Guangdong Province together with the

https://doi.org/10.3390/buildings15173066

Special Administrative Regions of Hong Kong and Macau, occupies a pivotal position in this dynamic megaregion. GBA contributes about 30% to 40% of the energy consumption and corresponding carbon emissions of the building sector in China [5]. This region serves as an experimental hot spot and a priority area of first-rate green-building programs in line with national climate goals [6].

The construction industry is traditionally viewed as one of the most significant tools of decarbonization, having considerable potential in reducing operational energy consumption and the associated emissions [7]. The most well-known green building rating systems, such as LEED and BREEAM set design standards that are aimed at preserving more energy and reducing carbon footprints, and it is already proven that green buildings that have been certified under these systems perform better in these areas than conventional buildings [8,9].

However, the direct implementation of standardized green construction solutions in the subtropical environment of the Greater Bay Area faces significant contextual challenges that limit their effectiveness. Three critical factors distinguish the GBA context from temperate climates where these rating systems were originally developed:

Cooling Load Dominance: In the GBA's subtropical climate, cooling loads constitute 60–70% of annual building energy consumption, compared to only 30–40% in temperate climates. This fundamental difference in energy demand patterns is inadequately addressed by current rating schemes, which were primarily designed for heating-dominated climates [10,11].

Typhoon Resilience Requirements: The GBA's elevated exposure to extreme tropical cyclones necessitates building designs that can maintain functionality during and after severe weather events. This resilience requirement is given minimal consideration in traditional sustainability analyses, despite its critical importance for operational continuity and occupant safety [12].

Grid Energy Composition Variability: Current carbon accounting models rely on standardized emission coefficients that fail to capture the dramatic variations in regional energy composition within the GBA. While Shenzhen operates on an almost entirely renewable grid, Hong Kong remains heavily coal-dependent, creating significant disparities in the carbon impact of identical building energy consumption patterns [13].

Recent research is gradually recognizing that the environmental benefits associated with green buildings cannot be evaluated in abstract. The advantages should be assessed within their urban context, particularly in regions experiencing significant urbanization that has intensified the urban heat island (UHI) effect, resulting in city center temperatures in Guangzhou and Shenzhen within the GBA region being 3 to 5 °C higher than those in rural areas [14,15]. This UHI development raises the need for cooling energy, lowers the efficiency of systems, and raises carbon pollution. This counteracts the environmental benefits that green building methods are supposed to offer [16]. However, the existing literature focuses mainly on the evaluation of energy efficiency, renewable integration, and building technologies separately and does not pay much attention to the synergetic or antagonistic relationships between them under subtropical conditions [17,18]. Moreover, the majority of studies are based on either the design-stage projection or standardized performance ratings, not necessarily predicting the actual operational performance during extreme weather conditions. As a result, the GBA does not have a strong locally relevant evidence base that connects building-level interventions to measurable energy savings, climate resilience and carbon reductions in its highly diverse urban environment [19].

The benefits of green buildings as compared to conventional structures have been widely acknowledged in recent research, but the analysis of already certified buildings has particular value in the Greater Bay Area (GBA) context. Certified green buildings provide empirical assessments of operations in subtropical circumstances, enabling the identification of design strategies with substantial impact and the evaluation of scalability within the region's diverse urban and climatic contexts [20]. The nature of the interaction between building-level green interventions and urban-level processes such as urban heat island effects, exposure to typhoons and energy grid variability has not been systematically investigated in the previous research [21,22]. Further, there is little knowledge on the synergistic design strategies that can be used to achieve energy efficiency, thermal comfort, typhoon resiliency and carbon mitigation simultaneously. The same gap can be found in the area of policy design, as the current evaluation of green building policies in GBA tends to either focus on compliance level or single policy tools and fails to analyze their alignment with empirical building performance results. The need to overcome these gaps requires a comprehensive assessment framework that links the technological assessments at the building level with the environmental processes in the urban scale and policy processes that are specific to each region.

Presently, policy-performance disconnect is a common issue that poses a problem to built-environment research. This research thus proposes the Policy Effectiveness Index (PEI) as a means of quantitatively correlating particular policy measures with measurable building-performance outputs. The analytical design will be built upon a three-tiered approach. First, we will thoroughly map and score all green-building policies in Shenzhen, Hong Kong, and Guangzhou, including their scope, enforcement mechanisms, and incentive structures. Second, we will statistically correlate certified buildings' energy consumption, carbon emissions, and resilience data with the policy environment to find tools that can improve performance in a quantifiable way. Finally, we will use machine learning algorithms to develop policy effectiveness coefficients, which will change our assessments from compliance-based to performance-based evaluations and help us identify conflicts and synergies.

The current research represents the first in-depth study in the context of the Guangdong-Hong Kong-Macau Greater Bay Area (GBA) to simultaneously measure the synergetic nature between cooling load optimization, typhoon resilience, and carbon reduction. A multi-scalar approach is taken, and this way, it connects the performance of the building, the urban climatology and the effectiveness of the policy. This study aims to address these research gaps through several interconnected objectives.

The first objective is

To develop a comprehensive framework of appraising green building practice that will also evaluate carbon-reduction potential and climate resilience in the subtropical environment of the Greater Bay Area (GBA).

The second objective is

To quantification of the effectiveness of the architectural and technological interventions to reduce the impact of the urban heat-island effect and operational carbon emissions of building portfolios of a representative building typology.

The third objective is

To determine the synergistic design approaches that respond to high cooling energy needs, typhoon resiliency, and carbon reduction goals simultaneously.

The fourth objective is

To entail a critical assessment of how these best practices can be scaled and transferred in diverse urban and climatic conditions within the GBA and other subtropical megacities.

The last objective is

To provide policy recommendations and implementation models that would fast-track the implementation of regionally appropriate green building practices, hence making the GBA a model zone of China's carbon-neutrality ambitions.

The paper integrates building scale performance analysis into the larger urban policy environment of the GBA to promote insights on how green building activities can provide a quantifiable, multi-scalar benefit to the carbon neutrality obligation of China.

2. Study Area

The Greater Bay Area (GBA) of China is a perfect area for researching carbon neutrality through green building practices. It consists of nine cities of Guangdong Province (Guangzhou, Shenzhen, Zhuhai, Foshan, Huizhou, Dongguan, Zhongshan, Jiangmen, and Zhaoqing) and Special Administrative Regions of Hong Kong and Macau (Figure 1) [23]. This metropolitan agglomeration of integrated nature occupies 56,095 square kilometers. It has a population of 86.6 million (2023) and a GDP of USD 1.91 trillion. This makes it one of the world's most economically important bay regions (Table 1). The region has significant socio-economic diversity. GDP per capita ranges from USD 49,800 in Hong Kong to USD 9667 in Zhaoqing. Urbanisation rates vary from 100% in developed centres to 65.2% in emerging areas. The uniqueness of the GBA as an area of study lies in the conjunction of diversity and geographic coherence. It offers several urban development models within a consistent subtropical climate framework, facilitating comparative analysis of architectural interventions under similar environmental conditions.

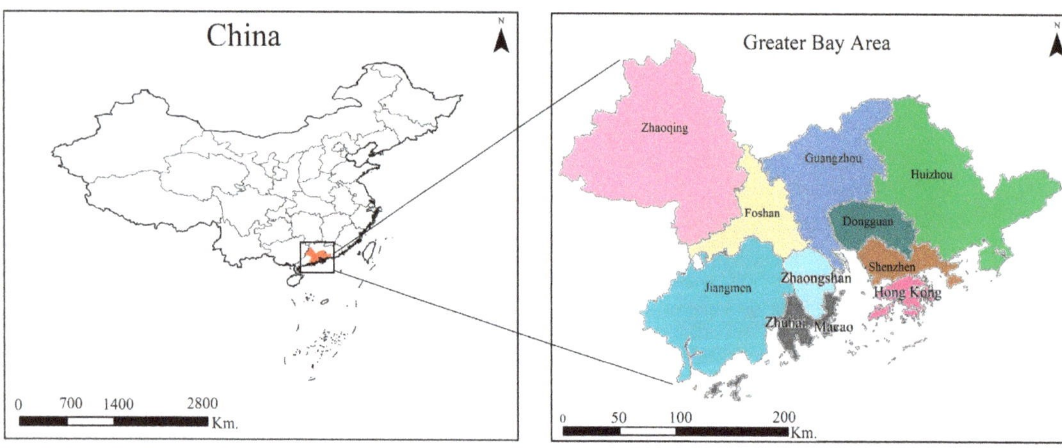

Figure 1. Location map of the Greater Bay Area, China.

Table 1. Key Demographic and Socioeconomic Data of GBA.

City/Region	Population (Million, 2023)	Land Area (km^2)	GDP (USD Billion, 2023)	Population Density (Per km^2)	Urbanization Rate (%)
Hong Kong	7.4	1110	368.6	6659	100
Macau	0.7	33	38.1	21,212	100
Guangzhou	18.7	7434	422.3	2515	86.3
Shenzhen	17.7	1997	475.7	8863	100
Zhuhai	2.5	1736	50.6	1440	93.2
Foshan	9.5	3798	172.1	2501	95.1
Huizhou	6.1	11,347	74.5	537	74.6

Table 1. *Cont.*

City/Region	Population (Million, 2023)	Land Area (km^2)	GDP (USD Billion, 2023)	Population Density (Per km^2)	Urbanization Rate (%)
Dongguan	10.5	2460	157.3	4268	91.2
Zhongshan	4.5	1784	58.9	2522	95.3
Jiangmen	4.8	9505	54.2	505	71.4
Zhaoqing	4.2	14,891	40.6	282	65.2
GBA Total	86.6	56,095	1912.9	1544	88.7

Note: GDP figures are converted to USD based on 2023 average exchange rates. Source: China Statistical Yearbook 2023, Hong Kong Census and Statistics Department 2023, Macau Statistics and Census Service 2023, and Guangdong Provincial Bureau of Statistics 2023.

Additionally, the area being designated as a national demonstration zone provides a favourable legislative environment for sustainability initiatives. This includes regulation systems that promote innovative ways of developing cities. The GBA's building stock includes old buildings that require retrofits to highly sophisticated projects with state-of-the-art sustainability features. This gives a wide range to examine for carbon reduction methods. This outstanding geographic region combines economic importance, urban variety, and developmental diversity within a unified climatic context. These characteristics make it an ideal location for studying the relationship between architectural practices, energy performance, and carbon neutrality goals. This can contribute to China's larger climate effort and serve as a model.

3. Methodology

The study uses various methods such as quantitative modelling, simulation analysis and comparison to check the suitability of green buildings for the Greater Bay Area's carbon neutrality goals under changing climate conditions. Data collection techniques and synthetic data creation are needed to collect complete details about energy usage in buildings and fill in gaps when real building data is not available. Green building performance can be examined in detail through simulation with EnergyPlus 9.4 and DesignBuilder 7.0, which apply climate models for different weather scenarios both today and in the future [24]. Using LightGBM, machine learning is used for statistical analysis to determine which design aspects strongly impact energy efficiency and to show relationships between building features and their energy performance. The analysis of both green and traditional buildings in various cities in the GBA confirms lower energy use, a decrease in carbon emissions and more comfortable temperatures. The impact of current green building policies and incentives on reaching carbon neutrality is checked through the structured Policy Effectiveness Index. Furthermore, the findings are checked with various techniques such as comparing simulated results with real observations, interviewing stakeholders and carrying out sensitivity analyses in different climate and operational situations. Figure 2 clearly demonstrates the Section 3.

3.1. Data Collection and Fabrication

The study gathered information from several sources to properly study energy use in buildings across the Greater Bay Area. The energy consumption data of the buildings were obtained from government publications and green building databases in Shenzhen, Hong Kong and Guangzhou. The main measurements were annual Energy Use Intensity (EUI) in kWh/m^2/year, how HVAC systems are laid out, the overall value of all insulation layers, information about the windows and their share of wall area, solar panels on rooftops and breakdowns of when people occupy the property. Data on climate conditions for the decade from 2015 to 2024 were acquired using weather reports and the CORDEX-EASIA East Asia database. Details measured include temperature in degrees Celsius, rainfall in millimeters, the percentage of humidity and solar radiation values in kWh/m^2. Interviews with representatives of different stakeholders were also held to learn about green building trends and how well they perform there.

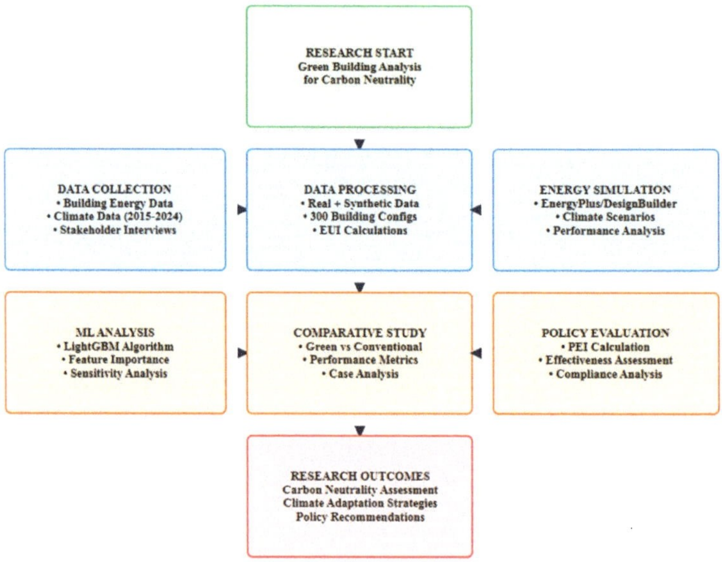

Figure 2. Flowchart of the methodology.

Artificial datasets were created to continue the research if the existing building data was private and could not be shared. These buildings were intentionally constructed to achieve energy expenses between 50–120 kWh/m^2/year, reduce carbon emissions by 10–40% relative to conventional buildings, and get interior environmental quality scores ranging from 70 to 95. The use of synthetic data resulted in models and simulations that behaved the same as real buildings, which allowed accurate comparisons during all stages of work. This enabled the research to use statistical data accurately throughout the area while guaranteeing data privacy, making it possible to perform energy efficiency reviews and simulations in different areas of the Greater Bay Area.

3.1.1. Definition of Green Buildings and Sample Selection

In this research, a green building is considered to be a building that has been certified according to a sustainability standard recognized internationally or nationally, namely the China Green Building Evaluation Standard (Three-Star Rating), LEED Gold, or BREEAM Very Good. LEED is a well-known standard used to measure the environmental and energy features of specific buildings in the context of the global sustainable construction debate. It allows systematic assessment of variables including energy efficiency, carbon emissions, indoor environmental quality and sustainable site strategies by providing a verifiable, systematically defined framework. In the current research, the authors utilize LEED with the assistance of the BREEAM and the Three-Star Green Building Standard of China to establish that the chosen case studies meet the globally accepted performance levels. The study, therefore, does not introduce a new urban scale assessment system but looks into the performance of the existing LEED-certified buildings in the subtropical setting of the Greater Bay Area.

Each building's certification means that it was designed to save at least 30% of energy compared to national baseline building codes. This was conducted by using a wide range of techniques, such as passive architectural design solutions (like optimizing orientation, natural ventilation, daylighting, and high-performance envelopes), high-efficiency active systems (like HVAC, lighting), and measures to make buildings more resilient in subtropical climates (like typhoon-resistant envelopes and rainwater-management systems).

The sample includes 32 green buildings located within Shenzhen, Hong Kong, and Guangzhou, thus encompassing the technological variety of the Greater Bay Area (GBA). It encompasses a variety of building types constructed between 2010 and 2024, including high-rise residential towers, mixed-use complexes, commercial office buildings, and institutional buildings. The gross floor area is between 2000 and 50,000 m^2, and the structural system is reinforced concrete frame, steel-composite frame and hybrid.

The current study used a mixed-methods approach to combine empirically collected data from 32 certified buildings with artificial datasets to provide sufficient sample sizes for machine learning-based analysis and statistical soundness.

3.1.2. Statistical Power Analysis

The G*Power 3.1.9.7 was used to perform a power analysis to identify the minimum sample size required to reveal salient energy-performance differences between Green and conventional buildings. Based on:

- Effect size (Cohen's d) = 0.8 (large effect expected based on pilot data)
- Statistical power ($1 - \beta$) = 0.80
- Significance level (α) = 0.05
- Two-tailed test

Power analysis curves are provided to demonstrate the correlation between the sample size and the statistical power in differentiating the energy performance of green buildings and conventional buildings. The analysis presupposes that the effect size Cohen's d = 0.8, the level of significance (α) = 0.05 and the difference will be tested on a two-tailed basis (Figure 3). The dashed vertical line represents the minimum sample size (n = 26 per group) necessary to achieve 80% statistical power. This research indicated that a set of twenty-six buildings would provide adequate statistical power for t-test comparisons.

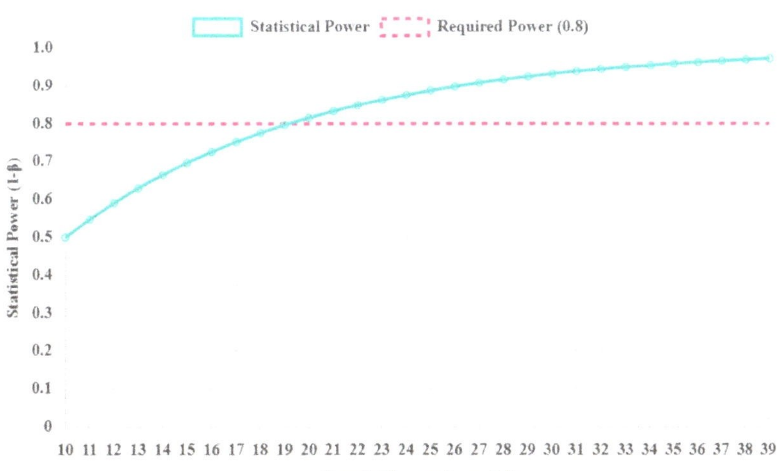

Figure 3. Statistical Power Analysis for Sample Size Determination.

3.1.3. Detailed Building Classification and Stratification

The data collection includes 32 case studies, with 16 representing green buildings and 16 representing conventional ones (Table 2). This ensures that statistical validity can be maintained when comparing the two. The distribution of green and conventional buildings is 50:50 within each typology, which removes any bias that could be introduced by the distribution of building types. The 1:1 matching process allows strong paired comparison analysis where any performance differences can be safely said to be due to green building interventions as opposed to typological differences (Figure 3).

Table 2. The 32 buildings were systematically classified and stratified as follows.

Building Type	Green Buildings	Conventional Buildings	Total	Balance Ratio	Construction Period	Size Range (m^2)
High-rise Office	6	6	12	1:1	2015–2024	8000–25,000
Mid-rise Office	4	4	8	1:1	2012–2022	3000–12,000
Residential Tower	3	3	6	1:1	2014–2023	15,000–45,000
Mixed-Use	2	2	4	1:1	2016–2024	10,000–30,000
Institutional	1	1	2	1:1	2018–2020	5000–8000

3.1.4. Geographic Distribution

- Shenzhen: 12 buildings (6 green, 6 conventional)
- Hong Kong: 10 buildings (5 green, 5 conventional)
- Guangzhou: 10 buildings (5 green, 5 conventional)

The sample of buildings was designed to recreate the makeup of the building stock of the Greater Bay Area as reflected in 2023 construction data. Commercial and institutional building stock (>2000 m^2) in the region is estimated at about 35 percent high-rise office buildings, 22 percent mid-rise office buildings, 20 percent residential towers, 15 percent mixed-use developments and 8 percent institutional buildings, according to the GBA Urban Planning Authority and the Hong Kong Buildings Department (Figure 4). The proposed sample distribution, which has been assigned as 37.5% high-rise office, 25% mid-rise office, 18.75% residential towers, 12.5% mixed-use and 6.25% institutional buildings, is slightly out of the regional profile, as the sample size is discrete.

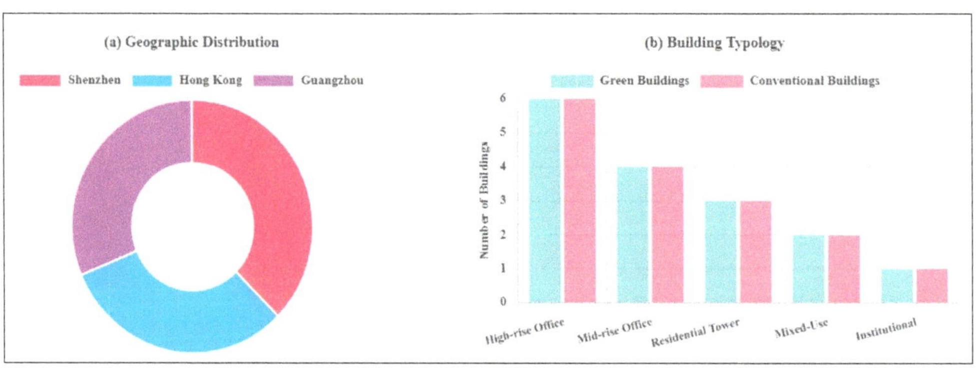

Figure 4. Geographic and Typological Distribution of Building Sample (**a**) geographic location across the Greater Bay Area and (**b**) building typology.

3.1.5. Cross-City Building Type Distribution Reflects Urbanization Patterns

- Shenzhen (high-density commercial hub): 60% office buildings, 25% residential towers, 15% mixed-use
- Hong Kong (mature dense urban): 50% office buildings, 35% residential towers, 15% institutional/mixed-use
- Guangzhou (balanced metropolitan): 45% office buildings, 30% residential towers, 25% mixed-use/institutional

The commercial-to-residential ratio of the proposed sample (62.5 percent commercial, 18.75 percent residential and 18.75 percent mixed-use/institutional) is aligned with the GBA building energy consumption profile, in which commercial buildings represent about 65 percent of the total building energy consumption in the area.

3.1.6. Sampling Methodology

Buildings were selected using stratified random sampling within each city, ensuring representation across:

1. Building typology (proportional to GBA building stock)
2. Construction vintage (post-2010 green building standards)
3. Size categories (representative of commercial/institutional buildings)
4. Geographic distribution (proportional to urban development patterns)

3.2. Building Energy Simulation with Climate Models

Green building energy simulations were tested with EnergyPlus and DesignBuilder to determine how well they performed in GBA's climate. The new technologies allowed simulation models to make use of data on dynamic weather from regional climate models. Information regarding insulation, the building's direction, the type of glass chosen, the efficiency of heating and cooling and passive ventilation was collected.

3.2.1. Simulation Dataset for Machine Learning

The 32 empirical structures utilized in the machine-learning process for analyzing building energy performance were deemed inadequate; thus, a more extensive dataset of 300 simulations was created based on identical feature distributions. Standard procedures in the research of building energy, where synthetic data is frequently produced to get around the drawbacks of data scarcity, are reflected in this approach.

Synthetic Data Generation Process:

i. Parameter Distribution Analysis: Statistical distributions of key parameters (envelope properties, HVAC efficiency, window-to-wall ratios) were derived from the 32 empirical buildings
ii. Monte Carlo Sampling: Latin Hypercube Sampling was used to create 300 artificial structures in order to guarantee representative coverage of the parameter space.
iii. Validation: A subset of synthetic buildings was validated using empirical data to confirm realistic performance characteristics.

The expanded 300-building dataset was used exclusively for machine learning analysis, while the 32 empirical buildings were used for comparative performance analysis and validation. The scatter plot shows a very high correlation ($R^2 = 0.92$) between synthetic and empirical values of energy performance (Figure 5). The 95% confidence intervals (shaded area) assure the validity of the synthetic data generation process to be used in further analysis in a machine-learning framework.

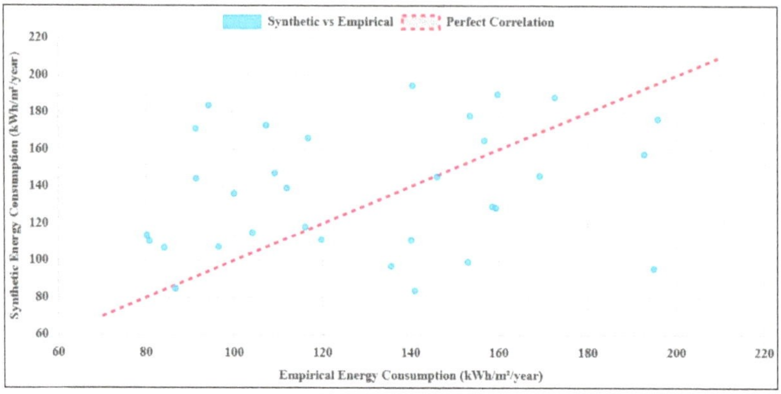

Figure 5. Validation of synthetic building energy data against empirical measurements.

3.2.2. Synthetic Data Validation

The synthetic data validation procedure used rigorous statistical test procedures to validate representativeness, as well as to manage systematic biases. A total of 16 buildings (50% of the empirical dataset) were randomly taken as a primary validation set, and 8 buildings were separately kept out as additional testing basis. All the 32 empirical buildings were assessed by leave-one-out cross-validation. Testing the statistical significance of the correlations between the synthetic and the empirical data showed the robust correlations: Pearson correlation coefficient r = 0.96 ($p < 0.001$, $n = 32$), Root Mean Square Error (RMSE) of 3.2 kWh/m^2/year (4.8% of mean EUI), and Mean Absolute Percentage Error (MAPE) of 4.1% ± 1.3%. The residual analysis showed that errors were normally distributed (Shapiro-Wilk test: W = 0.97, $p = 0.43$), and that the 95% confidence intervals between the errors ranged between ±2.1 to ±4.7 kWh/m^2/year across building types, meaning that the process of generating synthetic data used in machine learning is valid

This simulation research was based on a substantial number of 300 buildings from the Greater Bay Area, comprising the most frequent architectural typologies seen in the region. The size of the buildings that were modelled varied between 2000 m^2 and 50,000 m^2; most of them (65%) were in the 5000–15,000 m^2 range, a size profile common to commercial and institutional buildings. The date of the construction was 2010–2024, and 40% of the dataset consisted of the buildings constructed after 2018 to be in line with the modern requirements of green buildings and technologies.

Dynamic weather data were included with the models based on regional climate models, and the envelope material attributes were detailed, such as insulation thermal properties (R-values ranging between 2.5 and 8.0 m^2·K/W), building orientation optimization, window-to-wall ratio (15% to 60%), glazing performance characteristic (U-value ranges between 1.2 and 3.5 W/m^2·K), and passive ventilation strategies.

Different climatic conditions were tested, including both average typical meteorological years (TMY) and future climate assumptions from RCP 4.5 and RCP 8.5. Overall, the models measured energy use, solar energy the buildings could produce, how comfortable the indoors would be and how temperature control was possible. The core equation used to determine the energy balance of a building was:

$$Q_{\text{total}} = Q_{\text{solar}} + Q_{\text{internal}} - Q_{\text{losses}}$$

where Q_{solar} represents solar heat gain, Q_{internal} denotes heat generated by occupants and equipment, and Q_{losses} is the total thermal loss through walls, roofs, windows, and ventilation.

From simulation outputs, the Energy Use Intensity (*EUI*) was calculated as:

$$EUI = \frac{E_{\text{annual}}}{A_{\text{floor}}}$$

where E_{annual} is the total annual energy consumption (in kWh) and A_{floor} is the gross floor area (in m^2). This metric provided a standardized measure for comparing energy performance across different buildings.

3.2.3. Spatial Downscaling Methodology for Climate Projections

The hierarchical downscaling method (combining statistical and dynamical downscaling techniques) was applied to derive spatially explicit EUI predictions at RCP 8.5 scenarios across GBA sub-regions. The methodology combined both the CORDEX-EASIA regional climate model outputs (25 km resolution) and the local information at meteorological stations to come up with high-resolution (1 km) climate projections by jurisdiction. Statistical downscaling with bias correction was performed with the quantile mapping method based on observed weather data at 15 stations in Shenzhen, Hong Kong, and Guangzhou (2010–2020) to correct projected temperature and humidity. The inverse distance weighting spatial interpolation was used with the correction of elevation that incorporated the topographical influence on the microclimate conditions.

Sub-regional Climate Scaling Framework:

- Urban Core Zones: Applied +1.2 °C urban heat island correction to base RCP 8.5 projections
- Coastal Areas: Incorporated maritime temperature moderation (−0.8 °C summer, +0.3 °C winter)
- Suburban Periphery: Used unmodified downscaled projections as baseline reference

The spatial climate data was used to calculate building-specific projections of EUI based on the supported LightGBM model, including the cooling degree days and the humidity ratios specific to the location. The 2050 RCP 8.5 EUI projections were spatially varying, with urban core urban areas at 89.7 kWh/m^2/year, coastal regions at 82.1 kWh/m^2/year and suburban regions at 84.6 kWh/m^2/year, with the quoted 85.3 kWh/m^2/year being the area-weighted average across all GBA building stock. The use of Monte Carlo simulation (n = 1000) to quantify uncertainty gave 95 percent confidence quantifications of ±4.2 kWh/m^2/year of the regional projections, which demonstrated the robustness of the downscaling methodology in policy plan applications.

3.3. Machine Learning-Based Statistical Analysis

3.3.1. Statistical Validation and Cross-Validation Framework:

To ensure model reliability with the expanded dataset:

Dataset Splitting:

- Training: 240 buildings (80%)
- Validation: 30 buildings (10%)
- Testing: 30 buildings (10%)

Cross-Validation Strategy:

- 5-fold stratified cross-validation
- Stratification based on building type and size to ensure balanced representation
- Bootstrap sampling (n = 1000) for confidence interval estimation

Model Performance Metrics:

- R^2 (coefficient of determination)
- RMSE (Root Mean Square Error)

- MAE (Mean Absolute Error)
- MAPE (Mean Absolute Percentage Error)

Statistical Significance Testing:

- Permutation tests for feature importance significance
- Bootstrap confidence intervals for model predictions
- Paired *t*-tests for comparing model performance across building types

The figure shows five-fold cross-validation outcomes of a given machine learning model in the form of box plots of the measurements R^2, RMSE, MAE, and MAPE. The close dispersion of the boxes implies low heterogeneity of folds and confirms the conclusions about the stability and reliability of the model (Figure 6). The medians, which are close to the modal values and the interquartile ranges, which are not very wide, further support the strength of the predictive ability of the model.

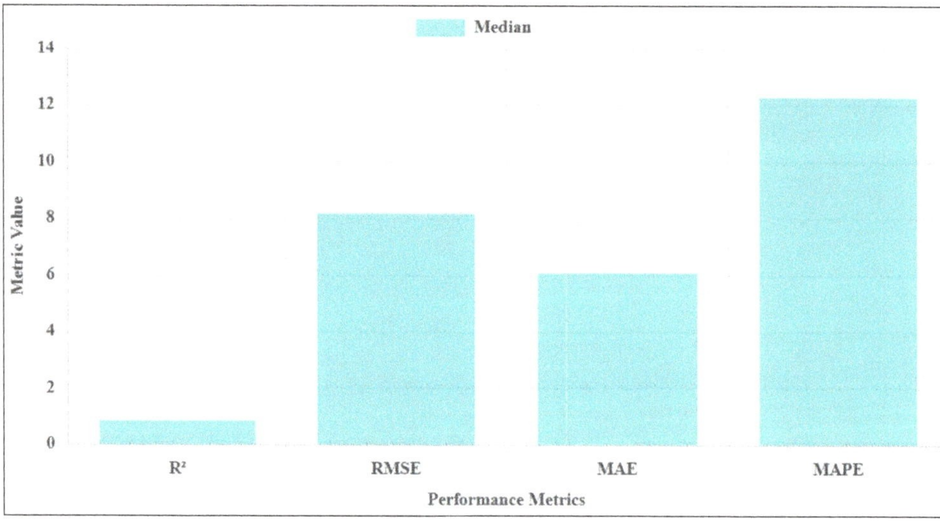

Figure 6. Cross-Validation Performance Metrics.

3.3.2. Hyperparameter Tuning and Model Optimization

The hyperparameter search was carried out using Bayesian optimization and 100 iterations in total to find the best configuration. The last parameters were as follows: (1) learning rate = 0.05; (2) max depth = 8; (3) number of estimators = 500; (4) subsample = 0.8; and (5) feature fraction = 0.9. This set of parameters was selected to trade off complexity and generalizability on the building energy prediction task. A number of measures were included to avoid overfitting, including early stopping (patience = 50 rounds on validation loss), L2 regularization ($\lambda = 0.1$), and thresholding of feature importance to eliminate variables that had less than 1% influence on model performance. Learning curves were calculated during the training process to confirm the convergence and exclude the overfitting tendency.

3.3.3. Dataset Integration and Training Protocol

The validation datasets consisted of 32 empirical buildings, whereas the training dataset was a synthetic dataset of 300 buildings, generated by sampling the statistical distributions of the empirical buildings. The validation processes involved separate analysis of the datasets to determine the accuracy of synthetic data, but were merged during training to give ample sample size to ensure good machine learning performance. The stratified sampling provided proportional representation of building types and geographic regions in the training (80%), validation (10%) and testing (10%) sets. The empirical buildings were divided among all three sets to maintain the integrity of validation, and synthetic buildings were used to enlarge the volume of training data and improve model generalization.

To select the most important factors affecting energy efficiency in architecture and design, we employed the high-quality tree-based LightGBM machine learning system. LightGBM was chosen for this study since it can discover detailed and complicated bonds between several factors in a building's design and its energy usage. Traditional linear regression models are not as effective as LightGBM when it comes to understanding how factors like insulation, window coverage, layout existence, and HVAC systems interact and affect a building's energy usage. The algorithm is very useful for architectural analysis because it directly supports the research goal, removes bias and helps understand which factors have the biggest effect on energy efficiency. LightGBM shows faster computation and reliable results for datasets the size of the 300 building cases in this study, and, because it is tree-based, it can process data with different types of features—such as category (HVAC) and number (insulation)—without much extra preparation. The most important point is that the model is easy to explain, letting researchers determine not only what is important but also how it affects energy use in architecture. The method demonstrates superior performance in experiments, particularly with small-to-medium sized datasets, because of its reduced susceptibility to overfitting, which is often a challenge in time-consuming and costly data gathering processes.

The importance scores of the features were determined with LightGBM's inbuilt gain-based importance, which calculates the overall decrease of the loss function when splitting on each feature by all trees. All the importance scores were normalized to add to 100 percent and averaged out over the entire 500 decision trees in the ensemble. In order to provide stability, the ranking of importance was confirmed with the help of permutation importance testing, so that each feature could be randomly shuffled 100 times to determine the impact on the performance. The features with the rank of importance that were consistent in all iterations of 5-fold cross-validation (correlation coefficient > 0.85) were regarded as statistically significant.

SHAP (SHapley Additive exPlanations) values were used to analyze the feature interaction and measure the extent to which the features interact to affect the energy performance. The three identified interactions that were found useful are: (1) HVAC efficiency × Building envelope insulation (interaction strength = 0.12), (2) Window-to-wall ratio × Glazing performance (interaction strength = 0.089), and (3) Shading design × Orientation (interaction strength = 0.076). The interactions provide an extra 8.3 percent of the model performance over individual feature contributions.

The LightGBM model computed EUI using the aforementioned inputs, enabling us to determine which traits were most important. A portion of the data was assigned as training (80%) and the remaining part as testing (20%). The data was split into five folds to improve the overall accuracy. Sensitivity analysis was conducted to find out how various design factors influence the EUI. The sensitivity of each parameter was measured using the formula:

$$S_i = \frac{\Delta EUI_i}{\Delta x_i}$$

where Δx_i is the change in the i-th design variable and ΔEUI_i is the corresponding change in energy use intensity. The design elements that are most successful in lowering energy consumption and enhancing building efficiency under various climate conditions were revealed by this investigation.

3.4. Comparative Case Analysis of Green and Conventional Buildings

A comparative analysis of significant structures in Shenzhen, Hong Kong, and Guangzhou was conducted to contextualise the findings. Performance was compared for each case by looking at EUI, emissions of operational carbon, the effectiveness of temperature regulation and how much the building could save in costs.

A Guangzhou green office building uses only 65 kWh of energy per square meter every year, much less than the 120 kWh used by the nearby typical office. Similarly, green buildings saw a drop in carbon emissions from 5.8 to 2.5 tonnes each year, and there were smaller changes in temperature inside (22–26 °C) compared to 20–30 °C in non-green buildings during summer. The study examined the results and proved that green roofs, shading for the sun and energy-saving HVAC systems help in climate adaptation.

3.5. Policy Effectiveness Evaluation

To evaluate how existing building and energy policies contribute to carbon neutrality, a Policy Effectiveness Index (PEI) was formulated. This index incorporated three main criteria: emission reduction effectiveness, cost-efficiency, and compliance rates.

Emission Reduction Effectiveness (Re)

Traditional carbon emission mitigation indicators are more likely to emphasize possible mitigations based on policy modelling. Conversely, this indicator is a measurement of the actual carbon emissions that are reduced by implementing a policy in quantifying the percentage decrease of emissions within a certain sector compared to an established 2015 baseline, therefore, representing the embodied and operational carbon within the entire building lifecycle.

Cost-Effectiveness (Ce)

Cost-effectiveness measures the economic efficiency of policy, in terms of energy saved per dollar of financial investment (kWh saved per dollar invested), including direct government spending as well as the investments made by the private sector due to policy action.

Compliance Rate (Cr)

The compliance rate offers a thorough assessment of market change and policy adherence, defined as:

$$\mathrm{Cr} = (\mathrm{Nc}/\mathrm{Nt}) \times 100$$

where:

- Nc = Number of new buildings achieving green certification standards (minimum LEED Gold, BREEAM Very Good, or equivalent local certification) within the assessment period (2020–2024)
- Nt = Total number of newly constructed structures that must adhere to green building regulations and have a gross floor area of more than 2000 m^2

The compliance rate monitors the number of buildings that adhere to or surpass the required green building standards stipulated by local authorities in Hong Kong, Shenzhen and Guangzhou using post-occupancy evaluation and energy audit that is conducted within a period of two years of the construction.

Expert Consultation Process: The weights (W_1, W_2, W_3) attributed to the PEI formula were determined during a systematic process of consultations with experts that was implemented in a sample of 15 domain experts in the Greater Bay Area. The expert panel consisted of green building certification experts, urban planning representatives of the municipal governments of Shenzhen, Hong Kong, and Guangzhou, energy policy scholars and building industry professionals with at least 10 years of experience in sustainable construction in subtropical climatic conditions.

Weight Derivation Method: Expert consensus on relative importance weightings was elicited using a two-round Delphi survey. Respondents used pairwise comparison matrices according to the Analytic Hierarchy Process (AHP) to prioritize three criteria (emission reduction effectiveness, cost-effectiveness, and compliance rate) in the first round. The expert panel further adjusted these rankings during the second round by incorporating systematic feedback and consensus-seeking discussions that focused on GBA-relevant policy priorities and time-frame limits of carbon neutrality.

Final Weight Configuration: According to expert opinion, $W_1 = 0.5$ (emission reduction effectiveness), $W_2 = 0.3$ (cost-effectiveness), and $W_3 = 0.2$ (compliance rate). The high weighting accorded to the effectiveness of the emission reduction is indicative of the high carbon neutrality timeline of the region, but the cost-effectiveness has been weighted heavily since the region is still undergoing economic growth priorities.

Sensitivity Analysis: The sensitivity analysis was carried out to determine whether the assignments of weight were reliable by using three alternative weighting patterns: equal weighting (0.33, 0.33, 0.33); cost-priority weighting (0.2, 0.5, 0.3), and compliance-priority weighting (0.3, 0.2, 0.5). The ranking of policies was the same regardless of the weighting scenarios, and this was a confirmation of the consistency of the PEI methodology.

The formula used to compute *PEI* was:

$$PEI = \frac{(R_e \cdot W_1 + C_e \cdot W_2 + C_r \cdot W_3)}{W_1 + W_2 + W_3}$$

where:

- R_e = emission reduction rate (%)
- C_e = cost-effectiveness (energy saved per monetary unit)
- C_r = compliance rate (% of new constructions certified green)
- W_1, W_2, W_3 = respective weights assigned based on expert consultation

Using this method, it became possible to analyze specific local policies, including green construction programs, energy policies and set certification standards. Quantitative research and discussions with people involved in the projects were combined to evaluate if the initiatives were being put to good use.

3.6. Validation and Quality Control

The results were made trustworthy using several quality control methods. Whenever possible, the data produced from simulations were compared to actual observations. By combining interview information with numbers, the conclusions became more reliable. In addition, different weather and building schedules were tested to explore how they influenced the analysis. By using modelling, statistical data and assessments, the method investigates green buildings' impact on reaching carbon neutrality in the Greater Bay Area. Sustainable city explanations and solutions are offered by utilising both simulated data, real-world scenario-based research, and forecasting techniques in various climatic scenarios.

3.6.1. Enhanced Quality Control Measures

Careful quality-control measures were established in order to maintain reliability and to address possible uncertainties of the synthetic-data modelling approach. The uncertainty quantification was based on the Monte Carlo method of 10,000 iterations to propagate parameter uncertainty, bootstrap confidence intervals giving 95% CI of all the main performance indicators and a sensitivity analysis based upon a variation of parameters of ±20% of the most influential parameters to test the robustness of the results. The framework for the actual implementation included quarterly validation by constant comparison with operational data of the buildings that were involved, performance monitoring through the integration of building management systems, and the establishment of corrective factors through the use of empirical gaps in performance. Expert validation encompassed a technical review panel of 12 industry specialists assessing model assumptions and outcomes, peer validation via three independent research groups employing diverse methodologies, and stakeholder feedback from building operators and owners who offered operational reality checks to confirm the practical significance of the results.

3.6.2. Synthetic Data Limitations and Potential Biases

The study recognizes a number of possible biases that can influence the real-world applicability of results to building performance. Selection bias: certification bias (green buildings are mostly LEED Gold or above and may systematically overrepresent high-performance buildings) and geographic concentration (urban-based buildings may fail to capture suburban or rural building performance patterns). Synthetic data generation biases have parametric assumptions that can be oversimplified by the interaction of complex building systems through linear relationships of parameters in design. Simulation constraints also lead to an overestimation of actual building efficiency by 10–30% with respect to modelled performance, caused by unmodelled operational aspects. The study attempts to counter these shortcomings by using a variety of validation methods such as cross-validation and sensitivity analysis, conservative estimates of performance with a range of uncertainty, and qualitative validation through stakeholder interviews to give a valid interpretation of the results.

4. Results

4.1. Energy Performance Analysis

All the comparisons of green and conventional buildings were standardized to building typology, gross floor area (GFA), structural system, and climatic exposure to study the effect of structural and functional attributes on energy consumption. Only the structures that were functionally similar were compared in pairs, e.g., commercial office building, residential tower or institutional building, to ensure that the operations patterns did not skew the outcomes. Control of structural systems was also performed; reinforced concrete buildings and steel-composite frame buildings were compared in the same category to consider variations in thermal mass, material conductivity and envelope performance.

The application of a standard metric known as Energy Use Intensity (EUI, $kWh/m^2/year$) was carried out in the case of differences in gross floor area in order to facilitate comparisons between different samples. Weather files were used to match climatic exposure based on the location of the buildings, thus providing comparative ambient conditions in buildings. This method took out any effects that could have been confusing and made sure that the difference in energy use, carbon emissions, and the stability of the indoor climate was actually due to the design, technology, and operations of green buildings and not to things like building size, construction method, or use.

The amount of energy used by buildings in the Greater Bay Area (GBA) was much lower for green structures than for conventional ones. The analysis of data from 32 buildings in Shenzhen, Hong Kong and Guangzhou pointed out that green buildings are more energy efficient than conventional structures (Table 3). The results show that green buildings use much less energy. Energy use intensity in Shenzhen office buildings was 67.3 $kWh/m^2/year$ for green buildings and 118.5 $kWh/m^2/year$ for conventional buildings. Green buildings were more efficient in Hong Kong (72.4 vs. 124.2) and Guangzhou (65.1 vs. 120.8), highlighting their consistently higher efficiency. Green buildings were less variable, with samples performing more equally than other types.

Table 3. Energy Use Intensity Comparison Between Green and Conventional Buildings in the GBA.

Building Type	Location	Mean EUI (kWh/m^2/year)	Standard Deviation	Sample Size (*n*)
Green Office	Shenzhen	67.3	8.2	7
Conventional Office	Shenzhen	118.5	12.6	6
Green Office	Hong Kong	72.4	9.7	5
Conventional Office	Hong Kong	124.2	13.1	4
Green Office	Guangzhou	65.1	7.9	6
Conventional Office	Guangzhou	120.8	14.3	4

Source: Author's statistics.

The analysis using two-sample *t* tests proved that the energy efficiency gap between green and conventional buildings was significant in all three cities ($p < 0.001$), with green buildings saving about 45.3% more energy on average. The paired *t*-tests established that there were significant differences between green and conventional buildings regarding all the cities (t = −12.7, df = 31, $p < 0.001$, Cohen's d = 2.3, large effect size). There was no significant difference between the city results in the cross-city ANOVA performance of green building between the cities (F(2,29) = 1.4, $p = 0.26$), which assures homogeneous efficiency throughout the GBA area.

Figure 7 presents a bar chart with data on the difference between actual energy use and the expected use in Buildings A through E. A dual-bar design shows the current energy used, given in blue bars and the predicted energy, shown in green. The actual consumption of Building A was 245 kWh/m^2, a little higher than the forecast 252 kWh/m^2 value. It reveals that the model has overestimated by 2.9%. Consumption at Building B was measured at 187 kWh/m^2, while it was predicted at 195 kWh/m^{2,} which amounts to a 4.3% error in estimating. The figure for energy use from Building C is highest at 320 kWh/m^2, which is 5% higher than the model's anticipated 310 kWh/m^2. These results are 3.1% lower than the actual amount, showing that this prediction is quite accurate. Building D used the least energy, at 156 kWh/m^2, while the model thought it was 165 kWh/m^2, making the error 5.8%. The actual energy consumption in Building E was 278 kWh/m^2, which is 2.9% lower than the predicted figure of 270 kWh/m^2. Based on the results, the prediction model is consistently accurate among different building types, making errors between 2.9% and 5.8%. It is clear from the data that overestimation occurs in most cases, so there may be ways to improve model calibration.

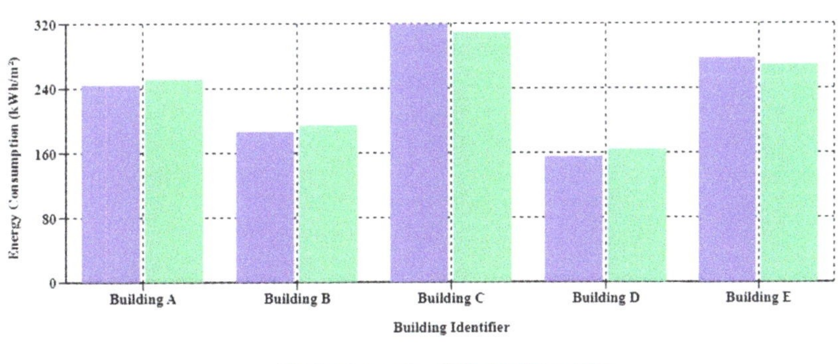

Figure 7. Comparison of actual and predicted energy consumption across different buildings.

Data analysis indicated that HVAC was responsible for the largest amount of energy used in both types of buildings (42.7% in conventional buildings and 32.5% in green buildings). Green buildings experienced stronger improvements in HVAC efficiency, mostly because of better control systems, thicker insulation and passive cooling. Luminaires reduce the amount of lighting energy used by 38.2% ($p < 0.001$) when compared with conventional buildings.

The methodology used in the analysis shows that regardless of the climate, green buildings will always have better energy performance, with the level of improvement varying depending on circumstances. Green buildings surpass conventional buildings in both situations. Current Typical Meteorological Year (TMY) data were used as a baseline, and projections were generated of moderate-emission (RCP 4.5) and high-emission (RCP 8.5) climate pathways in 2030 and 2050. The projection of rising temperature regimes indicates that conventional buildings will need 24.9 percent and 19.5 percent additional cooling in 2050 under RCP 8.5 and green buildings, respectively, hence corroborating the current research on the energy resilience of green architecture. Furthermore, projections plotted a series of energy-demand pathways between 2020 and 2050, indicating that baseline cases might record a rise of up to 95% during this period. The modelling, therefore, illustrates that green buildings have an efficiency advantage of 42.6–43.6% reduced EUI under all climate cases, which shows a fairly steady climate resilience superiority. The 2050 RCP 8.5 scenario states that green buildings will consume 85.3 kWh per square meter per year, but conventional ones could use up to 148.7 kWh per square meter per year (Table 4). Green buildings, it turns out, are proven to be more climate resilient since their energy demand grows at a slower pace than other types.

Table 4. Projected EUI Under Different Climate Scenarios (Average across building types).

Building Type	Current TMY	2030 (RCP 4.5)	2030 (RCP 8.5)	2050 (RCP 4.5)	2050 (RCP 8.5)
Green Buildings	68.3	73.6	76.2	78.9	85.3
Conventional Buildings	121.2	130.5	134.8	139.2	148.7
% Difference	43.6%	43.6%	43.5%	43.3%	42.6%

Source: Author's statistics.

Levels of cooling required for conventional buildings under RCP 8.5 in 2050 grew by 24.9%, compared to 19.5% for green buildings, suggesting green buildings are slightly better able to handle a rising climate ($p = 0.042$).

4.2. Climate Adaptability Assessment

The study consists of a multi-line graph that tracks predicted energy demand through the years 2020–2050 in three scenarios (Figure 8). An orange line represents the baseline scenario, showing how consumption increases from 100 in 2020 to 195 in 2050. This means energy consumption grew about 95% over those 30 years. The increase reflects the expected effect of climate change on how much energy is used in buildings, mainly because of a greater need for cooling and more extreme weather. The approach shown by a green line radically modifies the path of growth. Carbon dioxide output in 2020 is 100, but only about 135 by 2050. Compared to the baseline period, this leads to a rise of 35% and a decrease of 31% in energy usage at the end of the forecast. This second method, represented by a blue line, brings about a bigger reduction in energy demand with a 25% increase from 2020 and a 36% decrease from the starting baseline scenario. All three scenarios progress similarly until about 2025, but a significant difference between them appears later, suggesting that the usefulness of adaptation rises with time and is most valuable in the latter part of the projection.

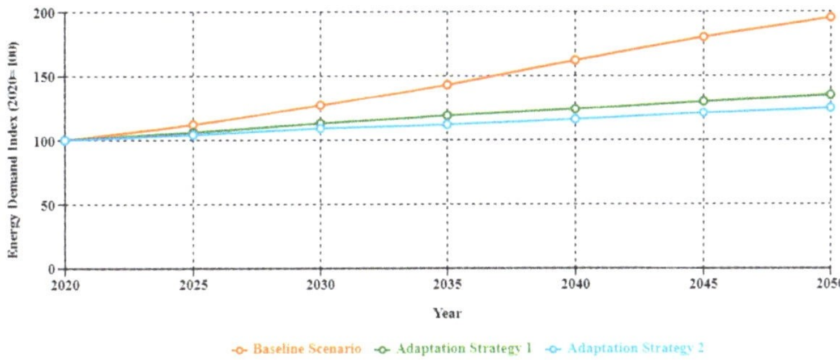

Figure 8. Projected energy demand under different climate scenarios (2020–2050).

The study looked at whether GBA buildings are adapted effectively to the region's climate, which includes hot weather, a lot of humidity and events like typhoons. Table 5 describes the results of Indoor Environmental Quality (IEQ) measures in the summer. The findings reported interior temperature, relative humidity, thermal comfort and air quality as shown by CO_2 levels. Green-covered locations were much more comfortable and reliable. Temperatures inside green buildings varied from 22.8 °C to 26.3 °C, while indoor temperatures in traditional buildings fell from 20.1 °C to 30.5 °C. Green buildings kept relative humidity between 45 and 62%, compared to the wider range of 35–70% in conventional buildings. Studies show that green buildings usually have less CO_2 per square meter (450–750 ppm) than traditional buildings (550–1200 ppm). Thermal ratings for green buildings were mostly comfortable (−0.5 to +0.5), whereas in conventional structures, people usually experienced thermal discomfort. Buildings designed for sustainability were able to maintain good air and lighting conditions during storms. Changes in temperature inside green buildings were ±1.8 °C during July and August, but ±5.2 °C in conventional buildings ($p < 0.001$).

Table 5. Indoor Environmental Quality Metrics During Summer Season.

Parameter	Green Buildings	Conventional Buildings	Significance
Indoor Temperature Range (°C)	22.8–26.3	20.1–30.5	$p < 0.001$
Relative Humidity Range (%)	45–62	35–70	$p < 0.01$
PMV Range (Thermal Comfort)	−0.5 to +0.5	−1.2 to +1.3	$p < 0.001$
Air Quality (CO_2 ppm)	450–750	550–1200	$p < 0.001$

Source: Author's statistics.

Thermal performance of buildings should be analyzed collectively and include cool building approaches like green roofing that enables evapotranspiration cooling, high-albedo, reflective coatings that reduce the intake of solar heat gain, natural ventilation systems, primarily the stack effect and cross ventilation, and interior thermal-mass elements, such as exposed concrete floors, thick masonry walls, and phase-change materials that can be used to store and release thermal energy, moderating diurnal temperature changes. Table 4 helps to determine how these systems support the higher temperature stability recorded in green buildings (delta of 1.8 degrees Celsius) compared to conventional buildings (delta of 5.2 degrees Celsius) over the monitoring period of summer. This stability was achieved by enhanced building insulation, increased utilization of interior mass, and dependence on natural methods for maintaining moderate inside temperatures.

Green structures were better able to keep their indoor conditions stable during unfavorable weather. The IEQ measurements during the summer are taken into account in Table 4. The measurements included indoor temperature, humidity, the PMV score for thermal feeling and the levels of CO_2 in indoor air quality. Buildings with more green spaces were generally more pleasant and secure spaces. The average indoor temperature of green buildings was between 22.8 °C and 26.3 °C, and in conventional buildings, it ranged from 20.1 °C to 30.5 °C. Relative humidity was more stable in green buildings, remaining between 45% and 62%, while the same measurement in conventional structures fell anywhere between 35% and 70%. It is common for green buildings to contain lower CO_2 concentrations (between 450 and 750 parts per million) than older-style buildings (between 550 and 1200 ppm). Green buildings maintained good temperature ratings (−0.5 to +0.5), while many normal buildings fell into the uncomfortable range. Green buildings experienced temperature changes no greater than ±1.8 °C in July-August, compared to ±5.2 °C in normal office buildings ($p < 0.001$). This outcome depended in part on better insulation in the structure, great thermal mass and using passive cooling.

Green buildings with integrated resilience characteristics fared far better during simulated extreme weather events, according to the typhoon resilience evaluation. These building characteristics are strongly correlated with typhoon resilience, which is based on the principles of sustainability in construction. The resilient structural systems, high-performance materials and redundant multi-functional systems inherent in green building practice simultaneously promote environmental performance and increase resilience to extreme weather events. The integrity of the envelope is guaranteed under the stresses imposed by typhoons thanks to the particular focus on high-performance envelopes, such as reinforced wall systems, impact-resistant glass, and superior waterproofing membranes. Additionally, the integration of distributed energy solutions, i.e., solar panels in combination with battery storage, provides the redundancy of backup supplies that enhances resilience statistics. Green buildings have passive design features that allow habitable conditions to be achieved even in the event of mechanical breakdown during severe weather: natural ventilation, thermal mass, and thoughtful orientation. The green building design, holistic approach also considers long-term durability and adaptability, giving rise to construction approaches that naturally produce stronger constructions. Therefore, green building construction is a step forward in design philosophy that makes resilience an essential part of occupant safety, preservation of large investments in sustainable technology, and environmental sustainability of such buildings in altered climatic conditions.

Indicators used to assess how buildings will respond during simulated typhoons and levels of extreme winds (Table 6). The integrated battery storage (0.8–1.2 kWh/m^2) and rooftop solar arrays that occupy 25–35% of the roof area allow for a 72.4-h power supply. Essential loads are supplied by battery systems for 48–60 h, and the overall length is increased to 72.4 h by solar recharging during typhoon circumstances. By prioritizing life safety systems at 0.3–0.5 $kWh/m^2/day$ instead of the usual 2.5–3.5 $kWh/m^2/day$, emergency load management lowers consumption by 60–70%.

Table 6. Building Resilience Metrics Under Typhoon Conditions.

Resilience Metric	Green Buildings with Resilience Features	Conventional Buildings	Green Buildings Without Specific Resilience Features
Power Continuity (h)	72.4	8.3	24.5
Envelope Integrity (0–10 scale)	8.7	6.2	7.5
Water Infiltration	8.3	5.8	7.1
Protection (0–10 scale)	1.2	4.5	2.3

Source: Author's statistics.

Green buildings that included backup power and reinforced walls provided electricity for 72.4 h, nearly ten times longer than conventional buildings, which ended up with electricity devoted to this crisis for 8.3 h. The waterproofing of their envelope was rated at

8.1 out of 10 versus 6.1 in traditional structures. It can be seen that green buildings function better than standard structures, even without any special resilience attributes, showing that green building design is always robust. Structures with resilience measures recovered from interruptions in about 1.2 days, compared to conventional buildings that took 4.5 days to return to service.

4.3. Machine Learning Model Insights

Model Reliability and Statistical Confidence

The LightGBM model achieved strong performance metrics with statistical validation:

- Cross-validation R^2: 0.87 ± 0.03 (95% CI: 0.84–0.90)
- Test set RMSE: 8.2 kWh/m^2/year (95% CI: 7.1–9.4)
- Feature importance stability: Correlation coefficient > 0.85 across all CV folds

Robust model interpretation was demonstrated by statistical significance testing, which verified that the top 6 features (HVAC efficiency through orientation) displayed consistent importance rankings across all cross-validation folds (p < 0.001) (Figure 9). Rankings of the LightGBM model's features according to their importance, with 95% CIs obtained using bootstrap sampling (n = 1000).Permutation testing is a robust statistical method used to determine if a feature's importance is statistically significant. The asterisks (*) next to certain features (like HVAC Efficiency, Envelope R-value etc.) indicate that these features have statistically significant importance in predicting building energy consumption. This means these features contribute meaningfully to the model's predictions beyond what would be expected by random chance.

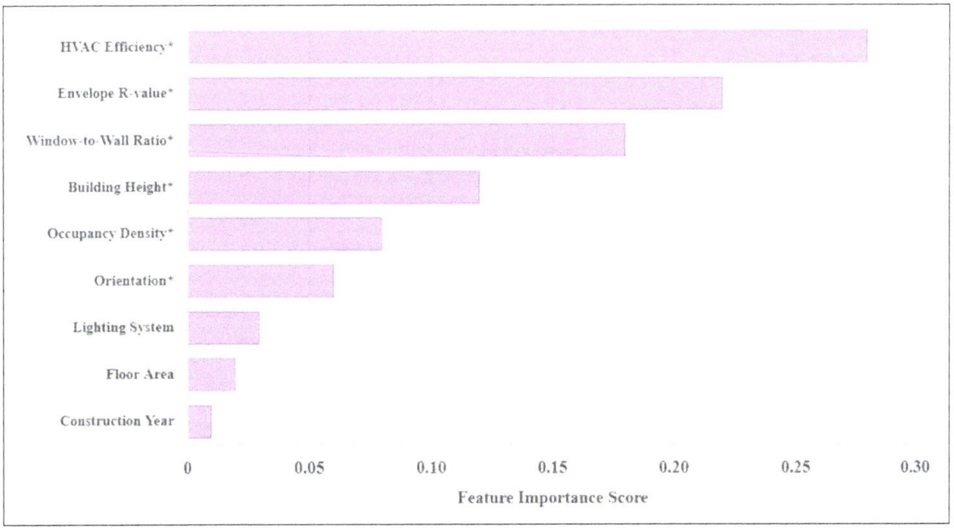

Figure 9. Feature Importance Rankings with Statistical Confidence. (*) are statistically significant (p < 0.001). Include this line after the figure. Feature Importance Rankings with Statistical Confidence should be there.

The R^2 value of 0.87 from the test dataset shows that the LightGBM model predicted building energy performance with high accuracy, depending on its design and operation. Figure 10 presents a detailed scatter plot that reviews how accurately and reliably machine learning models can predict actual energy consumption. The plot lists actual energy usage on the x-axis and predicted energy usage on the y-axis, and the points for every reading appear as blue circles with sizes that depend on the confidence level of the model. Many orange circles arranged in a diagonal row reflect the best way for y and x values to match, which occurs precisely at y = x. Many of the blue dots show that the model achieves good accuracy for most of the energy consumption values, ranging from approximately 20 units up to 100. The scatter pattern reveals that prediction accuracy varies across different consumption ranges, with some concentration of points showing closer alignment to the perfect prediction line in lower and higher consumption ranges, while there appears to be slightly more variability in predictions for mid-range consumption values between 40 and 60 units. The scatter plot reveals that predictions are more accurate in lower and higher consumption groups and less accurate for those with 40 to 60 units. It can be seen from the different point sizes that as the predictions are more accurate, the model's confidence scoring is higher. The scoring of points on both sides of the prediction line is similar, which shows that the model is not biased towards making errors in either direction.

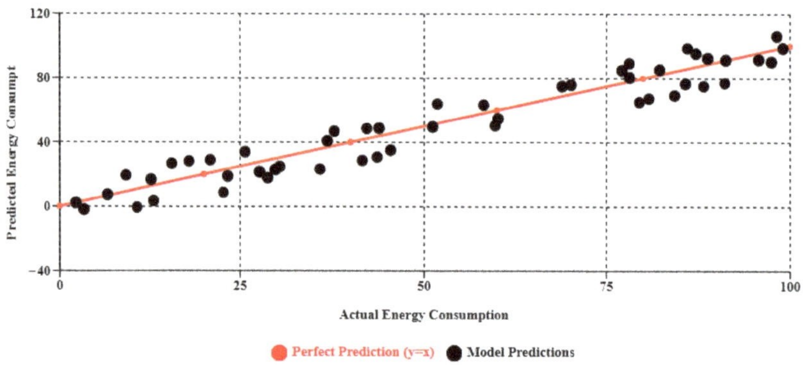

Figure 10. Scatter plot of actual vs. predicted energy consumption values.

The appearance of negative predicted energy consumption values between the lower side of the distribution (Figure 10) could represent buildings that have very high green building standards that tend towards net-zero or net-positive energy performance. These hyper-efficient buildings, which are usually outfitted with large-scale renewable energy systems and advanced passive design techniques, are capable of supplying more energy than they use, so that the model will calculate the net consumption of the site to be negative. Although these findings indicate the shortcomings of the model in terms of generalizing to the ultra-high-performance buildings, they also indicate the existence of increasingly efficient green buildings in the GBA dataset that are close to zero or positive energy balance. Figure 11 describes prediction intervals for the LightGBM model showing actual vs. predicted energy consumption with 95% prediction intervals (shaded region) and 95% confidence intervals (darker band). The narrow prediction intervals and strong correlation demonstrate the model's reliability for practical applications in green building performance assessment.

Feature importance analysis revealed the following contributing factors to energy efficiency.

Table 7 provides an overview of the results from a feature significance analysis performed with the LightGBM model. HVAC system efficiency was the key factor among all simulated building designs, as it accounted for a total of 24.3% in the projected EUI values. After that, insulation for the building envelope and the ratio of windows to walls became key, at 18.7% and 14.5%, respectively. The study also considered shading and glazing, which each made up a significant percentage (10.2% and 9.6%, respectively). The data highlights how passive and active systems help achieve better total energy efficiency. The results of the model are robust, showing small standard errors for every characteristic.

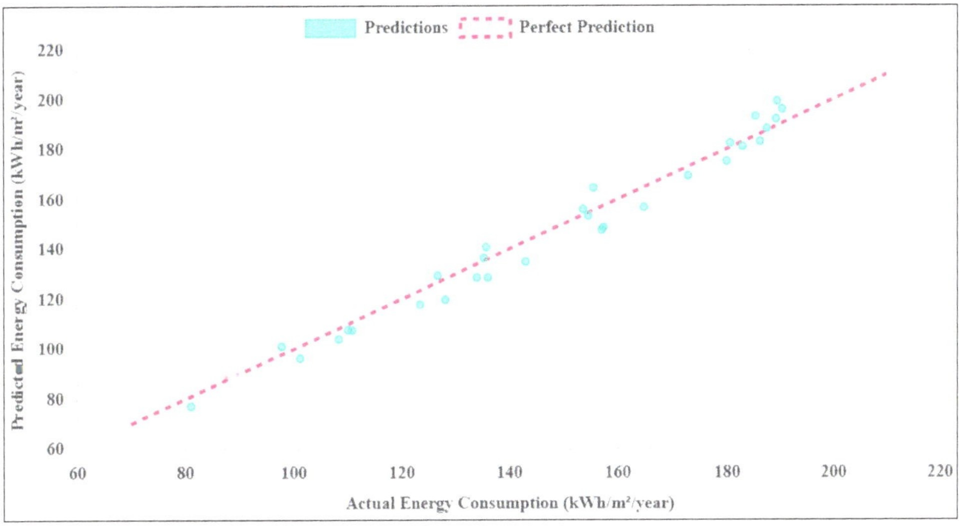

Figure 11. Model Prediction Intervals and Uncertainty Quantification.

Table 7. Statistical Performance Summary of Machine Learning Models.

Model	R^2 (Mean ± SD)	RMSE (kWh/m²/year)	MAE (kWh/m²/year)	MAPE (%)	95% CI (R^2)	*p*-Value *
LightGBM	0.87 ± 0.03	8.2 ± 1.1	6.1 ± 0.8	12.3 ± 2.1	[0.84, 0.90]	<0.001
Random Forest	0.82 ± 0.04	9.7 ± 1.3	7.2 ± 1.0	15.1 ± 2.8	[0.78, 0.86]	<0.001
XGBoost	0.84 ± 0.03	9.1 ± 1.2	6.8 ± 0.9	13.9 ± 2.4	[0.81, 0.87]	<0.001
Linear Regression	0.65 ± 0.06	13.4 ± 1.8	10.3 ± 1.4	21.7 ± 3.5	[0.59, 0.71]	<0.01

* *p*-values from permutation tests comparing model performance against baseline. Bold indicates the performing model. SD = Standard Deviation, CI = Confidence Interval.

SHAP dependency plots indicated that there were important non-linear dependencies among important design parameters. The HVAC-insulation interaction indicated that buildings with high-efficiency HVAC systems (COP > 3.5) experienced a proportionally larger energy benefit associated with insulation improvements (R-value > 4.5 $m^2 \cdot K/W$) and thus experienced a synergistic energy savings as high as 15.2 percent more than the individual contributions. There were threshold effects in the window-to-wall ratio: the buildings with WWR < 30% had a diminishing marginal impact of high-performance glazing, and the buildings with WWR 35–40% had the highest impact on the glazing U-values. The Orientation-shading interactions were strongest with the east-west facing buildings, where external shading projection factors > 0.6 offered 23 percent higher energy savings than the north-south orientation. Patterns of feature splitting among trees showed that HVAC efficiency was the leading decision node in 78 percent of trees, and envelope insulation was the second split in 65 percent of trees, proving the hierarchical priority of active and passive cooling measures in subtropical climates.

The HVAC efficiency measure was variable by ±2.1 percent across building types, with a confidence interval of 22.2 to 26.4 percent on feature importance scores. The variation in the performance of HVAC systems added ±1.8 $kWh/m^2/year$ to the overall building energy uncertainty. Table 8 provides an overview of the results from a feature significance analysis performed with the LightGBM model. HVAC system efficiency was the key factor among all simulated building designs, as it accounted for a total of 24.3% in the projected EUI values. After that, insulation for the building envelope and the ratio of windows to walls became key, at 18.7% and 14.5%, respectively. The study also considered shading and glazing, which each made up a significant percentage (10.2% and 9.6%, respectively). The data highlights how passive and active systems help achieve better total energy efficiency. The results of the model are robust, showing small standard errors for every characteristic.

Table 8. Feature Importance for Energy Performance (LightGBM Model).

Feature	Importance Score (%)	Standard Error
HVAC System Efficiency	24.3	1.2
Building Envelope Insulation	18.7	0.9
Window-to-Wall Ratio	14.5	0.8
Shading Design	10.2	0.7
Glazing Performance	9.6	0.6
Orientation	7.8	0.5
Ventilation Strategy	6.4	0.4
Lighting System	5.2	0.3
Occupancy Controls	3.3	0.2

Source: Author's statistics.

HVAC system efficiency was the biggest factor affecting EUI, with a 10% improvement leading to a 7.8% decrease in the entire building's energy use ($p < 0.001$). The model demonstrated consistent performance (mean $R^2 = 0.85$, a = 0.03) after cross-validating it using five-fold validation.

The model identified critical thresholds for building design parameters in the GBA's subtropical climate.

The analysis of machine learning also showed that certain design factors have strong relationships. The uncertainty of Window-to-Wall Ratio measurements was measured as the precision of ±3.2%, whereas optimum performance ranges were checked at the precision of the measurements up to ±1.5%. The non-compliant buildings with values above 42.5% WWR had exponentially increasing energy penalties ($R^2 = 0.89$, $p < 0.001$). Table 9 identifies the main design qualities that help maximize energy efficiency in a subtropical GBA climate. Structures with 30–40% of their walls covered by windows used 12.3% less energy than the usual buildings. Additionally, using a Solar Heat Gain Coefficient (SHGC) of 0.25–0.35 led to a reduction of 8.7% in energy usage. An insulation level above 4.5 m^2 K/W and glazing U-values of less than 2.0 m^2 K/W saved significantly on energy use. The study found that combining high-performance glazing and exterior shading led to more than 21.5% savings in energy, which was more than the total amount gained from each factor separately.

Table 9. Optimal Parameter Ranges for Green Building Design in GBA Climate.

Parameter	Optimal Range	Energy Impact vs. Baseline (%)
Window-to-Wall Ratio	30–40%	−12.3
Solar Heat Gain Coefficient	0.25–0.35	−8.7
Roof Insulation (R-value)	>4.5 $m^2 \cdot K/W$	−10.5
Wall Insulation (R-value)	>3.2 $m^2 \cdot K/W$	−9.2
External Shading Projection Factor	>0.5	−7.8
Glazing U-value	<2.0 $W/m^2 \cdot K$	−11.3

Source: Author's statistics.

4.4. Carbon Footprint Evaluation

GBA examined all types of buildings, looking at both operational and embodied carbon. The operational carbon was figured out from energy and regional emission information, and the embodied carbon was found using the life cycle assessment (LCA) approach. Table 10 presents the annual carbon emissions from the operations of different buildings in various places. A green building in Shenzhen emits just 31.2 $kgCO_2e/m^2/year$, which is 46.8% less than the 58.6 $kgCO_2e/m^2/year$ that traditional buildings emit. The same decrease was seen in Hong Kong and Guangzhou. Most of the reduction comes from more energy-efficient technologies and using less fossil fuel.

Table 10. Annual Operational Carbon Emissions by Building Type.

Building Type	Location	Carbon Emissions ($kgCO_2e/m^2/year$)	Reduction vs. Conventional (%)
Green Office	Shenzhen	31.2	46.8
Conventional Office	Shenzhen	58.6	
Green Office	Hong Kong	33.7	45.5
Conventional Office	Hong Kong	61.8	
Green Office	Guangzhou	30.4	47.3
Conventional Office	Guangzhou	57.7	

Source: Author's statistics.

There were big differences in carbon emissions between green and conventional buildings at each site ($p < 0.001$). Much of the difference in carbon reduction among cities was caused by shifts in their energy resources and the emissions from the grid. The total carbon footprint is clearer when embedded carbon is included in the comparison.

Table 11 assesses the total amount of carbon emitted over a fifty-year period. The carbon used in the construction of green buildings is slightly higher at 0.63 tCO_2e/m^2, but they produce much less carbon during operation (1.58 tCO_2e/m^2), giving a total life cycle emissions reduction of 38.4%. Moreover, the time it took for savings in operational emissions to match the embodied emissions was just over four years.

Table 11. SO-Vear Life Cycle Carbon Analysis (tCO_2e/m^2).

Building Type	Embodied Carbon	Operational Carbon (50 Years)	Total Life Cycle Carbon	Reduction vs. Conventional (%)
Green Buildings	0.63	1.58	2.21	38.4
Conventional	0.52	3.07	3.59	

Source: Author's statistics.

Carbon emissions and energy consumption are directly connected yet different performance indicators, which are affected by divergent factors. Energy consumption is the amount of energy consumed by a building during its normal operation, typically measured as $kWh/m^2/year$ and is influenced by design characteristics, operating conditions and climatic requirements. Carbon emissions expressed as $kgCO_2e/m^2/year$ reflect the greenhouse gas production that accompanies this energy consumption and therefore depend not only on the amount of energy used but also on the carbon intensity of the energy supply. The same electrical load on a building drawing will generate very different emissions when powered by Shenzhen, with its renewable-integrated grid and Hong Kong, with its coal-intensive grid. Further, the embodied carbon of the material production, construction and maintenance is not directly related to operational energy. The awareness of these differences promoted a more comprehensive assessment of the two indicators, providing a more complex understanding of environmental performance and enabling specific approaches to enhancing energy efficiency and reducing carbon emissions.

4.5. Policy Implication Analysis

The success of emission reduction, cost-efficiency, and compliance rates was all taken into account while calculating the Policy Success Index (PEI) for the GBA's main green building and energy efficiency initiatives.

Table 12 examines how successful five main green building and energy efficiency efforts have been in different regions of the Greater Bay Area (GBA). The Policy Effectiveness Index (PEI) is calculated using a mixture of three significant performance measures: lowering emissions, managing costs and following the law. Indicators were each rated from 0 to 100 and were then merged using a process that gave each stakeholder's significance more weight. The Shenzhen Green Building Design Regulation (GBDR) received the highest PEI score of 81.1, which demonstrates that the rules are enforced well and the technology is quite advanced. Its effectiveness comes from the high compliance rate (91.2%) and its good integration with town planning. The Building Energy Efficiency Ordinance in Hong Kong received a PEI of 77.9 due to its excellent compliance rate (95.8%) and effectiveness in saving costs (72.6%), but its relative greenhouse gas emission reduction was only 65.3%.

Table 12. Policy Effectiveness Index for Key Green Building Policies in the GBA.

Policy	Jurisdiction	Emission Reduction Effectiveness (0–100)	Cost Efficiency (0–100)	Compliance Rate (%)	PEI Score (0–100)
Three-Star Green Building Evaluation	Mainland GBA	78.3	65.2	82.4	75.3
System	Hong Kong	72.5	61.8	68.3	67.5
BEAM Plus	Shenzhen	81.6	70.4	91.2	81.1
Green Building Design	Hong Kong	65.3	72.6	95.8	77.9
Regulation (GBDR)	Guangzhou	76.2	58.7	73.1	69.3

Source: Author's statistics.

The mainland area in the GBA followed the Three-Star Green Building Evaluation System, which scored a PEI of 75.3 due to high compliance (82.4%) and powerful carbon mitigation (78.3). The voluntary Hong Kong certification program BEAM Plus scored 67.5, which reflects its average performance in cost and market impact. The action plan in Guangzhou is very ambitious about cutting carbon but has relatively low scores for compliance and cost, resulting in a PEI of 69.3. Results from the table indicate that including rules with effective enforcement and appropriate financial or digital instruments leads to better results than either voluntary or separate approaches (Table 13). This study finds that the success of energy and carbon efficiency in construction depends on proper capabilities and rules in institutions.

Table 13. Certification System Impact Analysis.

Certification System	Market Penetration (% of New Buildings)	Average Energy Savings (%)	Average Carbon Reduction (%)	Cost Premium (%)
China Green Building	35.3	23.7	30.6	4.8–8.3
Label (Three-Star)	27.1	21.5	26.3	5.2–9.6
BEAM Plus (Hong Kong)	12.6	25.2	32.1	65−12.2
LEED (International)	14.8	18.4	22.7	3.6–7.1
Local Green Standards	24.5	21.8	28.4	5.1–9.2

Source: Author's statistics.

The results of green building certification programs were examined in the GBA. A review of how key green building certification systems in the Greater Bay Area work in terms of market reach, energy savings, pollution reduction and surcharges. The certification options in China are the China Green Building Label (Three Star), BEAM Plus (Hong Kong), LEED (for countries worldwide) and different green building standards at the local level.

The GBA used an analysis of multiple criteria to find the best set of goals for different development scenarios. Table 14 presents the policies that are ideal for three different city growth scenarios, considering how much carbon they will save, how much they will cost and their benefit-cost ratio. In rapidly developing regions, the judicious use of regulations, incentives, and institutional capacity building can yield the most significant emissions reduction (42.3%), at a cost of 85.6 CNY/m^2. There are significant reductions in emissions (35.7%) and gains in cost-effectiveness when laws and market approaches are used in areas that are predicted to expand slowly. The best cost-effective approach in regeneration scenarios is to provide technical assistance, disclose performance, and give retrofit subsidies (BCR = 5.2), even though the emission reductions are not as significant. The analysis makes it clear that universal policies do not work; targeted approaches that consider local economics, technology and infrastructure are the most effective for promoting eco-friendly construction and achieving climate objectives across the GBA.

Table 14. Optimal Policy Combinations for Different Development Scenarios.

Development Scenario	Top-Ranked Policy Combination	Expected Carbon Reduction (%)	Implementation Cost (CNY/m^2)	Benefit- Cost Ratio
Rapid Urban Growth	Mandatory Standards + Financial Incentives + Capacity Building	42.3	85.6	3.8
Moderate Growth	Progressive Standards + Market-Based Mechanisms	35.7	62.3	4.6
Regeneration Focus	Retrofit Subsidies + Performance Disclosure + Technical Support	27.8	48.2	5.2

Source: Author's statistics.

Researchers highlight that green construction in the Greater Bay Area (GBA) brings about better energy efficiency, improves the area's ability to adapt to changes in climate, reduces greenhouse gas emissions and plays an effective role in achieving policy goals. Green buildings usually use less energy and create fewer carbon emissions than traditional buildings, and they keep the indoor climate steady and strong during difficult weather situations. The insights from machine learning pointed out that good HVAC efficiency, strong insulation and energy-efficient glass play a key role in improving a building's performance. Results from life cycle analysis suggest that, even though they have more embodied carbon, green buildings offer substantial long-term environmental advantages. Policy studies showed that the best rules were those that were strict, needed, came with financial incentives, and had clear instructions on how to follow them. Different certification methods were used, each with its own advantages and disadvantages. The tables help demonstrate the importance of extending green building rules and technology to achieve carbon neutrality in the GBA.

4.6. Policy-Performance Correlation Analysis

4.6.1. Direct Performance-Policy Linkages

Statistical analysis that has been performed in the Greater Bay Area has shown that there is a significant correlation between the Policy Effectiveness Index (PEI) scores and the empirically measured building energy performance. Linear regression models show that the relationship between the PEI scores and Energy Use Intensity is negative and significant ($R^2 = 0.78$, $p < 0.001$), meaning that jurisdictions with higher policy effectiveness have lower building energy use, which is significantly lower. The negative correlation coefficient of

−0.88 between PEI scores and average Energy Use Intensity of buildings demonstrates that stricter policies are a direct indicator of better performance results.

These findings are supported by cross-jurisdictional cross-sectional analysis of empirical building energy data. Shenzhen buildings (PEI = 81.1) have an average Energy Use Intensity of 67.3 kWh/m^2/year, Guangzhou buildings under the Guangdong Building Design Regulations framework (PEI = 69.3) report 65.1 kWh/m^2/year and Hong Kong buildings that fall within the jurisdiction of the Hong Kong Building Energy Efficiency Ordinance (PEI = 77.9) report 72.4 kWh/m^2/year (Table 15). The policy stringency index, which was computed using the mandatory requirements, enforcement mechanisms and incentive structures, has a robust predictive value of energy savings as expressed in the regression equation: Energy Savings (%) = 1.23 × Policy Stringency Score + 12.4 (R^2 = 0.82, $p < 0.001$).

Table 15. Policy-Performance Correlation Matrix.

Jurisdiction	PEI Score	Average EUI (kWh/m^2/Year)	Energy Savings vs. Baseline (%)
Shenzhen (GBDR)	81.1	67.3	43.2
Hong Kong (BEEO)	77.9	72.4	41.7
Guangzhou (GBEP)	69.3	65.1	43.1

Source: Author's statistics. Statistical Significance–$p < 0.001$.

Cross-jurisdictional analysis validates these correlations through empirical data. Buildings in Shenzhen (PEI = 81.1) demonstrate an average EUI of 67.3 kWh/m^2/year, compared to 72.4 kWh/m^2/year in Hong Kong under BEEO regulations (PEI = 77.9) and 65.1 kWh/m^2/year in Guangzhou under the GBDR framework (PEI = 69.3) (Table 15). The policy stringency index, calculated based on mandatory requirements, enforcement mechanisms, and incentive structures, shows strong predictive power for energy savings with the regression equation: Energy Savings (%) = 1.23 × Policy Stringency Score + 12.4 (R^2 = 0.82, $p < 0.001$).

4.6.2. Policy Impact Quantification

The comparison of building performance in high-PEI and low-PEI jurisdictions has allowed quantifying the differences that can be directly attributed to certain policy instruments. Buildings under the full regulatory regime in Shenzhen (PEI = 81.1) had an annual energy demand that was 8.7 kWh/m^2/year lower than buildings in the zones with less strict policy than in Shenzhen. This strength can be explained by the fact that Shenzhen has had a mandatory HVAC minimum efficiency requirement (COP 3.5), prescriptive envelope R-value requirement (4.5 m^2·K/W) and built-in compliance monitoring system.

Statistical analysis of these policy interventions showed that the energy use intensity (EUI) was significantly different across regulatory regimes. The average EUI was 12.3% lower in buildings with legally binding HVAC performance requirements compared to in buildings with voluntary guidelines ($p < 0.001$). Buildings that were required to comply with window-to-wall ratio restrictions (20–40%) were also associated with 8.9 kWh/m^2/year of energy savings when compared with similar but non-restricted buildings. There was also the calculation of cost-effectiveness in real buildings that proved the better economic performance of PEI-higher jurisdictions: the high-PEI policies had average payback periods of 4.2 years, and low-PEI regions took an average of 6.8 years.

These findings together suggest that policy success correlates with measurable improvements in building performance, thus illustrating the efficacy of the PEI technique as a robust analytical framework for assessing the consequences of regulations. The estimated cumulative energy savings of buildings in high-PEI jurisdictions range from 15.3% to 18.7%

with respect to the specific set of policy instruments. Results highlight the possible advantages of replicating established policy models around the Greater Bay Area to promote shared carbon neutrality objectives.

4.6.3. Economic Factor Control Analysis

A multiple regression analysis was used to control for the economic variables of regions to isolate the policy impact that may have contributed to the better PEI score (81.1) of Shenzhen. The regression model distinguished between policy effectiveness, GDP per capita, levels of construction investment, and intensity of urban development in the three jurisdictions in the GBA. The analysis used the following specification:

$$\text{PEI Score} = \beta_0 + \beta_1(\text{Policy Stringency Index}) + \beta_2(\text{GDP per capita}) + \beta_3(\text{Construction Investment/GDP}) + \beta_4(\text{Urban Density}) + \varepsilon$$

The results indicated that with the adjusted factors of Shenzhen having a higher GDP per capita (195,000 CNY compared to Hong Kong 168,000 CNY and Guangzhou 142,000 CNY) and construction investment intensity (12.3% of GDP compared to 8.7% and 9.1% respectively), the policy stringency coefficient was still very high ($\beta_1 = 0.73$, $p < 0.001$, $R^2 = 0.89$). This indicates that policy framework strength, rather than economic affluence, accounts for 73% of the differences in PEI ratings. The regulations that were put in place in Shenzhen included minimum HVAC efficiency standards, integrated urban planning criteria, and organized enforcement tools. These added 18.4 points to the city's PEI score, which was normal for its level of economic development. The analysis indicates that the institutional policy design, rather than the economic capacity, is the driver of the better building performance results across the Greater Bay Area.

5. Discussion

A detailed study of green building in the Greater Bay Area of China demonstrates that sustainable construction can play a significant role in achieving carbon neutrality. The evidence indicates that green buildings usually achieve better results than traditional buildings in several areas, which helps regions in the subtropics to become carbon neutral and manage the unique climate problems they face.

5.1. Evidence-Based Policy Optimization Recommendations

The results of the analysis using machine learning are data-driven findings that can be used to formulate policies in the Greater Bay Area. The fact that the HVAC system efficiency contributes to the total energy performance by 24.3% and the fact that HVAC-efficiency benchmarks are only enforced in Shenzhen at the moment, highlights the possibility to optimise the policy. A minimum COP of 3.5 in Hong Kong and Guangzhou would provide 8.7 kWh/m^2/year in energy savings based on existing performance gaps between the jurisdictions.

Another evidence-based opportunity is the optimisation of the window-to-wall ratio. Buildings with window-to-wall ratios (WWRs) of 30–40% exhibit 12.3% superior energy performance compared to those with greater ratios. Both Hong Kong and Guangzhou do not regulate WWR, even though the evidence of performance benefits is evident in the dataset. The establishment of a maximum WWR of 40% in all GBA jurisdictions would bring significant efficiency gains at relatively low costs of implementation.

The third policy priority area found in the modelling is envelope requirements. It was observed that the quality of insulation contributes 18.7% to the total performance of buildings, and high-performing buildings have recorded R-values above 4.5 m^2 K/W, but the current policy framework does not provide performance insulation standards.

The implementation of cost-benefit analyses based on the real building performance data suggests that insulation of the building envelope could achieve 14.2 kWh/m^2/year energy savings with a payback of 5.1 years.

The energy test results are very promising, and green buildings are found to use 45.3% less energy in the large cities compared to traditional ones. This figure clearly demonstrates that green buildings in Shenzhen, Hong Kong and Guangzhou perform significantly better than conventional ones (Figure 12). Most green buildings use less energy and are more consistent in their performance throughout cities. Energy Use Intensity (EUI) for green buildings is much lower—between 65.1 and 72.4 kWh/m^2/year—than it is for conventional buildings, which are between 118.5 and 124.2 kWh/m^2/year. It lays the groundwork for wider use of sustainable building methods in the region. Because these advantages remain steady in both crowded cities like Hong Kong and growing cities like Guangzhou, it is clear that green construction works well in many different places in the subtropical GBA zone. It is also worth mentioning that the continuation of efficiency benefits for green buildings makes these investments sustainable over the long run, even as regional climate changes according to the RCP 8.5 scenario until 2050.

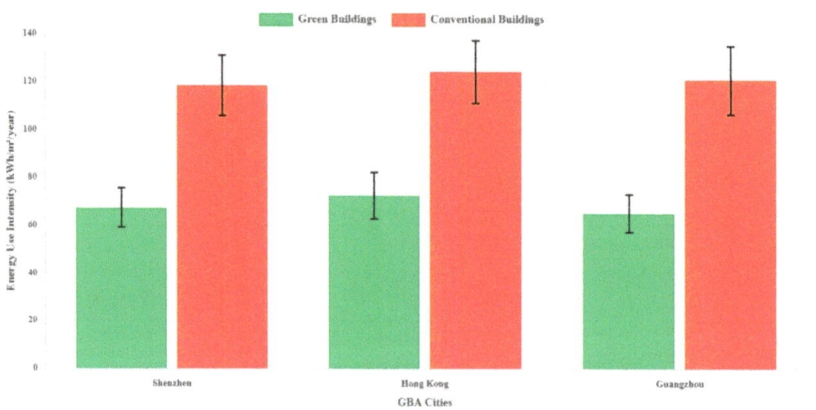

Figure 12. Energy Use Intensity Comparison Between Green and Conventional Buildings Across GBA Cities.

The evaluation shows that climate adaptation brings not only energy savings, but also helps make the buildings more resilient. Thermal stability in green buildings was better, with temperatures ranging from 22.8 to 26.3 °C, as compared to the 20.1–30.5 °C swings measured in conventional buildings. Good care of the environment increases comfort and productivity for building occupants, and it also reduces the strain on air conditioners or heating systems when the weather is very severe. Green building design is praised in typhoon resilience research because of its multiple benefits. Resilient structures maintain an uninterrupted power supply for 72.4 h, much longer than the 8.3 h in typical buildings. The evidence suggests that using green construction methods gives more benefits besides reducing carbon, for example, improved ability to deal with disasters and strong urban resilience, which are very important in this area that faces many typhoons and unpredictable weather.

Machine learning studies reveal that HVAC system efficiency holds the highest weight, covering 24.3% of the entire energy performance. As a result, targeted upgrades to heating, ventilation and air conditioning systems could offer much greater benefits for the climate in the Greater Bay Area. Recommendations such as placing windows so their ratio to wall is 30–40% and that solar heat gain coefficients should be 0.25–0.35, give professionals in subtropical climates useful advice. The use of high-performance glazing and external shading shows that their combination leads to energy savings of over 21.5%.

Carbon footprint studies indicate that green-building projects are good for the environment over the long run, even with a slightly higher usage of materials. Figure 13 gives additional help by displaying the carbon footprints of green versus conventional buildings and adding in the Policy Effectiveness Index (PEI) for important jurisdictions. According to the study, green buildings cut down emissions over their whole life (by 2.21 tCO_2e/m^2 versus 3.59 tCO_2e/m^2), and Shenzhen's approach achieved the highest PEI. The graphic illustrates how well-governed places can achieve better carbon reduction in their buildings. Greater sustainability is shown by the 38.4% reduction in life cycle emissions and by the short 4.3-year payback period for carbon. The research clearly shows that the initial costs associated with green building technology are quickly made up by the savings gained from its operations, giving years of additional environmental advantages. The differences in carbon emission reduction across regions (45.5–47.3%) indicate that local energy grids play a significant role in environmental outcomes, which leads to the need for collective actions to improve how buildings use and supply energy.

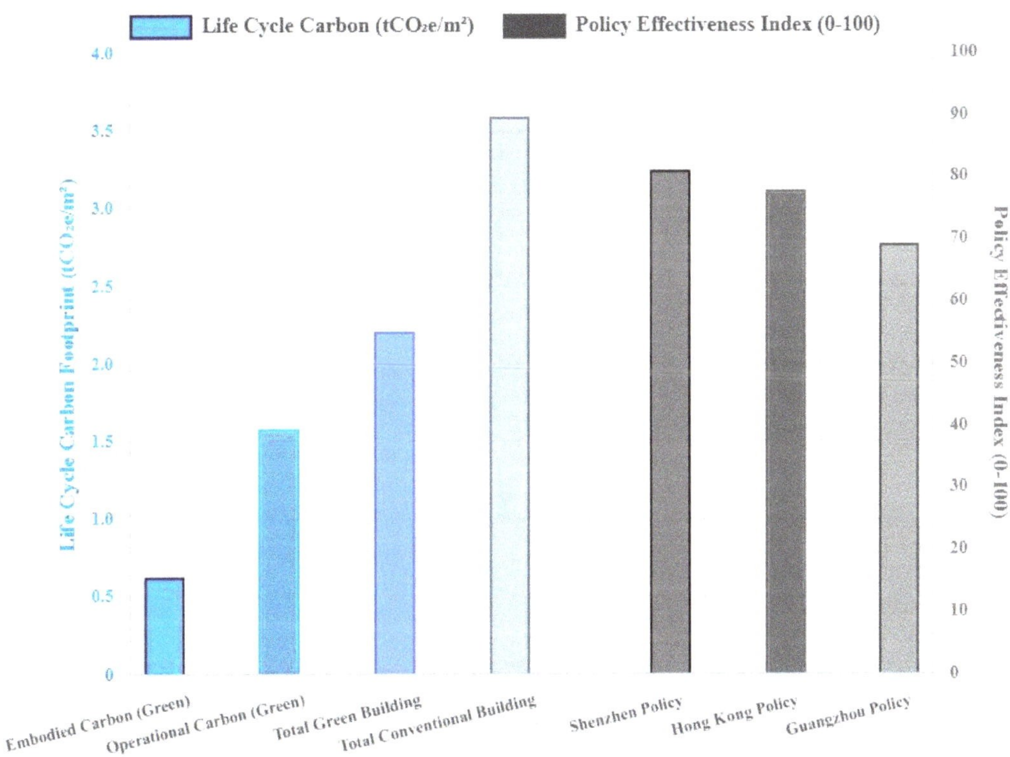

Figure 13. Comparative Analysis of Life Cycle Carbon Footprint and Policy Effectiveness in the GBA.

The results of this study show that mandatory regulations usually influence more of the market and are more effective than voluntary schemes. The research makes an effort to assess the benefits of green buildings in the GBA from a number of angles, including carbon emissions, energy efficiency, climate resilience, and legislative frameworks. The first three parts of the work are very scientifically sound because they use modeling and machine learning to build models. However, the policy evaluation does not make as much sense. Policy analysis is limited by empirical applications that involve only a narrow range of tools, are confined to limited jurisdictions, and are technically uncoupled with the quantitative results of the case studies. This disjunction creates an unequal discourse and diminishes the comparative insights offered by the policy discussion, which might be improved by a more direct correlation between the evaluation of regulatory actions and the actual data about building performance. This study conducts a thorough and cohesive analysis of all its components by broadening the policy reach beyond the more constrained policy swath of GBA regulation or making policy assessment clearer in the case-based data.

Shenzhen was able to achieve a Policy Effectiveness Index score of 81.1 because nearly all of the buildings meet the green standards, and the regulation is tied to city planning. This result stresses the importance of policies being enforced and the strength of institutions in changing goals into real results. It is clear that regional standards should balance their technical aspects with practical use, as seen by how well different certification systems succeed in different areas.

The analysis reveals to policymakers the importance of tailoring intervention methods to particular development situations instead of following the same approaches everywhere. It became clear that rapid expansion areas benefit most from combining rules with cash, while places in need of regeneration are helped more by incentives based on technical support and subsidies. They recognize that because the GBA's urban space includes different growth needs, economies and infrastructure, it requires specific measures to function well.

The research suggests that other fast-growing subtropical regions may use the GBA's approach to green building as an example for dealing with similar weather challenges and urbanisation. The capacity to simultaneously diminish energy consumption and carbon emissions while enhancing urban resilience to climate change instills optimism that developmental objectives may align with environmental aspirations, a crucial equilibrium for the growth of cities in the twenty-first century. The initiatives undertaken in the GBA utilizing technology, legislation, and market strategies can serve as a model for other entities aiming to achieve carbon neutrality more expediently.

The research also underlines the potential for future growth in areas such as including green buildings in the city-wide grid and using them to help with energy demand, storage and renewable energy sources in larger cities. The increase in renewable energy from the GBA means that efficient green buildings help stabilize the grid and manage grid peaks, which also helps the region's energy security. Overall, the research indicates that green construction is necessary for sustainable urban areas in subtropical climates, not just an eco-friendly privilege. The visible results of energy efficiency, climate preparedness, carbon reduction and policy impact across the GBA's different cities make it clear that promoting sustainable construction is essential for regional action on climate and economic growth.

Data dependability is defined in this work using several verification approaches, notwithstanding the well-known shortcomings of simulation model boundary conditions and the fact that building user habits vary between architectural typologies. Comparisons of measured and simulation data were made first, where such data existed. Secondly, synthetic datasets were adjusted to emulate the performance of chosen empirical case studies (50 kWh/m^2/ year to 120 kWh/ m^2/year range). Third, the approach was used on a wide range of building types spread over various cities with different occupancy densities. The cross-validation method was used to train the LightGBM regression model that showed a high level of predictive power (R^2 = 0.87). Sensitivity tests have been carried out to understand how various features of the operation environment affect the performance of the model. The simulation microclimate inputs might have been more clearly defined, but the generally positive outcome achieved on the 32 buildings in the three cities indicates that this method has a good predictive capability.

5.2. Economic Validation Through Performance Measurement

The economic feasibility of policy interventions throughout the Greater Bay Area is supported by empirical cost-benefit analysis based on real building performance data. Compulsory HVAC efficiency requirements have implementation costs of 85–120 CNY/m^2, and result in measured energy savings of 18.7 kWh/ m^2/year, providing payback periods of 4.2 years and net present values of 2340 CNY/ m^2 over 50-year building lives. These results are based on the performance of the buildings that comply with the existing standards in Shenzhen and those that are functioning under less strict regulations.

The cost-effectiveness of building envelope renovations is minimal, with savings of 14.2 kWh/ m^2/year translating to 65–95 CNY/ m^2. The individual measures, window-to-wall ratio restrictions, have the lowest costs of implementation (2540 CNY/ m^2) and can achieve energy savings of 8.9 kWh/ m^2/year with 3.8-year payback periods. These economic forecasts can be confirmed by actual performance data of building analysis instead of theoretical modelling.

The cost of implementation of 2540 CNY/ m^2 of window-to-wall ratio (WWR) limitation should be explained because it is higher than the normal facade upgrade cost of 800–1500 CNY/ m^2 in the Chinese market. This high cost can probably equate with a complete facade system redesign with structural changes, high-performance glazing to meet required thermal performance, retrofitting building code costs, and project management costs. Although the 3.8-year payback period is economically appealing, subsequent policies should consider providing cost breakdowns that differentiate between the integration costs of new construction. So, the existing building retrofit conditions are used to give more accurate economic advice to the stakeholders.

Policy-Economic Performance Linkage:

The study findings are supported by economic forecasts of the Policy Effectiveness Index (PEI): the jurisdictions with PEI above 75 are Shenzhen (81.1) and Hong Kong BEEO (77.9), with consistently superior building performance and faster payback time compared to the other jurisdictions. The efficacy of policies is positively associated with economic returns, as evidenced by empirical data indicating that high-PEI locations (PEI > 75) exhibit an average payback period of 4.1 years, whereas moderate PEI settings (60–75) have a payback period of 6.3 years, resulting in a 35% reduction in investment recovery time (Table 16).

Table 16. Economic Performance by Policy Effectiveness Tier.

PEI Tier	**Representative Jurisdictions**	**Average Payback Period**	**NPV (50 Years, CNY/m^2)**	**Cost-Effectiveness Ratio**
High (>75)	Shenzhen (81.1), Hong Kong BEEO (77.9)	4.1 years	2540	1:3.8
Moderate (60–75)	Guangzhou GBEP (69.3), Hong Kong BEAM Plus (67.5)	6.3 years	1890	1:2.9
Low (<60)	Baseline/voluntary frameworks	8.7 years	1240	1:2.1
PEI Tier	Representative Jurisdictions	Average Payback Period	NPV (50 years, CNY/m^2)	Cost-Effectiveness Ratio

The construction activities of high-PEI policy frameworks have better economic performance on all analyzed parameters. The average savings in total lifecycle costs in Shenzhen's regulatory environment amount to 156 CNY/m^2/year, which is twofold higher than in the regulatory environments with a moderate policy (89 CNY/m^2/year), which corresponds to a 75% increase in financial performance. This economic benefit is an indication of the holistic aspect of high-PEI policies, which targets several performance parameters at the same time and generates synergistic cost savings that exceed the summation of isolated interventions.

Regional Economic Optimization:

The proposed regional approach to policy harmonization would introduce Greater Bay Area-wide standards, including a minimum HVAC COP of 3.5, a maximum window-to-wall ratio (WWR) of 40%, a minimum envelope R-value of 4.5 $m^2 \cdot K/W$ and a requirement to incorporate climate resilience characteristics. The implementation of these evidence-based standards within all three jurisdictions would provide system-wide energy savings of 27.6 $kWh/m^2/year$ at a total cost of 110–160 CNY/m^2 with an overall benefit-cost ratio of 4.2:1 using measured performance data in PEI-high jurisdictions. The harmonized standards would also remove existing economic inefficiencies whereby the same building will be able to deliver different performances merely because of the regulation, and thus, provide stable economic returns across the Greater Bay Area, as well as maximize the potential of carbon neutrality.

This research paper adds a solid empirical base, although a number of methodological limitations should be mentioned. The simulations are based on standard weather data and manufacturer-specified performance parameters, but they do not perfectly represent the variability of the way buildings are actually operated. Besides, the behavior of the occupants, specifically thermal comfort preferences and behavioral patterns, differs among occupants of green buildings and conventional buildings and is not completely captured by the conventions of conventional models. Lastly, future empirical validation will be necessary to use the machine-learning insights generated through the 32-building dataset to extrapolate to broader building typologies and operating conditions within the heterogeneous urban fabric of the GBA.

6. Conclusions

This paper conducts a detailed analysis of the operating efficiency of the certified green buildings in the subtropical Greater Bay Area (GBA), in terms of energy efficiency, climate adaptability, carbon footprint, and policy efficiency. The analytical variables are gross floor area, typology of functions, method of construction, and climatic exposure. Normalized performance measures have been used to facilitate rigorous comparisons, such that the differences will indicate design and technological improvements, and not scale or usage patterns. The analysis shows that green office buildings in the sample experienced a 43–46% decrease in Energy Use Intensity (EUI) compared to conventional buildings, and with it a 45–47% decrease in operational carbon emissions, even when the effects of future climatic conditions were projected. Statistical normalization based on EUI ($kWh/m^2/year$) ensured that these numbers can not be explained by the difference in occupancy, operation hours, or volume of buildings.

The results also explain that certification status is not enough to ensure high performance. Comparing each case separately shows that normally constructed buildings that are geared toward passive design and natural air can save more energy than green-certified manually cooled buildings in some situations. This evidence supports the need for contextualized, climate-responsive design in subtropical environments where cooling loads prevail over total energy consumption. The paper admits limitations, especially in the lack of explicit microclimate modelling of factors like the shading of the surrounding environment, vegetation, and urban heat island enhancement. Nevertheless, a measure of 32 buildings gave a good correlation (R^2 = 0.92) of the comparative framework and the validity of relative performance trends.

The analysis of thirty-two building datasets of the domestic scale was conducted, twenty-four (75%) of them were used to train the models, and eight (25%) were used to validate the models. The robustness was determined through the five-fold cross-validation, thus guaranteeing the generalizability of the model, whereas sensitivity analyses identified HVAC efficiency, envelope insulation, window-to-wall ratio, and shading configuration as the most significant parameters. A sensitivity analysis showed that an improvement of the HVAC system efficiency by 10 percent provides an average 7.8 percent decrease in EUI. Moreover, the synergistic energy reductions due to combinations of parameter optimizations were over 20 percent. These results increase the faith in the predictive power of the model and identify key drivers of enhancing the performance of green buildings.

The results have direct implications for the policy of urban planning and climate adaptation. The major contribution of HVAC efficiency (24.3 percent of total energy performance) helps to meet the mandatory minimum COP of 3.5, whereas best window-to-wall ratios of 30–40 percent and envelope insulation (R-values > 4.5 $m^2 \cdot K/W$) provide a set of regulatory targets with 5.1 year payback. The analysis of typhoon resilience depicting 72.4 h vs. 8.3 h power continuity between green versus conventional buildings gives quantitative standards of disaster preparedness. The effectiveness of policy is directly related to building performance results, and high-PEI jurisdictions (>75) enjoy 35 percent quicker investment recovery and 75 percent greater lifecycle cost savings due to integrated regulatory methods that incorporate mandatory standards, financial incentives and robust enforcement. GBA-wide harmonization of regional standards would provide 27.6 $kWh/m^2/year$ energy savings at a 4.2:1 benefit-cost ratio, proving that the Shenzhen model can be replicated in other subtropical jurisdictions following faster decarbonization.

The research will conclude with a controlled and evidence-based study that may be used to quantify the real-life consequences of green building operations in fast-emerging subtropical megacities. The methodology and results of the research can be directly applied in other comparable urban settings with other climatic challenges, such as Southeast Asian mega-cities (Bangkok, Jakarta, Manila), Brazilian mega-cities (Sao Paulo, Rio De Janeiro), and new African urban centers (Lagos, Kinshasa). The subtropical climatic zone includes more than 2 billion urban dwellers around the world, and the experience of the GBA as a pilot of the carbon neutrality strategies is of primary international importance.

The transferability is not limited to technical performance measures but also to governance structures, as PEI, as a methodology, offers a uniform way of assessing the effectiveness of policies in various institutional settings. The study demonstrates that performance outcomes rely on the synergy of design, contextual flexibility, and successful policy frameworks, rather than only on certification status, therefore offering a sustainable paradigm for urban development replication. Because subtropical megacities across the globe are currently facing the triple pressures of high rates of urbanization, climate fragility, and decarbonization demands, the integrated policy and performance of green building in the GBA can not only offer technical standards but also governance strategies toward carbon-neutral cities.

Future studies should focus on longitudinal pre- and post-retrofit studies, microclimate-sensitive modelling techniques, and region-specific performance data to make findings more practical for global policymakers and urban planners working on creating climate-resilient cities. The conceptual framework developed in this research stands the Greater Bay Area as a model of evidence-based sustainable city building, whose lessons can be applied directly to the burgeoning community of subtropical megacities with a goal to achieve carbon neutrality by 2050.

Author Contributions: Conceptualization, X.F. and F.X.; methodology, X.F.; software, C.L.; validation, X.F., C.L. and F.X.; formal analysis, C.L. and X.F.; investigation, X.F.; resources, X.F.; data curation, F.X.; writing—original draft preparation, F.X.; writing—review and editing, C.L.; visualization, X.F.;

supervision, X.F.; project administration, F.X.; funding acquisition, F.X. All authors have read and agreed to the published version of the manuscript.

Funding: This study is a phased result of the Guangdong Provincial Department of Education's key scientific research platforms and projects for general universities in 2023: Guangdong, Hong Kong, and Macau Cultural Heritage Protection and Innovation Design Team funded project (Grant Number: 2023WCXTD042). The funders had no role in study conceptualization, data curation, formal analysis, methodology, software, decision to publish, or preparation of the manuscript. There was no additional external funding received for this study.

Data Availability Statement: The datasets used and analyzed during the current study are available from Xinshu Feng (fengxinshu9987@xhsysu.edu.cn) on reasonable request.

Conflicts of Interest: The authors declare no conflicts of interest.

References

1. Wang, Y.; Guo, C.H.; Chen, X.J.; Jia, L.Q.; Guo, X.N.; Chen, R.S.; Zhang, M.-S.; Chen, Z.-Y.; Wang, H.D. Carbon peak and carbon neutrality in China: Goals, implementation path and prospects. *China Geol.* **2021**, *4*, 720–746. [CrossRef]
2. He, J.K. Global low-carbon transition and China's response strategies. *Adv. Clim. Change Res.* **2016**, *7*, 204–212. [CrossRef]
3. Luo, J.; Yu, J. Facing the policy challenges of climate change: Assessing China's strategy and actions in international environmental crisis communication. *Commun. Humanit. Res.* **2024**, *33*, 205–212. [CrossRef]
4. Kuhn, B.M. China's commitment to the sustainable development goals: An analysis of push and pull factors and implementation challenges. *Chin. Political Sci. Rev.* **2018**, *3*, 359–388. [CrossRef]
5. Zhou, Y.; Li, K.; Liang, S.; Zeng, X.; Cai, Y.; Meng, J.; Shan, Y.; Guan, D.; Yang, Z. Trends, drivers, and mitigation of CO_2 emissions in the guangdong–hong kong–macao greater bay area. *Engineering* **2023**, *23*, 138–148. [CrossRef]
6. Wang, R.; Ren, C.; Liao, C.; Huang, Y.; Liu, Z.; Cai, M. Sectoral carbon emission prediction and spatial modeling framework: A local climate zone-based case study of the Guangdong-Hong Kong-Macao Greater Bay Area. *Sustain. Cities Soc.* **2024**, *114*, 105756. [CrossRef]
7. Wang, F.; Harindintwali, J.D.; Yuan, Z.; Wang, M.; Wang, F.; Li, S.; Yin, Z.; Huang, L.; Fu, Y.; Chen, J.M. Technologies and perspectives for achieving carbon neutrality. *Innovation* **2021**, *2*, 100180. [CrossRef]
8. Zhao, C.; Ju, S.; Xue, Y.; Ren, T.; Ji, Y.; Chen, X. China's energy transitions for carbon neutrality: Challenges and opportunities. *Carbon Neutrality* **2022**, *1*, 7. [CrossRef]
9. Zhang, D.; Wang, J.; Lin, Y.; Si, Y.; Huang, C.; Yang, J.; Huang, B.; Li, W. Present situation and future prospect of renewable energy in China. *Renew. Sustain. Energy Rev.* **2017**, *76*, 865–871. [CrossRef]
10. Awadh, O. Sustainability and green building rating systems: LEED, BREEAM, GSAS and Estidama critical analysis. *J. Build. Eng.* **2017**, *11*, 25–29. [CrossRef]
11. Zhang, Y.; Bai, X.; Mills, F.P.; Pezzey, J.C. Rethinking the role of occupant behavior in building energy performance: A review. *Energy Build.* **2018**, *172*, 279–294. [CrossRef]
12. Fang, G. Typhoon Wind Modeling and Flutter Fragility Analysis of Long-Span Bridges in Coastal Regions of China. Ph.D. Thesis, Clemson University, Clemson, SC, USA, 2019.
13. Wang, Z.; Xiao, Z.; Tam, C.Y.; Pan, W.; Chen, J.; Hu, C.; Ren, C.; Wei, W.; Yang, S. The projected effects of urbanization and climate change on summer thermal environment in Guangdong-Hong Kong-Macao Greater Bay Area of China. *Urban Clim.* **2021**, *37*, 100866. [CrossRef]
14. Zhou, S.; Li, M.; Xie, J. Evaluating urban–rural gradients and urban forms in metropolitan areas: A local climate zone approach with future spatial simulation. *Sustain. Cities Soc.* **2024**, *112*, 105636. [CrossRef]
15. Liu, Y.; Li, S.; Wang, Y.; Zhang, T.; Peng, J.; Li, T. Identification of multiple climatic extremes in metropolis: A comparison of Guangzhou and Shenzhen, China. *Nat. Hazards* **2015**, *79*, 939–953. [CrossRef]
16. Oh, J.H.; Kim, S.S. Prefabricated Envelope Green Remodeling Potential of Public Office Buildings in Korea. *Buildings* **2024**, *14*, 2182. [CrossRef]
17. Bibri, S.E.; Krogstie, J. Environmentally data-driven smart sustainable cities: Applied innovative solutions for energy efficiency, pollution reduction, and urban metabolism. *Energy Inform.* **2020**, *3*, 29. [CrossRef]
18. Orikpete, O.F.; Ikemba, S.; Ewim, D.R.E. Integration of renewable energy technologies in smart building design for enhanced energy efficiency and self-sufficiency. *J. Eng. Exact Sci.* **2023**, *9*, 16423-01e. [CrossRef]
19. Wang, Y.; Yao, Y.; Chen, S.; Ni, Z.; Xia, B. Spatiotemporal evolution of urban development and surface urban heat island in Guangdong-Hong Kong-Macau greater bay area of China from 2013 to 2019. *Resour. Conserv. Recycl.* **2022**, *179*, 106063. [CrossRef]
20. Duarte, C.; Raftery, P.; Schiavon, S. Development of whole-building energy models for detailed energy insights of a large office building with green certification rating in singapore. *Energy Technol.* **2018**, *6*, 84–93. [CrossRef]

21. Li, L.; Chan, P.W.; Deng, T.; Yang, H.L.; Luo, H.Y.; Xia, D.; He, Y.Q. Review of advances in urban climate study in the Guangdong-Hong Kong-Macau greater bay area, China. *Atmos. Res.* **2021**, *261*, 105759. [CrossRef]
22. Liang, H.; Bian, X.; Dong, L. Towards net zero carbon buildings: Accounting the building embodied carbon and life cycle-based policy design for Greater Bay Area, China. *Geosci. Front.* **2024**, *15*, 101760. [CrossRef]
23. Patchell, J. *China's Greater Bay Area: Agglomeration, External Economies, Governance and Urbanization*; Routledge: Oxfordshire, UK, 2023.
24. Han, T.; Huang, Q.; Zhang, A.; Zhang, Q. Simulation-based decision support tools in the early design stages of a green building—A review. *Sustainability* **2018**, *10*, 3696. [CrossRef]

buildings

MDPI

Article

Machine Learning in the Design Decision-Making of Traditional Garden Space Renewal: A Case Study of the Classical Gardens of Jiangnan

Lina Yan [1], Liang Zheng [2], Xingkang Jia [1], Yi Zhang [3] and Yile Chen [2,*]

1 College of Arts, Shanghai Zhongqiao Vocational and Technical University, No. 3888 Caolang Road, Jinshan District, Shanghai 201514, China; linayan@shzq.edu.cn (L.Y.); jiaxingkang@shzq.edu.cn (X.J.)
2 Faculty of Humanities and Arts, Macau University of Science and Technology, Avenida Wai Long, Taipa, Macau 999078, China; zhengliang@must.edu.mo
3 Shanghai GOODLINKS International Design Group, No. 258 Yangzhai Road, Changning District, Shanghai 200051, China; fish8@126.com
* Correspondence: chenyile@must.edu.mo

Abstract: This research takes the Suzhou Gardens, a World Cultural Heritage Site, as the object of study and investigates a rapid scheme generation approach for garden restoration and expansion projects, assisting designers in making scientific decisions. Considering the limitations of current garden design, which is inefficient and relies on human experience, this study proposes an intelligent generation framework based on a conditional generative adversarial network (CGAN). In constructing the CGAN model, we determine the spatial characteristics of the Suzhou Gardens and, combined with historical floor plan data, train the network. We then design an optimization strategy for the model training process and finally test and verify the generative space scheme. The research results indicate the following: (1) The CGAN model can effectively capture the key elements of the garden space and generate a planar scheme that conforms to the traditional space with an accuracy rate reaching 91.08%. (2) This model can be applied to projects ranging from 200 to 1000 square meters. The generated results can provide multiple scheme comparisons for update decisions, helping managers to efficiently select the optimal solution. (3) Decision-makers can conduct space utilization analyses and evaluations based on the generated results. In conclusion, this study can help decision-makers to efficiently generate and evaluate the feasibility of different design schemes, providing intelligent support for decision-making in urban renewal plans.

Keywords: urban landscape renewal; artificial intelligence; design decision-making; Jiangnan gardens

1. Introduction

1.1. Research Background

Urban renewal is crucial for revitalizing historic cities, and effective site planning is key to ensuring successful redevelopment [1]. In recent years, local governments have paid increasing attention to the renewal of landscape spaces within historic conservation areas. Spatial design is crucial for the spatial renewal of these historically protected areas. The traditional gardens of Jiangnan are outstanding representatives of Chinese classical garden art and an important part of the world's cultural heritage [2].

However, the acceleration of urbanization and changing social demands are challenging the protection and renewal of Jiangnan gardens. Within Suzhou's historic and cultural

https://doi.org/10.3390/buildings15142401

city conservation area, conducting spatial renewal while preserving its historical features has become a critical issue for decision-makers (Figure 1).

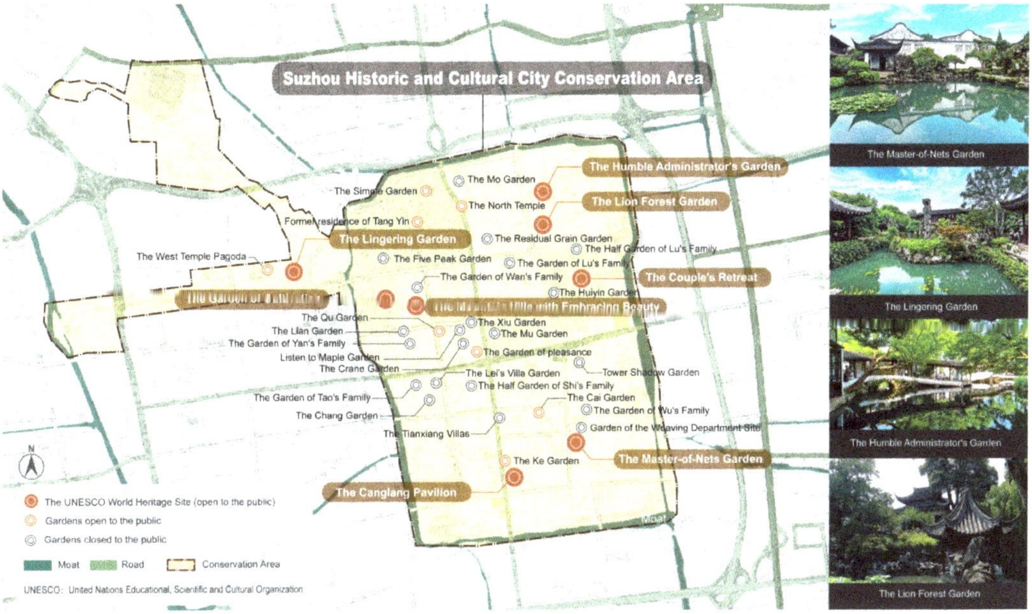

Figure 1. Map of the historic and cultural city conservation area in Suzhou, China.

At present, decisions regarding the renewal design of Jiangnan gardens still rely on the experience and judgment of designers, lacking systematic and scientific methodological support. Most urban renewal decision-making methods often result in homogeneous spaces that lack local characteristics [3]. The traditional garden design process is often time-consuming and fails to fully consider the complex characteristics of the space. The challenge for designers lies in effectively preserving characteristics such as spatial scale, types of landscape elements, rules of landscape combination, and the layout of landscape space. Furthermore, due to the lack of quantitative analysis tools, many renewal plans fail to fully conserve the spatial essence of Jiangnan gardens, resulting in unreasonable design decisions and possibly even destruction of the original cultural values and landscape continuity.

Therefore, efficiently and accurately extracting and applying the characteristics of garden spaces in the renewal of historic protected areas has become a key issue in improving the efficiency and quality of design decisions.

1.2. Literature Review

Researchers have recently begun to adopt machine learning methods to quantitatively analyze the characteristics of garden spaces. The rapid development of machine learning technology has provided new solutions in this field. Some scholars have applied computer vision technology to the recognition of spatial types; for example, the intelligent recognition and classification of the types of images of architectural ruins in Jiangnan were achieved using machine learning [4,5]. Intelligent recognition of surface damage in Jiangnan garden architecture was performed based on the computer vision technology of machine learning [6]. An analysis framework for spatial indicators in Chinese classical gardens has been constructed and studied based on spatial syntax and machine learning [7]. The modular information fusion model of garden landscapes was studied using machine learning algorithms [8]. Research on the characteristics of traditional garden roads and the intelligent generation of space, etc., has also been conducted [9]. In terms of spatial layout, some scholars have studied the layout characteristics of the geometric space of Jiangnan gardens using machine learning [10,11]. In particular, the application of GAN technology in landscape space has gradually matured. For example, the conditional generative adversarial network (CGAN) was applied to the generation of landscape plan layouts [12,13], and GAN technology has been used for the determination of spatial area and prediction of garden landscapes, etc. [14].

In terms of design decision-making, integrating machine learning technology into the garden decision-making process is shaping the current research frontier. In recent years, scholars have focused on the application of artificial intelligence in landscape space design and management, such as garden planning systems, site selection, and spatial quality assessment [15]. For example, intelligent planning and design were achieved through the integration of multi-source data [16,17]. The site selection of landscape space was accomplished using machine learning [18]. Research has been conducted on spatial renewal, transformation design, quality assessment based on machine learning, etc. [19–21].

In conclusion, the application of machine learning technology in the landscape renewal design process has brought about major breakthroughs in this field. By relying on the power of automated data processing methods, the efficiency of design analysis can be significantly improved. In particular, the conditional generative adversarial network (CGAN) can automatically generate garden layouts. However, existing CGAN applications primarily focus on generic urban green spaces, with several domain-specific gaps persisting in machine learning implementations: (1) there is no specialized framework for the design principles of Jiangnan gardens; (2) they lack the integration of small-scale historical spatial scales (element types, combinations); (3) and the distribution of spatial element categories is imbalanced. These limitations constitute the primary directions for future research.

1.3. Problem Statement and Objectives

This study considers the renewal of historically protected areas in Jiangnan gardens as a case study. Addressing the limitations of traditional garden design—including its heavy reliance on empirical expertise, prolonged scheme generation cycles, and existing gaps in machine learning applications for this domain—we pioneer a CGAN-based framework for the spatial analysis and generation of Jiangnan gardens. Leveraging CGAN technology to decode critical spatial patterns, our approach enables rapid scheme generation for garden restoration/expansion, effectively streamlining design decision-making timelines. This study aims to explore the application of machine learning technology in the decision-making process regarding the spatial renewal design of traditional gardens in the Jiangnan region. We construct a data-driven auxiliary design framework, with the expectation of providing decision support for the protection and innovation of world cultural heritage.

Our objectives were as follows:

(1) Rapid generation of typologically diverse schemes adhering to spatial characteristics;
(2) Preservation of landscape features at fine-grained heritage spatial scales;
(3) Quantitative evaluation of spatial element compositional ratios.

By harnessing the ability of CGANs to decode complex spatial relationships, our approach uncovers latent patterns beyond the reach of traditional methods. Furthermore, at the design decision-making level, it facilitates a paradigm shift from empirical approaches to data-driven science in landscape research, delivering robust technical support for the contemporary reinterpretation and design decisions of historic spaces.

Through our planning practice and research, the following six issues are mainly discussed:

(1) What types of landscape spaces exist in traditional Jiangnan gardens?

(2) What are the specific methods and processes of CGAN technology in the generation of traditional landscape spaces?
(3) How accurate are the data training process and test results?
(4) Compared with the traditional design process, what advantages does the process of generating floor plans with CGAN have in design decision-making?
(5) Can multiple schemes be output to meet the decision-makers' needs for scheme comparison and selection?
(6) Does the utilization rate of the generated planar space meet the normative requirements of the design decision?

2. Materials and Methods

2.1. The Characteristics of Garden Space

Traditional Jiangnan gardens, renowned for their poetic integration of architecture, nature, and philosophy, employ meticulously designed landscape spaces to create harmony and layered visual experiences [22,23]. Generally, based on the characteristics of landscape space elements, the structure of their main spatial types can be classified into the following types [24–26].

(1) Water-Island Space

This is a space with water as its scenery; for example, central ponds, rivers, and streams. The curved waterways guide the movement, with zigzag bridges amplifying spatial depth.

(2) Rockery Sculpture Space

The garden mainly features artificial mountains, using porous Taihu Lake rocks to simulate peaks or mountain ranges; for example, Lion Grove Garden's labyrinthine rock formations.

(3) Pavilion Corridor Space

The space is composed of pavilions, terraces, and towers in the garden; for example, open-sided structures for contemplation. The corridor space that winds through the garden serves as a walkway and spatial divider.

(4) Botanical Spaces

These usually include floral focus areas, such as lotus ponds in summer, osmanthus groves in autumn, and prunus mume (plum) clusters in winter, etc. Also present are bamboo groves planted near walls to cast moving shadows, and tall trees (such as ginkgo) providing shade, mid-layer shrubs (camellias), and ground covers like moss or ferns (Figure 2).

Traditional Jiangnan gardens artfully integrate aquatic, rockery, architectural, and botanical spaces to create harmonious, layered landscapes that blend nature with poetic philosophy [27]. This study respects the traditional classification of spatial types and employs these four spatial types: rockery sculpture space, water-island space, pavilion gallery space, and botanical spaces.

Figure 2. Four typical spatial types of classical gardens in the Jiangnan region.

2.2. CGAN Method

2.2.1. Framework and Principles

This study adopts a CGAN (conditional generative adversarial network) as the core framework for garden floor plan generation. CGAN is an extension of GAN (generative adversarial network), and it is mainly applied in image generation tasks [28].

Compared to CGAN, diffusion models and Transformer models exhibit distinct advantages and limitations in generative tasks, primarily differing in generative mechanisms, training stability, and application scenarios. Diffusion models produce high-quality samples through stepwise denoising, excelling in output diversity but suffering from high computational costs and slow generation speeds [29]. Transformer models rely on self-attention mechanisms to achieve the global modeling of structured data, yet their computational complexity increases quadratically with input size, and they are prone to overfitting on small datasets [30]. CGANs enable rapid generation via adversarial training, but face issues such as mode collapse and training instability in complex scenarios [31] (Table 1).

Table 1. Key technical comparisons.

Model Type	Strengths	Limitations	Suitability for Garden Floor Plans
Diffusion Models	High-quality details	High computational cost; slow generation	Limited (real-time iteration required)
Transformers	Global context modeling	Quadratic complexity; data-hungry	Overkill (local patterns > global links)
CGAN	Fast generation; local feature capture	Mode collapse risks	Optimal: Balances speed and spatial granularity

CGAN remains the optimal methodological framework due to three practical advantages: (1) single-pass inference enables the real-time generation essential for design iteration, outperforming diffusion models in speed; (2) convolutional inductive bias efficiently captures the local spatial patterns critical to garden layouts, avoiding the computational overhead of Transformer models; (3) architectural priors ensure robustness with limited data and reduce overfitting risks, which balances speed, feature localization, and adaptability in resource-constrained scenarios. Therefore, the CGAN was selected as the methodological framework for this study, offering the optimal balance of speed, localized feature capture, and practical adaptability within resource-constrained scenarios.

The CGAN framework consists of a generator and a discriminator (Figure 3). Among them, the function of the generator is to generate images, while that of the discriminator is to determine whether the generated images are real images. Unlike ordinary GANs, CGANs require that additional conditions be provided, and these conditions can be images of any form [32]. By introducing conditions (TrainA), CGAN can generate more targeted and personalized images.

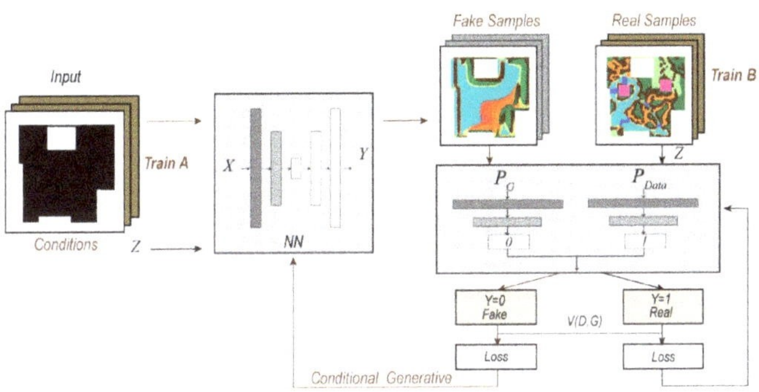

Figure 3. The framework and principle of CGAN.

During the CGAN training process, the generator and discriminator are trained through adversarial learning. The generator deceives the discriminator (TrainB) by learning how to generate images that match the condition (TrainA), while the discriminator prevents deception by learning to distinguish between the real images and generated images [33]. During this cyclic training process, the generator and the discriminator keep iterating until the generator can generate sufficiently realistic images, such that the discriminator is unable to determine which images are real. After the training is completed, by adjusting the input conditions, the attributes of the generated images can be controlled, thereby obtaining images with specific features. This study, based on the principle of the CGAN generative adversarial network model and taking the spatial characteristics of different garden artistic conceptions as conditions, generated garden floor plans with different landscape spatial characteristics.

The goal of this study is to enable the machine to automatically generate floor plans with the spatial characteristics of garden landscapes after learning a large amount of sample data through CGAN. This not only retains the planar characteristics of the traditional garden landscape space but also serves as an intelligent tool for generating garden schemes.

2.2.2. Research Methods and Procedures

The method of generating the spatial floor plan of garden landscapes using the CGAN comprises a new attempt to combine the protection of garden cultural heritage with AI. This method mainly consists of five process steps, namely, data collection, data processing, model training, model evaluation, and model application (Figure 4). The specific method contents are as follows.

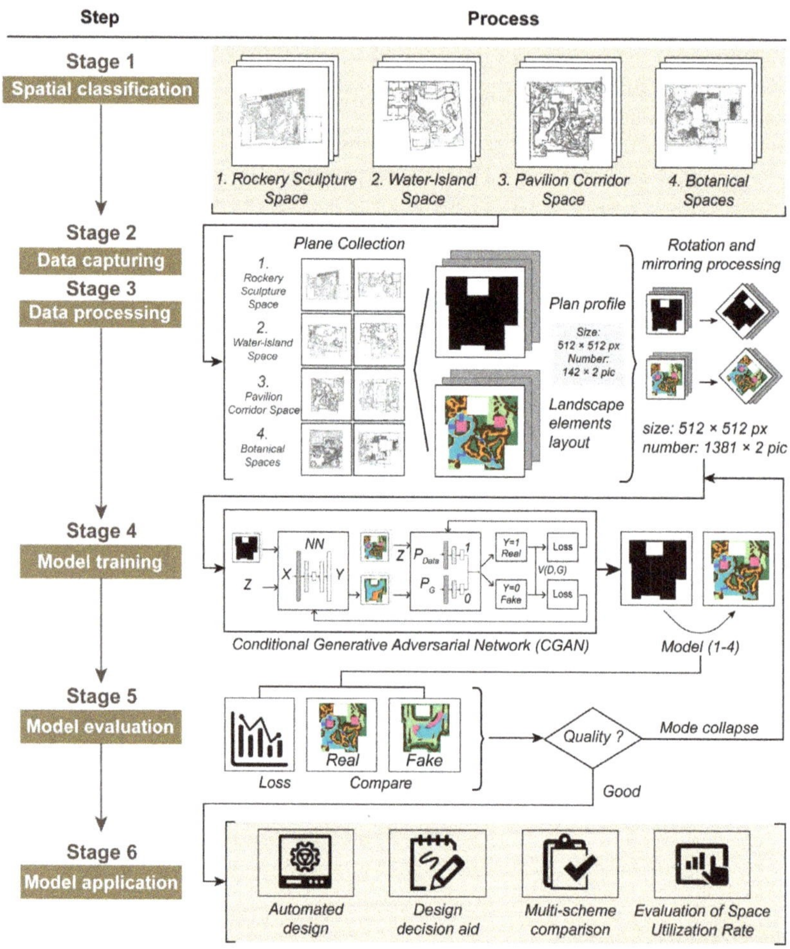

Figure 4. Research methods and procedures.

(1) Data classification: Spatial type division is carried out in accordance with traditional landscape features. According to the characteristics of landscape space elements in traditional Jiangnan classical gardens, their spatial types are classified as: rockery sculpture space, water-island space, pavilion corridor space, and botanical spaces.

(2) Data collection: At the same scale, spatial floor plans with corresponding features are extracted from the original floor plans of classical gardens according to different types of landscapes as the basic samples for training.

(3) Data processing: Targeted processing is conducted based on the characteristics of the samples. Firstly, the sample size is increased. The sample size of the landscape space is limited, and it is necessary to increase the training sample size through technical means. Secondly, the contour map is drawn, which will serve as an input for the generation condition in the CGAN learning process. Secondly, the landscape space elements are classified and colored to enable the machine to effectively recognize the

content and features of the drawing during the learning process. Finally, the size of the training images is unified to improve the efficiency of training.

(4) Model training: Paired training samples are placed in the CGAN training framework for machine learning. The CGAN model framework requires paired samples for training. The "contour atlas" and the corresponding "spatial element atlas" obtained above are paired sample sets. Each of the contour maps corresponds to a spatial element map for paired training. The purpose is to obtain the spatial meta-layout primitives under contour control; that is, to input one contour map and derive a corresponding spatial element layout map scheme.

(5) Model Evaluation: To verify the training effect of the model, this study conducted evaluations in three aspects. First, the loss value of CGAN was observed in each iteration, and test images were generated during the iteration process. Secondly, the generated test samples were compared with the original samples to evaluate whether there was any deviation in the effect of the generated samples. Finally, a professional satisfaction survey was conducted between the samples generated by the test and the original samples to obtain a professional assessment of the spatial effect. After the above three rounds of evaluation, if the effect was not satisfactory, it was necessary to adopt cyclic optimization training, repeatedly adjusting parameters, increasing the number of iterations, and employing other methods to improve the effect of model training. If the training effect was satisfactory, the next application stage could be entered.

(6) Model Application: The results generated through the training of this process method are applied in actual design as an auxiliary tool for the generation of garden landscape space.

2.3. Data Collection and Processing

The collection of floor plans: Based on the four types of spaces (rockery sculpture space, water-island space, pavilion corridor space, botanical spaces), a total of 142 floor plans of the corresponding landscape space types were selected from the classical garden floor plans. The screening criteria for the 142 initial samples were:

(1) Area selection: Between 200 and 1000 m^2, to control the scale rationality of spatial generation. Excluded were instances like the 8 m^2 floor plan courtyard in the Lingering Garden.

(2) Scope definition: It must possess relatively clear spatial boundaries; for example, the scope of architectural spaces is defined by building boundaries, and plant spaces are defined by plant courtyards.

(3) Distinct element characteristics: Each spatial type must contain identifiable landscape elements. For instance, a water-island space must include water bodies, and a rockery space must contain rocks. Excluded were spaces dominated by a single element, such as a 200 m^2 plant space filled solely with bamboo, or an area consisting entirely of water bodies without other landscape elements.

To ensure the proportion and scale of the original floor plan, the spatial range was cut at the same scale of 1:100. Among the characteristics were 27 mountain and rock types, 38 water feature types, 35 architectural types, and 42 flower and tree types. The size of the space obtained after interception ranged from 200 to 1000 m^2. The space obtained after training could be applied in the renewal of micro and small spaces in urban historic protection blocks.

Increase the sample size: Obviously, the 142 images were not sufficient for model training. Therefore, we expanded the sample size. By rotating the original samples by 90°, 45°, or mirroring them, the number of samples could be increased [34]. After each sample was rotated and mirrored in all directions, it could be extended to 6 to 10 samples. The garden floor plans after rotation and mirroring did not undergo any changes such as scaling, distortion, or deletion, nor was the original layout content of the floor plan altered. The final total sample size reached 1381, which met the training requirements of the samples. Among them, there were 235 mountain and rock types, 380 water feature types, 346 architectural types, and 420 flower and tree types.

Draw the spatial contour: After the training quantity was met, it was necessary to draw the contour of the sample. The specific operation filled the garden floor plan obtained by the above organization with black according to the outline to obtain the "outline atlas." The contour map was used as the input condition of the CGAN.

Classification and coloring of spatial elements: To enable the machine to effectively read the complex information on the drawing surface, color separation processing was performed on the different landscape elements in the floor plan (one color represented one landscape element). According to the spatial elements of the landscape, a total of 12 color types were classified. The buildings are red (R255, G0, B0), the pavilions and structures are rose red (R255, G0, B255), the corridors are yellow (R255, G255, B0), the bridges are dark blue (R0, G0, B255), the water is light blue (R0, G255, B255), and the trees and shrubs are green (R0, G255, B0). The ground cover plants are dark green (R9, G124, B37), the garden paths are orange (R255, G150, B0), the rockery stones are brown (R106, G57, B0), the platforms are cyan (R0, G117, B169), and the open spaces in the square are pinkish green (R172, G213, B152). The revetments are light yellow (R245, G245, B156). The classification and annotation of colors not only help machines identify the differences in image content, but also improve the feature learning of machines for different types of landscape spaces [35]. After color-marking each floor plan, the "spatial element layout atlas" corresponding to the abovementioned "outline atlas" was obtained (Figure 5).

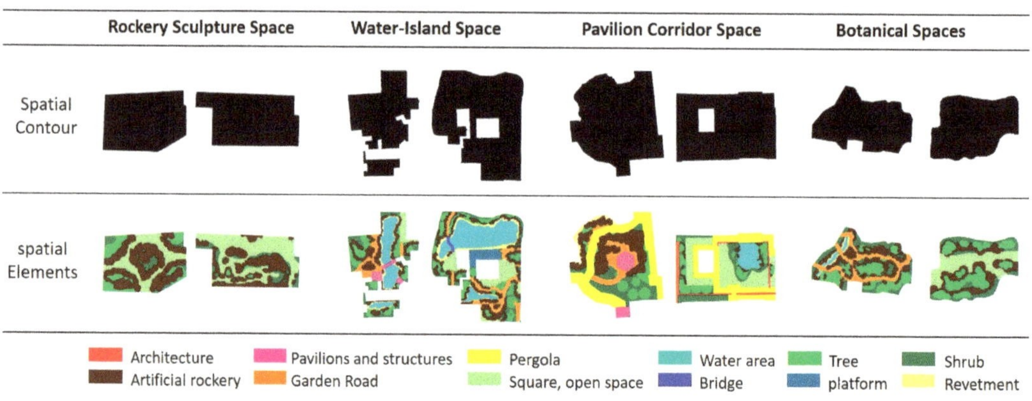

Figure 5. The sample images used in the research.

Unify image size: To avoid generation errors caused by differences in sample image sizes, the above two image sets were uniformly processed into 512 × 512 pixels, 96 dpi resolution, and 24-bit depth images.

3. Model Training and Testing

3.1. Model Training Process

The sorted paired samples were input into the CGAN for training (Appendix A). The "contour atlas" was the conditional sample, and the "spatial element layout atlas" was the generation sample to train the layout generation of spatial elements under the control of the contour conditions. To ensure the effectiveness of the training process, it was necessary to constantly observe the loss logs during the training process and the test images in the iterations.

Based on prior training experience, 300–500 epochs are typically sufficient to complete the training process [36]. Given the complexity of landscape garden spaces, 500 epochs were selected to ensure effective machine learning. To preserve the generic features learned during training (edges, textures) while significantly reducing computational load and memory consumption, freezing was implemented at epoch 150. Additionally, model snapshots were saved every 50 epochs to facilitate progress monitoring (Appendix B). This configuration ensured training validity, retained critical learned features, and enabled efficient process inspection (Table 2).

Table 2. Key parameters of model training.

Parameter Type	Input
Picture size	512
Training Epochs (niter)	500
Freezing Epochs (niter decay)	150
Checkpoint Interval (save epoch freq)	50

3.1.1. Loss Value Log Data

During the model training process, loss logs were generated simultaneously, and the loss values in them reflected the training quality of the generative adversarial network [37]. A loss value with a continuous downward trend indicated that the samples generated by the model during the learning process were constantly approaching the real samples. A loss value no longer decreasing or fluctuating indicated that the learning of the model reached a bottleneck and an over-fitting situation occurred [38].

In this study, the corresponding loss value line graphs were obtained by statistically analyzing the loss values of each iteration of the generator and discriminator during the model training process of the four landscape types (Figure 4). As can be seen from Figure 5, the loss curves of all four model variants exhibited a sustained decline throughout training. Critically, after ~480 epochs, losses stabilized within a low-value regime (<1.0) with minimal fluctuations ($\pm$0.05), indicating convergence of both discriminative and generative components. The trained models were deemed effective.

Among them, the models of the water-island space (a) and the rockery sculpture space (b) had a relatively fast convergence speed, and the downward trend of the loss value curve was obvious. The reason is that the spatial form characteristics of mountains and water are relatively obvious, and there are significant differences among the various elements, so the efficiency of model learning is relatively high. Although the loss value curves of the pavilion corridor space (c) and the botanical spaces (d) also declined significantly, the convergence process of the models fluctuated greatly. Due to the complexity of the buildings and plant spaces, difficulties emerged in the learning process of the model, especially in the architectural space, the complex architectural contours (curved eaves, interconnected pavilions, corridors and roof lines), and the multi-scale spatial dependence (the infiltration of vegetation through the corridor frame). These complexities led to unstable gradients during the training process, causing significant fluctuations in the learning process and a slower convergence speed. Nevertheless, all four groups of models gradually converged after nearly 500 epochs and achieved good learning results in terms of data (Figure 6).

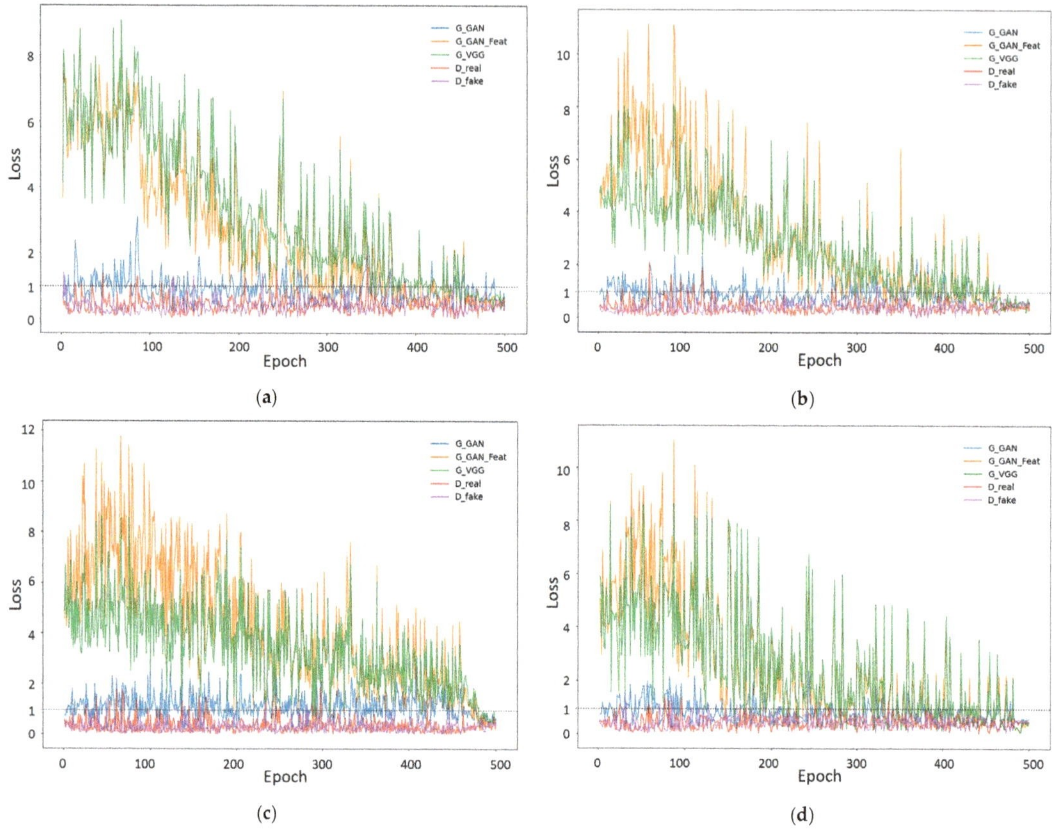

Figure 6. The learning curves of different landscape model trainings: (**a**) the loss value of the water-island space model; (**b**) the loss value of the rockery sculpture space model; (**c**) the loss value of the pavilion corridor space model; (**d**) the loss value of the botanical spaces model.

3.1.2. Accuracy of the Iterative Process Model

In addition to observing the model's learning status from the loss value data, it was also necessary to constantly observe whether the test images during the iteration process gradually became clear and realistic, as well as whether there were any conditions such as blurriness or distortion. This allowed us to determine whether the model converged or whether the machine learning parameters needed to be readjusted.

During the model training process, one check image was set to be output every 10 epochs to determine changes in the model's learning accuracy. As shown in Figure 7, the effect generated by the model in the 10th epoch is very blurry. By the 100th epoch, there is already a relatively clear spatial outline. Starting from the 200th epoch, the effect begins to approach the real picture. Until the 500th epoch, the effect is almost consistent with the real picture. This indicates that the model has basically achieved the training objective after 500 epochs of training. The results produced by the generator have learned the characteristics of these types of garden spaces (Figure 7).

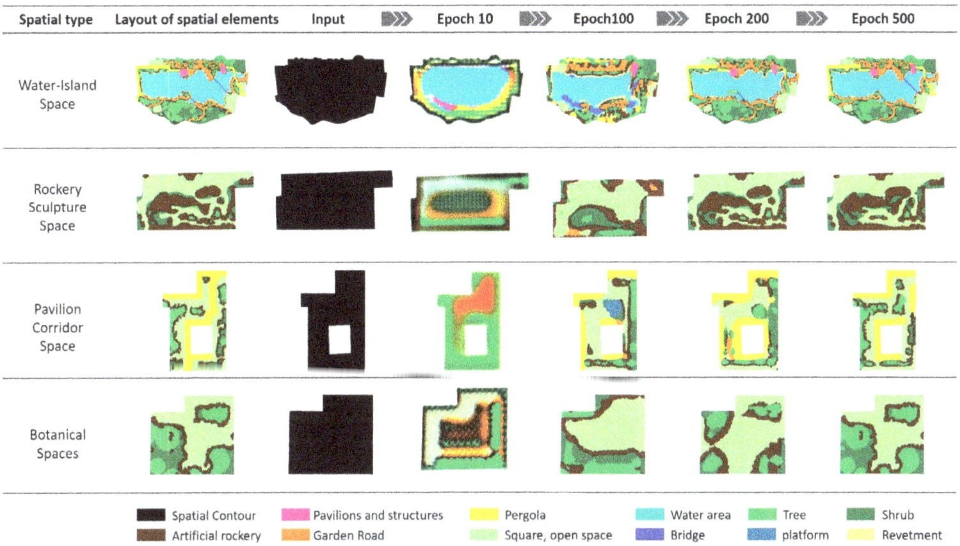

Figure 7. Changes in model accuracy during the iterative process.

In terms of image similarity, the training effects of the water-island space, the rockery sculpture space, and the botanical spaces were relatively good. In the 100th epoch, the spatial forms are quite different from the real pictures, and there were many errors starting from the 200th epoch; however, the results from the generator were very close to the real pictures. The morphological elements of the pavilion corridor space were rather complex, and the efficiency of the model in learning the spatial morphological rules was relatively slow. The results at 200 epochs still exhibited many errors when compared with the real pictures. However, with a continuous increase in the number of training sessions, the spatial generation effect of the architectural type was almost the same after 500 epochs of training as that of the real picture.

In summary, after 500 epochs of training by CGAN, the four sets of loss value log data for the spatial floor plan samples of the four landscape types finally began to converge at the 480th epoch. The results generated by its generator were almost consistent with the real samples. This indicated that training the model to generate the "layout diagram of landscape space elements" through "outline" was effective. The model completed the feature learning of the landscape space and had a good effect. The model training was completed.

3.2. Model Accuracy Test and Comparison

After the model was trained, it was further tested using new samples to evaluate its performance. After the new untrained samples were contoured, they were placed as input samples into the trained CGAN program. The generated results were compared with the original real samples to test the similarity and accuracy of the generated results. The testing comprised two components: shape contour similarity (assessed via image pixel matching) and spatial element semantic accuracy (measured by compositional deviation of element types).

As shown in Figure 8, after four groups of new landscape samples were placed in the trained CGAN model, the output results were found to be extremely similar to those of the real samples. For shape contour similarity, a Python 3.8.3 script was developed to calculate image pixel matching. By calculating the pixel differences between the two images, the specific locations and similarity percentages of the morphological differences between the two were obtained more intuitively.

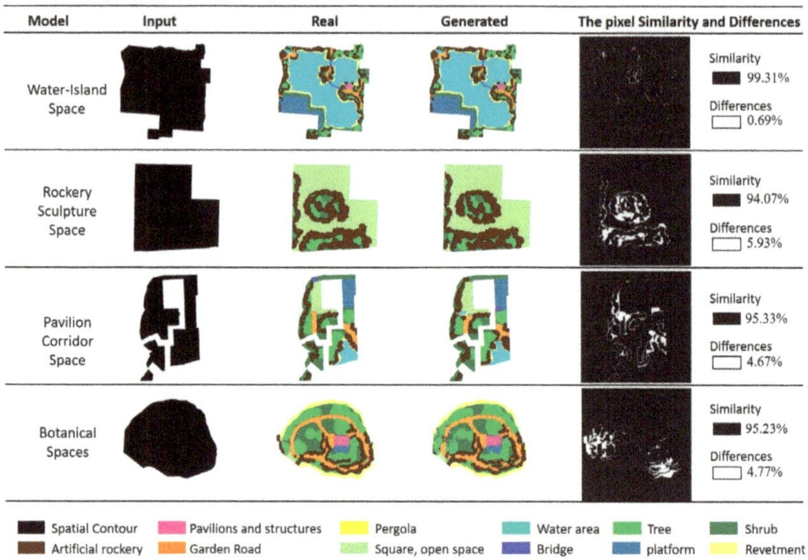

Figure 8. Model test similarity comparison.

Among the spaces, the similarity of the water-island space was the highest, reaching 99.31%. The morphological differences occurred in the contour lines, and the differences were extremely small. The pavilion corridor space and botanical spaces showed similarities of 95.33% and 95.23%, respectively. The differences in pavilion corridor space were located at the positions of water areas and some roads, while the morphological differences in botanical spaces mainly lay in the plants and roads at the left and right ends. The overall morphological similarity was very high. In contrast, the morphological differences in the mountain and rock scenes were relatively greater, but the similarity still reached 94.07%. The main difference lay in the treatment of the contours of the mountain and rocks (Figure 8). Based on the training logs, similarity analysis was conducted on the images generated by the four model groups at epoch 480 (model convergence), yielding an overall average similarity of 91.08%.

The accuracy of the landscape semantics of the floor plan generated by CGAN and the original floor plan was determined by calculating the color proportions of landscape element types. As shown in Figure 9, among the four groups of typical plans, the types and quantities of elements generated by CGAN were consistent with the original floor plan types.

Among the 10 spatial types in the first group of water-island spaces, the difference between the elements generated by CGAN and the original image ranged from 0.00% to 0.08%. Among them, the shrub element generated by CGAN was 3.99%, which was 0.04% higher than the 3.95% of the original image; the tree was 0.03% different, the artificial rockery was 0.07% different; and the water area was 0.02% different. Garden road, square, and open space differed by 0.08%, while revetment, pavilions and structures, and platform differed by only 0.01%. Similarly, in the comparison of the four types of elements in the second group, rockery sculpture space, the difference in the proportion of elements was 0.00% to 0.25%. The comparison differences of the nine element types in the third group, pavilion corridor space, ranged from 0.00% to 1.44%. The comparison differences among the eight element types in the fourth group of botanical spaces ranged from 0.00% to 1.30% (Figure 9).

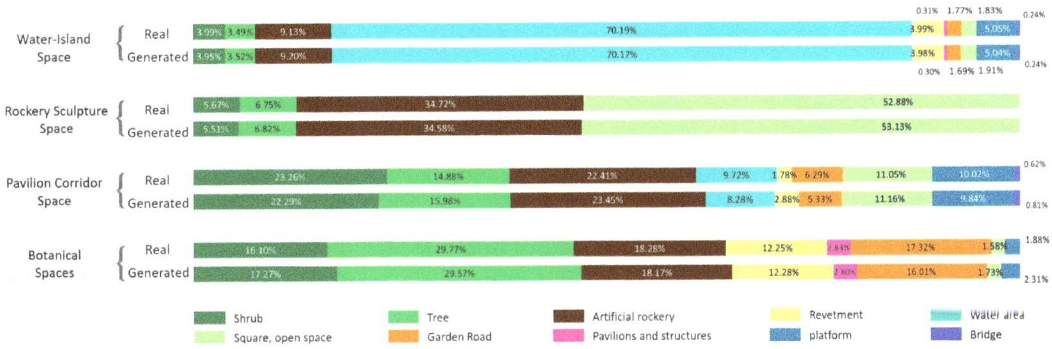

Figure 9. Semantic comparison of spatial elements in the model testing.

It can be seen from this that the proportion of planar element types generated by CGAN was extremely different from the original image. This reflects the relatively high accuracy of the CGAN generation floor plan in terms of element semantics.

3.3. Model Robustness Test

To verify whether the model can generate multiple schemes, conducting a robustness test of the CGAN model is an important step [39]. From the test model samples, one representative floor plan from each of the four groups of landscape space types was selected as the test object. When changing the noise input of the model, the four groups of landscape space models generated a variety of different morphological schemes.

As can be seen from Figure 10, the spatial elements generated by each group of models exhibited differences while conforming to the theme. Among them, the pavilion corridor space changed significantly. The reason is that there are many types of spatial elements around the building, especially the integration of water elements, which enrich this entire space. Among the different schemes for generating water features, the main characteristics of the water surface were retained, and the positions of the pavilions and garden paths were also changed accordingly. However, there were slight flaws in the generation of bridges in Effects 3 and 4 of the water-island space generation, indicating that the learning of bridges in water-island space required improvement. The test results of the water-island space showed a relatively stable state, and the generated planar forms were quite similar. The results of multiple schemes for botanical spaces varied significantly. Pavilions were generated in Effects 1 and 2, while no pavilions were generated in Effects 3 and 4 (Figure 10). These locally occurring problems can be solved by increasing the sample size or the number of training sessions.

In summary, the robustness test results of the four groups of CGAN models were relatively ideal. The models generated multiple schemes that conformed to the topic from the same contour norm space. These generated schemes demonstrated rich creativity in various aspects, such as the proportion, position, form, and outline of spatial elements. This model can generate diverse design schemes or inspirations with the spatial characteristics of Jiangnan gardens for architects in different spatial scenarios, thus creating more flexible and personalized garden space forms.

Meanwhile, the author also tested the model's generation of random contour samples. The randomly drawn planar contours were used as input conditions for the four groups of trained models to observe the generated effects. The results of the random generation experiment are shown in Figure 11. Under the randomly drawn outline, the model generated the element layout content corresponding to the topic.

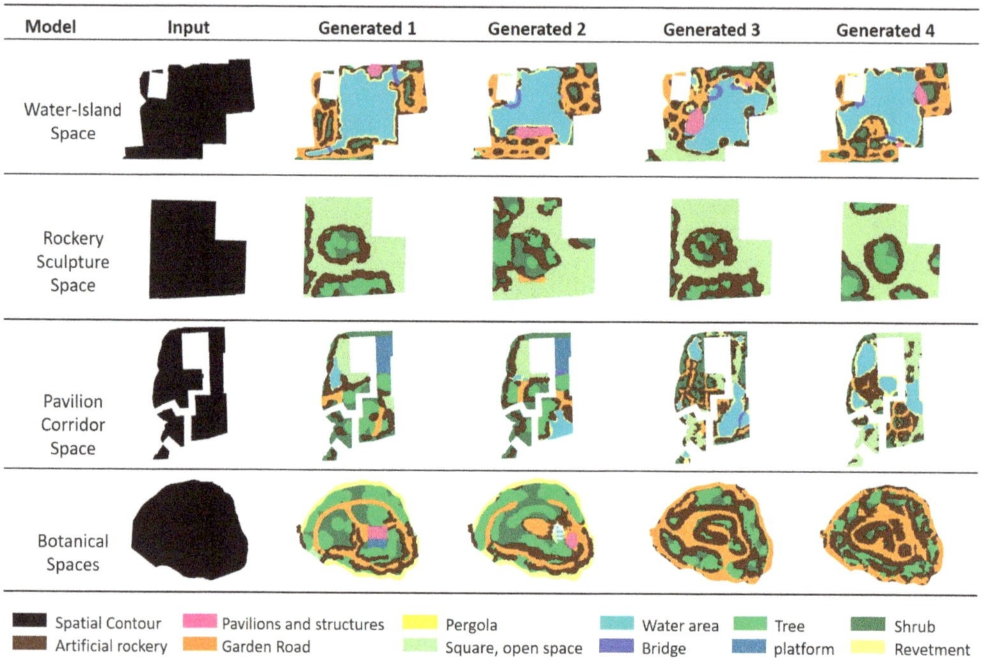

Figure 10. Multiple model generations.

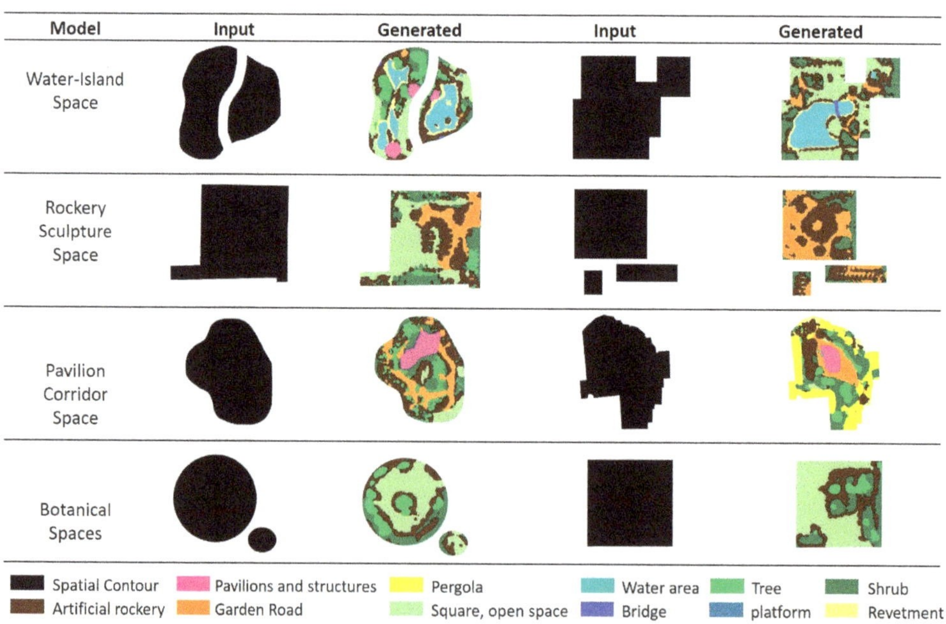

Figure 11. The morphological scheme generated by the model under the randomly drawn planar contour.

In terms of water-island space generation, the model demonstrated stable performance within the contour range. It was capable of generating different water body forms that conformed to the size of the space. In terms of the generation of rockery sculpture space and botanical spaces, the model combined flowers and trees with mountain stones to generate various forms of mountain stone and flower and tree combination spaces. It automatically adjusted the proportion of the rockery and plants according to the rock, flower, or tree theme. In terms of pavilion corridor space generation, the generation effect of element shapes was relatively poor. However, the layout framework between the building and its various elements could still be observed from it. It formed an architectural space with a pavilion as the core, surrounded by trees or embraced by corridors. The problem of random generation results of architectural models can be solved by generating more schemes and selecting a more reasonable option (Figure 11).

Overall, after the robustness test of the model and the random contour sample experiment, we determined that the generated floor plans conformed to the basic effect of creating a garden landscape space. In the future, the clarity of architectural forms must be further optimized. These test results prove the reliability and practical value of the model. It is capable of effectively generating layout schemes that meet the morphological characteristics of various landscape spaces.

4. Discussion: Application in Design Decision-Making

4.1. Simplify the Decision-Making and Design Process of the Scheme

The well-trained CGAN has shown significant application potential in the field of landscape floor planning and design. With the help of the trained CGAN model, it can assist designers or decision-makers in quickly generating planar schemes that conform to the local garden spatial structure, reducing the scheme design time.

The design and renovation of the Suzhou Gardens Conservation Area, a World Cultural Heritage site, demand that designers possess extensive professional knowledge and design experience. Designers or decision-makers require time to evaluate the characteristics of traditional garden spaces and must transform this knowledge into practical design experience. During the design process, preliminary preparations must be integrated with the extraction of traditional spatial features, and the scheme undergoes multiple revisions before finalization. Extracting these traditional spatial features necessitates a deep understanding of and rich experience with spatial composition, element types, combination patterns, and the proportion of elements. Consequently, the traditional design process is not only intricate but also highly dependent on the designer's accumulated experience, thus typically resulting in a lengthy design cycle of one to three months or even longer (Figure 12).

However, the generation mechanism of CGAN actuates the inheritance of the complex spatial characteristics of gardens through algorithmic operations. Its visual output not only decodes the spatial topological rules of regional gardens but also illustrates the spatial composition patterns under the influence of regional cultural contexts. By deploying the well-trained CGAN model, designers can spend less energy on studying the characteristics of classical gardens. It only takes 5 to 10 s to generate 30 to 50 different scheme floor plans using the CGAN model trained in this study. This has shortened the design cycle from the traditional 3 months to 10 s. This represents a shift from human-scale deliberation (weeks/months) to machine-speed synthesis (seconds), fundamentally transforming the design paradigm (Figure 13).

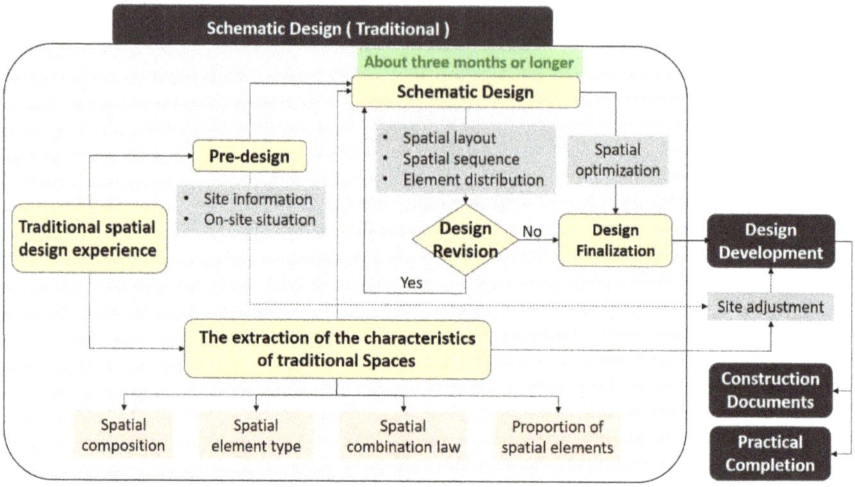

Figure 12. The traditional scheme design process.

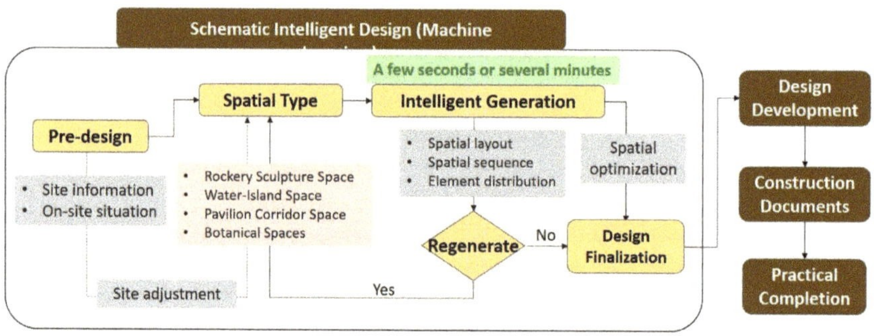

Figure 13. The design process generated by the CGAN scheme.

CGAN is not only a research medium for the spatial cognition of traditional culture, but also an efficient decision-making tool for scheme design. It can play a key role in shortening the scheme design cycle in historic site protection and renewal.

4.2. Multi-Scheme Comparison of the Planning Site

The model trained in this study supports the generation of multiple types of schemes. Designers or decision-makers can compare and select multiple schemes based on the characteristics and types of the site. This model can significantly shorten the design cycle while ensuring that the scheme maintains coherence with regional traditions in terms of spatial organization logic. With the help of this trained CGAN model, four different types of landscape spaces can be generated according to the characteristics of the spatial environment (rockery sculpture space, water-island space, pavilion gallery space, and botanical spaces).

(1) Different schemes are generated for comparison and selection at the same site.

The spatial size of the training samples for this model ranges from 200 to 1000 square meters. The floor plans generated by CGAN can be applied in the renewal of micro and small spaces in urban historic protection blocks; for example, micro-space updates (200–500 m^2), similar in size to pocket parks, street corners, and tiny green spaces. Small-scale space updates (500–800 m^2) are similar to community plazas and courtyards. The medium-scale space update (>800–1000 m^2) is similar to neighborhood parks. Decision-makers can select different types of models based on the size of the site space to generate multiple landscape space floor plans on the same site. Thus, decisions and comparisons can be made based on the actual site requirements (Figure 14).

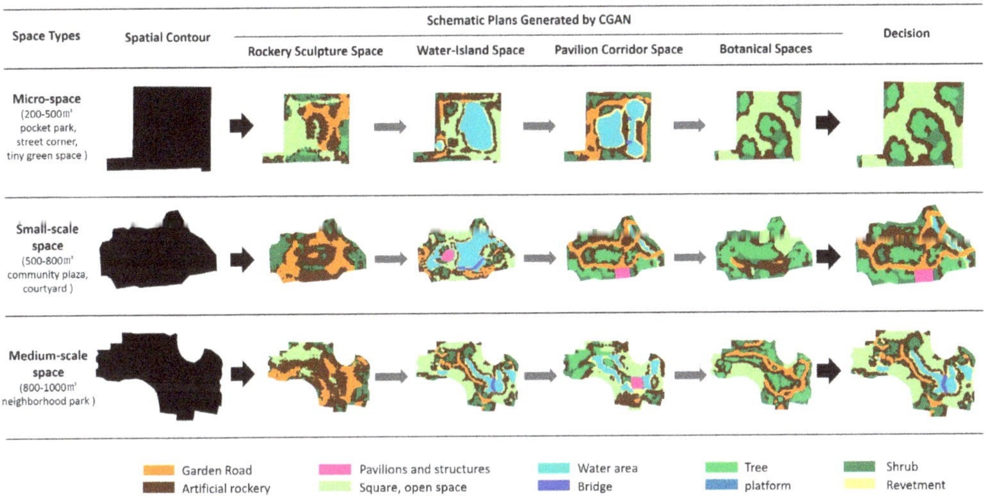

Figure 14. In the same site, different spatial types of schemes can be generated for comparison and selection.

(2) Different venues generate different scheme comparisons.

Against the complex background of urban renewal and public space planning, decision-makers often face the core predicament of the synchronous and overall planning of multiple sites. They must conduct simultaneous design evaluations for multiple spaces, such as pocket parks, community squares, block green spaces, etc. The traditional method relies on manual sequential analysis, resulting in low efficiency and difficulty in unifying standards. Meanwhile, the design floor plans for different venues are proposed by different teams, with significant differences in style, function, or cost. Therefore, decision-makers lack a standardized comparison framework and find it difficult to quantify the advantages and disadvantages (Figure 15).

With the CGAN, the specific difficulties of multi-site collaborative planning can be solved. The CGAN model trained in this study can automatically generate design schemes according to different spatial types with a unified style and differentiated landscape functions. The CGAN transforms multi-site planning from "empirical manual labor" to "multi-type intelligent generation." This not only effectively leverages the decision-makers' experience and efficiency but also intuitively assists in the generation of design decisions through visual floor plans. This technology can provide a scalable new paradigm for the fragmented space renewal of high-density historic and culturally protected areas.

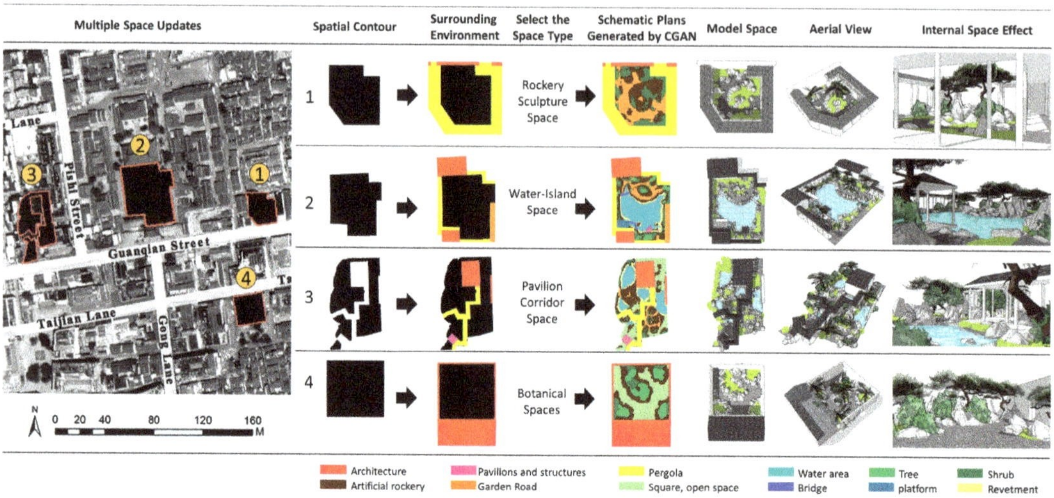

Figure 15. Different spatial types of schemes generated in multiple sites.

4.3. Evaluation of Space Utilization Rate

The scheme floor plan generated by CGAN marks landscape types in different colors (comprising 12 landscape elements such as architecture, pavilions and structures, pergolas, water areas, trees, etc.). The land-use ratio of each landscape element in the scheme can be calculated through the color proportions. Decision-makers can evaluate the space utilization rate of the generated floor plan by comparing it with the park design specifications based on the actual site type. Finally, the best scheme with relatively rich landscape elements that meets the design specifications is selected.

When multiple schemes are executed simultaneously, decision-makers can first create different spaces according to the size and outline of the site (rockery sculpture space, water-island space, pavilion gallery space) and spatial generation (botanical spaces). Subsequently, all the generated landscape elements and technical indicators are calculated individually. This way, the types and proportions of the landscape elements and the space utilization rate of each generated result can be obtained. Finally, by comparing the space utilization rate with the land-use indicators in the park design specifications, the scheme floor plan that meets the design specifications can be identified (Figure 16).

We take three project sites of different scales as examples. Among the generation results for the micro-space, the pavilion corridor space was the most abundant and conformed to the park design specifications. According to China's Code for the Design of Public Park (GB 51, 192–2016) [40], for small parks with an area of less than 2 hectares, the proportion of landscape greening (including water areas) should exceed 65%, the proportion of buildings should be less than 1.0%, and the land for garden paths and paving should be between 15% and 30%. The pavilion corridor space conformed best to the specifications (the proportion of landscape greening was 82.61%, the proportion of buildings was 0%, and the proportion of garden paths and paving was 17.39%).

Figure 16. The space utilization rate is calculated based on the landscape elements, from which the optimal scheme can be decided.

Meanwhile, according to the Park Design Code (2016), for community parks with an area of less than 2 hectares, the proportion of landscape greening (including water body areas) must be > 65%, and the proportion of buildings should be <3% (with the proportion of management buildings less than 0.5%, the proportion of service buildings less than 2.5%, and the land for garden paths and paving between 15% and 30%). Among the four results for the small-scale space, the water-island space agreed most with the norms (the proportion of landscape greening was 68.53%, the proportion of buildings 2.47%, and the proportion of garden paths and paving 29.01%).

Furthermore, among the results for the medium-scale space, the pavilion gallery space was the most abundant. It complied with the park design standards (landscape greening accounted for 71.69%, buildings for 2.91%, and garden paths and paving for 25.37%).

Thus, decision-makers can compare the space utilization rates of the different generated results, refer to the design specifications, and form the best design decision. In practical applications, they can select the appropriate spatial type. Based on the output results of the model and the requirements, they can prioritize the schemes with rich landscape elements and a high space utilization rate and evaluate and adjust the floor plan to meet the needs of urban renewal.

4.4. Satisfaction with the Generated Floor Plan

4.4.1. Investigation Methods and Verification

To further evaluate the satisfaction of designers with respect to the garden space floor plans generated using the CGAN, the researchers selected one group of typical garden spaces from each of the four spatial types, forming a total of four groups of comparison schemes (each group included one original scheme and one CGAN-generated scheme), for a total of eight design schemes as the investigation samples. The satisfaction survey was conducted from five evaluation perspectives: the rationality of the layout, the adaptability of the scale, the richness of the content, the aesthetic quality, and the sense of layering in the landscape (Figure 17).

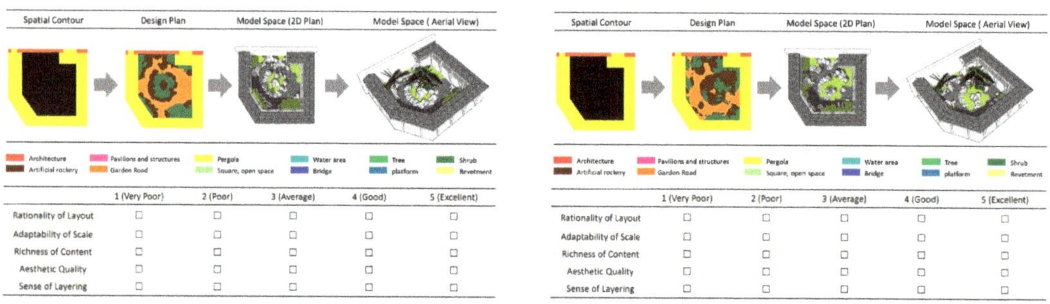

(a) (b)

Figure 17. Comparison of rockery sculpture spaces: (**a**) Floor Plan A; (**b**) Floor Plan B.

The survey was scored using a 5-point Likert scale: 5 points = very satisfied/very much in agreement; 4 points = relatively satisfied/relatively in agreement; 3 points = average/ acceptable; 2 points = not very satisfied/not quite in agreement; and 1 point = dissatisfied/ disagree. Under double-blind testing conditions (without revealing the source of the scheme), designers and decision-makers were invited to independently score the scheme based on their professional judgment and personal preferences (Appendix C). Comparing the score differences between each group of original schemes and the CGAN-generated schemes, the designers' acceptance of the AI-generated schemes was analyzed.

A total of 310 questionnaires were distributed in this study. After screening and eliminating 4 invalid questionnaires, 306 valid questionnaires were finally obtained. To ensure the reliability and consistency of the research data, we conducted reliability and validity analyses of the questionnaire using the SPSS 21 statistical software. Reliability is used to measure the stability of questionnaire results and the authenticity of the measured features. Cronbach's alpha coefficient is the most commonly used internal consistency evaluation index, which can effectively reflect the reliability level of the questionnaire [41]. The value of the standardized Cronbach's alpha coefficient ranges from 0 to 1; the closer the coefficient is to 1, the higher the consistency within the questionnaire and the more reliable the results. It is generally believed that a questionnaire with a standardized Cronbach's alpha coefficient value greater than 0.6 is sufficiently reliable [42].

The reliability results of this questionnaire are shown in Table 3. The Cronbach's alpha coefficient of all measurement dimensions was 0.924, which is greater than the standard of 0.6, indicating that the questionnaire had good overall reliability (Table 3).

Table 3. Reliability statistics.

Cronbach's Alpha	Cronbach's Alpha Based on Standardized Items	No. of Items
0.924	0.923	40

Furthermore, to ensure the validity of the questionnaire measurement results, the validity of the questionnaire was also tested. Validity reflects the extent to which the measurement tool can accurately assess the characteristics of the target. The validity level was evaluated by calculating the KMO (Kaiser–Meyer–Olkin) value. The value range of this index is 0–1. The closer the value is to 1, the better the structural validity of the questionnaire, and the more truly the measurement results can reflect the characteristics of the research object [43]. In this study, the validity analysis of the questionnaire mainly applied the KMO and Bartlett sphere tests to illustrate whether the internal structure of the questionnaire was valid. The KMO and Bartlett values are considered the best indicators of validity. When the KMO test coefficient is > 0.5 and the Bartlett significance level is <0.05, the questionnaire has structural validity [44].

According to the results in Table 4, the KMO value of this questionnaire was 0.902, Bartlett's approx. Chi-square was 5728.149, the degree of freedom (df) was 780, and Bartlett's significance (Sig) was 0.000, which is less than 0.05, indicating that this questionnaire has good authenticity and validity (Table 4).

Table 4. KMO and Bartlett's test results.

Kaiser–Meyer–Olkin Measure of Sampling Adequacy	Bartlett's Test of Sphericity		
	Approx. Chi-Square	df	Sig.
0.902	5728.149	780	0.000

4.4.2. Analysis of Investigation Results

(1) Overall satisfaction comparison

According to the statistical results of the total average value of each group of spatial evaluations, the average satisfaction score for all floor plans was 3.871, which is greater than the level of 3 (general). Among them, the average satisfaction scores generated by the CGAN model were mostly higher than the average scores for the original floor plan. In the floor plans generated by CGAN, the average satisfaction score for the water-island space was the highest at 4.080, while the satisfaction score for botanical spaces was relatively low, at 3.659. In the original floor plan, the average satisfaction score for the water-island space was the highest, at 4.006, while for the botanical spaces it was the lowest, at 3.628 (Table 5).

Table 5. Item statistics.

	Name of Shape Group	Source	Mean	Std. Deviation	No. of Items
1	Rockery sculpture space	Authentic garden space	3.882	1.007	306
		CGAN model's design	3.712	1.078	306
2	Water-island space	Authentic garden space	4.006	0.969	306
		CGAN model's design	4.083	0.943	306
3	Pavilion corridor space	Authentic garden space	3.943	0.974	306
		CGAN model's design	4.063	0.966	306
4	Botanical spaces	Authentic garden space	3.620	1.156	306
		CGAN model's design	3.659	1.132	306

According to the comparison scores for the different type groups, the average scores of the three shape groups in the floor plan generated by the CGAN model were all higher than the scores of the original floor plan. In the rockery sculpture space comparison group, the average satisfaction score of the CGAN-generated planar floor plan was slightly lower than that of the original planar floor plan (3.882) and the CGAN-generated planar floor plan (3.712) (Figure 18).

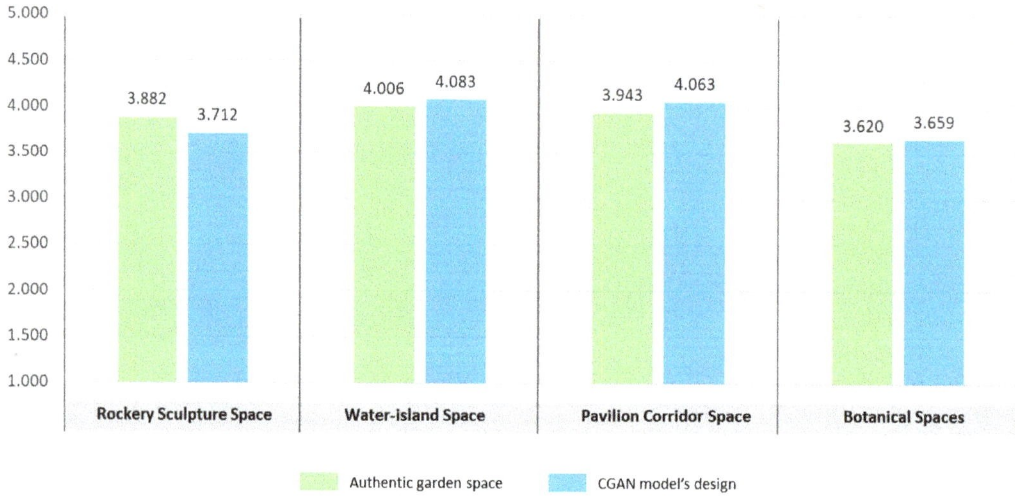

Figure 18. Average satisfaction scores in the four groups of floor plans.

(2) Analysis of performance in each dimension

Rationality of Layout: In the CGAN scheme, the water-island space (4.02) and the space around the building (4.00) scored the highest, and the performance in the rock space (3.62) was relatively weak.

Adaptability of Scale: The water-island space (CGAN scheme: 4.20) and the space around the building (CGAN scheme: 4.20) performed the best, indicating that CGAN had more advantages in coordinating the spatial proportion.

Richness of Content: The water-island space (CGAN: 4.14) received the highest score, while the flower and tree space (CGAN: 3.73) and the rock space (CGAN: 3.82) received lower scores.

Aesthetic Quality: Among all types, the water-island space (CGAN scheme: 3.96) and the space around the building (CGAN scheme: 3.92) had relatively high aesthetic evaluations, but the flower and tree space (CGAN scheme: 3.49) and the rock space (CGAN scheme: 3.52) performed relatively weakly.

Sense of Layering: The CGAN scheme performed well in terms of sense of layering in the water-island space (4.09) and the space around the building (4.07), but needs to be strengthened in the rock space (3.72) and in the design of spatial depth (Figure 19).

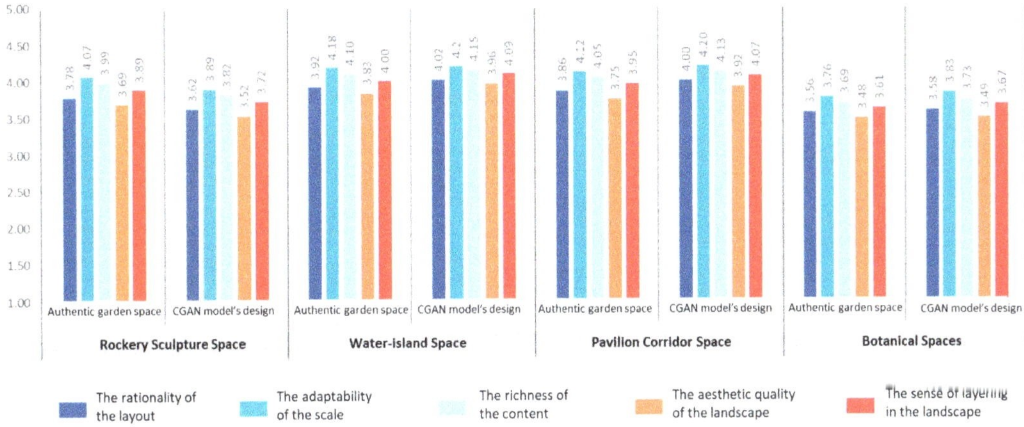

Figure 19. Satisfaction scores in five dimensions for the four groups of floor plans.

The comprehensive analysis indicated that the garden floor plan schemes generated by CGAN performed exceptionally well in terms of overall satisfaction, with an average score (3.879) higher than the authentic garden space (3.863).

Among them, the CGAN design schemes for the water-island space and the pavilion corridor space had the highest satisfaction rates (4.083 and 4.063, respectively). In dimensions such as scale coordination (4.20), content richness (4.14), and layering (4.09), CGAN has prominent advantages, demonstrating the potential of AI in optimizing spatial proportion and landscape structure. However, satisfaction with the CGAN scheme for the rockery sculpture space (3.712) was slightly lower than that for the original design (3.882), and it performed poorly in terms of layout rationality and aesthetic appeal. This indicates that simulating complex natural forms with the algorithm needs to be improved. Furthermore, the overall evaluation of the botanical spaces was relatively low (CGAN scheme: 3.659), reflecting the limitations of the existing models in terms of capturing the content richness and visual aesthetics of the botanical spaces.

Future research will focus on optimizing the layout logic of rock and mountain spaces and enhancing the content richness of flower and tree spaces. At the same time, it will maximize CGAN's technical advantages in water areas and the spaces around buildings to comprehensively promote the application of AI-assisted design in landscape planning.

5. Conclusions

This research studied the Suzhou Gardens, a World Cultural Heritage Site, with the aim of exploring innovative methods using artificial intelligence technology to assist in decision-making regarding the design of garden spaces. It provides efficient and scientific tools for the protection and renewal of traditional gardens by urban decision-makers restricted by the limitations of current garden design, such as reliance on manual experience, low efficiency, and insufficient continuity of traditional styles. This study proposes an intelligent generation framework based on CGAN to enhance the feasibility and innovativeness of garden space design decisions.

The main innovations of this research are reflected in the following three aspects: Firstly, we analyzed the core characteristics of traditional garden spaces. According to these characteristics, the four main spatial types that constitute Suzhou Gardens were systematically summarized, including rockery sculpture space, water-island space, pavilion gallery space, and botanical spaces. Meanwhile, color annotations were applied to the 12 main landscape space elements. Secondly, a deep learning architecture based on CGAN was designed. This model adopted the adversarial training mechanism, including a generator and a discriminator. In terms of the training strategy, the "contour atlas" was taken as the conditional sample and the "spatial element layout atlas" as the generation sample to train the layout generation of spatial elements under the control of contour conditions. By observing the loss logs (loss values) during the training process and the test images in the iteration, we ensured that the generated results not only conformed to the data distribution but also maintained spatial rationality. Finally, we applied the proposed model to generate design decisions. Application methods such as the scheme design process assisted by machine learning, the multi-scheme comparison and selection mode, and the space utilization rate evaluation were proposed, providing designers with an efficient and multidimensional design decision-making basis.

The main conclusions of this study are as follows:

(1) The CGAN model can effectively capture the key layout characteristics of the garden space. Four different types of landscape spaces, namely, rockery sculpture space, water-island space, pavilion gallery space, and botanical spaces, can be generated according to the characteristics of the spatial environment.
(2) This model can be applied to projects with space sizes ranging from 200 to 1000 square meters. The plans generated by CGAN can be applied in the renewal of micro and small spaces in urban historic conservation blocks.
(3) After 500 epochs of training, the CGAN model can accurately learn the organizational rules of garden spaces, and the average accuracy rate of the generated planar schemes reached 91.08%.
(4) The efficiency of the scheme generation was significantly improved. Compared with traditional design, which relies on experience and takes 1 to 3 months, 30 to 50 floor plans can be generated within 3 to 10 s on the same site. This has greatly improved the design efficiency.
(5) CGAN can provide multiple scheme comparisons for decision-making in urban historic space renewal, helping managers to select the optimal solution in a short period of time.
(6) The land-use proportions of 12 spatial elements can be quickly calculated based on the color of the element annotations. From this, the design specifications can be referred to for a quantitative assessment of the generated space utilization rate. Among the generated floor plans, the proportion of landscape greening basically met the specification requirements (>65%).
(7) The floor plan generated by CGAN was slightly higher than the original floor plan in terms of the overall average satisfaction. The average satisfaction with the CGAN scheme (3.879) was higher than with the authentic garden space (3.863).

This study can help decision-makers efficiently generate and evaluate the feasibility of different design schemes, providing intelligent support for decision-making in urban renewal plans. However, this model remains limited in depicting small-scale details, and the data and algorithms need to be further optimized. The following deficiencies remain:

(1) The detailed processing ability of the model needs to be improved to generate clear and realistic image results.
(2) The model still faces certain challenges when dealing with the planar contours of curved surfaces and requires further improvement and optimization.
(3) There is room for improvement in the clarity and cleanliness of the generated images.
(4) The layout logic of building and rockery spaces must be optimized, the content richness of botanical spaces enhanced, and CGAN's advantages in water areas consolidated.

In summary, generative adversarial networks (GANs) have shown great potential in the field of landscape design decision-making due to their powerful image generation capabilities. The significance of this research relates to providing an intelligent technical path for the protective development of historic gardens. In particular, the digital inheritance of traditional construction wisdom can be achieved through machine learning methods. On the other hand, the decision support system constructed in this study can assist managers in quickly evaluating the feasibility of different strategies in the early stage of scheme design. It can effectively balance the contradiction between protection and development.

In the future, researchers will further improve and optimize the CGAN model to enhance its robustness and stability. Existing planar generation results can also be combined with 3D model generation tools to generate 3D model spaces, offering the functional expansion from floor plan generation to 3D model generation. We can explore methods of collaborative design decision making between models and designers to further improve the design quality and achieve better human–computer interactions. This technology can be further extended to the field of protection and renewal of historic districts in other cities, providing intelligent support for urban planning decisions.

Supplementary Materials: The following supporting information can be downloaded at: https://www.mdpi.com/article/10.3390/buildings15142401/s1.

Author Contributions: Conceptualization, L.Y. and Y.C.; methodology, L.Y. and L.Z.; software, L.Y.; validation, L.Y. and L.Z.; formal analysis, L.Y., Y.Z. and L.Z.; investigation, L.Y., Y.Z. and X.J.; resources, L.Y., Y.Z. and X.J.; data curation, L.Y. and Y.Z.; writing—original draft preparation, L.Y.; writing—review and editing, L.Y. and Y.C.; visualization, L.Y. and X.J.; supervision, Y.C.; project administration, Y.C.; funding acquisition, L.Z. All authors have read and agreed to the published version of the manuscript.

Funding: This research was supported by the Faculty Research Grants funded by Macau University of Science and Technology (grant number: FRG-25-041-FA). The funders had no role in study conceptualization, data curation, formal analysis, methodology, software, decision to publish, or preparation of the manuscript. There was no additional external funding received for this study.

Data Availability Statement: The datasets used and analyzed during the current study are available from the corresponding author upon reasonable request.

Conflicts of Interest: Author Yi Zhang was employed by the company Shanghai GOODLINKS International Design Group. The remaining authors declare that the research was conducted in the absence of any commercial or financial relationships that could be construed as a potential conflict of interest.

Abbreviations

The following abbreviations are used in this manuscript:

CGAN	Conditional Generative Adversarial Network
GAN	Generative Adversarial Network

Appendix A

Machine learning environment configuration: The operating system was Windows 10 (X64), the Cuda version was 11.7, the deep-learning framework was Pytorch, the graphics card was NVIDIA GeForce RTX 3060 (32 G) (NVIDIA, Santa Clara, CA, USA), and the processor was 12th Gen Intel(R) Core(TM) i5-12490F (3.00 GHz) (Intel, Santa Clara, CA, USA).

Appendix B

Main code (train and test):

```
# -*- coding: utf-8 -*-
import os
name = '***you name'
dataroot = './datasets/*** you name/'
size = '512'
cmd = 'python train.py --name {} --dataroot {} --no_instance --label_nc 0 --loadSize {} --fineSize {} --resize_or_crop none --niter 500 --niter_decay 150 --save_epoch_freq 50 --no_flip'.format(name, dataroot, size, size)
#cmd = 'python test.py --name {} --dataroot {} --no_instance --label_nc 0 --loadSize {} --fineSize {} --resize_or_crop none --how_many 2000'.format(name, dataroot, size, size)
os.environ["PYTORCH_CUDA_ALLOC_CONF"] = "max_split_size_mb:256"
res = os.popen(cmd)
output_str = res.read()
print(output_str)
print(' Task completed!')
# The code related to the CGAN model is in Supplementary Materials.
```

Appendix C

Appendix C.1

AI-Assisted Design Generation-Satisfaction Survey on Jiangnan Classical Garden Spatial Layout

Dear participant,

This survey aims to evaluate public satisfaction with AI-generated Jiangnan classical garden spatial layouts. Please rate the following designs based on your impressions.

The research team has categorized the designs into four landscape features: Rockeries & Mountains, Water & Islands, Architecture, Plants & Flowers. The satisfaction survey for the design proposals will be conducted from the following five evaluation dimensions: Layout rationality, Scale adaptability, Content richness, Landscape aesthetic quality, Spatial layering perception.

Each group includes two design schemes: One is a real classical garden design. The other is AI-generated.

To ensure objectivity, the source (human/AI) of each scheme will not be disclosed. Please evaluate them solely based on your satisfaction.

Survey Details: Anonymous: Only basic demographic data (e.g., gender, age) will be collected. Data Protection: Encrypted storage, compliant with China's PIPL and EU's GDPR.

Thank you for your participation!

Rating Scale (1–5): 5 = Very Satisfied (Excellent), 4 = Satisfied (Good), 3 = Neutral (Average), 2 = Dissatisfied (Poor), 1 = Very Dissatisfied (Very Poor). Please respond based on your genuine preferences.

Group 1: Comparison of rockery sculpture space

1. Rockery sculpture space (Floor Plan A)

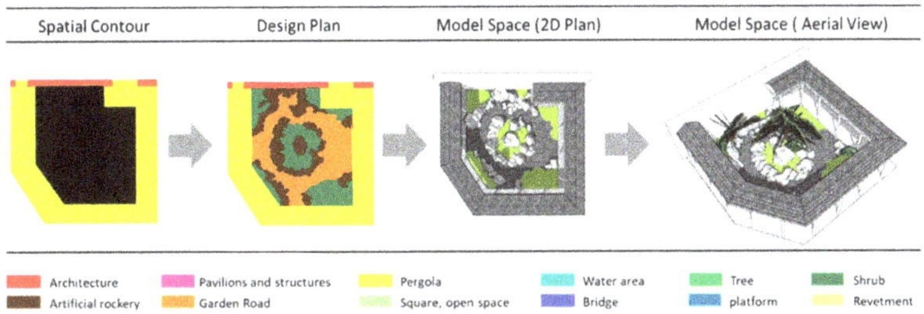

	1 (Very Poor)	2 (Poor)	3 (Average)	4 (Good)	5 (Excellent)
Rationality of Layout	☐	☐	☐	☐	☐
Adaptability of Scale	☐	☐	☐	☐	☐
Richness of Content	☐	☐	☐	☐	☐
Aesthetic Quality	☐	☐	☐	☐	☐
Sense of Layering	☐	☐	☐	☐	☐

2. Rockery sculpture space (Floor Plan B)

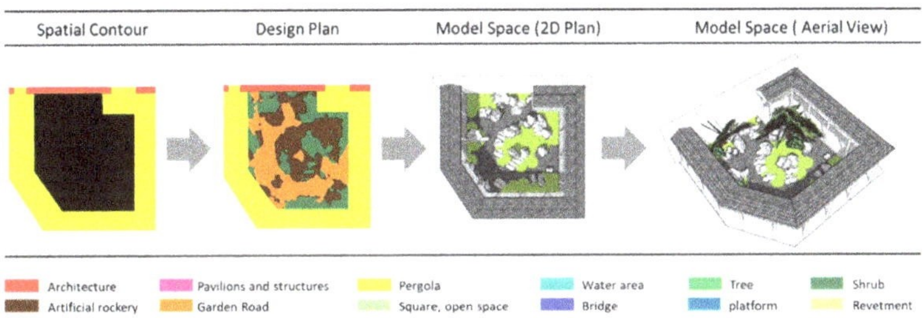

	1 (Very Poor)	2 (Poor)	3 (Average)	4 (Good)	5 (Excellent)
Rationality of Layout	☐	☐	☐	☐	☐
Adaptability of Scale	☐	☐	☐	☐	☐
Richness of Content	☐	☐	☐	☐	☐
Aesthetic Quality	☐	☐	☐	☐	☐
Sense of Layering	☐	☐	☐	☐	☐

Group 2: Comparison of water-island space

3. Water-island space (Floor Plan A)

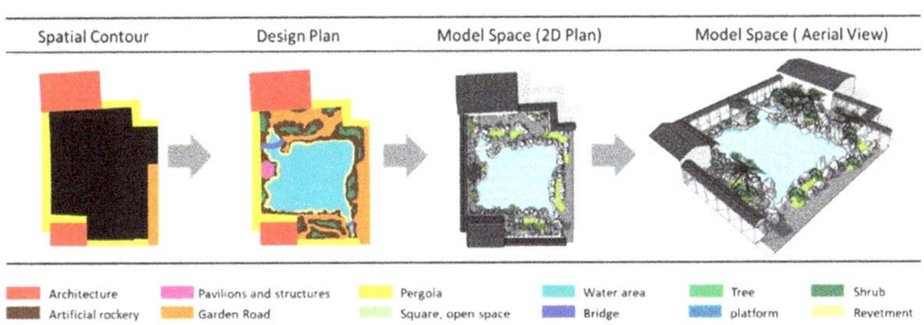

	1 (Very Poor)	2 (Poor)	3 (Average)	4 (Good)	5 (Excellent)
Rationality of Layout	☐	☐	☐	☐	☐
Adaptability of Scale	☐	☐	☐	☐	☐
Richness of Content	☐	☐	☐	☐	☐
Aesthetic Quality	☐	☐	☐	☐	☐
Sense of Layering	☐	☐	☐	☐	☐

4. Water-island space (Floor Plan B)

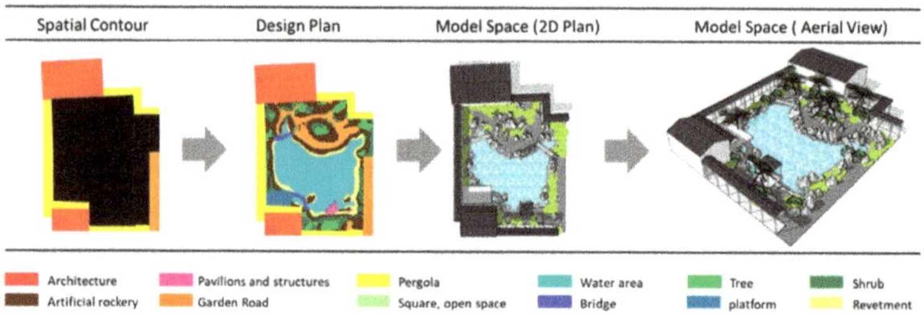

	1 (Very Poor)	2 (Poor)	3 (Average)	4 (Good)	5 (Excellent)
Rationality of Layout	☐	☐	☐	☐	☐
Adaptability of Scale	☐	☐	☐	☐	☐
Richness of Content	☐	☐	☐	☐	☐
Aesthetic Quality	☐	☐	☐	☐	☐
Sense of Layering	☐	☐	☐	☐	☐

Group 3: Comparison of pavilion corridor space

5. Pavilion corridor space (Floor Plan A)

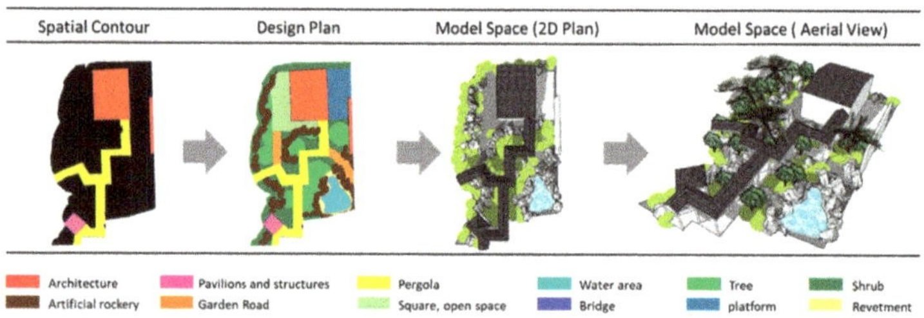

	1 (Very Poor)	2 (Poor)	3 (Average)	4 (Good)	5 (Excellent)
Rationality of Layout	☐	☐	☐	☐	☐
Adaptability of Scale	☐	☐	☐	☐	☐
Richness of Content	☐	☐	☐	☐	☐
Aesthetic Quality	☐	☐	☐	☐	☐
Sense of Layering	☐	☐	☐	☐	☐

6. Pavilion corridor space (Floor Plan B)

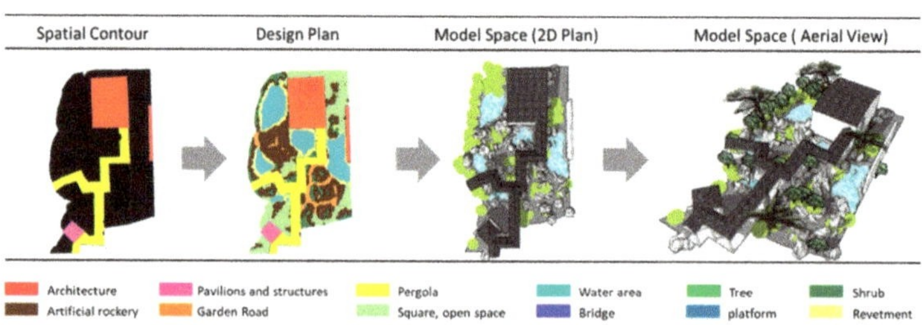

	1 (Very Poor)	2 (Poor)	3 (Average)	4 (Good)	5 (Excellent)
Rationality of Layout	☐	☐	☐	☐	☐
Adaptability of Scale	☐	☐	☐	☐	☐
Richness of Content	☐	☐	☐	☐	☐
Aesthetic Quality	☐	☐	☐	☐	☐
Sense of Layering	☐	☐	☐	☐	☐

Group 4: Comparison of botanical spaces

7. Botanical spaces (Floor Plan A)

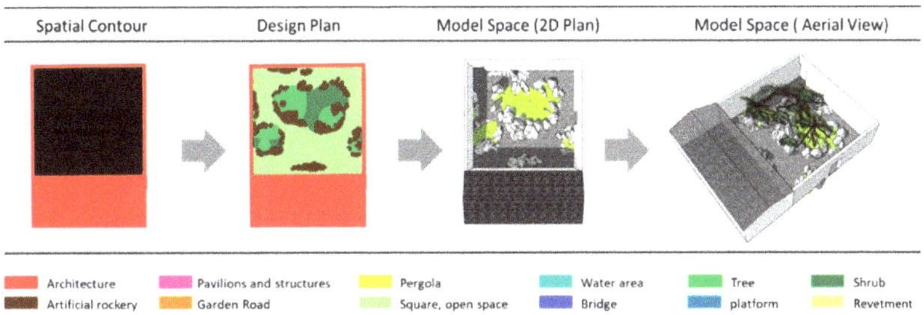

	1 (Very Poor)	2 (Poor)	3 (Average)	4 (Good)	5 (Excellent)
Rationality of Layout	☐	☐	☐	☐	☐
Adaptability of Scale	☐	☐	☐	☐	☐
Richness of Content	☐	☐	☐	☐	☐
Aesthetic Quality	☐	☐	☐	☐	☐
Sense of Layering	☐	☐	☐	☐	☐

8. Botanical spaces (Floor Plan B)

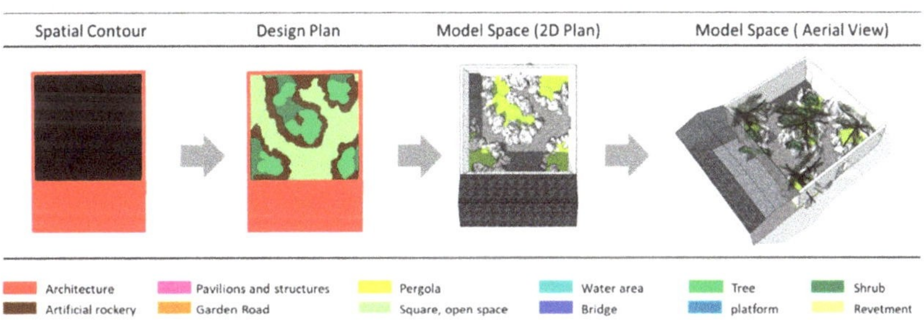

	1 (Very Poor)	2 (Poor)	3 (Average)	4 (Good)	5 (Excellent)
Rationality of Layout	☐	☐	☐	☐	☐
Adaptability of Scale	☐	☐	☐	☐	☐
Richness of Content	☐	☐	☐	☐	☐
Aesthetic Quality	☐	☐	☐	☐	☐
Sense of Layering	☐	☐	☐	☐	☐

9. Gender: ☐ Male, ☐ Female

10. Age group: ☐ under 18, ☐ 18~25, ☐ 26~30, ☐ 31~40, ☐ 41~50, ☐ 51~60, ☐ Over 60.

Appendix C.2

Table A1. Summary of survey results.

Types	Evaluation	Mean	Std. Deviation
1. Rockery sculpture space (A)	Rationality of Layout	3.78	1.046
	Adaptability of Scale	4.07	0.946
	Richness of Content	3.99	0.968
	Aesthetic Quality	3.69	1.067
	Sense of Layering	3.89	1.010
2. Rockery sculpture space (B)	Rationality of Layout	3.62	1.084
	Adaptability of Scale	3.89	1.033
	Richness of Content	3.82	1.033
	Aesthetic Quality	3.52	1.131
	Sense of Layering	3.72	1.108
3. Water-island space (A)	Rationality of Layout	3.92	1.000
	Adaptability of Scale	4.18	0.905
	Richness of Content	4.10	0.923
	Aesthetic Quality	3.83	1.013
	Sense of Layering	4.00	1.003
4. Water-island space (B)	Rationality of Layout	4.02	0.946
	Adaptability of Scale	4.20	0.903
	Richness of Content	4.14	0.903
	Aesthetic Quality	3.96	1.022
	Sense of Layering	4.09	0.938
5. Pavilion corridor space (A)	Rationality of Layout	3.86	1.008
	Adaptability of Scale	4.12	0.916
	Richness of Content	4.05	0.950
	Aesthetic Quality	3.75	1.031
	Sense of Layering	3.95	0.965
6. Pavilion corridor space (B)	Rationality of Layout	4.00	0.978
	Adaptability of Scale	4.20	0.871
	Richness of Content	4.13	1.006
	Aesthetic Quality	3.92	1.001
	Sense of Layering	4.07	0.973
7. Botanical spaces (A)	Rationality of Layout	3.56	1.156
	Adaptability of Scale	3.76	1.162
	Richness of Content	3.69	1.167
	Aesthetic Quality	3.48	1.163
	Sense of Layering	3.61	1.132
8. Botanical spaces (B)	Rationality of Layout	3.58	1.160
	Adaptability of Scale	3.83	1.097
	Richness of Content	3.73	1.137
	Aesthetic Quality	3.49	1.140
	Sense of Layering	3.67	1.125

References

1. Wang, H.; Shen, Q.; Tang, B.S.; Lu, C.; Peng, Y.; Tang, L. A framework of decision-making factors and supporting information for facilitating sustainable site planning in urban renewal projects. *Cities* **2014**, *40*, 44–55. [CrossRef]
2. Yang, H. *A Treatise on the Garden of Jiangnan: A study on the Art of Chinese Classical Garden*, 1st ed.; Springer Nature: Berlin/Heidelberg, Germany, 2022. [CrossRef]
3. Li, T.; Pintong, S. A Study on Innovative Strategies for Spatial Renewal Design in the Context of Rapid Urbanization: A Case Study of Nanjing Menxi Area. *Int. J. Sociol. Anthropol. Sci. Rev.* **2025**, *5*, 471–480. [CrossRef]
4. Huang, Y.; Yang, S. Machine Learning Model for Building Type Classification of Cultural Heritage Sites along Jiangnan Canal: A Comparative Study of Historical and Modern Images. *Int. J. Des. Soc.* **2024**, *18*, 77. [CrossRef]
5. Zhang, R.; Zhao, Y.; Kong, J.; Cheng, C.; Liu, X.; Zhang, C. Intelligent recognition method of decorative openwork windows with sustainable application for Suzhou traditional private gardens in China. *Sustainability* **2021**, *13*, 8439. [CrossRef]

6. Yan, L.; Chen, Y.; Zheng, L.; Zhang, Y. Application of computer vision technology in surface damage detection and analysis of shedthin tiles in China: A case study of the classical gardens of Suzhou. *Herit. Sci.* **2024**, *12*, 72. [CrossRef]
7. Chen, X.; Yu, H.; Xiong, R.; Ye, Y. Construction of an analytical framework for spatial indicator of Chinese classical gardens based on Space Syntax and machine learning. *Landsc. Archit.* **2024**, *31*, 123–131. [CrossRef]
8. Zeng, Q.; Xu, J.; Chen, J. The Modular Information Fusion Model of Landscape Based on Machine Learning Algorithm. In Proceedings of the 2021 3rd International Conference on Artificial Intelligence and Advanced Manufacture, Wuxi, China, 10–12 September 2021; pp. 1695–1698. [CrossRef]
9. Yan, L.; Chen, Y.; Zheng, L.; Zhang, Y.; Liang, X.; Zhu, C. Intelligent Generation Method and Sustainable Application of Road Systems in Urban Green Spaces: Taking Jiangnan Gardens as an Example. *Int. J. Environ. Res. Public Health* **2023**, *20*, 3158. [CrossRef]
10. Sun, C.; Jiang, Z.; Yu, B. How to interpret Jiangnan gardens: A study of the spatial layout of Jiangnan gardens from the perspective of fractal geometry. *Herit. Sci.* **2024**, *12*, 353. [CrossRef]
11. Liu, Y.; Fang, C.; Yang, Z.; Wang, X.; Zhou, Z.; Deng, Q.; Liang, L. Exploration on machine learning layout generation of Chinese private garden in Southern Yangtze. In *Proceedings of the 2021 DigitalFUTURES: The 3rd International Conference on Computational Design and Robotic Fabrication (CDRF 2021) 3*; Springer: Singapore, 2021; pp. 35–44. [CrossRef]
12. Chen, R.; Zhao, J.; Yao, X.; Jiang, S.; He, Y.; Bao, B.; Wang, C. Generative design of outdoor green spaces based on generative adversarial networks. *Buildings* **2023**, *13*, 1083. [CrossRef]
13. Pan, X.; Lin, Q.; Ye, S.; Li, L.; Guo, L.; Harmon, B. Deep learning based approaches from semantic point clouds to semantic BIM models for heritage digital twin. *Herit. Sci.* **2024**, *12*, 65. [CrossRef]
14. Cui, Y. Research on garden landscape reconstruction based on geographic information system under the background of deep learning. *Acta Geophys.* **2023**, *71*, 1491–1513. [CrossRef]
15. Xing, Y.; Gan, W.; Chen, Q. Artificial intelligence in landscape architecture: A survey. *Int. J. Mach. Learn. Cybern.* **2025**, 1–26. [CrossRef]
16. Chen, X. Environmental landscape design and planning system based on computer vision and deep learning. *J. Intell. Syst.* **2023**, *32*, 20220092. [CrossRef]
17. Tao, L. Machine Learning Based Landscape Garden Plan Analysis and Rendering Application. *J. Inf. Sci. Eng.* **2025**, *41*, 279–295. [CrossRef]
18. Zhao, B.; Zheng, H.; Cheng, X. A Machine Learning Approach to Predict Site Selection from the Perspective of Vitality Improvement. *Land* **2023**, *13*, 2113. [CrossRef]
19. Xu, L. Value Assessment and Renovation Design of Buildings in the Perspective of Urban Renewal: Case Studies in China and Italy. 2023. Available online: https://hdl.handle.net/11577/3474288 (accessed on 24 June 2025).
20. Wu, T.; Lin, D.; Chen, Y.; Wu, J. Integrating Street View Images, Deep Learning, and sDNA for Evaluating University Campus Outdoor Public Spaces: A Focus on Restorative Benefits and Accessibility. *Land* **2025**, *14*, 610. [CrossRef]
21. Ye, Y.; Zeng, W.; Shen, Q.; Zhang, X.; Lu, Y. The visual quality of streets: A human-centred continuous measurement based on machine learning algorithms and street view images. *Environ. Plan. B Urban Anal. City Sci.* **2019**, *46*, 1439–1457. [CrossRef]
22. Yang, B.; Volkman, N.J. From traditional to contemporary: Revelations in Chinese garden and public space design. *Urban Des. Int.* **2010**, *15*, 208–220. [CrossRef]
23. Gu, L. Trends in Chinese Garden-making: The Qianlong Emperor (1711–99) and Gardens of Jiangnan. *Gard. Hist.* **2018**, *46*, 184–195.
24. Rinaldi, B.M. *The Chinese Garden: Garden Types for Contemporary Landscape Architecture*, 1st ed.; Walter de Gruyter: Berlin, Germany, 2012. [CrossRef]
25. Zheng, J. Art and the shift in garden culture in the Jiangnan Area in China (16th–17th Century). *Asian Cult. Hist.* **2013**, *5*, 1. [CrossRef]
26. Kai, G. *Literati Gardens of the Jiangnan Region: Characters and Mutations*, 1st ed.; Routledge: London, UK; New York, NY, USA, 2022; pp. 102–120. [CrossRef]
27. Liu, K.; Wang, Y.; Yang, R.; Xian, Z.; Takeda, S.; Zhang, J.; Xing, S. Interpreting the space characteristics of everyday heritage gardens of Suzhou, China, through a space syntax approach. *J. Asian Archit. Build. Eng.* **2024**, 1–26. [CrossRef]
28. Mirza, M.; Osindero, S. Conditional generative adversarial nets. *arXiv* **2014**, arXiv:1411.1784. [CrossRef]
29. Dosovitskiy, A.; Beyer, L.; Kolesnikov, A.; Weissenborn, D.; Zhai, X.; Unterthiner, T.; Dehghani, M.; Minderer, M.; Heigold, G.; Gelly, S.; et al. An Image is Worth 16x16 Words: Transformers for Image Recognition at Scale. *arXiv* **2020**, arXiv:2010.11929. [CrossRef]
30. Gupta, K.; Achille, A.; Lazarow, J.; Davis, L.; Mahadevan, V.; Shrivastava, A. Layout Generation and Completion with Self-attention. *arXiv* **2021**, arXiv:2006.14615. [CrossRef]
31. Liu, B.; Zhu, Y.; Song, K.; Elgammal, A. Towards Faster and Stabilized GAN Training for High-fidelity Few-shot Image Synthesis. *arXiv* **2021**, arXiv:2101.04775. [CrossRef]

32. Zhang, H.; Sindagi, V.; Patel, V.M. Image de-raining using a conditional generative adversarial network. *IEEE Trans. Circuits Syst. Video Technol.* **2019**, *30*, 3943–3956. [CrossRef]
33. Ding, X.; Wang, Y.; Xu, Z.; Welch, W.J.; Wang, Z.J. Ccgan: Continuous conditional generative adversarial networks for image generation. In Proceedings of the International Conference on Learning Representations, Virtual Event, Austria, 3–7 May 2021; Available online: https://openreview.net/pdf?id=PrzjugOsDeE (accessed on 6 July 2025).
34. Yang, J.; Liu, J.; Xie, J.; Wang, C.; Ding, T. Conditional GAN and 2-D CNN for bearing fault diagnosis with small samples. *IEEE Trans. Instrum. Meas.* **2021**, *70*, 3525712. [CrossRef]
35. Zhang, D.; Islam, M.M.; Lu, G. A review on automatic image annotation techniques. *Pattern Recognit.* **2012**, *45*, 346–362. [CrossRef]
36. Abu-Srhan, A.; Abushariah, M.A.; Al-Kadi, O.S. The effect of loss function on conditional generative adversarial networks. *J. King Saud Univ.-Comput. Inf. Sci.* **2022**, *34*, 6977–6988. [CrossRef]
37. Zheng, L.; Chen, Y.; Yan, L.; Zheng, J. The Impact of High-Density Urban Wind Environments on the Distribution of COVID-19 Based on Machine Learning: A Case Study of Macau. *Buildings* **2023**, *13*, 1711. [CrossRef]
38. Wu, L.; Tian, F.; Xia, Y.; Fan, Y.; Qin, T.; Jian-Huang, L.; Liu, T.Y. Learning to teach with dynamic loss functions. In Proceedings of the Advances in Neural Information Processing Systems, Montréal, QC, Canada, 3–8 December 2018; p. 31. [CrossRef]
39. Thekumparampil, K.K.; Khetan, A.; Lin, Z.; Oh, S. Robustness of conditional gans to noisy labels. In Proceedings of the Advances in Neural Information Processing Systems, Montréal, QC, Canada, 2–8 December 2018; p. 31. [CrossRef]
40. *GB 51192-2016*; Ministry of Housing and Urban-Rural Development of the People's Republic of China. Code for the Design of Public Park. China Building Industry Publishing: Beijing, China, 2016; p. 8.
41. Shrestha, N. Factor analysis as a tool for survey analysis. *Am. J. Appl. Math. Stat.* **2021**, *9*, 4–11. [CrossRef]
42. Everitt, B.; Dunn, G. *Applied Multivariate Data Analysis*; Arnold: London, UK, 2001; Volume 2. [CrossRef]
43. Tabachnick, B.G.; Fidell, L.S.; Ullman, J.B. *Using Multivariate Statistics*; Pearson: Boston, MA, USA, 2013; Volume 6, pp. 497–516.
44. Verma, J.P. *Data Analysis in Management with SPSS Software*; Springer Science & Business Media: Berlin/Heidelberg, Germany, 2012. [CrossRef]

buildings

MDPI

Article

Building Instance Extraction via Multi-Scale Hybrid Dual-Attention Network

Qingqing Hu [1], Yiran Peng [2], Chi Zhang [1], Yunqi Lin [1], KinTak U [2,*] and Junming Chen [1,*]

1 Faculty of Humanities and Arts, Macau University of Science and Technology, Macau 999078, China; 3230005309@student.must.edu.mo (Q.H.); 3240003803@student.must.edu.mo (C.Z.); 2250028261@student.must.edu.mo (Y.L.)

2 Faculty of Innovation Engineering, Macau University of Science and Technology, Macau 999078, China; 3230002514@student.must.edu.mo

* Correspondence: ktu@must.edu.mo (K.U.); jmchen@must.edu.mo (J.C.)

Abstract: Accurate building instance segmentation from high-resolution remote sensing images remains challenging due to complex urban scenes featuring occlusions, irregular building shapes, and heterogeneous textures. To address these issues, we propose a novel Multi-Scale Hybrid Dual-Attention Network (MS-HDAN), which integrates a dual-stream encoder, multi-scale feature extraction, and a hybrid attention mechanism. Specifically, the encoder is designed with a Local Feature Extraction Pathway (LFEP) and a Global Context Modeling Pathway (GCMP), enabling simultaneous capture of structural details and long-range semantic dependencies. A Local-Global Collaborative Perception Enhancement Module (LG-CPEM) is introduced to fuse the outputs from both streams, enhancing contextual representation. The decoder adopts a hierarchical up-sampling structure with skip connections and incorporates a dual-attention module to refine boundary-level details and suppress background noise. Extensive experiments on benchmark urban building datasets demonstrate that MS-HDAN significantly outperforms existing state-of-the-art methods, particularly in handling densely distributed and structurally complex buildings. The proposed framework offers a robust and scalable solution for real-world applications, such as urban planning, where precise building segmentation is crucial.

Keywords: building instance; remote sensing; deep learning; dual attention

1. Introduction

Building instance segmentation from high-resolution remote sensing images, typically with a spatial resolution finer than 1 m per pixel, has emerged as a fundamental technique. It plays a crucial role in the broader digital revolution of the construction sector, where the integration of remote sensing, Building Information Modeling (BIM), and digital twin technologies is reshaping urban planning and management practices. Accurately identifying building boundaries enables application scenarios such as 3D city model reconstruction and population density analysis. Traditional building extraction methods typically rely on manually constructed spectral–texture features or geometric cues such as best-fit plane estimation and elevation differences, which are then combined with machine learning classifiers to complete the recognition task [1]. Although these approaches are effective in relatively regular scenes, they struggle to cope with the complexity and diversity of urban environments. With convolutional neural network (CNN) gradually established as the core technical framework, the application of deep learning has changed the field.

The improved architectures based on U-Net (ResUNet, Attention U-Net, etc.) achieve performance optimization by integrating cross-layer skip connections with attention control modules. Although such networks have advantages in capturing local details, they still face technical constraints in modeling long-range spatial correlations between buildings. This challenge is particularly evident in densely built-up areas, which are typically characterized by a building coverage ratio exceeding 60%. The Transformer architecture breaks through this limitation with the global context-awareness paradigm, and the Multi-layer Perceptron (MLP) [2] architecture opens up a new modeling path through the feature channel mixing mechanism [3].

Despite the rapid advancement of deep learning techniques, accurate building instance extraction from high-resolution remote sensing imagery remains a challenging task. This difficulty arises not only in urban areas but also in non-urban settings, where scale differences, occlusions, and diverse background conditions remain unresolved issues. First, architectural entities in urban areas exhibit significant variability in shape, scale, and spectral appearance, requiring models to dynamically adapt their receptive fields to capture diverse structures effectively. Second, vegetation occlusion, shadow interference, and adjacency to other buildings often cause building contours to break or remain incomplete, leading to missed detections. Third, due to the inherent class imbalance between sparse building targets and extensive background areas, models must possess fine-grained feature discrimination capabilities to maintain boundary precision. While recent studies have attempted to address these challenges individually—such as employing attention mechanisms for modeling long-range dependencies [4], deformable convolutions for adaptive feature sampling [5], and multi-scale fusion strategies for handling size variations [6]—their effectiveness remains limited in complex urban environments due to the lack of a holistic, integrated solution that simultaneously addresses geometric adaptability, contextual reasoning, and structural refinement.

To address the challenges, we propose a framework named the Multi-Scale Hybrid Dual-Attention Network (MS-HDAN). Built upon a U-Net-style encoder–decoder architecture, MS-HDAN introduces a dual-stream encoder that includes a Local Feature Extraction Pathway (LFEP) and a Global Context Modeling Pathway (GCMP). The LFEP utilizes pre-activation residual blocks to preserve edge and building outline information. At the same time, the GCMP employs a Gated Dual-Attention BiFormer Block (GDABB) to capture long-range semantic relations. To enhance interaction between the two streams, we design a Local-Global Collaborative Perception Enhancement Module (LG-CPEM), which fuses semantic and structural features through a combination of standard and deformable convolutions. It improves adaptability to varying building geometries while maintaining semantic consistency. Regarding the decoder, an attention-guided reconstruction module incorporates spatial and channel attention information to refine boundaries and restore spatial resolution. The hierarchical skip connections ensure the propagation of multi-scale context and structural information.

The main contributions of this paper include the following:

1. We propose a Multi-Scale Hybrid Dual-Attention Network (MS-HDAN) that integrates local detail modeling with global semantic reasoning for accurate building instance segmentation.
2. We introduce the Gated Dual-Attention BiFormer Block (GDABB) for dynamic global feature selection, and the Local-Global Collaborative Perception Enhancement Module (LG-CPEM) for adaptive feature alignment under geometric variations.
3. A dual-attention guided reconstruction module is designed to refine building boundaries and enhance spatial detail consistency.

The structure of this paper is as follows. Section 2 reviews related work on building instance extraction. Section 3 details the proposed Multi-Scale Hybrid Dual-Attention Network (MS-HDAN). Section 4 presents the experimental setup, results, and analysis. Section 5 provides further discussion, and Section 6 concludes the paper.

2. Related Work

2.1. Deep Learning and Attention Mechanisms for Building Extraction

The iterative upgrading of deep learning technology has improved the accuracy level of building extraction from remote sensing images. As the founding framework of pixel-level semantic segmentation, fully convolutional networks (FCNs) construct the technical foundation of end-to-end prediction [7]. On this basis, U-Net [8] deepened the feature expression ability by using a symmetric encoding and decoding architecture and a cross-layer skip connection mechanism, establishing itself as the benchmark model system for building extraction. ResUNet [9] optimizes the deep network training process by integrating the residual learning module, and attention U-Net [10] utilizes a spatial attention gate mechanism to enhance the feature focusing ability of the target region.

In recent years, Transformer-based models have gradually attracted attention in remote sensing applications. Swin Transformer adopts a hierarchical self-attention structure to effectively capture long-range dependencies in high-resolution remote sensing images and improve the feature discrimination of ground targets [3,11]. However, the Transformer architecture is inefficient in extracting local features and requires many computational resources. In order to solve this problem, by deeply fusing the local perception advantages of the convolution operation and the global modeling ability of self-attention, the CNN-Transformer hybrid network effectively improves the accuracy and robustness of remote sensing image interpretation [12]. The attention mechanism exhibits a powerful feature focusing ability in complex terrain scenes, providing a new solution to the problem of feature redundancy commonly found in remote sensing images. The Dual Attention Network (DANet) [4] introduced the simultaneous use of position and channel attention for scene segmentation. HANet establishes a multi-level attention architecture to enable cross-scale feature interaction and adaptive fusion [13]. RADANet (On the Road Enhanced Deformable Attention Network) combines the geometric modeling benefits of deformable convolutions with the dynamic weighting properties of the attention mechanism through an optimization system tailored for linear objects, such as road networks [14]. Remote sensing image building extraction benefits most from the attention module, which uses dynamic channel feature enhancement. Squeeze-and-excitation block-based [15] channel attention mechanisms have demonstrated effectiveness in building extraction by focusing on important feature channels. The Convolutional Block Attention Module (CBAM) [16] added spatial attention to enhance channel-wise feature recalibration.

2.2. Deformable Convolutions and Hybrid Architectures for Building Extraction

Deformable convolutional networks solved the problem of fixed geometric transformations in standard CNNs by learning dynamic sampling locations [5]. This capability is particularly valuable for building extraction, where structures exhibit diverse shapes and orientations. Refine-UNet [17] demonstrated how Atrous Spatial Pyramid Pooling (ASPP) could complement deformable convolutions to capture multi-scale building features. Recent work has explored the integration of deformable operations with attention mechanisms. The Feature-Fusion Segmentation Network (FFS-Net) [18] showed that combining deformable convolutions with attention could improve boundary accuracy for irregular objects. However, these approaches typically apply deformable operations at fixed network stages, while our RM module integrates them throughout the feature hierarchy for continuous geometric adaptation. The resurgence of MLP-based architectures, exemplified by MLP-Mixer [2], offered new perspectives on feature mixing without convolution or self-attention. Gated MLP variants like gMLP [19] introduced data-dependent projections to control information flow. These architectures demonstrated competitive performance on image classification while maintaining computational efficiency. However, their application to dense prediction tasks, such as building extraction, remains largely unexplored. The proposed GateMLP module differs from previous MLP approaches in several key aspects. First, it operates on feature patches rather than individual tokens, thereby preserving spatial relationships that are crucial for boundary delineation. Second, the gating mechanism incorporates sigmoid activation and element-wise multiplication, enabling finer control over feature propagation. Third, it integrates seamlessly with convolutional features through residual connections, combining the strengths of both paradigms. The last decade has witnessed remarkable progress in building extraction methodologies, from early CNN-based approaches to recent hybrid architectures. While existing methods have addressed various aspects of the problem—geometric adaptation through deformable convolutions, context modeling via attention mechanisms, and multi-scale feature fusion—they often treat these components separately.

In summary, existing methods have addressed some of the existing problems, such as geometric adaptation via deformable convolutions, context modeling via attention mechanisms [20], and handling size changes via multi-scale feature fusion [21]. However, these methods treat each component separately and lack a systematic solution to simultaneously optimize local detail preservation, global context integration, and adaptive feature selection. Despite achieving high performance in specific scenarios, existing building extraction methods can reach accuracy levels above 85–90% when evaluated by metrics such as IoU or F1-score. However, their limitations become evident in complex urban environments, where buildings with diverse scales, shapes, and structural appearances together with occlusions and background interferences significantly reduce robustness. For example, the dramatic changes in geometry, scale distribution, and spectral performance of buildings still pose challenges to the adaptability of existing models. At the same time, vegetation occlusion, shadow interference and the influence of adjacent buildings may also cause contour breakage and missed detection problems. Therefore, improving the location accuracy of building boundaries and the integrity of the overall area in complex urban environments is still a key problem to be solved.

3. Method

When semantic segmentation of urban buildings is performed on remote sensing images, traditional image segmentation methods often exhibit insufficient feature expression capabilities in complex scenes and high-resolution images. At the same time, these methods also have limitations in modeling spatial semantic information, especially in multi-scale feature fusion and long-distance dependency, making it challenging to capture the structural and semantic information. To solve these problems, we propose a Multi-Scale Hybrid Dual-Attention Network (MS-HDAN).

Let $x \in \mathbb{R}^{H \times W \times C}$ represent an urban remote sensing image, where $H \times W$ is the resolution and C is the number of channels. Our goal is to predict the pixel-wise label for segmentation. Our model is built upon the classic U-Net architecture, utilizing an encoder–decoder framework. It integrates multi-scale feature extraction and a hybrid dual-attention mechanism to enhance the representation of contextual information and improve segmentation accuracy in complex urban scenes. Specifically, the encoder is designed as a dual-stream structure, where the main branch focuses on semantic features and the auxiliary branch provides complementary structural details. After feature extraction, the decoder employs a stepwise up-sampling strategy, combined with skip connections from the encoder, to reconstruct the spatial resolution. Additionally, a dual-attention mechanism is integrated into the decoder path to enhance feature fusion and improve the reconstruction of structural details, such as building edges. This design significantly improves the segmentation accuracy of complex urban scenes in remote sensing images. The flowchart is shown in Figure 1.

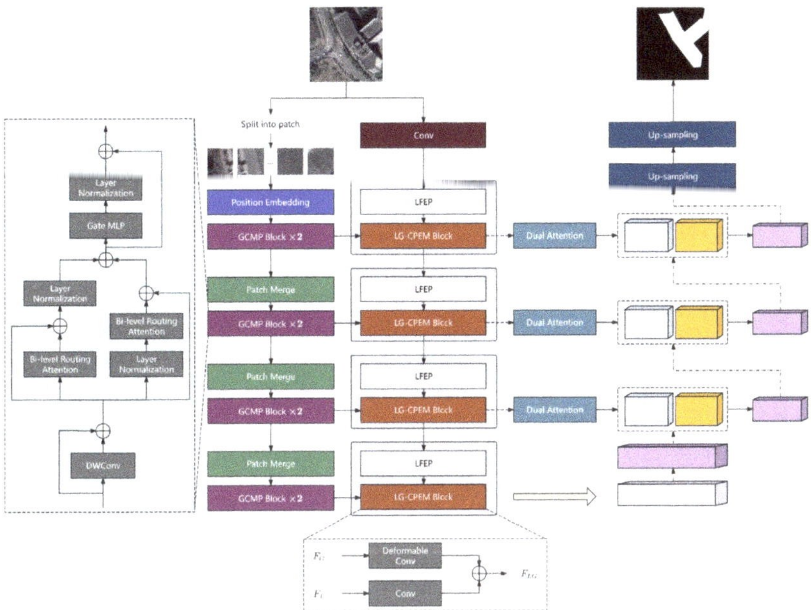

Figure 1. The overall model structure of MS-HDAN.

3.1. Dual-Stream Encoder

To effectively extract diverse features from urban remote sensing images, we designed the dual-stream encoder of MS-HDAN. First, we propose a dual-stream encoder composed of two processing pathways: a Local Feature Extraction Pathway (LFEP) and a Global Context Modeling Pathway (GCMP). The LFEP is designed to capture spatial patterns and structural textures based on convolutional operations. The GCMP captures long-range dependencies and semantic coherence across the remote sensing images. Among them, the results of each layer will be fused through the designed Local-Global Collaborative Perception Enhancement Module (LG-CPEM) for feature fusion.

Firstly, we convert each remote sensing image $x \in \mathbb{R}^{H \times W \times C}$ into a sequence of image patches with size $P \times P$. Each patch $M_i \in \mathbb{R}^{C \times P \times P}$ is then flattened and formed into the sequence matrix $X_M \in \mathbb{R}^{N \times C \times P \times P}$. $N = \frac{H \times W}{P^2}$ is the number of patches. Secondly, we project the image patches X_M into the D-dimension embedding space using a learnable linear projection. The calculation is as follows:

$$Z = \{X_M^1 W, X_M^2 W, \ldots, X_M^N W\} + E_p \tag{1}$$

where $W \in \mathbb{R}^{P \times P \times C \times D}$ is the projection. $Z \in \mathbb{R}^{N \times D}$ denotes the embedding patches. E_p is the learnable position embeddings.

Here, we input Z and x into the dual-stream encoder. Specifically, Z is fed into the GCMP, and x is fed into the LFEP. The overall structure can be expressed as

$$F = E_{DS}(Z, x) \tag{2}$$

where E_{DS} denotes the dual-stream encoder. It consists of N blocks:

$$F_L^0 = x \tag{3}$$

$$F_G^0 = Z \tag{4}$$

$$F_L^i = f_L(F_L^{i-1}) \tag{5}$$

$$F_G^i = f_G(F_G^{i-1}) \tag{6}$$

$$F^i = f_{fuse}^i(F_L^i, F_G^i) \tag{7}$$

where f_L represents the LFEP and f_G represents the GCMP. f_{fuse} represents the LG-CPEM.

3.1.1. Global Context Modeling Pathway

To enhance the model's ability to model long-distance dependencies and global semantics in complex urban scenes, we introduce the GCMP based on the Gated Dual-Attention BiFormer Block (GDABB). It emphasizes informative features for improving discriminative representation. Specifically, GDABB is an improved version of BiFormer. BiFormer demonstrates strong potential in modeling long-range dependencies. However, the MLP has limited expressive capability when dealing with complex feature relationships. At the same time, GateMLP can adaptively adjust the information flow according to the input by introducing a dynamic gating mechanism. It enhances the feature selectivity and expressive capability. Meanwhile, the conventional normalization strategy can be stably trained, but it is insufficient in modeling complex feature interactions. For this reason, we introduce the dual-attention mechanism to explore the relationship between features from multiple perspectives to enrich the semantic expression.

Given the input Z, the GCMP can be expressed as

$$F_G = PM^i(H_{GDB}^i(Z))i = 1,2,3 \tag{8}$$

PM denotes the patch merge. H_{GDB} denotes the GDABB. i denotes the i-round of the structure.

To better illustrate the GCMP, we now provide a detailed description of the Gated Dual-Attention BiFormer Block (GDABB), which serves as the fundamental unit of GCMP. Given that feature representation Y, the GDABB first applies a depthwise convolution branch to extract localized spatial patterns. Then, the Bi-Level Routing Attention is employed to model the global dependencies. Also, to further enhance the expressive diversity of global context modeling, we propose a dual-way parallel attention mechanism. Specifically, we adopt a structurally symmetric dual-path attention mechanism. This scheme aims to model global dependencies from both pre-normalization and post-normalization perspectives, thereby achieving complementary modeling of attention representations.

$$Y_1 = Y + DWConv_{3\times3}(Y) \tag{9}$$

where $DWConv$ denotes the depthwise convolution.

$$Y_2 = Y_1 + BRA(LN(Y_1)) + LN(BRA(Y_2)) \tag{10}$$

where BRA represents the Bi-Level Routing Attention. LN represents the layer normalization.

BRA is a dynamic, query-aware sparse attention mechanism. It takes the query Q, key K, and value V as input, and then calculates the adjacency matrix A. Secondly, a routing index matrix (I) is set to contain the top-k connections for each region. Then, it gathers the key and value to ensure high computational efficiency during sparse attention.

$$A = QK^T \tag{11}$$

$$I = TopK(A) \tag{12}$$

$$K_G = gather(K, I) \tag{13}$$

$$V_G = gather(V, I) \tag{14}$$

where K_G and V_G are the gathered key and value. $TopK$ denotes the operation of selecting the top-k strongest connections.

Finally, we apply the attention based on the gathered key and value.

$$O_A = Att(Q, K_G, V_G) + DWConv(V) \tag{15}$$

$$Att(Q, K_G, V_G) = \text{Softmax}\left(\frac{QK_G^T}{\sqrt{d_k}}\right)V_G \tag{16}$$

where d_k denotes the dimensionality of the key and query embeddings. $DWConv$ represents the depthwise separable convolution.

After calculating the attention, a gated feedforward network is introduced to fuse the features adaptively. This step is formulated as

$$O = GateMLP(LN(O_A)) \tag{17}$$

$$GateMLP(X) = (DWConv(GELU(XW_1 + b_1)) \odot \sigma(XW_1 + b_g))W_2 + b_2 \tag{18}$$

$$\text{GELU}(x) = x \cdot \Phi(x) = x \cdot \frac{1}{2}\left(1 + \text{erf}\left(\frac{x}{\sqrt{2}}\right)\right) \tag{19}$$

Here, O is the output of the BRA. $\odot$ denotes element-wise multiplication. σ denotes the activation function. $\Phi(x)$ denotes the cumulative distribution function of the standard normal distribution, and erf is the Gaussian error function.

3.1.2. Local Feature Extraction Pathway

To enhance the model's ability to extract fine-grained features that correspond to subtle pixel-level variations in remote sensing images, the Local Feature Extraction Pathway (LFEP) is designed. Specifically, we adopt the Pre-activation ResNet (PRN) block as the backbone structure to extract discriminative, detailed features in local regions. Unlike the standard ResNet structure, Pre-activation ResNet places the batch normalization and activation function (ReLU) before the convolution operation, effectively mitigating the problem of gradient vanishing. It makes the local feature expression more stable and training more efficient.

Given the input x, the backbone structure can be expressed as

$$F_L = LFEP(x) \tag{20}$$

$$LFEP(x) = PRN^3(Conv(x)) \tag{21}$$

where PRN^3 denotes the PRN being performed three times.

$$PRN(x) = x + Conv_{3\times3}(ReLU(BN(x))) \tag{22}$$

where BN denotes the batch normalization, and $ReLU$ is the activation function.

3.1.3. Local-Global Collaborative Perception Enhancement Module

To effectively fuse local details with global semantic information, we propose the Local-Global Collaborative Perception Enhancement Module (LG-CPEM). This module enhances the discriminative ability of the overall feature representation by efficiently fusing the global and local information from both paths. Specifically, LG-CPEM introduces a differentiated feature enhancement strategy. On the one hand, features output from the global path are introduced into a standard 3×3 convolution to preserve the semantic backbone. On the other hand, the deformable convolution is applied to the features of the local path to enhance its modeling capability. The dual-path features are processed and then fused by element-by-element addition to obtain a synergistically enhanced multi-scale feature representation. This fusion approach not only enables complementary perception at the semantic and structural levels but also enhances computational efficiency and structural flexibility.

Given the extracted F_L from the LFEP and F_G from GCMP, the overall process can be stated as

$$F_{LG} = Conv_3(F_L) + DeformConv_3(F_G) \tag{23}$$

where $DeformConv$ denotes deformable convolution.

3.2. *Attention-Guided Decoder*

The decoder structure of traditional U-Net generally suffers from insufficient information fusion as well as insufficient detail recovery. So, we construct an attention-guided decoder. Specifically, we adopt a skip-connection based on dual attention (DA) for feature enhancement. The decoder not only adopts the up-sampling features from the previous layer but also jointly utilizes the output features from the fusion of local and global features (F^i) provided by the encoder. Firstly, the fused feature F^i is feature-enhanced by the DA to achieve a more accurate and contextually consistent reconstruction. Then, the feature is up-sampled as a result of decoding feature D^i in the previous layer. The process can be formulated as follows:

$$F_S^i = DA(F^i) \tag{24}$$

where the F_S^i denotes the enhanced features from the dual-attention mechanism.

$$D^i = UpSampling(Concat(D^{i+1}, F_S^i)) \tag{25}$$

DA consists of the spatial attention (SPA) and the channel attention (CA). For spatial attention, when given the feature A, it first generates the feature maps B, C, and D, then performs the matrix multiplication and applies the softmax to generate the spatial attention map. Here, D is used for the feature weighing of SPA. For channel attention, when given the feature A, it first performs the matrix multiplication between A and the transpose of A. Then, applies the softmax to generate the channel attention map. Here, A is used for feature weighting in CA. After weighting the two types of attention based on their features, feature summation is then performed. The overall process can be expressed as

$$DA(x) = SPA(x) + CA(x) \tag{26}$$

$$s_{ji} = \frac{\exp(B_i \cdot C_j)}{\sum_{i=1}^{N} \exp(B_i \cdot C_j)} \tag{27}$$

$$SPA(x) = \alpha \sum_{i=1}^{N} s_{ji} D_i + A_j \tag{28}$$

$$x_{ji} = \frac{\exp(A_i \cdot A_j)}{\sum_{i=1}^{C} \exp(A_i \cdot A_j)} \tag{29}$$

$$CA(x) = \beta \sum_{i=1}^{C} x_{ji} A_i + A_j \tag{30}$$

where α and β are the parameters. N denotes the number of pixels and C denotes the channel number.

4. Experiments

4.1. Dataset Details

Two public datasets are used for experimental evaluation: Wuhan and Shanghai, taken from the buildings instances of typical cities in China (BITC) [22]. Among these, there are 1231 images of Shanghai and 1448 images of Wuhan. The size of each image is 500×500 and its spatial resolution is 0.29 m.

4.2. Evaluation Metrics

For evaluation, we employ three standard metrics: accuracy, F1-score, and mean intersection over union (mIoU) [23]:

$$Accuracy = \frac{TP + TN}{TP + TN + FP + FN} \tag{31}$$

$$F1\text{-}score = \frac{2 \cdot Precision \cdot Recall}{Precision + Recall} \tag{32}$$

$$mIoU = \frac{1}{K} \sum_{k=1}^{K} \frac{TP_k}{TP_k + FP_k + FN_k} \tag{33}$$

where TP, TN, FP, and FN represent the number of true positives, true negatives, false positives, and false negatives, respectively. K denotes the number of classes, and TP_k, FP_k, and FN_k represent the true positives, false positives, and false negatives for the k-th class.

4.3. Performance Evaluation

To conduct a comprehensive evaluation of the model's performance, we performed experiments on the proposed method using the Shanghai-BITC and Wuhan-BTTC datasets.

Table 1 presents the quantitative results of the Shanghai-BITC dataset, comparing the performance of the five segmentation models. Overall, our proposed method consistently achieves higher performance across all metrics, reaching an accuracy of 0.9451, an F1-score of 0.9445, and an mIoU of 0.8445, which highlights its improved segmentation ability and model generalization. Compared to other mainstream models, TransUNet also performs better (accuracy: 0.9329, F1-score: 0.9325, mIoU: 0.8325), verifying the effectiveness of fusing CNN and Transformer architectures in improving semantic understanding and spatial perception. DeepLabV3 benefits from its Atrous Spatial Pyramid Pooling (ASPP) structure, which effectively enhances multi-scale context capture, achieving a mean IoU (mIoU) of 0.7841. UNeXt consistently demonstrates improvements over the base U-Net model across all evaluation metrics. Although these differences are not statistically significant, they reveal a positive trend, suggesting that the optimized decoding structure and jump connection mechanism help preserve information. Although SwinUNet introduces a hierarchical window attention mechanism, it does not surpass TransUNet and DeepLabV3 in all indices, especially in mIoU (0.7513), which is only slightly higher than that of the basic U-Net, indicating that the balance between local and global features still has some deficiencies. These results collectively demonstrate the effectiveness of the proposed method, particularly in feature extraction and boundary maintenance.

Table 2 presents the results of the Wuhan-BITC dataset. The accuracy, F1-score, and mIoU of all models on this dataset are generally lower than their performance on the Shanghai-BITC dataset. Among them, our proposed method still achieves the highest accuracy (0.9232), F1-score (0.9221), and mIoU (0.7638). DeepLabV3 also shows stable performance on Wuhan-BITC, with an accuracy of 0.9139 and an mIoU of 0.7427. In contrast, TransUNet's performance slightly declines on this dataset, with mIoU dropping from 0.8325 to 0.7033. SwinUNet achieves an mIoU of 0.7056, which is slightly higher than that of TransUNet, but still significantly lower than its performance on the Shanghai-BITC dataset. The performance of UNeXt and U-Net is further weakened, especially the mIoU of UNeXt, which decreases to 0.6613, likely due to weaker boundary representation.

In summary, the proposed method maintains consistently strong performance across two datasets with different distributions and complexities, highlighting its robustness and generalization ability.

Table 1. Comparison of segmentation models on Shanghai-BITC dataset.

Model	Accuracy	F1-Score	mIoU
U-Net [8]	0.9034	0.9014	0.7466
UNeXt [24]	0.9121	0.9111	0.7693
DeepLabV3 [25]	0.9208	0.9187	0.7841
TransUNet [26]	0.9329	0.9325	0.8325
SwinUNet [27]	0.9046	0.9031	0.7513
Our proposed method	0.9451	0.9445	0.8445

Table 2. Comparison of segmentation models on Wuhan-BITC dataset.

Model	Accuracy	F1-Score	mIoU
U-Net [8]	0.8972	0.8922	0.6919
UNeXt [24]	0.8789	0.8762	0.6613
DeepLabV3 [25]	0.9139	0.9121	0.7427
TransUNet [26]	0.9017	0.8971	0.7033
SwinUNet [27]	0.9025	0.8993	0.7056
Our proposed method	0.9232	0.9221	0.7638

4.4. Visualization Results

Figures 2 and 3 display the image segmentation visualization results for the two remote sensing datasets, Shanghai-BITC and Wuhan-BITC, respectively. These figures contain the segmentation outputs of the original image, the ground truth labels, the five mainstream models, and the proposed method. In the Shanghai-BITC dataset (Figure 2), the traditional U-Net and its variant UNeXt are prone to target sticking or breaking, and the edge details are not well preserved. DeepLabV3 has a slight advantage in extracting the backbone structure, but it still suffers from deformation or missegmentation in fine areas. SwinUNet and TransUNet alleviate these problems to some extent, particularly in reconstructing the general contour. Primarily, it is more complete in reconstructing the general contour. However, the proposed method can better restore building edges, slender structures, and narrow gap areas, and the generated prediction maps are closer to the real labels in terms of structural continuity and geometric alignment. In the Wuhan-BITC dataset (Figure 3), the scenario is more complex, leading to a significant increase in the difficulty of model prediction. The overall observation reveals that U-Net and UNetX exhibit significant background leakage and boundary collapse in several areas, and their performance is volatile in small target regions. DeepLabV3 retains larger regions but has limited ability to recognize fine edges. SwinUNet and TransUNet perform slightly better in the third and fourth rows of the image, identifying the general building layouts. However, residual errors are still visible, mainly around complex boundaries and textured regions, where misclassifications and noise are more likely to occur. Experimental results across multiple scenes demonstrate that the proposed method consistently maintains higher accuracy and mIoU compared with other models, indicating improved robustness.

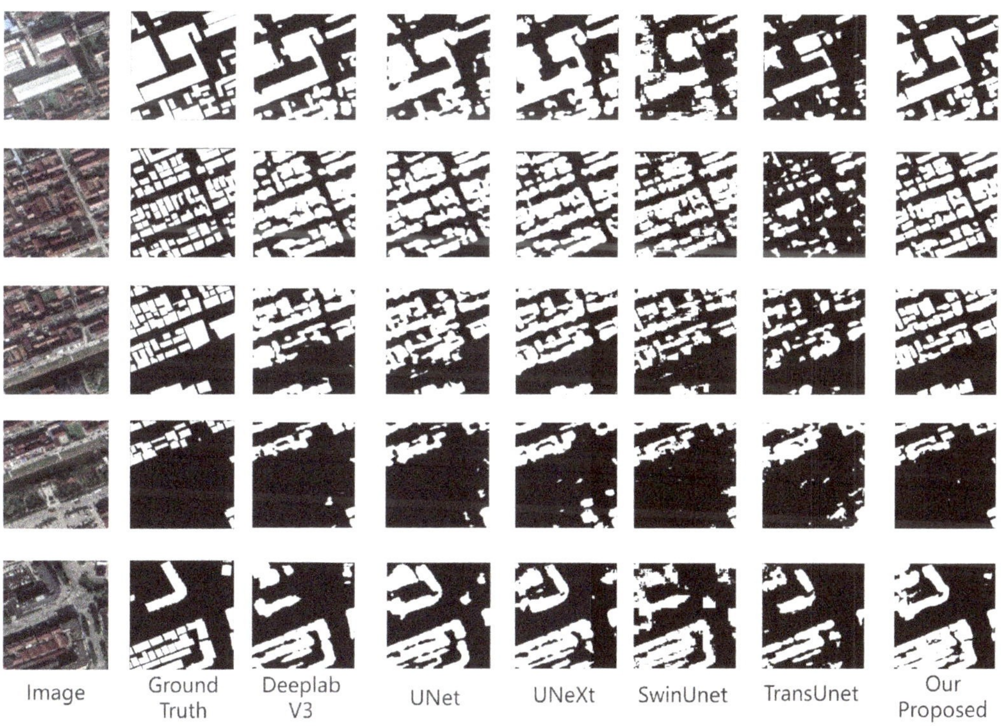

Figure 2. Visualization results for the Shanghai-BITC dataset.

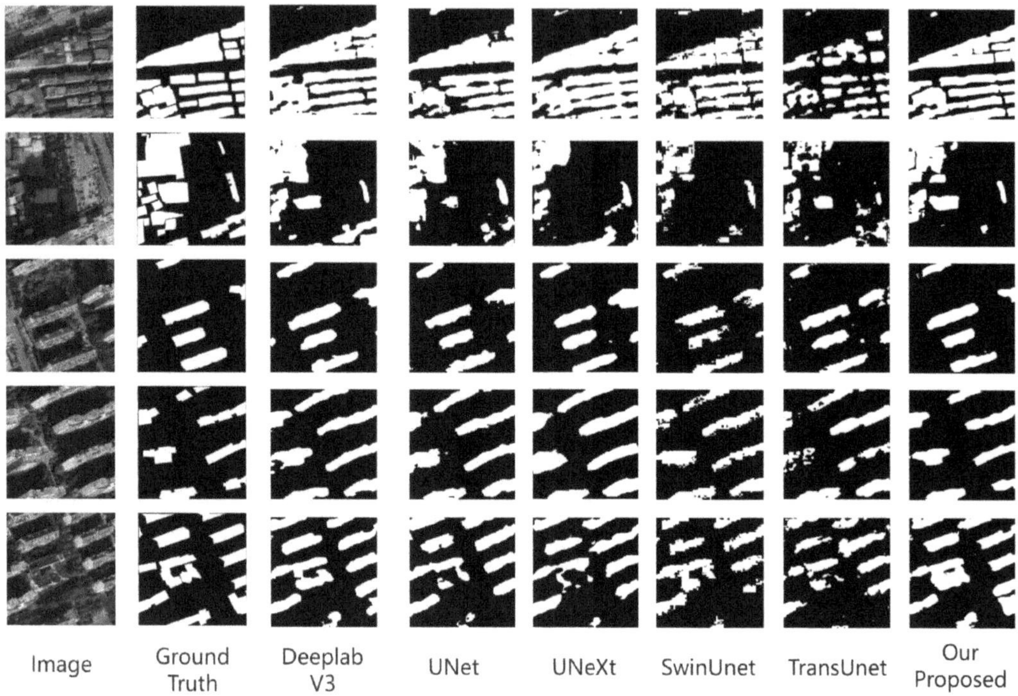

Figure 3. Visualization results for the Wuhan-BITC dataset.

4.5. Confusion Matrices

To further analyze the classification performance of the models at the pixel level, Figures 4 and 5 show the confusion matrix results of each segmentation model on the Shanghai-BITC and Wuhan-BITC datasets, respectively. The proposed method performs best on the Shanghai-BITC dataset (Figure 4). In contrast, the Class 1 recognition rates of U-Net and UNeXt are 0.80 and 0.81, both lower than the 0.85 achieved by the proposed method, particularly at the intersection of background and foreground. TransUNet and DeepLabV3 show relatively consistent performance across datasets, with smaller performance fluctuations compared to other models. In particular, TransUNet demonstrates a stronger ability to recognize background regions. In contrast, SwinUNet's performance is relatively unstable, indicating that there are still obvious omissions in the detection of target regions, such as buildings. On the Wuhan-BITC dataset (Figure 5), the overall classification performance is degraded, reflecting the challenges of this dataset, such as its complex background and fuzzy target boundaries. Nevertheless, the proposed method still maintains high accuracy in both prediction classes, reaching 0.94 for Class 0 and 0.82 for Class 1. These results are higher than those of other methods, particularly in classifying foreground buildings, and demonstrate stronger robustness.

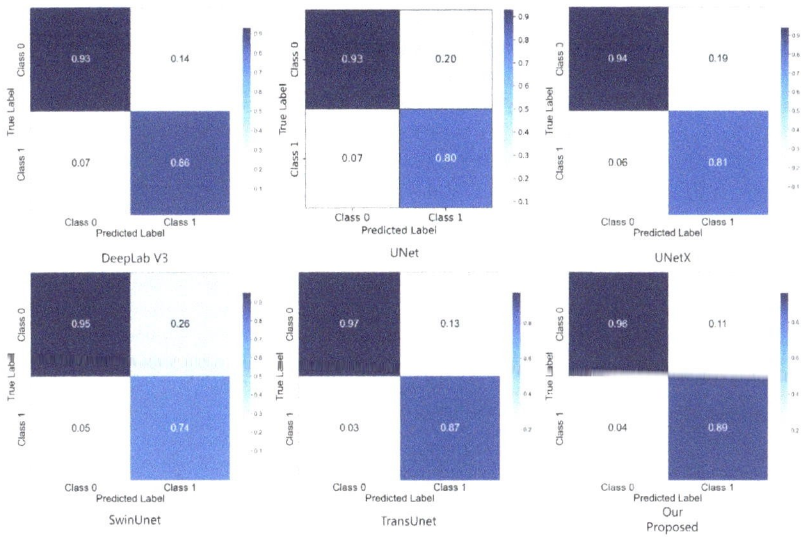

Figure 4. Confusion matrix results for the Shanghai-BITC dataset.

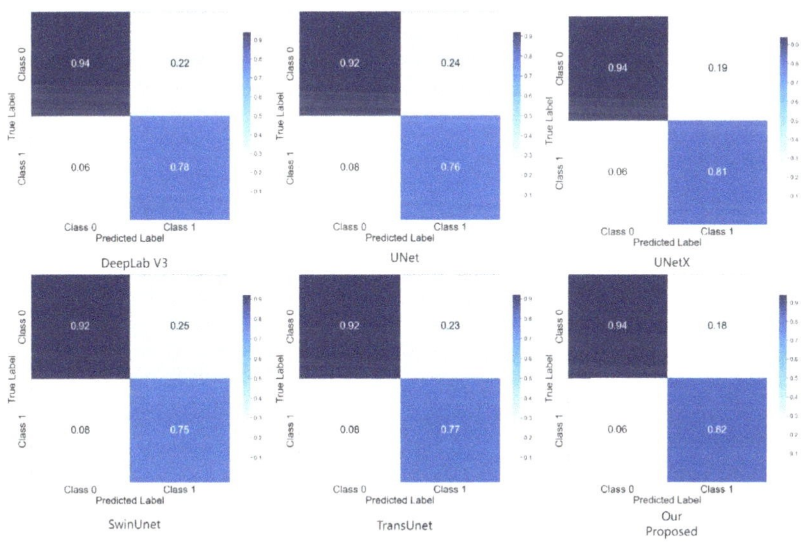

Figure 5. Confusion matrix results for the Wuhan-BITC dataset.

4.6. Ablation Experiment

To verify the impact of the key modules in the proposed method on the overall performance, we conduct ablation experiments on the Shanghai-BITC and Wuhan-BITC datasets. Here, two main modules are examined: (1) the GDAB (Gated Dual-Attention BiFormer) block, which enhances the relationship between features from multiple perspectives to enrich semantic expression; and (2) the DA (dual-attention) mechanism, which improves the discrimination of feature expression. Tables 3 and 4 show the comparison results of the complete model (our proposed) and its two ablation versions (w/o GDAB, w/o DA) on the Shanghai-BITC and Wuhan-BITC datasets, respectively. As shown in the table, the complete model achieves the highest results in all three metrics on both the Shanghai-BITC and Wuhan-BITC datasets. This improvement should be interpreted as a consistent trend rather than a statistically proven significance, since no statistical hypothesis testing (e.g., *p*-value analysis) was conducted. When the GDAB module is removed, the model's mIoU on the Shanghai-BITC and Wuhan-BITC datasets decreases by 2.41% and 1.23%, respectively. These results suggest that GDAB contributes to improving boundary accuracy and reducing false detections. In contrast, the F1-score decreases when the DA module is removed, indicating that the mechanism plays a crucial role in enhancing feature discrimination capability and comparability. Overall, the observed differences support the effectiveness of the proposed modules.

Table 3. Ablation study results on the Shanghai-BITC dataset.

Method	Accuracy	F1-Score	mIoU
Our proposed	0.9451	0.9447	0.8458
w/o GDAB	0.9363	0.9354	0.8217
w/o DA	0.9344	0.9345	0.8216

Table 4. Ablation study results on the Wuhan-BITC dataset.

Method	Accuracy	F1-Score	mIoU
Our proposed	0.9232	0.9221	0.7638
w/o GDAB	0.9182	0.9161	0.7515
w/o DA	0.9191	0.9177	0.7571

5. Discussion

5.1. Model Effectiveness in Complex Urban Scenes

With the acceleration of urbanization, high-resolution remote sensing images are playing an increasingly important role in urban planning, land use, and disaster monitoring [28]. As the core element of urban space, accurate segmentation of buildings is crucial for constructing high-quality urban information maps [29]. However, traditional semantic segmentation methods often suffer from insufficient feature expressiveness and limited spatial semantic modeling when faced with complex urban scenes and architectural structures of varying scales.

Our proposed MS-HDAN addresses these issues by combining local detail preservation with global semantic reasoning, leading to more coherent boundary predictions without relying on a single type of feature representation. Rather than restating the technical details of each module, here we emphasize that the overall framework improves both contextual understanding and structural fidelity in challenging urban scenarios.The experimental results demonstrate that MS-HDAN consistently outperforms existing mainstream methods, thereby validating its adaptability and robustness in complex urban scenes. In particular, the model shows improvements in densely built-up areas and regions with ambiguous boundaries, where maintaining regional consistency is especially difficult.

5.2. Potential Impact on Urban Decision Making

MS-HDAN has shown outstanding performance in high-precision building semantic segmentation, providing more accurate and automated data support for urban management and decision making [30]. In practical applications, our proposed method can be widely used for tasks such as extracting urban spatial information, conducting building surveys, monitoring urban expansion, and evaluating planning. For example, in the process of urban renewal, renovation of old residential areas, and infrastructure expansion, decision-makers can rely on model segmentation results to quickly identify building distribution density, boundary morphology, and structural characteristics, thereby developing more targeted and scientific intervention strategies.

By introducing the dual-stream structure and dual-attention mechanism, MS-HDAN models the structural details and semantic contours of buildings in complex urban environments more accurately, effectively alleviating the misjudgment and missed detection problems of traditional methods in densely populated urban areas. Therefore, this method shows potential practical value and may provide useful technical references for urban intelligent governance and data-driven spatial decision making, although further validation in real-world applications is needed.

5.3. Limitations and Future Work

Although the proposed MS-HDAN has demonstrated significant advantages in the semantic segmentation of urban buildings, certain limitations remain. Firstly, the complex structure of the model significantly increases the overall parameter count and computational cost compared to traditional segmentation models. This may become a limiting factor in resource-constrained edge devices or application scenarios that require high real-time performance. Secondly, model training still heavily relies on high-quality, pixel-level annotated data (i.e., remote sensing images with manually delineated building boundaries serving as ground truth). In practical remote sensing applications, the cost of obtaining building labels is high, and annotation accuracy is easily affected by human subjectivity, which can significantly impact the performance of cross-regional promotion and large-scale deployment.

Given the above problems, future research work can be carried out from the following directions: on the one hand, model lightweight strategies can be explored, such as structural pruning, knowledge distillation or neural architecture search (NAS), which can significantly reduce model complexity to meet the actual needs of edge computing and real-time reasoning on the premise of ensuring segmentation performance; on the other hand, multi-source remote sensing data (such as hyperspectral images, LiDAR, or multi-phase images) can be introduced for fusion modeling, further enhancing the model's perception and segmentation capabilities for complex terrain structures, occluded areas, and multi-scale targets. In addition, future segmentation results can be deeply integrated with tasks such as urban functional zoning, building type recognition, or 3D modeling, providing more accurate and multidimensional data support for intelligent decision-making systems, including urban management, disaster response, and spatial planning.

6. Conclusions

In this paper, we propose a novel Multi-Scale Hybrid Dual-Attention Network (MS-HDAN) for accurate building instance segmentation in high-resolution remote sensing images. To address the challenges posed by complex urban scenes, including geometric variability and semantic ambiguity, MS-HDAN integrates a dual-stream encoder that captures both local structural details and global contextual semantics. Beyond its methodological contributions, the approach can provide valuable support for applications such as BIM-based modeling, GIS data updates, and urban spatial monitoring. Potential users include urban planning authorities, competent city developers, and agencies involved in disaster response and land-use management. The method may also be aligned with emerging standards for urban spatial data exchange, although further investigation is needed. Some limitations remain, including the model's relatively high complexity and reliance on pixel-level annotations. Future work may explore lightweight architectures, multi-source data integration, and closer linkage with planning and construction workflows. Overall,

MS-HDAN provides an incremental improvement over existing approaches, with potential value for integration into digital urban management practices.

Author Contributions: Conceptualization: Q.H. and Y.P.; data curation: C.Z. and Y.L.; formal analysis: Q.H. and C.Z.; investigation: Y.P. and Y.L.; methodology: Q.H. and Y.P.; project administration: K.U. and J.C.; resources: Q.H. and Y.P.; software: Y.P.; visualization: Q.H., Y.P. and Y.L.; writing—original draft: Q.H., Y.P., Y.L., C.Z., J.C. and K.U.; writing—review and editing: Q.H., Y.P., K.U., J.C. and K.U. All authors have read and agreed to the published version of the manuscript.

Funding: This research work is funded by the project FRG-25-064-FA and granted by the Research Fund of Macao University of Science and Technology (FRG-MUST).

Data Availability Statement: The datasets used in this study are publicly available and can be accessed from the following repositories: the buildings instances of typical cities in China (BITC) dataset is available at https://aistudio.baidu.com/datasetdetail/250828 (accessed on 25 June 2025).

Conflicts of Interest: The authors declare no conflicts of interest.

References

1. Raghavan, R.; Verma, D.C.; Pandey, D.; Anand, R.; Pandey, B.K.; Singh, H. Optimized building extraction from high-resolution satellite imagery using deep learning. *Multimed. Tools Appl.* **2022**, *81*, 42309–42323. [CrossRef]
2. Tolstikhin, I.O.; Houlsby, N.; Kolesnikov, A.; Beyer, L.; Zhai, X.; Unterthiner, T.; Yung, J.; Steiner, A.; Keysers, D.; Uszkoreit, J.; et al. Mlp-mixer: An all-mlp architecture for vision. *Adv. Neural Inf. Process. Syst.* **2021**, *34*, 24261–24272.
3. Liu, Z.; Lin, Y.; Cao, Y.; Hu, H.; Wei, Y.; Zhang, Z.; Lin, S.; Guo, B. Swin transformer: Hierarchical vision transformer using shifted windows. In Proceedings of the IEEE/CVF International Conference on Computer Vision, Montreal, BC, Canada, 11–17 October 2021.
4. Fu, J.; Liu, J.; Tian, H.; Li, Y.; Bao, Y.; Fang, Z.; Lu, H. Dual attention network for scene segmentation. In Proceedings of the IEEE Conference on Computer Vision and Pattern Recognition, Long Beach, CA, USA, 15–20 June 2019.
5. Dai, J.; Qi, H.; Xiong, Y.; Li, Y.; Zhang, G.; Hu, H.; Wei, Y. Deformable convolutional networks. In Proceedings of the IEEE International Conference on Computer Vision, Venice, Italy, 22–29 October 2017.
6. Zhao, H.; Shi, J.; Qi, X.; Wang, X.; Jia, J. Pyramid scene parsing network. In Proceedings of the IEEE Conference on Computer Vision and Pattern Recognition, Honolulu, HI, USA, 21–26 July 2017.
7. Long, J.; Shelhamer, E.; Darrell, T. Fully convolutional networks for semantic segmentation. In Proceedings of the IEEE Conference on Computer Vision and Pattern Recognition, Boston, MA, USA, 7–12 June 2015.
8. Ronneberger, O.; Fischer, P.; Brox, T. U-net: Convolutional networks for biomedical image segmentation. In Proceedings of the International Conference on Medical Image Computing and Computer-Assisted Intervention, Munich, Germany, 5–9 October 2015; Springer: Berlin/Heidelberg, Germany, 2015; pp. 234–241.
9. Liu, Y.; Shen, J.; Yang, L.; Bian, G.; Yu, H. ResDO-UNet: A deep residual network for accurate retinal vessel segmentation from fundus images. *Biomed. Signal Process. Control* **2023**, *79*, 104087. [CrossRef]
10. Oktay, O.; Schlemper, J.; Folgoc, L.L.; Lee, M.; Heinrich, M.; Misawa, K.; Mori, K.; McDonagh, S.; Hammerla, N.Y.; Kainz, B.; et al. Attention u-net: Learning where to look for the pancreas. *arXiv* **2018**, arXiv:1804.03999. [CrossRef]
11. He, X.; Zhou, Y.; Zhao, J.; Zhang, D.; Yao, R.; Xue, Y. Swin transformer embedding UNet for remote sensing image semantic segmentation. *IEEE Trans. Geosci. Remote Sens.* **2022**, *60*, 4408715. [CrossRef]
12. Yin, M.; Chen, Z.; Zhang, C. A CNN-transformer network combining CBAM for change detection in high-resolution remote sensing images. *Remote Sens.* **2023**, *15*, 2406. [CrossRef]
13. Han, C.; Wu, C.; Guo, H.; Hu, M.; Chen, H. HANet: A hierarchical attention network for change detection with bitemporal very-high-resolution remote sensing images. *IEEE J. Sel. Top. Appl. Earth Obs. Remote Sens.* **2023**, *16*, 3867–3878. [CrossRef]
14. Dai, L.; Zhang, G.; Zhang, R. RADANet: Road Augmented Deformable Attention Network for Road Extraction From Complex High-Resolution Remote-Sensing Images. *IEEE Trans. Geosci. Remote Sens.* **2023**, *61*, 5602213. [CrossRef]
15. Hu, J.; Shen, L.; Sun, G. Squeeze-and-excitation networks. In Proceedings of the IEEE Conference on Computer Vision and Pattern Recognition, Salt Lake City, UT, USA, 18–22 June 2018.
16. Woo, S.; Park, J.; Lee, J.; Kweon, I.S. Cbam: Convolutional block attention module. In Proceedings of the European Conference on Computer Vision, Munich, Germany, 8–14 September 2018.
17. Qiu, W.; Gu, L.; Gao, F.; Jiang, T. Building Extraction From Very High-Resolution Remote Sensing Images Using Refine-UNet. *IEEE Geosci. Remote Sens. Lett.* **2023**, *20*, 6002905. [CrossRef]

18. Liu, X.; Peng, Y.; Lu, Z.; Li, W.; Yu, J.; Ge, D.; Xiang, W. Feature-Fusion Segmentation Network for Landslide Detection Using High-Resolution Remote Sensing Images and Digital Elevation Model Data. *IEEE Trans. Geosci. Remote Sens.* **2023**, *61*, 4500314. [CrossRef]
19. Rajagopal, A.; Nirmala, V. Convolutional gated MLP: Combining convolutions and gMLP. In Proceedings of the International Conference on Big Data, Machine Learning and Applications, Orlando, FL, USA, 6–9 December 2021.
20. Zuo, R.; Zhang, G.; Zhang, R.; Jia, X. A deformable attention network for high-resolution remote sensing images semantic segmentation. *IEEE Trans. Geosci. Remote Sens.* **2021**, *60*, 4406314. [CrossRef]
21. Chen, J.; Yi, J.; Chen, A.; Jin, Z. EFCOMFF-Net: A multiscale feature fusion architecture with enhanced feature correlation for remote sensing image scene classification. *IEEE Trans. Geosci. Remote Sens.* **2023**, *61*, 5604917. [CrossRef]
22. Wu, K.; Zheng, D.; Chen, Y.; Zeng, L.; Zhang, J.; Chai, S.; Xu, W.; Yang, Y.; Li, S.; Liu, Y.; et al. A dataset of building instances of typical cities in China. *Chin. Sci. Data* **2021**, *6*, 182–190. [CrossRef]
23. You, D.; Wang, S.; Wang, F.; Zhou, Y.; Wang, Z.; Wang, J.; Xiong, Y. EfficientUNet+: A building extraction method for emergency shelters based on deep learning. *Remote Sens.* **2022**, *14*, 2207. [CrossRef]
24. Valanarasu, J.M.J.; Patel, V.M. Unext: Mlp based rapid medical image segmentation network. In Proceedings of the International Conference on Medical Image Computing and Computer-Assisted Intervention, Singapore, 18–22 September 2022; Springer: Berlin/Heidelberg, Germany, 2022; pp. 23–33.
25. Chen, L.C.; Papandreou, G.; Schroff, F.; Adam, H. Rethinking atrous convolution for semantic image segmentation. *arXiv* **2017**, arXiv:1706.05587. [CrossRef]
26. Chen, J.; Lu, Y.; Yu, Q.; Luo, X.; Adeli, E.; Wang, Y.; Lu, L.; Yuille, A.L.; Zhou, Y. Transunet: Transformers make strong encoders for medical image segmentation. *arXiv* **2021**, arXiv:2102.04306. [CrossRef]
27. Cao, H.; Wang, Y.; Chen, J.; Jiang, D.; Zhang, X.; Tian, Q.; Wang, M. Swin-unet: Unet-like pure transformer for medical image segmentation. In Proceedings of the European Conference on Computer Vision, Tel Aviv, Israel, 23–27 October 2022.
28. Al Shafian, S.; Hu, D. Integrating machine learning and remote sensing in disaster management: A decadal review of post-disaster building damage assessment. *Buildings* **2024**, *14*, 2344. [CrossRef]
29. Dabove, P.; Daud, M.; Olivotto, L. Revolutionizing urban mapping: Deep learning and data fusion strategies for accurate building footprint segmentation. *Sci. Rep.* **2024**, *14*, 13510. [CrossRef] [PubMed]
30. Arulananth, T.; Kuppusamy, P.; Ayyasamy, R.K.; Alhashmi, S.M.; Mahalakshmi, M.; Vasanth, K.; Chinnasamy, P. Semantic segmentation of urban environments: Leveraging U-Net deep learning model for cityscape image analysis. *PLoS ONE* **2024**, *19*, e0300767. [CrossRef] [PubMed]

buildings

MDPI

Article

Spatially Heterogeneous Effects of Microscale Built Environments on $PM_{2.5}$ Concentrations Based on Street View Imagery and Machine Learning

Tian Hu [1,*], Ke Wu [1], Yarui Wu [2] and Lei Wang [3,*]

[1] School of Architecture and Civil Engineering, Xi'an University of Science and Technology, Xi'an 710054, China; 23204110050@stu.xust.edu.cn
[2] College of Geomatics, Xi'an University of Science and Technology, Xi'an 710054, China; wuyarui@xust.edu.cn
[3] School of Architecture, Tianjin University, Tianjin 300072, China
* Correspondence: ht1015@xust.edu.cn (T.H.); wanglei2021@tju.edu.cn (L.W.)

Abstract: $PM_{2.5}$ pollution is a significant environmental problem in global urbanization. However, traditional macro-scale studies are constrained by data resolution limitations, failing to accurately characterize the microscale built environment or thoroughly investigate its spatially heterogeneous effects on $PM_{2.5}$ concentrations. To address this gap, this study constructs a multidisciplinary framework of "Street View Imagery element extraction–spatial heterogeneity modeling–planning strategy optimization" with Xi'an as the case. Leveraging machine learning techniques, the study employs the ResNet50 deep learning model and the ADE20K dataset to precisely extract ten microscale built environment factors from tens of thousands of street view images. Combined with the High-resolution and High-quality Ground-level $PM_{2.5}$ Dataset for China, Ordinary Least Squares (OLS), Geographically Weighted Regression (GWR), and Multiscale Geographically Weighted Regression (MGWR) models were used to systematically reveal the impacts of the microscale built environment on $PM_{2.5}$ concentrations. Ten built environment factors were identified with varying spatial heterogeneity in their effects on the $PM_{2.5}$ concentrations, as follows: (1) factors with positive effects, in descending order of strength, include building, wall, fence, tree, sky, and grass; (2) factors with negative effects, in descending order of strength, include sidewalk, plant, and car; (3) compared with other factors, the road factor showed a relatively weaker effect. This research provides decision-making support for targeted urban planning and environmental protection, while offering valuable references for air pollution control in other cities.

Keywords: street view imagery; built environment; $PM_{2.5}$; MGWR; spatial heterogeneity

1. Introduction

Air pollution is a major global public health issue. $PM_{2.5}$ (particulate matter with an aerodynamic diameter of 2.5 μm or less) is one of the primary air pollutants and a key indicator of air quality, posing a serious threat to public health [1,2]. When inhaled, $PM_{2.5}$ particles can penetrate deep into the lungs and even enter the bloodstream, leading to numerous health complications. Studies have shown that exposure to $PM_{2.5}$ can result in a range of diseases, including respiratory diseases [3,4], cardiovascular diseases [5,6], cancer [7–9], pulmonary diseases [10,11], neurological disorders [12,13], and even an increased risk of premature death [14–16]. Worldwide, $PM_{2.5}$ is responsible for millions of deaths [17–19]. Therefore, the issue of global $PM_{2.5}$ pollution remains critical, and urgent and effective measures are required to address it.

Air pollution has become one of the most severe environmental issues in China [20]. In recent years, the Chinese government has implemented a series of policies to mitigate $PM_{2.5}$ pollution [21–24]. The State Council of the People's Republic of China issued the Clean Air Action Plans in 2013, 2018, and 2023, including the Air Pollution Prevention and Control Action Plan, the Three-Year Action Plan for Winning the Blue Sky Defense Battle, and the Action Plan for Continuous Improvement of Air Quality. These policies aim to significantly reduce the total emissions of major air pollutants, coordinate reductions in greenhouse gas emissions, and further decrease $PM_{2.5}$ concentrations [25–27]. Although these Clean Air Action Plans have yielded marked improvements in reducing $PM_{2.5}$ pollution levels and enhancing air quality since 2013 [21,28–30], China's annual average $PM_{2.5}$ concentration still exceeds the World Health Organization's current annual guideline value of 5 $\mu g/m^3$ [31]. Consequently, $PM_{2.5}$ pollution continues to pose a serious challenge for China.

The urban built environment refers to the human-made surroundings that support human activities, including buildings, roads, and other infrastructure [32]. Exploring the effect of the built environment on $PM_{2.5}$ concentrations and identifying effective strategies to reduce $PM_{2.5}$ pollution have long been central concerns in the fields of urban planning, city management, environmental protection, and public health. Numerous domestic and international studies have revealed a complex relationship between various elements of the urban built environment and $PM_{2.5}$ concentrations. These elements include urban landscape patterns [33], urban morphology [34–36], and land use [37], all of which significantly affect $PM_{2.5}$ concentrations. Although substantial research has been conducted on how the built environment affects $PM_{2.5}$ concentrations, most studies have relied on two-dimensional data to quantify the urban built environment. For example, most remote sensing imagery is based on a top-down perspective and focuses on large-scale monitoring data, making it unsuitable for detailed analysis of the microscale built environment. Zhou et al. employed mobile monitoring combined with two-dimensional built environment indicators, revealing the spatiotemporal heterogeneity of road-level $PM_{2.5}$ and its relationship with the microscale built environment [38]. However, at present, research on the spatially heterogeneous effects of the microscale built environment on $PM_{2.5}$ concentrations from a three-dimensional perspective remains limited. Therefore, selecting appropriate data sources and research methods is essential for thoroughly investigating the relationship between microscale built environments and $PM_{2.5}$ concentrations.

In recent years, with the advancement of computer vision and machine learning technologies, Street View Imagery has emerged as a novel data source, providing new opportunities for urban research and planning [39–41]. With advantages such as high resolution, easy accessibility, and broad spatial coverage, Street View Imagery has been widely used in studies on the green view index [42–44], urban vitality [45,46], spatial perception [47], land surface temperature [48,49], urban color [50–52], urban economy [53], carbon emissions [54], urban visual environments [55], housing prices [56,57], and air pollution [58–60]. Qi et al. revealed that Street View Imagery has significant potential for predicting street-level air pollution [61]. Furthermore, over the past decade, Street View Imagery has been extensively employed to analyze urban street characteristics and the built environment [62]. Therefore, this study employs Street View Imagery as a distinctive data source to examine the relationship between micro-scale urban built environments and $PM_{2.5}$ concentrations.

As a representative high-density city in China and a heavily polluted city in the Fenwei Plain, Xi'an is frequently affected by air pollution due to its complex urban built environment. Accordingly, this study integrates knowledge from urban planning, environmental science, and computer science to establish a multidisciplinary framework for studying $PM_{2.5}$ concentrations. By combining Street View Imagery, machine learning techniques, and spatial regression models, the research aims to help urban planners and administrators gain more targeted insights into the built environment factors affecting air pollution in Xi'an. This study seeks to address the following questions:

(1) What is the effect of the micro-scale urban built environment on $PM_{2.5}$ concentrations?

(2) Can the Multiscale Geographically Weighted Regression (MGWR) model accurately explain the effect of the microscale built environment on $PM_{2.5}$ concentrations?

(3) What is the strength of the effect of microscale built environment factors on $PM_{2.5}$ concentrations?

2. Methods

2.1. Research Framework

The research framework, as illustrated in Figure 1, can be divided into the following stages:

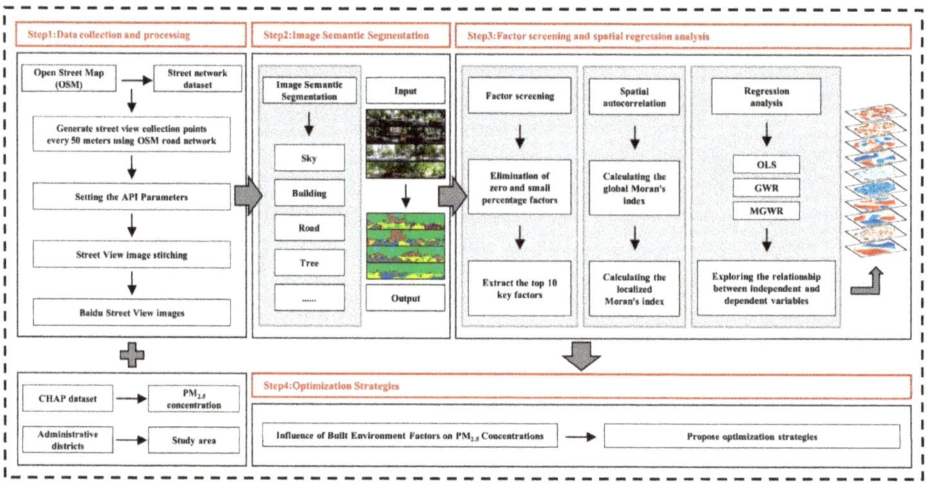

Figure 1. Research framework.

(1) Acquire OSM road network data, generate Street View collection points every 50 m along the OSM road network, set API parameters to collect Baidu Street View Imagery data, and stitch the images. Obtain $PM_{2.5}$ concentration data and administrative division data to ensure data accuracy and completeness.

(2) Employ the ResNet50 deep convolutional neural network architecture combined with the ADE20K dataset to perform semantic segmentation on Street View Imagery, quantifying various elements within the images.

(3) Screen key factors with effects on $PM_{2.5}$ concentrations. Utilize spatial autocorrelation methods to calculate global Moran's I and local Moran's I, exploring the aggregation and distribution patterns of $PM_{2.5}$ concentrations at both regional and local scales within the study area.

(4) Compare the Ordinary Least Squares (OLS), Geographically Weighted Regression (GWR), and Multi-Scale Geographically Weighted Regression (MGWR) models to select the optimal model for comprehensively explaining the spatially heterogeneous effects of urban built environments on $PM_{2.5}$ concentrations.

(5) Summarize the spatial heterogeneity characteristics between urban built environments and $PM_{2.5}$ concentrations in Xi'an, providing optimization strategies for urban planning and environmental protection in the study area.

2.2. Study Area

Xi'an, located in the central part of the Guanzhong Plain, is bounded by the Qinling Mountains to the south, the Wei River to the north, and embraced by the Lishan Mountains on its eastern periphery. Its geographical coordinates span from 107°40′ E to 109°49′ E and from 33°42′ N to 34°45′ N. The city has a warm-temperate semi-humid continental monsoon climate, characterized by distinct seasons with an annual average temperature ranging from 13.5 °C to 14.9 °C. By the end of 2023, Xi'an administered 11 districts and 2 counties, covering a total area of 10,096.89 km^2 and supporting a permanent population of 13.0782 million. Xi'an, serving as an international comprehensive transportation hub city, a critical node city of the Belt and Road Initiative, a national central city of China, a major industrial city in Northwest China, and the core city of the Guanzhong Plain urban agglomeration, occupies a pivotal strategic position. However, as the provincial capital of Shaanxi and situated within the Fenwei Plain (a heavily polluted region), Xi'an faces severe $PM_{2.5}$ pollution challenges.

The study area is the central urban area enclosed by the Xi'an Ring Expressway, a region characterized by complex built environments, diverse urban morphologies, high population density, intense traffic flow, mixed land use, and pronounced street canyon effects. The area also features a distinctive checkerboard-radial road network. Furthermore, comprehensive coverage of Baidu Street View Imagery data in this region enhances the accuracy and reliability of research outcomes. Figure 2 shows the overview map of the study area. This research selects the $PM_{2.5}$ pollution problem of Xi'an, a typical polluted city in the Fenwei Plain, for the study, which has significant representativeness and urgency.

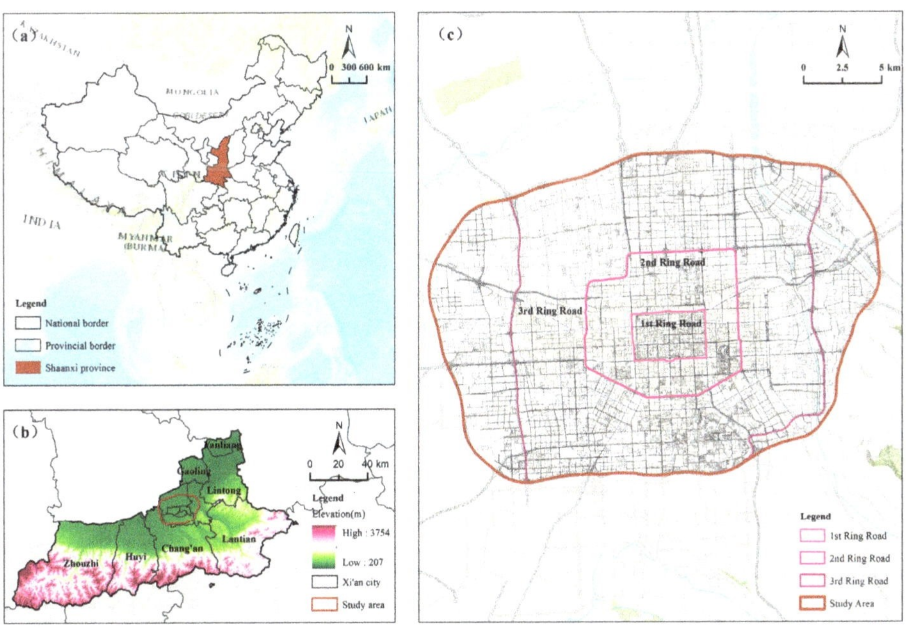

Figure 2. Overview of the study area: (**a**) location of Shaanxi Province within China; (**b**) location of the study area within Xi'an City; (**c**) boundary of the study area. Note: Administrative boundary vector data are sourced from the National Geomatics Center of China (Approval No. GS (2024) 0650).

2.3. Street View Imagery Data Collection

Street View Imagery, as a data source capable of visually reflecting built environment characteristics, has been widely applied in urban planning, environmental protection, and other fields. Google Street View (GSV), a globally renowned map service, provides extensive coverage in over 90 countries and regions. However, due to policy restrictions, Google services are unavailable in mainland China, and GSV has almost no coverage in this region. Tencent Street View (TSV) has covered most cities in mainland China since 2011, but its data updates have been discontinued due to business adjustments [63]. In contrast, Baidu Street View (BSV) covers 652 cities in mainland China, with a total road imagery length of 2.295 million kilometers, covering more than 80% of Chinese cities. Its data quality is comparable to GSV, and it has been widely utilized in various research fields [39].

This study selected Baidu Maps (https://map.baidu.com) as the data source for Street View Imagery, focusing on the area within the Xi'an Ring Expressway. The specific data collection process is as follows: Firstly, road network data of the study area were downloaded from OpenStreetMap (https://www.openstreetmap.org/). Using ArcGIS 10.8 software, Street View Imagery sampling points were set at 50 m intervals along the road network, with a total of 45,404 sampling points established in the study area. Secondly, Python 3.9 software was utilized to invoke Baidu Maps' API, with camera angle parameters set to 0°, 90°, 180°, and 270° to collect Street View Imagery from the front, back, left, and right directions for each sampling point. The images had a resolution of 600 × 480 pixels, and the vertical viewing angle was set to 7°. After data collection, data cleaning and removal of invalid data were performed, resulting in 133,860 valid Street View Imagery. Finally, the front, back, left, and right images of each sampling point were stitched to generate 33,465 panoramic Street View Images for subsequent analysis. Table 1 presents the parameter settings for Baidu Maps' panoramic static images.

Table 1. Baidu map panoramic static imagery parameter settings.

Parameter	Description	Default	Set Value
Heading	Horizontal visual angle of imagery, range [0, 360]	0	0/90/180/270
Pitch	Vertical visual angle of imagery, range [0, 90]	0	7

2.4. Semantic Segmentation Model and Dataset

ADE20K is the largest open-source dataset for semantic segmentation and scene parsing released by the MIT computer vision team. The dataset includes 150 categories of elements, such as sky, road, car, and plant, and has been widely used for semantic segmentation of Street View Imagery [64]. ADE20K comprises three types of data: training data (20,210 images), validation data (2000 images), and test data (3000 images) [65]. ResNet50 is a commonly used convolutional neural network architecture widely applied in image recognition tasks. The model architecture consists of an encoder and a decoder. The CSAILVision framework, released on GitHub (https://github.com/CSAILVision/semantic-segmentation-pytorch, accessed on 30 September 2025) by the MIT Computer Science and Artificial Intelligence Laboratory (CSAIL), was adopted in this study. This project is a mature PyTorch framework based on the ResNet50 model pre-trained on the ADE20K dataset, which has been extensively cited and applied in the computer vision community. This ensures the reproducibility and reliability of the approach. As shown in Table 2, the performance of the ResNet50 model indicates its effectiveness in segmenting Street View Imagery. Figure 3 illustrates the process of Street View Imagery collection and semantic segmentation.

Table 2. ResNet50 model performance metrics in street view imagery processing.

Architecture	Multiscale Testing	Mean IoU	Pixel Accuracy (%)	Overall Score	Inference Speed (fps)
ResNet50dilated	NO	41.26	79.73	60.50	8.3
+ PPM_deepsup	Yes	42.14	80.13	61.14	2.6

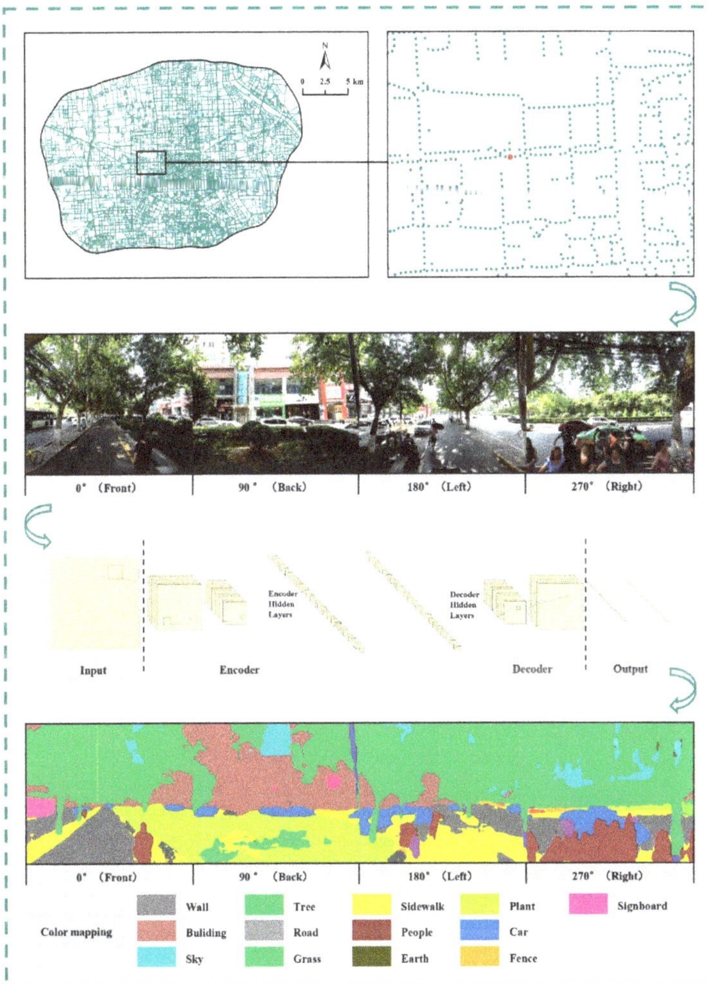

Figure 3. Street view imagery collection and image semantic segmentation process.

2.5. $PM_{2.5}$ Concentration Data

This study selected the High-resolution and High-quality Ground-level $PM_{2.5}$ Dataset for China [66,67], provided by the National Tibetan Plateau Data Center (http://data.tpdc.ac.cn). This dataset offers $PM_{2.5}$ data from 2000 to the present, with a ten-fold cross-validation coefficient of determination (R^2) of 0.92 and a root mean square error (RMSE) of 10.76 $\mu g/m^3$. Covering the entire territory of China, the dataset has a spatial resolution of 1 km and temporal resolutions of daily, monthly, and annual scales, with units expressed in $\mu g/m^3$.

Compared to traditional ground-based monitoring station data, the ChinaHighAirPollutants (CHAP) dataset offers advantages such as extensive coverage, a long time series, high spatial resolution, and low cost. In recent years, it has been widely applied in multiple fields, including air pollution research [68–70], environmental science [6,71–73], and medical research [74–76]. Furthermore, the validity and accuracy of this dataset have been extensively studied and verified.

According to data released by the Ministry of Ecology and Environment of the People's Republic of China in 2023, Xi'an ranked as the fifth-lowest among 168 key cities in China in terms of air quality [77]. Therefore, the 2023 annual average $PM_{2.5}$ pollution data were selected for analysis in this study. The selection of this specific timeframe aims to better elucidate the spatial heterogeneity effects of Xi'an's microscale built environment on $PM_{2.5}$ concentrations, thereby providing a scientific basis for improving air quality in Xi'an.

2.6. Spatial Autocorrelation Analysis

Spatial autocorrelation analysis is commonly employed in research fields such as environmental science, urban planning, and ecology [78–81]. This study calculated the Global Moran's I to analyze the spatial clustering and dispersion patterns of $PM_{2.5}$ concentrations within the study area. The value of Global Moran's I ranges from −1 to 1, where values greater than 0 indicate spatial positive autocorrelation, values less than 0 indicate spatial negative autocorrelation, and values close to 0 suggest random spatial distribution. The calculation formula for Global Moran's I is as follows:

$$I = \frac{n \sum_{i=1}^{n} \sum_{j=1}^{n} w_{ij} \sum_{i=1}^{n} (x_i - \overline{x})(x_j - \overline{x})}{\sum_{i=1}^{n} \sum_{j=1}^{n} w_{ij} \sum_{i=1}^{n} (x_j - \overline{x})^2} \quad (1)$$

In the formula, I represents the Global Moran's I; x_i and x_j denote the $PM_{2.5}$ concentrations in grids i and j, respectively; $\overline{x}$ signifies the mean $PM_{2.5}$ concentration; W_{ij} denotes the spatial weight matrix; and n represents the total number of samples.

Building on the previous analysis, this study further calculated the Local Moran's I to examine the spatial clustering of high and low $PM_{2.5}$ concentrations within the study area. The Local Moran's I values can be categorized into four types of clusters: the high–high cluster, high–low cluster, low–low cluster, and low–high cluster. The calculation formula for Local Moran's I is as follows:

$$I_i = \frac{(x_i - \overline{x}) \sum_{j-1.j \neq i}^{n} W_{ij}(x_j - \overline{x})}{S_i^2} \quad (2)$$

$$S_i^2 = \frac{\sum_{j-1,j \neq i}^{n} (x_j - \overline{x})^2}{n-1} - (\overline{x})^2 \quad (3)$$

In the formula, I_i denotes the Local Moran's I, $S_i{}^2$ represents the variance of $PM_{2.5}$ concentrations, and the definitions of other variables remain consistent with those in Equation (2).

2.7. Spatial Regression Analysis

In urban planning, environmental science, and related research fields, fishnet grids are commonly employed for spatial partitioning to facilitate data integration and overlay analysis [59,82]. Consequently, this study utilized ArcGIS 10.8 software to generate a fishnet tool, dividing the study area into 200 m × 200 m grids. Ten urban built environment factor datasets were then assigned to the corresponding grids. During spatial analysis, these fishnet grids served as the fundamental spatial unit, laying the groundwork for subsequent spatial regression analysis.

To analyze the effects of microscale built environment factors on $PM_{2.5}$ concentrations, three spatial regression models were used in this study. First, the OLS model was used to establish the base regression model. Second, the GWR model was used to explain the spatial non-stationarity. Finally, the MGWR model was used to explain the spatial heterogeneity of the effect factors.

(1) OLS Regression Model

The OLS regression model has been widely adopted in urban planning, environmental science, and related fields [83,84]. As a foundational global linear regression model, it does not account for spatial variations. In this study, the OLS model was initially employed to examine whether significant linear relationships exist between built environment factors and $PM_{2.5}$ concentrations. The OLS model is formulated as:

$$Y = X\beta + \epsilon \tag{4}$$

In the formula, Y represents the dependent variable, X denotes the matrix of explanatory variables, β signifies the coefficient vector, and ε is the random error term vector.

(2) GWR Regression Model

Due to the limitation of the OLS regression model in neglecting spatial heterogeneity, the GWR model is a modification of the OLS model. The GWR model permits regression coefficients to vary spatially by employing a uniform bandwidth, thereby capturing spatial non-stationarity. The GWR model is formulated as:

$$Y_i = \alpha_{0(u_i,v_i)} + \sum_{k=1}^{m} a_{k(u_i,v_i)} X_{k(u_i,v_i)} + \varepsilon_i \tag{5}$$

In the formula, Y_i represents the $PM_{2.5}$ concentration, (u_i, v_i) denotes the coordinates of the i-th sampling point, α_0 (u_i, v_i) is the model intercept, α_k signifies the regression coefficient of the k-th independent variable at (u_i, v_i), X_k (u_i, v_i) is the k-th attribute at location i, and ε_i is the random error term.

(3) MGWR Regression Model

The MGWR model is a modification of the GWR model. By assigning distinct bandwidths to each explanatory variable, it effectively addresses spatial non-stationarity and comprehensively reveals spatial heterogeneity across various effect factors. Research has demonstrated the MGWR model's notable advantages in fields like atmospheric pollution. For example, Duan et al. utilized the MGWR model to elucidate the spatial heterogeneity between three-dimensional urban landscape patterns and $PM_{2.5}$ pollution in Wuhan [68]. Wu et al. employed the MGWR model to analyze multi-scale interactions and spatial heterogeneity of natural and socioeconomic factors affecting $PM_{2.5}$ concentrations [85]. Zhou et al., focusing on the Niger River Basin in West Africa, applied the MGWR model to investigate relationships between $PM_{2.5}$ concentrations and natural/land-use factors [86]. These studies fully validate the effectiveness of the MGWR model in revealing spatial heterogeneity from different perspectives and regions. Therefore, the present study employs the MGWR 2.2 software, incorporating built environment factors as independent variables and $PM_{2.5}$ concentrations as the dependent variable, to explore their spatial heterogeneity. The MGWR model is formulated as:

$$Y_i = \alpha_{0(u_i,v_i)} + \sum_{k=1}^{m} a_{bwk(u_i,v_i)} X_{k(u_i,v_i)} + \varepsilon_i \tag{6}$$

The MGWR model is a modification of the GWR model. In the formula, Y_i represents the $PM_{2.5}$ concentration at the i-th sampling point, (u_i, v_i) denotes the coordinates of the i-th sampling point, m is the total number of sampling points, a_{bwk} is the k-th explanatory variable with an added bandwidth parameter bw, and each explanatory variable has a distinct spatial scale. ε_i is the random error term.

3. Results

3.1. Screening of Urban Built Environment Factors

This study employs the ResNet50 deep convolutional neural network architecture combined with the ADE20K dataset to perform semantic segmentation on 33,465 panoramic Street View Images of the study area, quantifying various elements within the images. The semantic segmentation results enable the calculation of the percentage of each element relative to the total elements in the Street View Imagery. The ADE20K dataset encompasses diverse scene elements, including 150 semantic categories covering indoor and outdoor elements such as architecture, natural landscapes, transportation facilities, and furniture [64,65]. Given that not every element can be considered an urban built environment factor affecting $PM_{2.5}$ concentrations, it is necessary to screen the elements in the semantic segmentation results of Street View Imagery.

This study adopts a three-stage approach to screen indicator factors, ensuring efficient and accurate identification of key factors. First, during the preliminary filtering stage, factors with zero or low proportions (such as Ship, Fan) were excluded from the 150 elements derived from Street View Imagery semantic segmentation results, resulting in the elimination of 125 invalid variables. Second, in the primary factor identification stage, elements were ranked in descending order based on their average proportions in Street View Imagery, and the top 10 key built environment factors with the highest proportions were extracted, including sky, building, road, tree, and car. Relevant studies also indicate that factors such as sky, building, tree, and car have an effect on $PM_{2.5}$ dispersion [68,87,88]. Finally, during the data analysis stage, descriptive statistical analysis was conducted on these 10 key built environment factors using IBM SPSS Statistics 26 software. Table 3 shows the descriptive statistics of the 10 built environment factors.

Table 3. Descriptive statistics of 10 built environment factors.

Number	Elements	Min	Max	Mean	STD
1	Sky	0.000	0.632	0.260	0.136
2	Building	0.000	0.892	0.250	0.173
3	Road	0.000	0.415	0.173	0.077
4	Tree	0.000	0.746	0.127	0.111
5	Car	0.000	0.344	0.045	0.052
6	Sidewalk	0.000	0.379	0.041	0.042
7	Wall	0.000	0.774	0.020	0.046
8	Plant	0.000	0.464	0.018	0.029
9	Fence	0.000	0.351	0.014	0.023
10	Grass	0.000	0.331	0.011	0.023

Note: The built environment factors in the table are sorted in descending order based on the mean value. Data were extracted from 33,466 street-view panoramas in the study area.

The research findings indicate significant differences among the 10 built environment factors in terms of minimum, maximum, mean, and standard deviation values. Sky, as a crucial component of urban landscapes, not only represents the openness of the sky but also affects surface temperature and air quality. In this study, the sky factor exhibited the highest proportion in Street View Imagery, with a mean value of 0.260. However, substantial fluctuations were observed across the study area, with a minimum value of

0.000, a maximum value of 0.632, and a standard deviation of 0.136. These findings suggest significant disparities in the sky factor across different neighborhoods within the study area. Such disparities may be attributed to factors such as neighborhood building density and street orientation.

Buildings, as an important part of the urban built environment, have a density that affects air quality, the urban heat island effect, ventilation conditions, and the comfort of residents. In this study, the building factor demonstrates a mean value of 0.250, which is second only to the sky factor, indicating its dominant presence in the Street View Imagery of the study area. The minimum and maximum values of 0.000 and 0.892 reveal substantial spatial heterogeneity in building distribution, with localized areas exhibiting high-density development patterns. A standard deviation of 0.173 underscores the complexity and diversity of urban building densities across the study area, which may contribute to microclimatic variations.

As an important part of the urban transportation system, the width, density, and traffic flow of roads affect urban traffic conditions, air quality, and accessibility. In this study, the mean value of the road factor is 0.173, which is lower than the sky and building, but still an important part of the urban built environment. The standard deviation is 0.077, which indicates that roads in the study area are characterized by an unbalanced distribution.

Trees, as critical components of the urban green space system, significantly contribute to improving urban air quality and influencing $PM_{2.5}$ dispersion and deposition. In this study, the tree factor exhibits a mean value of 0.127, with a maximum of 0.746, indicating that while street green spaces remain limited, certain areas demonstrate effective tree preservation and utilization, achieving relatively high coverage rates. However, a standard deviation of 0.111 reveals considerable data variability, reflecting the uneven spatial distribution of street green spaces across the study area.

In this study, the minimum value of the car factor is 0.000, the maximum value is 0.344, the mean value is 0.045, and the standard deviation is 0.052. As an important component of urban transportation, the effect of cars on air quality mainly involves two aspects: one is the bluff body effect, which physically obstructs airflow, and the other is their role as a source of $PM_{2.5}$ emissions. However, since Street View Imagery is static, the car factor extracted from it primarily reflects the former—the obstruction effect on airflow—while it cannot effectively distinguish between moving and parked vehicles, nor directly capture emission information.

Sidewalks, as critical components of urban transportation systems, have a significant effect on traffic accessibility through their quantity, density, and width. In this study, the sidewalk factor exhibits a mean value of 0.041, a maximum of 0.379, and a standard deviation of 0.042, indicating significant variations in sidewalk infrastructure across different neighborhoods.

Walls, as boundary elements of urban space, reflect wall facade visibility and spatial closure. In this study, the mean value of the wall is 0.020, and the standard deviation is 0.046. This shows the unevenness of the distribution of the wall factor in the study area. This inhomogeneity may be due to different urban planning and the choice of building materials in different areas.

Plants, as critical components of the urban green space system, have a significant effect on urban air quality and ecological quality through their species composition, quantity, and spatial distribution. In this study, the plant factor exhibits a mean value of 0.018 and a standard deviation of 0.029, indicating limited spatial distribution across the study area.

As an important tool for urban space management, the type, height, and density of fences affect the urban interface, the tidiness and safety of the urban landscape, and the travel behavior of residents. In this study, the mean value of fences is 0.014, which is low but still has some presence in the Street View Imagery. The maximum value of the fence factor is 0.351, indicating that some neighborhoods are densely fenced, which may be closely related to the needs of urban space management in some areas.

Grass, as a critical component of the urban green space system, has a significant effect on urban microclimates, surface temperatures, and air quality. In this study, the grass factor exhibited a minimum value of 0.000, a maximum value of 0.331, and an average value of 0.011. These results suggest that grass distribution within the study area is relatively sparse and uneven.

In summary, the descriptive statistical analysis of the ten built environment factors systematically reveals their distribution characteristics and data quality, accurately grasps the distribution patterns of the data, and identifies potential anomalies. Through this analysis, it ensures the accuracy and credibility of the research findings and provides a solid foundation for subsequent spatial heterogeneity analysis.

3.2. Spatial Distribution of $PM_{2.5}$ Concentrations

The study obtained $PM_{2.5}$ concentration data from the National Tibetan Plateau Data Center. The annual average $PM_{2.5}$ concentration in the study area ranged from 45.6 μg/m^3 to 52.7 μg/m^3 during 2023. Figure 4a illustrates the spatial distribution of $PM_{2.5}$ concentrations in Xi'an City (China) during 2023.

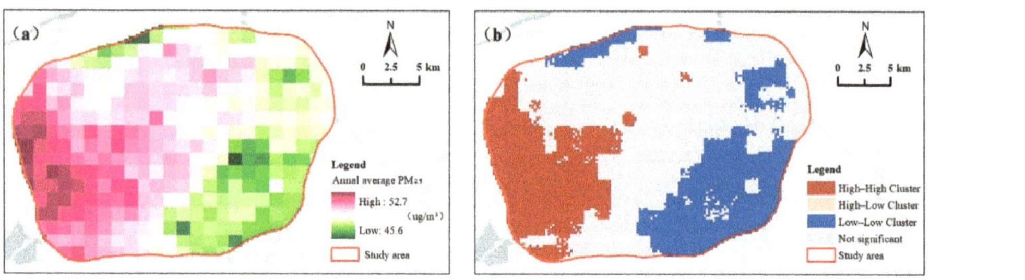

Figure 4. Spatial distribution and local spatial autocorrelation analysis of $PM_{2.5}$ concentrations in Xi'an City, 2023: (**a**) spatial distribution of $PM_{2.5}$ concentrations; (**b**) local spatial autocorrelation distribution of $PM_{2.5}$ concentrations.

In this study, spatial autocorrelation analysis of $PM_{2.5}$ concentration was conducted to reveal the characteristics of the spatial distribution of $PM_{2.5}$ concentration in the study area. The results of the global Moran's index and local Moran's index analysis are as follows. The results of the global Moran's index analysis showed that the Moran's I index was 0.97, indicating a positive spatial effect of $PM_{2.5}$ concentration. The p-value was 0.00 (p-value < 0.01), and the z-value was 148.11 (z-value > 2.58). It indicates that the distribution of $PM_{2.5}$ concentration in the study area shows a clustering trend. The results of the localized Moran index analysis showed that high and low values were spatially clustered, with high–high clusters mainly concentrated in the west side of the study area, and low–low clusters mainly concentrated in the southeast and northeast side of the study area. High–low clusters and low–high clusters are relatively rare in the study area. Figure 4b shows the local spatial autocorrelation distribution.

3.3. Spatial Regression Results

(1) OLS model regression results

In this study, the OLS model was first used to initially explore the linear relationships between the 10 variables and $PM_{2.5}$ concentrations in the microscale built environment. In the OLS model, the VIF values of the 10 variables were lower than 3.760, which indicated that there was no significant multicollinearity problem among the variables selected in this study. The results of this model showed an R^2 of 0.030, an adjusted R^2 of 0.029, and an AICc of 38506.261. In addition, the p-value of the Koenker (BP) statistic was 0.000 (p-value < 0.01), which indicated that the OLS modeling relationship was non-stationary and there was spatial heterogeneity in the variables. Therefore, further in-depth analysis using GWR modeling is necessary.

(2) GWR model regression results

This study was further analyzed using the GWR model, which is able to account for spatial heterogeneity. The results of this model showed an R^2 of 0.916, an adjusted R^2 of 0.908, and an AICc of 11,379.701, which suggests that this model is a better fit than the OLS model. However, the GWR model with the same bandwidth settings for all explanatory variables has some limitations in explaining spatial heterogeneity and needs to be further analyzed using the MGWR model.

(3) MGWR model regression results

MGWR is a further improvement of the GWR model, which allows different bandwidths to be set for different explanatory variables with higher accuracy and more precise refinement of the spatial heterogeneity of the different variables. The results of the MGWR model show an R^2 of 0.981 and an adjusted R^2 of 0.974, indicating a high goodness-of-fit. Compared with the OLS model (AICc = 32,970.558) and the GWR model (AICc = 11,379.701), the MGWR model exhibits a marked decrease in AICc to −3897.948. Meanwhile, the residual standard deviation (Sigma) decreases from 0.385 (GWR) to 0.163 (MGWR). These simultaneous reductions in AICc and Sigma suggest that the improved fit is due to the superior performance of the MGWR model rather than overfitting. The robustness of the model is further supported by its ability to assign distinct bandwidths to each variable, with specific variables operating at different spatial scales, such as Building (bandwidth of 43.000) representing a highly local influence and Grass (bandwidth of 9870.000) representing an essentially global influence.

This study compares the modeling performance of OLS, GWR, and MGWR models, respectively. The results show that the OLS model has the smallest adjusted R^2 value and the largest AICc value. The MGWR model has the largest adjusted R^2 value and the smallest AICc value. This indicates that the MGWR model has the best fit and is better than the GWR and OLS models. The MGWR model can aptly reveal the spatial heterogeneity of the relationship between the microscale built environment and $PM_{2.5}$ concentration. Table 4 shows the comparison of regression results of OLS, GWR, and MGWR models. Figure 5 illustrates the regression coefficient maps of the ten built environment factors derived from the MGWR model.

Table 4. Comparison of regression results for OLS, GWR, and MGWR models.

--	OLS Coefficients	GWR Coefficients				MGWR Coefficients			
Variables	Mean	Mean	Min	Max	Bandwidth	Mean	Min	Max	Bandwidth
Sky	1.250	−0.111	−6.569	3.961	1380.539	0.002	−0.486	0.836	47.000
Building	0.653	0.035	−3.288	2.575	1380.539	0.017	−0.852	3.140	43.000
Road	−1.809	0.320	−5.652	7.057	1380.539	0.000	−0.134	0.169	322.000
Tree	0.268	−0.271	−3.920	2.105	1380.539	0.003	−0.012	0.012	2890.000
Car	−1.225	0.492	−9.216	16.746	1380.539	−0.001	−3.654	1.463	43.000
Sidewalk	−4.313	0.131	−6.915	13.111	1380.539	−0.010	−4.198	0.698	43.000
Wall	−0.144	0.712	−3.829	8.230	1380.539	0.006	−0.029	0.045	1008.000
Plant	1.226	0.098	−25.415	12.417	1380.539	−0.009	−0.028	0.005	2503.000
Fence	−5.835	−0.460	−33.791	15.895	1380.539	0.004	−0.961	0.964	47.000
Grass	−8.254	0.718	−14.533	24.936	1380.539	0.001	−0.002	0.005	9870.000
R^2	0.030	0.916				0.981			
Adj.R^2	0.029	0.908				0.974			
AICc	38,506.261	11,379.701				−3897.948			

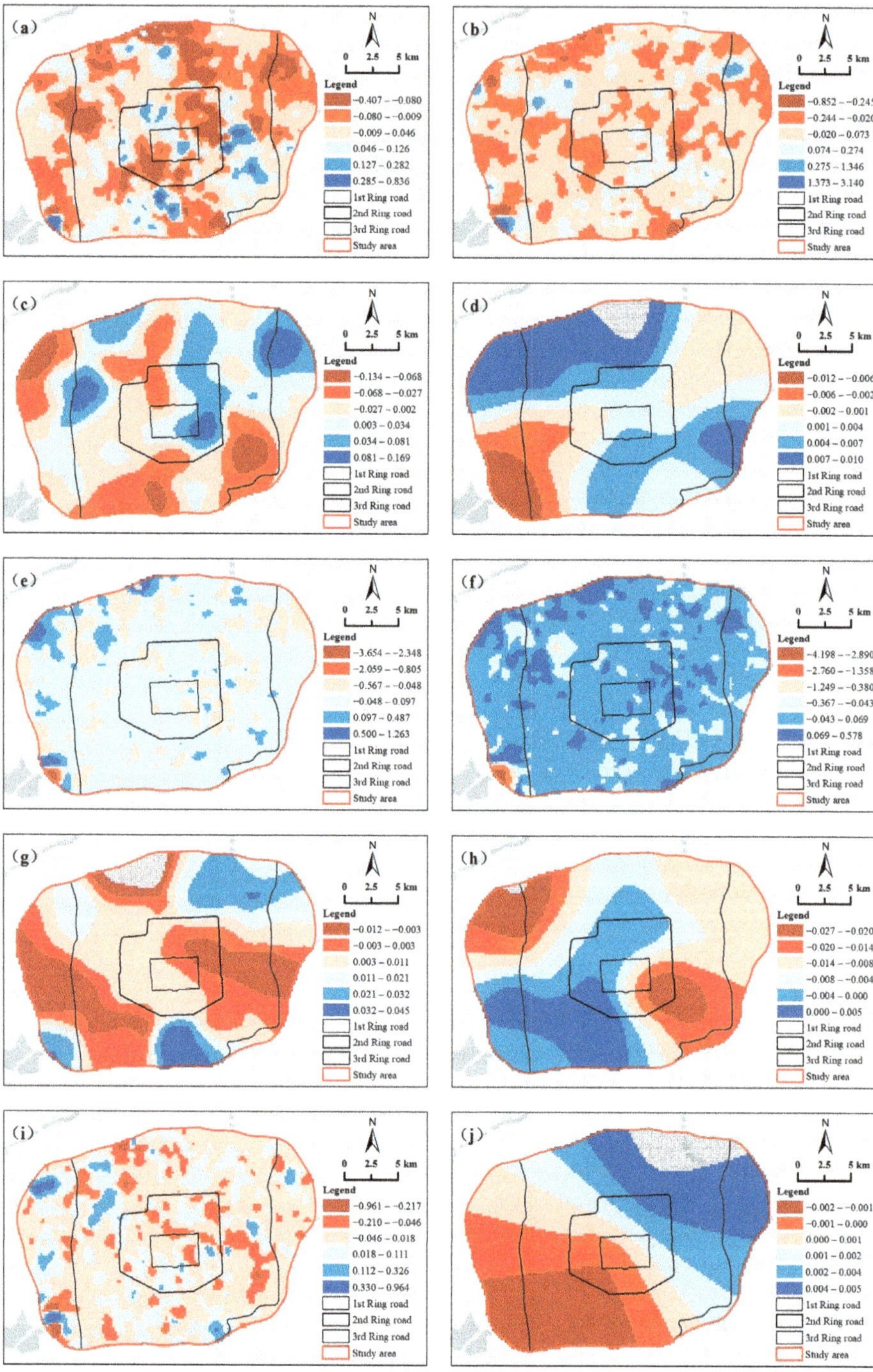

Figure 5. MGWR regression coefficient map for 10 built environment factors in Xi'an: (**a**) Sky; (**b**) Building; (**c**) Road; (**d**) Tree; (**e**) Car; (**f**) Sidewalk; (**g**) Wall; (**h**) Plant; (**i**) Fence; (**j**) Grass.

The MGWR model revealed the effects of 10 built environment factors on $PM_{2.5}$ concentrations, and there was significant spatial heterogeneity in these effects. The specific analysis of the regression coefficients of the MGWR model is presented below:

Firstly, Table 4 presents the optimal bandwidths for different factors within the MGWR model. By assigning factor-specific optimal bandwidths, the MGWR model optimizes its fitting performance. The results reveal substantial variations in bandwidths across these factors. Notably, factors such as road, tree, wall, plant, and grass exhibit relatively large bandwidths, spanning from 322.000 to 9870.000. This suggests that the effect of these variables on $PM_{2.5}$ concentrations manifests at broader spatial scales, with regression coefficients exhibiting weaker spatial heterogeneity. Conversely, factors including sky, building, car, sidewalk, and fence demonstrate relatively small bandwidths, ranging between 43.000 and 47.000. This indicates a localized effect of these factors on $PM_{2.5}$ concentrations, accompanied by stronger spatial heterogeneity in their regression coefficients.

Furthermore, Table 4 presents the mean, minimum, and maximum values of regression coefficients for each factor within the MGWR model. A detailed analysis of each factor's regression coefficients is provided as follows:

The sky factor represents the degree of openness of the sky. Its average coefficient is positive (average coefficient of 0.002), which indicates that the sky factor has a positive effect on $PM_{2.5}$ concentration. It reflects that the lack of vegetation cover in the open sky area causes $PM_{2.5}$ dispersion to be affected. The sky factor demonstrates a maximum coefficient of 0.836 and a minimum of −0.486, indicating regional variations in its effect across different areas.

The building factor demonstrates a positive effect on $PM_{2.5}$ concentrations (average coefficient of 0.017). This suggests that densely built environments may impede airflow, thereby obstructing $PM_{2.5}$ dispersion. The building factor exhibits a maximum coefficient of 3.140 and a minimum of −0.852, indicating spatial disparities in its effect across localized regions.

The average coefficient of the road factor is 0.000, with a minimum value of −0.134 and a maximum value of 0.169, which indicates that the road factor as a whole has a weak effect on the $PM_{2.5}$ concentration, but there are some differences in localized areas.

The average coefficient of the tree factor is positive (0.003, with minimum and maximum values of −0.012 and 0.012, respectively). This may be related to the limitations of Street View Imagery, which mainly captures horizontal perspectives of the built environment and may have limited ability to reflect differences in tree species or the vertical structure of tree canopies.

The average coefficient of the car factor is negative (−0.001), with minimum and maximum values of −3.654 and 1.463, respectively. This may reflect the effects of other factors on the car factor. Street View Imagery performs well in extracting static features of the built environment; therefore, the extracted car factor primarily reflects static characteristics, but it has limitations in capturing dynamic features, such as traffic flow or vehicle density.

The average coefficient of the sidewalk factor is negative (average coefficient of −0.010), with minimum and maximum values of −4.198 and 0.698, respectively. This suggests that the sidewalk factor has a negative effect on $PM_{2.5}$ concentrations. Sidewalk width and pedestrian friendliness enhancement may reduce short-distance motor vehicle use, and locally isolated sidewalks with greenery may reduce $PM_{2.5}$ concentrations.

The mean coefficient of the wall factor is positive (mean coefficient of 0.006), which indicates that the wall factor positively affects the $PM_{2.5}$ concentration. The minimum and maximum values of the wall factor were −0.029 and 0.045, respectively. This indicates that the effect between the two varies considerably in the study area, with localized continuous walls impeding ventilation and exacerbating pollutant accumulation.

The average coefficient of the plant factor was negative (mean coefficient of −0.009), which indicated that the plant factor had a negative effect on the $PM_{2.5}$ concentration, and three-dimensional greening plants would significantly inhibit $PM_{2.5}$, and the plant dust retention efficiency was affected by the hierarchical structure of the vegetation and the maintenance status. The minimum and maximum values of the plant factor were −0.028 and 0.005, respectively, which indicated that there were differences in the effects of the plant factor on $PM_{2.5}$ concentration.

The fence factor demonstrates a positive mean coefficient (mean coefficient of 0.004), with coefficient values ranging from −0.961 to 0.964. This indicates that fences exert a positive effect on $PM_{2.5}$ concentrations within the study area. Functioning as physical barriers, permeable fences enhance ventilation while solid fences exacerbate pollution accumulation.

The average coefficient of the grass factor was 0.001, with a minimum value of −0.002 and a maximum value of 0.005, which indicates that the grass factor has a weak effect on $PM_{2.5}$ concentrations, but there is some variation in localized areas.

In conclusion, the results of the MGWR model showed that the 10 built environment factors in the study area had significant effects on the spatial heterogeneity of $PM_{2.5}$ concentrations. Based on the positive and negative regression coefficients and the strength of their effects, they can be categorized into the following three categories. The first category—buildings, walls, and fences—has a strong positive effect on $PM_{2.5}$, and the strength of their effect may be related to pollutant retention in high-density built environments. Trees, sky, and grass have a weak positive effect on $PM_{2.5}$ concentrations, which may be associated with vegetation blocking diffusion or open space inversion effects. This may also be related to the limitation of Street View Imagery, which mainly captures horizontal perspectives and may not fully reflect differences in tree species, vegetation types, or canopy structures. The second category—sidewalks, plants, and cars—has certain negative regression effects on $PM_{2.5}$ concentrations. On one hand, this reflects the potential of pedestrian-friendly design, vegetation adsorption, and traffic management to mitigate $PM_{2.5}$ pollution. On the other hand, car factors extracted from Street View Imagery mainly capture static characteristics and are limited in representing dynamic features, such as traffic flow or vehicle density, which may affect the regression results. The third category—compared with other built environment factors, the overall effect of the road factor on $PM_{2.5}$ concentrations is relatively weak.

This study employed the MGWR model to investigate the spatial heterogeneity relationship between built environment factors and $PM_{2.5}$ concentrations. The following presents a detailed analysis of the spatial distribution patterns of these regression coefficients:

The sky factor exhibits an overall positive effect on $PM_{2.5}$ concentrations, yet with pronounced spatial heterogeneity. Areas demonstrating a significantly positive effect are primarily concentrated in the industrial heritage zone outside the Second Ring Road, specifically the old industrial bases surrounding the Xingfu Forest Belt, where dense factory clusters impede pollutant diffusion. Additionally, the urban–rural transition zone southwest of the Ring Expressway is also positively significant, likely due to village sprawl and fragmented open spaces in this region. Conversely, areas with significant negative effects are predominantly located in large parks and green spaces, such as Daming Palace National Heritage Park and Xingqing Palace Park within the Second Ring Road, and in science and education clusters. They are also found near numerous parks, including Xi'an World Horticultural Expo Park and Qujiang Lake Heritage Park, both inside and outside the Third Ring Road.

The building factor has a positive effect on the $PM_{2.5}$ concentration as a whole, but with significant spatial heterogeneity. The significant positive effect areas are mainly concentrated in the high-density built-up area of the city. They are distributed in the northwestern part of the Second Ring Road and in the densely built-up area from the Second Ring Road to the southeastern part of the Third Ring Road, peaking around the villages in the southwestern part of the Ring Expressway due to the agglomeration of low-rise contiguous building clusters. Significant negative effect areas are mainly concentrated around Changle Park in the eastern part of the Second Ring Road, parks along the Third Ring Road, and near a number of industrial parks in the western part of the Ring Expressway.

The road factor exhibits a relatively weak overall effect on $PM_{2.5}$ concentrations but demonstrates localized positive and negative effects. Notable positive effect zones primarily concentrate in the southeastern sections of the First and Second Ring Roads, as well as along the eastern and western stretches of the Third Ring Road. Conversely, significant negative effect areas are mainly distributed in the southern part of the Third Ring Road. Multiple urban parks, including Yongyang Park, Tang City Wall Ruins Park, Qujiang Lake Heritage Park, and Xi'an Botanical Garden, are located there, contributing to $PM_{2.5}$ mitigation. Additionally, negative effects are observed near the Dongfang Factory on the eastern Third Ring Road and industrial parks northwest of the Ring Expressway.

The tree factor generally exhibits a positive effect on $PM_{2.5}$ concentrations; however, spatial variations exist within the study area. Limited tree distribution and low green coverage in urban centers contribute to their limited effectiveness in reducing $PM_{2.5}$ concentrations. In contrast, areas southwest and northeast of the Ring Expressway, adjacent to regions such as Kunmingchi Qixi Park and Xi'an World Horticultural Expo Park, demonstrate a significant negative effect on $PM_{2.5}$ concentrations.

The car factor has a weak negative effect on the $PM_{2.5}$ concentration as a whole, but there is significant spatial heterogeneity. The significant positive effect area is mainly concentrated in the north side of the Third Ring Road and the northwestern part of the city ring road, and around the city highway near the industrial park. The significant negative effect area is mainly concentrated along the city ring road and the Ring Expressway.

The sidewalk factor had an overall negative effect on $PM_{2.5}$ concentrations, but there was significant spatial heterogeneity. Except for the urban ring road and along the Ring Expressway in the study area, the sidewalk factor showed a significant negative effect on $PM_{2.5}$ concentration in most parts of the study area.

The wall factor exhibits an overall positive effect on $PM_{2.5}$ concentrations. Except for the southwestern and southeastern regions of the study area, dense building facade areas demonstrate a significant positive effect on $PM_{2.5}$ concentrations across most regions.

Except for the urban–rural transition zone in the southwestern zone of the study area, most of the areas of the study area showed significant negative effects of plant factors on $PM_{2.5}$ concentrations. Among them, a significant negative effect was presented near several areas with better ecological environments, such as Changle Park, Xingqing Park, and Qinglong Temple Heritage Park.

The fence factor has an overall positive effect on $PM_{2.5}$ concentrations, but there is significant spatial heterogeneity. The area of significant positive effect is mainly concentrated in the vicinity of industrial parks in the western part of the Ring Expressway. The zone of significant negative effect is mainly concentrated in the north of the Third Ring Road and the southeast near the Oriental Factory.

The grass factor had an overall positive effect on $PM_{2.5}$ concentrations. However, there were some differences in the study area. The area of significant positive effect is mainly concentrated in the northeast of the study area. However, in the southwest of the study area, the grass factor in Yongyang Park, Tang City Wall Heritage Park, and Fengqing Park had a significant negative effect on $PM_{2.5}$ concentrations.

In summary, the distribution characteristics of regression coefficients for the ten built environment factors in the study area indicate significant differences in their effect intensity and spatial distribution on $PM_{2.5}$ concentrations. This finding not only reveals the complex nonlinear relationships between these ten built environment factors and $PM_{2.5}$ concentrations but also challenges the conventional homogeneous governance model under the "one city, one policy" framework. Consequently, subsequent strategies should be formulated with precise optimization measures based on spatial heterogeneity characteristics.

4. Discussion

4.1. Effects of Micro-Scale Built Environment on $PM_{2.5}$ Pollution

In this study, semantic segmentation of Street View Imagery was employed to quantify the morphological characteristics of the urban built environment, enabling the extraction of key indicator variables influencing $PM_{2.5}$ concentration distributions. Three regression models, OLS, GWR, and MGWR, were comparatively analyzed, and the MGWR model was selected as optimal for this study. The relationship between the urban built environment and $PM_{2.5}$ pollution was analyzed using MGWR, revealing the effect of significant spatial heterogeneity.

Firstly, the study findings indicate that buildings and walls exert a significant positive effect on $PM_{2.5}$ concentrations. At the microscale, the layout of buildings and walls can impede $PM_{2.5}$ diffusion by reducing air ventilation efficiency, which aligns with the research results of Wu et al. [89]. Generally, urban greening plays a certain mitigating role in $PM_{2.5}$ concentrations. However, this study discovered a positive effect between trees, grass, and $PM_{2.5}$ concentrations in localized areas of the study region. This finding has also been confirmed in relevant studies, indicating that the effect of urban vegetation on air quality is more complex than previously assumed. Trees and other vegetation may hinder ventilation and reduce air circulation, thereby affecting pollutant dispersion [90]. Urban greening at the street level may exacerbate air pollution [91]. The study results demonstrate that unreasonable sky view factors in densely built-up industrial areas and areas near the rural–urban fringe can impede air circulation and aggravate air pollution. This is consistent with the research findings of Fan et al., who pointed out a significant effect between sky view factors and $PM_{2.5}$ concentrations [92].

Secondly, the mitigating effect of urban parks on air pollution has been widely confirmed. This study further corroborates this conclusion, revealing a significant negative effect between plant factors near numerous parks within the study area and $PM_{2.5}$ concentrations. This finding is consistent with previous research, such as that conducted by Li et al., which demonstrated that increased vegetation coverage around urban parks significantly expands the $PM_{2.5}$ reduction range [93]. Notably, the car factor exhibits a weak negative correlation with $PM_{2.5}$ concentrations. This phenomenon can be attributed to several factors. On one hand, the city of Xi'an has witnessed a continuous increase in the number of clean energy vehicles in recent years, effectively reducing the intensity of vehicle exhaust emissions. On the other hand, the study area is situated within the Xi'an Ring Expressway, characterized by a high level of urbanization and well-developed infrastructure, including comprehensive charging facilities that facilitate the widespread adoption of clean energy vehicles. Relevant studies also indicate that the transition to electric vehicles can significantly reduce $PM_{2.5}$ concentrations and improve urban air quality [94]. Moreover, the car factor in this study was extracted from Baidu Street View Imagery. Street View Imagery can be used to detect the number and types of vehicles on the road, but it mainly records static features. Due to this characteristic, Street View Imagery has difficulty distinguishing moving vehicles from parked ones and cannot track vehicle movement or real-time emission data. This limitation has also been confirmed in studies by Biljecki et al. [39].

Furthermore, the road factor exhibits a relatively weak overall effect on $PM_{2.5}$ concentrations. However, it demonstrates a significant positive effect along the southeastern and eastern-western sections of the urban ring road. This phenomenon may be attributed to the ring road serving as a major urban artery, carrying substantial traffic volume, which leads to elevated $PM_{2.5}$ concentrations. This finding aligns with the research conducted by Wang et al., who observed a notable increase in $PM_{2.5}$ concentrations in areas with dense high-grade road networks, such as along ring roads [95]. Additionally, this study reveals a significant negative effect of the road factor on $PM_{2.5}$ concentrations in park-dense areas, which is consistent with the research findings of Du et al. [96].

4.2. Optimization Strategies

With the gradual transformation of urban construction in China toward high-quality development, $PM_{2.5}$ pollution mitigation has become a core issue in ecological environment governance. In response to the urgent need for air pollution prevention and control, the state has successively issued systemic policy documents to improve air quality. Notably, the 2025 Implementation Plan for Beautiful City Construction further focuses on key areas and prominent issues of the urban ecological environment, explicitly mandating the enhancement of urban ecological quality. As one of the first batch of pilot cities for urban renewal in China (designated in 2021) and the sole pilot city in Shaanxi Province, Xi'an confronts the dual imperative of reconciling $PM_{2.5}$ pollution mitigation with the optimization of its built environment during its ongoing new urbanization process.

In the face of various problems in urban development, urban renewal has become an important research topic of common concern for domestic and international academics [97]. Its core lies in the comprehensive optimization of urban economy, culture, environment, and other dimensions [98]. In this context, urban street space, as an important part of urban public space, is also the main place for residents' daily activities, and its environmental quality is directly related to the health and well-being of residents. Therefore, it is particularly important to conduct a refined quantitative analysis of urban environmental characteristics [99]. This can not only provide a scientific basis and data support for urban renewal and air pollution control but also provide decision support for optimizing urban spatial layout, improving air quality, and upgrading ecological environment quality.

Based on the bandwidths, regression coefficients, and spatial distribution characteristics of various factors revealed by the MGWR model, this study proposes the following optimization strategies. Firstly, regarding the differences in bandwidth among factors, it is recommended that local governments emphasize not only macroscale regional collaborative governance but also microscale interventions. The importance of tailoring strategies to local conditions should be highlighted, implementing differentiated governance measures to accommodate the complex spatial heterogeneity within cities.

Secondly, for factors such as buildings, walls, and fences, it is suggested to pay attention to optimizing the openness of urban spatial interfaces during urban renewal. In areas with dense buildings, it is suggested to reasonably manage building spacing to maintain ventilation. In areas with dense industrial buildings and along the Ring Expressway, it is suggested to optimize land-use layout and the configuration of open spaces. The addition of vertical green walls on building facades is also suggested to enhance ecological performance. For fences with high enclosures, promoting green-permeable fences is suggested to reduce facade closure, with particular attention to improving areas around the western industrial park along the Ring Expressway. Additionally, it is suggested to optimize the urban green structure to avoid excessive planting that may hinder pollutant dispersion.

Finally, for factors such as sidewalks, plants, and cars, it is suggested to focus on improving the pedestrian environment along urban ring roads and the Ring Expressway, enhancing sidewalk continuity. It is suggested to optimize the layout of the urban green space system to increase plants' capacity for pollutant absorption. It is also suggested to promote low-emission or cleaner vehicles, implement a public-transport-first strategy, and strengthen the management of motor vehicle emissions. At the same time, it is suggested to reinforce the road pollution monitoring network on urban ring roads and main roads to achieve refined pollution management.

In conclusion, most of the existing air pollution control strategies follow the macro governance framework of "one city, one policy". Although certain achievements have been made at the regional scale, they tend to overlook the micro differences in the built environment elements within cities. This study has promoted the transformation of thinking in $PM_{2.5}$ pollution control from "homogeneous control" to "differentiated control", systematically revealing the influence of factors such as the sky, building, and plant on the spatial heterogeneity of $PM_{2.5}$ concentration. This not only provides a scientific basis for avoiding the "one-size-fits-all" policy in environmental protection but also provides an optimization basis for the layout of buildings and green space systems in urban planning. It is also a response to the strategy of "precise pollution control, scientific pollution control, and law-based pollution control" in the "Action Plan for Continuous Improvement of Air Quality" issued by The State Council in 2023. Although this study focuses on Xi'an, the multidisciplinary research framework established here can be extended to other similar cities.

4.3. Research Limitations and Future Directions

This study primarily adopts a method combining Street View Imagery and machine learning to quantify the microscale built environment. Investigating the spatial heterogeneity effects of microscale built environments on $PM_{2.5}$ concentrations demonstrates a certain degree of innovation. However, this study still has the following limitations:

(1) Geographical Limitations: This study analyzes data from one city in one year. Differences in built environments across countries, cities, or years are not considered. Google Street View data are not available for Xi'an; therefore, Baidu Street View is used. This may limit cross-platform comparisons and the generalizability of the findings. Future research suggestions should validate the results in regions with multiple street view platforms.

(2) Update Frequency Limitations: This study uses commercial Street View platforms, whose irregular update frequency limits their applicability for continuous temporal analyses. To address this limitation, future studies could consider using continuous time-series or multi-year Street View Imagery, for example, by supplementing commercial data with self-collected images, to further investigate the dynamic variations of $PM_{2.5}$ concentrations.

(3) Multifactorial Integration Limitations: This study primarily relies on Street View Imagery, which captures only street-level visuals and does not account for data such as rooftop emissions or indoor sources. Future research should integrate multi-source datasets, particularly topography, meteorological, socio-economic, and other emission-related data, to construct a more comprehensive analytical model. This approach can enhance the robustness, accuracy, and comprehensiveness of the research findings and better capture the full range of determinants influencing $PM_{2.5}$ concentrations.

(4) Limitations of Semantic Segmentation and Feature Extraction: This study applies semantic segmentation to Street View Imagery using the ResNet50 deep learning model and the ADE20K dataset, and the overall model performance is robust. However, this approach still has certain limitations in capturing dynamic characteristics and fine-grained features. For example, when quantifying the car factor, Street View Imagery finds it difficult

to distinguish between parked and moving vehicles, which limits the accurate reflection of real-time vehicle emissions. When extracting vegetation factors, horizontal viewing angles limit the identification of canopy structures and vegetation types. Future research should address these limitations by integrating multi-source data, improving semantic segmentation algorithms, and enhancing the accuracy of feature recognition.

5. Conclusions

This study integrates multidisciplinary knowledge from urban planning, environmental science, computer science, and geographic information science (GIS) to overcome the limitations of a single disciplinary perspective and to form an integrated environment–technology–planning analysis paradigm for interdisciplinary research on $PM_{2.5}$ concentrations. By combining Street View Imagery with machine learning techniques, it addresses the limitations of traditional two-dimensional data at the microscale, enabling the quantification of urban built environments from three-dimensional and microscale perspectives. A research framework titled "Street View Imagery Element Extraction—Spatial Heterogeneity Modeling—Planning Strategy Optimization" was constructed, uncovering the spatially heterogeneous relationships between microscale built environments and $PM_{2.5}$ concentrations and demonstrating a certain degree of innovation. The key findings are summarized as follows:

(1) Ten microscale built environment factors in the study area exhibit spatially heterogeneous effects on $PM_{2.5}$ concentrations.

(2) This study compares OLS, GWR, and MGWR spatial regression models. The results indicate that the MGWR model has the highest goodness-of-fit, accurately explaining the spatially heterogeneous effects of microscale built environment factors on $PM_{2.5}$ concentrations.

(3) Significant differences exist in the spatially heterogeneous effects of the ten microscale built environment factors on $PM_{2.5}$ concentrations. Among them, positive effect factors ranked by strength are building, wall, fence, tree, sky, and grass; negative effect factors ranked by strength are sidewalk, plant, and car. In contrast, road factors have a relatively weaker overall effect on $PM_{2.5}$ concentrations.

This study systematically reveals the spatially heterogeneous effects of the microscale built environment on $PM_{2.5}$ concentrations through a multidisciplinary cross-cutting approach, which may provide new insights for $PM_{2.5}$ pollution management. The research results are conducive to formulating effective air pollution policies, optimizing urban planning, design, and management, and have certain reference significance for promoting ecological civilization construction, facilitating sustainable development, building healthy cities, and achieving United Nations Sustainable Development Goals 3.9 and 11.6.2.

Author Contributions: T.H. was involved in the overall conception and design of the study and led the drafting and finalization of the manuscript. K.W. participated in the formulation of the research plan, data collection, and analysis, and contributed to the initial draft of the manuscript. Y.W. made substantial intellectual revisions to the content. L.W. contributed to the review and revision of the manuscript, provided key suggestions, and jointly finalized the approved version. All authors have read and agreed to the published version of the manuscript.

Funding: This work was supported by the 'Scientists + engineers' team development project of YuLin High Tech Industrial Development Zone under Grant YGXKG-2023-105 and the National Key R & D Program under Grant 2022YFE0119200.

Data Availability Statement: Data will be provided upon request. Any questions should be addressed to the corresponding author.

Acknowledgments: The authors thank the National Tibetan Plateau Data Center for providing the High-resolution and High-quality Ground-level $PM_{2.5}$ Dataset for China.

Conflicts of Interest: The authors report there are no competing interests to declare.

References

1. Gu, B.; Zhang, L.; Van Dingenen, R.; Vieno, M.; Van Grinsven, H.J.; Zhang, X.; Zhang, S.; Chen, Y.; Wang, S.; Ren, C.; et al. Abating Ammonia Is More Cost-Effective than Nitrogen Oxides for Mitigating $PM_{2.5}$ Air Pollution. *Science* **2021**, *374*, 758–762. [CrossRef]
2. Ulpiani, G. On the Linkage between Urban Heat Island and Urban Pollution Island: Three-Decade Literature Review towards a Conceptual Framework. *Sci. Total Environ.* **2021**, *751*, 141727. [CrossRef]
3. Liu, K.; Hua, S.; Song, L. $PM_{2.5}$ Exposure and Asthma Development: The Key Role of Oxidative Stress. *Oxidative Med. Cell. Longev.* **2022**, *2022*, 3618806. [CrossRef]
4. Yan, M.; Ge, H.; Zhang, L.; Chen, X.; Yang, X.; Liu, F.; Shan, A.; Liang, F.; Li, X.; Ma, Z.; et al. Long-Term $PM_{2.5}$ Exposure in Association with Chronic Respiratory Diseases Morbidity: A Cohort Study in Northern China. *Ecotoxicol. Environ. Saf.* **2022**, *244*, 114025. [CrossRef]
5. Huang, W.; Zhou, Y.; Chen, X.; Zeng, X.; Knibbs, L.D.; Zhang, Y.; Jalaludin, B.; Dharmage, S.C.; Morawska, L.; Guo, Y.; et al. Individual and Joint Associations of Long-Term Exposure to Air Pollutants and Cardiopulmonary Mortality: A 22-Year Cohort Study in Northern China. *Lancet Reg. Health-West. Pac.* **2023**, *36*, 100776. [CrossRef]
6. Tian, Y.; Ma, Y.; Wu, J.; Wu, Y.; Wu, T.; Hu, Y.; Wei, J. Ambient $PM_{2.5}$ Chemical Composition and Cardiovascular Disease Hospitalizations in China. *Environ. Sci. Technol.* **2024**, *58*, 16327–16335. [CrossRef]
7. Hill, W.; Lim, E.L.; Weeden, C.E.; Lee, C.; Augustine, M.; Chen, K.; Kuan, F.-C.; Marongiu, F.; Evans, E.J.; Moore, D.A.; et al. Lung Adenocarcinoma Promotion by Air Pollutants. *Nature* **2023**, *616*, 159–167. [CrossRef] [PubMed]
8. Park, S.-H.; Yoon, S.-J.; Choi, S.; Jung, J.; Park, J.-Y.; Park, Y.-H.; Seo, J.; Lee, J.; Lee, M.-S.; Lee, S.-J.; et al. Particulate Matter Promotes Cancer Metastasis through Increased HBEGF Expression in Macrophages. *Exp. Mol. Med.* **2022**, *54*, 1901–1912. [CrossRef] [PubMed]
9. Li, M.-Y.; Liu, L.-Z.; Li, W.; Ng, C.S.H.; Liu, Y.; Kong, A.W.Y.; Zhao, Z.; Wang, S.; Qi, H.; Jia, H.; et al. Ambient Fine Particulate Matter Inhibits 15-Lipoxygenases to Promote Lung Carcinogenesis. *J. Exp. Clin. Cancer Res.* **2019**, *38*, 359. [CrossRef]
10. Hou, D.; Ge, Y.; Chen, C.; Tan, Q.; Chen, R.; Yang, Y.; Li, L.; Wang, J.; Ye, M.; Li, C.; et al. Associations of Long-Term Exposure to Ambient Fine Particulate Matter and Nitrogen Dioxide with Lung Function: A Cross-Sectional Study in China. *Environ. Int.* **2020**, *144*, 105977. [CrossRef]
11. Huang, Q.; Hu, D.; Wang, X.; Chen, Y.; Wu, Y.; Pan, L.; Li, H.; Zhang, J.; Deng, F.; Guo, X.; et al. The Modification of Indoor $PM_{2.5}$ Exposure to Chronic Obstructive Pulmonary Disease in Chinese Elderly People: A Meet-in-Metabolite Analysis. *Environ. Int.* **2018**, *121*, 1243–1252. [CrossRef]
12. Shi, L.; Wu, X.; Yazdi, M.D.; Braun, D.; Abu Awad, Y.; Wei, Y.; Liu, P.; Di, Q.; Wang, Y.; Schwartz, J.; et al. Long-Term Effects of $PM_{2.5}$ on Neurological Disorders in the American Medicare Population: A Longitudinal Cohort Study. *Lancet Planet. Health* **2020**, *4*, E557–E565. [CrossRef]
13. Ku, T.; Li, B.; Gao, R.; Zhang, Y.; Yan, W.; Ji, X.; Li, G.; Sang, N. NF-κB-Regulated microRNA-574-5p Underlies Synaptic and Cognitive Impairment in Response to Atmospheric $PM_{2.5}$ Aspiration. *Part. Fibre Toxicol.* **2017**, *14*, 34. [CrossRef] [PubMed]
14. Liu, M.; Saari, R.K.; Zhou, G.; Li, J.; Han, L.; Liu, X. Recent Trends in Premature Mortality and Health Disparities Attributable to Ambient $PM_{2.5}$ Exposure in China: 2005–2017. *Environ. Pollut.* **2021**, *279*, 116882. [CrossRef] [PubMed]
15. Nansai, K.; Tohno, S.; Chatani, S.; Kanemoto, K.; Kagawa, S.; Kondo, Y.; Takayanagi, W.; Lenzen, M. Consumption in the G20 Nations Causes Particulate Air Pollution Resulting in Two Million Premature Deaths Annually. *Nat. Commun.* **2021**, *12*, 6286. [CrossRef]
16. Wu, X.; Braun, D.; Schwartz, J.; Kioumourtzoglou, M.A.; Dominici, F. Evaluating the Impact of Long-Term Exposure to Fine Particulate Matter on Mortality among the Elderly. *Sci. Adv.* **2020**, *6*, eaba5692. [CrossRef]
17. Yin, S. Decadal Changes in $PM_{2.5}$-Related Health Impacts in China from 1990 to 2019 and Implications for Current and Future Emission Controls. *Sci. Total Environ.* **2022**, *834*, 155334. [CrossRef] [PubMed]
18. Shi, C.; Nduka, I.C.; Yang, Y.; Huang, Y.; Yao, R.; Zhang, H.; He, B.; Xie, C.; Wang, Z.; Yim, S.H.L. Characteristics and Meteorological Mechanisms of Transboundary Air Pollution in a Persistent Heavy $PM_{2.5}$ Pollution Episode in Central-East China. *Atmos. Environ.* **2020**, *223*, 117239. [CrossRef]
19. Cohen, A.J.; Brauer, M.; Burnett, R.; Anderson, H.R.; Frostad, J.; Estep, K.; Balakrishnan, K.; Brunekreef, B.; Dandona, L.; Dandona, R.; et al. Estimates and 25-Year Trends of the Global Burden of Disease Attributable to Ambient Air Pollution: An Analysis of Data from the Global Burden of Diseases Study 2015. *Lancet* **2017**, *389*, 1907–1918. [CrossRef]
20. Lu, X.; Zhang, S.; Xing, J.; Wang, Y.; Chen, W.; Ding, D.; Wu, Y.; Wang, S.; Duan, L.; Hao, J. Progress of Air Pollution Control in China and Its Challenges and Opportunities in the Ecological Civilization Era. *Engineering* **2020**, *6*, 1423–1431. [CrossRef]

21. Geng, G.; Liu, Y.; Liu, Y.; Liu, S.; Cheng, J.; Yan, L.; Wu, N.; Hu, H.; Tong, D.; Zheng, B.; et al. Efficacy of China's Clean Air Actions to Tackle $PM_{2.5}$ Pollution between 2013 and 2020. *Nat. Geosci.* **2024**, *17*, 987–994. [CrossRef]
22. Weng, Z.; Tong, D.; Wu, S.; Xie, Y. Improved Air Quality from China's Clean Air Actions Alleviates Health Expenditure Inequality. *Environ. Int.* **2023**, *173*, 107831. [CrossRef] [PubMed]
23. Huang, J.; Pan, X.; Guo, X.; Li, G. Health Impact of China's Air Pollution Prevention and Control Action Plan: An Analysis of National Air Quality Monitoring and Mortality Data. *Lancet Planet. Health* **2018**, *2*, e313–e323. [CrossRef]
24. Wen, Z.; Ma, X.; Xu, W.; Si, R.; Liu, L.; Ma, M.; Zhao, Y.; Tang, A.; Zhang, Y.; Wang, K.; et al. Combined Short-Term and Long-Term Emission Controls Improve Air Quality Sustainably in China. *Nat. Commun.* **2024**, *15*, 5169. [CrossRef]
25. Yue, H.; He, C.; Huang, Q.; Yin, D.; Bryan, B.A. Stronger Policy Required to Substantially Reduce Deaths from $PM_{2.5}$ Pollution in China. *Nat. Commun.* **2020**, *11*, 1462. [CrossRef]
26. Shu, Y.; Hu, J.; Zhang, S.; Schöpp, W.; Tang, W.; Du, J.; Cofala, J.; Kiesewetter, G.; Sander, R.; Winiwarter, W.; et al. Analysis of the Air Pollution Reduction and Climate Change Mitigation Effects of the Three-Year Action Plan for Blue Skies on the "2 + 26" Cities in China. *J. Environ. Manag.* **2022**, *317*, 115455. [CrossRef]
27. Xue, T.; Kang, N.; Wan, W.; Zhu, T. Health-Oriented Strategies Are Needed to Optimize China's 2025 Clean Air Action Plan. *Sci. Bull.* **2024**, *69*, 2007–2008. [CrossRef]
28. Zheng, Y.; Xue, T.; Zhang, Q.; Geng, G.; Tong, D.; Li, X.; He, K. Air Quality Improvements and Health Benefits from China's Clean Air Action since 2013. *Environ. Res. Lett.* **2017**, *12*, 114020. [CrossRef]
29. Chen, Z.; Wang, J.-N.; Ma, G.-X.; Zhang, Y.-S. China Tackles the Health Effects of Air Pollution. *Lancet* **2013**, *382*, 1959–1960. [CrossRef] [PubMed]
30. Feng, Y.; Ning, M.; Lei, Y.; Sun, Y.; Liu, W.; Wang, J. Defending Blue Sky in China: Effectiveness of the "Air Pollution Prevention and Control Action Plan" on Air Quality Improvements from 2013 to 2017. *J. Environ. Manag.* **2019**, *252*, 109603. [CrossRef]
31. World Health Organization. WHO Global Air Quality Guidelines: Particulate Matter (PM2.5 and PM10), Ozone, Nitrogen Dioxide, Sulfur Dioxide and Carbon Monoxide. Available online: https://www.who.int/publications/i/item/9789240034228 (accessed on 30 September 2025).
32. Bao, Y.; Huang, Z.; Guo, Q.; Liu, Y. Spatial Calculation of Urban Built Environment Stock: Progress and Prospects. *Natl. Remote Sens. Bull.* **2022**, *26*, 1909–1919. [CrossRef]
33. Meng, J.; Han, W.; Yuan, C. Seasonal and Multi-Scale Difference of the Relationship between Built-up Land Landscape Pattern and $PM_{2.5}$ Concentration Distribution in Nanjing. *Ecol. Indic.* **2023**, *156*, 111079. [CrossRef]
34. Fan, C.; Tian, L.; Zhou, L.; Hou, D.; Song, Y.; Qiao, X.; Li, J. Examining the Impacts of Urban Form on Air Pollutant Emissions: Evidence from China. *J. Environ. Manag.* **2018**, *212*, 405–414. [CrossRef]
35. Yuan, M.; Song, Y.; Huang, Y.; Hong, S.; Huang, L. Exploring the Association between Urban Form and Air Quality in China. *J. Plan. Educ. Res.* **2018**, *38*, 413–426. [CrossRef]
36. Shi, K.; Wang, H.; Yang, Q.; Wang, L.; Sun, X.; Li, Y. Exploring the Relationships between Urban Forms and Fine Particulate ($PM_{2.5}$) Concentration in China: A Multi-Perspective Study. *J. Clean. Prod.* **2019**, *231*, 990–1004. [CrossRef]
37. Lin, Y.; Yuan, X.; Zhai, T.; Wang, J. Effects of Land-Use Patterns on $PM_{2.5}$ in China's Developed Coastal Region: Exploration and Solutions. *Sci. Total Environ.* **2020**, *703*, 135602. [CrossRef]
38. Zhou, S.; Lin, R. Spatial-Temporal Heterogeneity of Air Pollution: The Relationship between Built Environment and on-Road $PM_{2.5}$ at Micro Scale. *Transp. Res. Part D Transp. Environ.* **2019**, *76*, 305–322. [CrossRef]
39. Biljecki, F.; Ito, K. Street View Imagery in Urban Analytics and GIS: A Review. *Landsc. Urban Plan.* **2021**, *215*, 104217. [CrossRef]
40. Liu, L.; Sevtsuk, A. Clarity or Confusion: A Review of Computer Vision Street Attributes in Urban Studies and Planning. *Cities* **2024**, *150*, 105022. [CrossRef]
41. Hou, Y.; Biljecki, F. A Comprehensive Framework for Evaluating the Quality of Street View Imagery. *Int. J. Appl. Earth Obs. Geoinf.* **2022**, *115*, 103094. [CrossRef]
42. Han, Y.; Zhong, T.; Yeh, A.G.O.; Zhong, X.; Chen, M.; Lu, G. Mapping Seasonal Changes of Street Greenery Using Multi-Temporal Street-View Images. *Sustain. Cities Soc.* **2023**, *92*, 104498. [CrossRef]
43. Dang, H.; Li, J. The Integration of Urban Streetscapes Provides the Possibility to Fully Quantify the Ecological Landscape of Urban Green Spaces: A Case Study of Xi'an City. *Ecol. Indic.* **2021**, *133*, 108388. [CrossRef]
44. Zhang, L.; Wang, L.; Wu, J.; Li, P.; Dong, J.; Wang, T. Decoding Urban Green Spaces: Deep Learning and Google Street View Measure Greening Structures. *Urban For. Urban Green.* **2023**, *87*, 128028. [CrossRef]
45. Doan, Q.C.; Ma, J.; Chen, S.; Zhang, X. Nonlinear and Threshold Effects of the Built Environment, Road Vehicles and Air Pollution on Urban Vitality. *Landsc. Urban Plan.* **2025**, *253*, 105204. [CrossRef]
46. Wu, C.; Ye, Y.; Gao, F.; Ye, X. Using Street View Images to Examine the Association between Human Perceptions of Locale and Urban Vitality in Shenzhen, China. *Sustain. Cities Soc.* **2023**, *88*, 104291. [CrossRef]
47. Tang, F.; Zeng, P.; Wang, L.; Zhang, L.; Xu, W. Urban Perception Evaluation and Street Refinement Governance Supported by Street View Visual Elements Analysis. *Remote Sens.* **2024**, *16*, 3661. [CrossRef]

48. Chen, K.; Tian, M.; Zhang, J.; Xu, X.; Yuan, L. Evaluating the Seasonal Effects of Building Form and Street View Indicators on Street-Level Land Surface Temperature Using Random Forest Regression. *Build. Environ.* **2023**, *245*, 110884. [CrossRef]
49. Klimenka, M.; Zhao, K.; Hilland, R.; Zhang, F.; Voogt, J.; Ratti, C. Instant Infrared: Estimating Urban Surface Temperatures from Street View Imagery. *Build. Environ.* **2025**, *267*, 112122. [CrossRef]
50. Han, X.; Yu, Y.; Liu, L.; Li, M.; Wang, L.; Zhang, T.; Tang, F.; Shen, Y.; Li, M.; Yu, S.; et al. Exploration of Street Space Architectural Color Measurement Based on Street View Big Data and Deep Learning—A Case Study of Jiefang North Road Street in Tianjin. *PLoS ONE* **2023**, *18*, e0289305. [CrossRef] [PubMed]
51. Chen, N.; Wang, L.; Xu, T.; Wang, M. Perception of Urban Street Visual Color Environment Based on the CEP-KASS Framework. *Landsc. Urban Plan.* **2025**, *259*, 105359. [CrossRef]
52. Xue, X.; Tian, Z.; Yang, Y.; Wang, J.; Cao, S.-J. Sustaining the Local Color of a Global City. *Nat. Cities* **2025**, *2*, 400–412. [CrossRef]
53. Li, W.; Sun, R.; He, H.; Yan, M.; Chen, L. Perceptible Landscape Patterns Reveal Invisible Socioeconomic Profiles of Cities. *Sci. Bull.* **2024**, *69*, 3291–3302. [CrossRef] [PubMed]
54. Shi, W.; Xiang, Y.; Ying, Y.; Jiao, Y.; Zhao, R.; Qiu, W. Predicting Neighborhood-Level Residential Carbon Emissions from Street View Images Using Computer Vision and Machine Learning. *Remote Sens.* **2024**, *16*, 1312. [CrossRef]
55. Liang, X.; Zhao, T.; Biljecki, F. Revealing Spatio-Temporal Evolution of Urban Visual Environments with Street View Imagery. *Landsc. Urban Plan.* **2023**, *237*, 104802. [CrossRef]
56. Suzuki, M.; Mori, J.; Maeda, T.N.; Ikeda, J. The Economic Value of Urban Landscapes in a Suburban City of Tokyo, Japan: A Semantic Segmentation Approach Using Google Street View Images. *J. Asian Arch. Build. Eng.* **2023**, *22*, 1110–1125. [CrossRef]
57. Luo, L.; Yang, X.; Li, J.; Song, Y.; Zhao, Z. Deciphering House Prices by Integrating Street Perceptions with a Machine-Learning Algorithm: A Case Study of Xi'an, China. *Cities* **2025**, *156*, 105542. [CrossRef]
58. Zhong, H.; Chen, D.; Wang, P.; Wang, W.; Shen, S.; Liu, Y.; Zhu, M. Predicting On-Road Air Pollution Coupling Street View Images and Machine Learning: A Quantitative Analysis of the Optimal Strategy. *Environ. Sci. Technol.* **2025**, *59*, 3582–3591. [CrossRef]
59. Deng, X.; Gao, F.; Liao, S.; Li, S. Unraveling the Association between the Built Environment and Air Pollution from a Geospatial Perspective. *J. Clean. Prod.* **2023**, *386*, 135768. [CrossRef]
60. Liu, Y.; Cao, Z.; Wei, H.; Guo, P. Decoding Prediction of $PM_{2.5}$ against Jointly Street-Tree Canopy Size and Running Vehicle Density Using Big Data in Streetscapes. *Urban Clim.* **2025**, *59*, 102282. [CrossRef]
61. Qi, M.; Hankey, S. Using Street View Imagery to Predict Street-Level Particulate Air Pollution. *Environ. Sci. Technol.* **2021**, *55*, 2695–2704. [CrossRef]
62. Ito, K.; Zhu, Y.; Abdelrahman, M.; Liang, X.; Fan, Z.; Hou, Y.; Zhao, T.; Ma, R.; Fujiwara, K.; Ouyang, J.; et al. ZenSVI: An Open-Source Software for the Integrated Acquisition, Processing and Analysis of Street View Imagery towards Scalable Urban Science. *Comput. Environ. Urban Syst.* **2025**, *119*, 102283. [CrossRef]
63. Wang, L.; Zhang, T.; He, J.; Kada, M. Cross-Platform Complementarity: Assessing the Data Quality and Availability of Google Street View and Baidu Street View. *Trans. Urban Data Sci. Technol.* **2025**, *4*, 22–47. [CrossRef]
64. Zhou, B.; Zhao, H.; Puig, X.; Xiao, T.; Fidler, S.; Barriuso, A.; Torralba, A. Semantic Understanding of Scenes through the ADE20K Dataset. *Int. J. Comput. Vis.* **2019**, *127*, 302–321. [CrossRef]
65. Zhou, B.; Zhao, H.; Puig, X.; Fidler, S.; Barriuso, A.; Torralba, A. Scene Parsing through ADE20K Dataset. In Proceedings of the 2017 IEEE Conference on Computer Vision and Pattern Recognition (CVPR), Honolulu, HI, USA, 21–26 July 2017; pp. 5122–5130.
66. Wei, J.; Li, Z.; Lyapustin, A.; Sun, L.; Peng, Y.; Xue, W.; Su, T.; Cribb, M. Reconstructing 1-Km-Resolution High-Quality $PM_{2.5}$ Data Records from 2000 to 2018 in China: Spatiotemporal Variations and Policy Implications. *Remote Sens. Environ.* **2021**, *252*, 112136. [CrossRef]
67. Wei, J.; Li, Z.; Cribb, M.; Huang, W.; Xue, W.; Sun, L.; Guo, J.; Peng, Y.; Li, J.; Lyapustin, A.; et al. Improved 1 Km Resolution $PM_{2.5}$ Estimates across China Using Enhanced Space–Time Extremely Randomized Trees. *Atmos. Chem. Phys.* **2020**, *20*, 3273–3289. [CrossRef]
68. Duan, H.; Cao, Q.; Wang, L.; Gu, X.; Ashrafi, K. Exploring the Relationships between 3D Urban Landscape Patterns and $PM_{2.5}$ Pollution Using the Multiscale Geographic Weighted Regression Model. *Prog. Phys. Geogr. Earth Environ.* **2024**, *48*, 368–388. [CrossRef]
69. Feng, Z.; Wang, X.; Yuan, J.; Zhang, Y.; Yu, M. Changes in Air Pollution, Land Surface Temperature, and Urban Heat Islands during the COVID-19 Lockdown in Three Chinese Urban Agglomerations. *Sci. Total Environ.* **2023**, *892*, 164496. [CrossRef]
70. Cao, W.; Zhou, W.; Yu, W.; Wu, T. Combined Effects of Urban Forests on Land Surface Temperature and $PM_{2.5}$ Pollution in the Winter and Summer. *Sustain. Cities Soc.* **2024**, *104*, 105309. [CrossRef]
71. Guo, J.; Zhou, J.; Han, R.; Wang, Y.; Lian, X.; Tang, Z.; Ye, J.; He, X.; Yu, H.; Huang, S.; et al. Association of Short-Term Co-Exposure to Particulate Matter and Ozone with Mortality Risk. *Environ. Sci. Technol.* **2023**, *57*, 15825–15834. [CrossRef] [PubMed]
72. Bachwenkizi, J.; Liu, C.; Meng, X.; Zhang, L.; Wang, W.; Van Donkelaar, A.; Martin, R.V.; Hammer, M.S.; Chen, R.; Kan, H. Fine Particulate Matter Constituents and Infant Mortality in Africa: A Multicountry Study. *Environ. Int.* **2021**, *156*, 106739. [CrossRef] [PubMed]

73. Tang, F.; Zeng, P.; Guo, Y.; Shen, Y.; Wang, L.; Liu, K.; Zhang, L. Decoding the Spatiotemporal Dynamics and Driving Mechanisms of Ecological Resilience in the Beijing-Tianjin-Hebei Urban Agglomeration: A Deep Learning Approach. *Urban Clim.* **2025**, *61*, 102436. [CrossRef]
74. Lin, L.; Yi, X.; Liu, H.; Meng, R.; Li, S.; Liu, X.; Yang, J.; Xu, Y.; Li, C.; Wang, Y.; et al. The Airway Microbiome Mediates the Interaction between Environmental Exposure and Respiratory Health in Humans. *Nat. Med.* **2023**, *29*, 1750–1759. [CrossRef]
75. Cai, M.; Zhang, S.; Lin, X.; Qian, Z.; McMillin, S.E.; Yang, Y.; Zhang, Z.; Pan, J.; Lin, H. Association of Ambient Particulate Matter Pollution of Different Sizes With In-Hospital Case Fatality Among Stroke Patients in China. *Neurology* **2022**, *98*, e2474–e2486. [CrossRef]
76. Cai, M.; Wei, J.; Zhang, S.; Liu, W.; Wang, L.; Qian, Z.; Lin, H.; Liu, E.; McMillin, S.E.; Cao, Y.; et al. Short-Term Air Pollution Exposure Associated with Death from Kidney Diseases: A Nationwide Time-Stratified Case-Crossover Study in China from 2015 to 2019. *BMC Med.* **2023**, *21*, 32. [CrossRef]
77. Ministry of Ecology and Environment of the People's Republic of China. Ministry of Ecology and Environment Announces National Ambient Air Quality Status for December 2023 and January–December 2023. Available online: https://www.mee.gov.cn/ywdt/xwfb/202401/t20240125_1064784.shtml (accessed on 30 September 2025).
78. Herfort, B.; Lautenbach, S.; Porto De Albuquerque, J.; Anderson, J.; Zipf, A. A Spatio-Temporal Analysis Investigating Completeness and Inequalities of Global Urban Building Data in OpenStreetMap. *Nat. Commun.* **2023**, *14*, 3985. [CrossRef]
79. Liu, Q. The Varying Driving Forces of $PM_{2.5}$ Concentrations in Chinese Cities: Insights from a Geographically and Temporally Weighted Regression Model. *Environ. Int.* **2020**, *145*, 106168. [CrossRef] [PubMed]
80. Han, C.; Zhou, L.; Zhou, T. How Does the Built Environment Affect Hotel Prices? A Study Using Multiscale GWR and Deep Learning. *J. Asian Arch. Build. Eng.* **2024**, *23*, 1717–1734. [CrossRef]
81. Zhang, L.; Fang, C.; Zhao, R.; Zhu, C.; Guan, J. Spatial–Temporal Evolution and Driving Force Analysis of Eco-Quality in Urban Agglomerations in China. *Sci. Total Environ.* **2023**, *866*, 161465. [CrossRef]
82. Li, Y.; Zhang, Y.; Wu, Q.; Xue, R.; Wang, X.; Si, M.; Zhang, Y. Greening the Concrete Jungle: Unveiling the Co-Mitigation of Greenspace Configuration on $PM_{2.5}$ and Land Surface Temperature with Explanatory Machine Learning. *Urban For. Urban Green.* **2023**, *88*, 128086. [CrossRef]
83. Jiang, Z.; Wu, C.; Chung, H. The 15-Minute Community Life Circle for Older People: Walkability Measurement Based on Service Accessibility and Street-Level Built Environment—A Case Study of Suzhou, China. *Cities* **2025**, *157*, 105587. [CrossRef]
84. Liu, L.; Liu, Y.; Cheng, F.; Yu, Y.; Wang, J.; Wang, C.; Nong, L.; Deng, H. Remote Sensing Estimation of Regional $PM_{2.5}$ Based on GTWR Model -a Case Study of Southwest China. *Environ. Pollut.* **2024**, *351*, 124057. [CrossRef] [PubMed]
85. Wu, T.; Zhou, L.; Jiang, G.; Meadows, M.E.; Zhang, J.; Pu, L.; Wu, C.; Xie, X. Modelling Spatial Heterogeneity in the Effects of Natural and Socioeconomic Factors, and Their Interactions, on Atmospheric $PM_{2.5}$ Concentrations in China from 2000–2015. *Remote Sens.* **2021**, *13*, 2152. [CrossRef]
86. Zhou, L.; Wu, T.; Pu, L.; Meadows, M.; Jiang, G.; Zhang, J.; Xie, X. Spatially Heterogeneous Relationships of $PM_{2.5}$ Concentrations with Natural and Land Use Factors in the Niger River Watershed, West Africa. *J. Clean. Prod.* **2023**, *394*, 136406. [CrossRef]
87. Wu, D.; Gong, J.; Liang, J.; Sun, J.; Zhang, G. Analyzing the Influence of Urban Street Greening and Street Buildings on Summertime Air Pollution Based on Street View Image Data. *ISPRS Int. J. Geo-Inf.* **2020**, *9*, 500. [CrossRef]
88. Wang, X.; Zhou, Z.; Xiang, Y.; Peng, C.; Peng, C. Effects of Street Plants on Atmospheric Particulate Dispersion in Urban Streets: A Review. *Environ. Rev.* **2024**, *2*, 114–130. [CrossRef]
89. Wu, C.; Eldesoky, A.H.; Morello, E. Understanding the Effect of Built-up and Green Spaces upon Air Quality at Multiple Spatial Scales: A Systematic Literature Review. *Landsc. Urban Plan.* **2025**, *257*, 105304. [CrossRef]
90. Vos, P.E.J.; Maiheu, B.; Vankerkom, J.; Janssen, S. Improving Local Air Quality in Cities: To Tree or Not to Tree? *Environ. Pollut.* **2013**, *183*, 113–122. [CrossRef]
91. Venter, Z.S.; Hassani, A.; Stange, E.; Schneider, P.; Castell, N. Reassessing the Role of Urban Green Space in Air Pollution Control. *Proc. Natl. Acad. Sci. USA* **2024**, *121*, e2306200121. [CrossRef]
92. Fan, S.; Zhang, M.; Li, Y.; Li, K.; Dong, L. Impacts of Composition and Canopy Characteristics of Plant Communities on Microclimate and Airborne Particles in Beijing, China. *Sustainability* **2021**, *13*, 4791. [CrossRef]
93. Li, P.; Chen, Y.; Niu, H.; Zhang, L.; Tang, Y.; Zhu, G.; Zhang, Z.; Ma, Y.; Wu, W. How to Evaluate the Reduction Effect of the Park on $PM_{2.5}$? Exploratory Application of the Maximum and Cumulative Perspective. *Sustain. Cities Soc.* **2024**, *116*, 105909. [CrossRef]
94. Wang, L.; Chen, X.; Zhang, Y.; Li, M.; Li, P.; Jiang, L.; Xia, Y.; Li, Z.; Li, J.; Wang, L.; et al. Switching to Electric Vehicles Can Lead to Significant Reductions of $PM_{2.5}$ and NO_2 across China. *One Earth* **2021**, *4*, 1037–1048. [CrossRef]
95. Wang, F.; Peng, Y.; Jiang, C. Influence of Road Patterns on $PM_{2.5}$ Concentrations and the Available Solutions: The Case of Beijing City, China. *Sustainability* **2017**, *9*, 217. [CrossRef]
96. Du, W.; Chen, L.; Wang, H.; Shan, Z.; Zhou, Z.; Li, W.; Wang, Y. Deciphering Urban Traffic Impacts on Air Quality by Deep Learning and Emission Inventory. *J. Environ. Sci.* **2023**, *124*, 745–757. [CrossRef]

97. Liu, Y.; Shen, L.; Ren, Y.; Zhou, T. Regeneration towards Suitability: A Decision-Making Framework for Determining Urban Regeneration Mode and Strategies. *Habitat Int.* **2023**, *138*, 102870. [CrossRef]
98. Liao, Z.; Liu, M. Critical Barriers and Countermeasures to Urban Regeneration from the Stakeholder Perspective: A Literature Review. *Front. Sustain. Cities* **2023**, *5*, 1115648. [CrossRef]
99. Zhang, F.; Salazar-Miranda, A.; Duarte, F.; Vale, L.; Hack, G.; Chen, M.; Liu, Y.; Batty, M.; Ratti, C. Urban Visual Intelligence: Studying Cities with Artificial Intelligence and Street-Level Imagery. *Ann. Am. Assoc. Geogr.* **2024**, *114*, 876–897. [CrossRef]

buildings

MDPI

Article

Prediction and Optimization of the Restoration Quality of University Outdoor Spaces: A Data-Driven Study Using Image Semantic Segmentation and Explainable Machine Learning

Xiaowen Zhuang [1,†], Zhenpeng Tang [2,†], Shuo Lin [2] and Zheng Ding [1,*]

1 College of Landscape Architecture and Art, Fujian Agriculture and Forestry University, Fuzhou 350002, China; 12319097012@fafu.edu.cn

2 College of Economics and Management, Fujian Agriculture and Forestry University, Fuzhou 350002, China; [illegible]@fafu.edu.cn (Z.T.); 12315089001@fafu.edu.cn (S.L.)

* Correspondence: 000q151039@fafu.edu.cn

† These authors contributed equally to this work.

Abstract: Evaluating the restoration quality of university outdoor spaces is often constrained by subjective surveys and manual assessment, limiting scalability and objectivity. This study addresses this gap by applying explainable machine learning to predict restorative quality from campus imagery, enabling large-scale, data-driven evaluation and capturing complex nonlinear relationships that traditional methods may overlook. Using Fujian Agriculture and Forestry University as a case study, this study extracted road network data, generated 297 coordinates at 50-m intervals, and collected 1197 images. Surveys were conducted to obtain restorative quality scores. The Mask2Former model was used to extract landscape features, and decision tree algorithms (RF, XGBoost, GBR) were selected based on MAE, MSE, and EVS metrics. The combination of optimal algorithms and SHAP was employed to predict restoration quality and identify key features. This research also used a multivariate linear regression model to identify features with significant statistical impact but lower features importance ranking. Finally, the study also analyzed heterogeneity in scores for three restoration indicators and five campus zones using k-means clustering. Empirical results show that natural elements like vegetation and water positively affect psychological perception, while structural components like walls and fences have negative or nonlinear effects. On this basis, this study proposes spatial optimization strategies for different campus areas, offering a foundation for creating high-quality outdoor environments with restorative and social functions.

Keywords: university campus; landscape optimization; image semantic segmentation; interpretable machine learning

1. Introduction

University campuses, as the physical carriers of higher education, bear the core mission of knowledge dissemination, value guidance, and character development. Their environmental attributes profoundly shape students' cognitive development, social behavior, and cultural identity [1]. A growing body of empirical research reveals that campus outdoor spaces, which include dynamic networks connecting the built environment, such as street systems, blue-green infrastructure, and multifunctional areas, hold substantial restorative potential [2,3]. These spaces can effectively alleviate academic stress, social anxiety, and psychological fatigue commonly experienced by university students [4]. Beyond serving as everyday settings for commuting, informal interactions, and contemplation, campus outdoor spaces function as essential containers of public life, carrying symbolic cultural significance that transcends their utilitarian roles. Moreover, the efficacy of outdoor spaces arises from the complex interplay between their physical attributes (e.g., vegetation coverage, facility layout) and user characteristics (e.g., behavioral patterns, psychological expectations) [5–8]. This systemic "people–scene–activity" relationship makes optimizing outdoor space experiences a key strategy for enhancing the overall quality of university education. Quantifying the coupling mechanisms between campus spatial features and students' multidimensional perceptions and identifying the psychological effects of natural elements embedded in outdoor environments can provide a robust empirical foundation for campus design.

While existing literature has begun to uncover the positive psychological effects of campus outdoor spaces, several notable gaps remain in current research perspectives [9–13]. First, many studies are based on universities with generalized academic profiles, failing to examine the unique spatial-psychological interaction patterns demanded by students from discipline-specific institutions. Second, methodological limitations persist. Traditional approaches often struggle to capture the nonlinear and dynamic associations between the complex visual features of campus environments and students' multidimensional perceptions. They are also insufficient for identifying threshold effects and changes in the influence strength of specific visual features. Third, previous studies have largely overlooked the spatial drivers underlying both high and low psychological ratings and have paid little attention to the internal functional heterogeneity of campus outdoor spaces. Consequently, the differentiated psychological impacts of functional zones, such as academic areas, core leisure spaces, and sports facilities, have not yet been systematically explored. These research limitations hinder the development of tailored and adaptive landscape planning practices for specific types of university campuses.

Conventional approaches to evaluating campus outdoor spaces rely heavily on small-scale, subjective surveys and manual observations, limiting scalability and overlooking complex nonlinear relationships between environmental features and psychological outcomes. This study applied machine learning (ML) to overcome these limitations, leveraging its ability to process large volumes of visual data, detect hidden patterns, and provide interpretable results through SHAP (Shapley additive explanations), thereby supporting both accurate prediction and actionable design insights. Focusing on a forestry university in China, we integrated Mask2Former-based image semantic segmentation, SHAP-based interpretable decision trees, and multivariate linear regression to examine how spatial features influence students' emotional uplift, social interaction, and psychological restoration. Landscape features were extracted from campus images, and survey-based psychological scores were collected. Decision tree models (RF, XGBoost, GBR) combined with SHAP identified key features and threshold effects, while regression analysis detected statistically significant but lower-salience elements. Psychological scores were then categorized into high, medium, and low groups, and k-means clustering was used to define functional zones, enabling the analysis of heterogeneous effects across campus areas.

This study integrated multiple methods to support the optimization of campus landscapes in forestry universities, tailored to students' psychological and social well-being. First, the geographic coordinates of selected campus locations were manually collected, and original landscape images were obtained via web-crawling techniques. The advanced Mask2former model was then used to extract semantic features from the images. Unlike previous studies that focused solely on building high-accuracy prediction models for psychological outcomes, our approach innovatively combines SHAP-based interpretable decision tree methods with multivariate linear regression analysis. While the former identifies the most influential semantic features, the latter uncovers additional variables with significant psychological impacts that machine learning models may have omitted. Our focus was not merely on prediction accuracy, but on understanding the role of critical landscape features in shaping students' psychological experiences.

Furthermore, the complementary use of SHAP-based decision trees and multivariate regression enhances analytical robustness. While interpretable decision trees can highlight key visual elements associated with the students' psychological outcomes, they often prioritize visually prominent features and may ignore subtler yet impactful landscape components. Multivariate linear regression compensates for this limitation by statistically identifying additional features with significant psychological relevance.

Our analytical perspective is also more comprehensive. Compared with single-dimensional measures of restoration, the selected psychological indicators, including emotional uplift (reflecting positive and negative affect), social interaction (indicating campus social networks), and psychological restoration (representing recovery from stress and fatigue), collectively offer a holistic view of the students' mental health. The heterogeneity analyses in this study contribute novel insights by identifying both positively and negatively associated landscape features across different score groups. Additionally, the functional zone-based clustering approach provides further guidance for spatial optimization strategies tailored to forestry universities.

In summary, this study offers empirical evidence to support the refined design of campus outdoor spaces in forestry universities. It breaks through the limitations of homogenized planning, reinforces the educational function of the environment, and provides both theoretical and practical references for building sustainable campuses that integrate pedagogical, ecological, and health objectives.

The remainder of this paper is structured as follows. Section 2 reviews relevant literature and theoretical foundations; Section 3 outlines the research framework; Section 4 presents the empirical results on how campus landscape features influence three psychological dimensions; Section 5 discusses the findings; and Section 6 concludes with key implications for campus design and future research directions.

2. Literature Review and Theory

2.1. Literature Review

2.1.1. Campus Outdoor Spaces and Student Mental Health

Campus outdoor spaces are composed of both fixed and semi-fixed elements and are widely recognized as key determinants of campus life quality [5]. Studies have shown that well-designed spatial environments can encourage students to linger and engage in social interaction, thereby enhancing their quality of life and mental well-being [1,14]. The environmental attributes of outdoor spaces are closely associated with the students' sense of satisfaction, well-being, and creativity [15,16].

Among these spaces, green areas offer distinct advantages in addressing common mental health challenges faced by students, such as alleviating academic stress and improving sleep quality [17], while also enhancing the campus's overall restorative potential [3,4,18]. In forestry universities, the diversity of green spaces not only meets the heterogeneous needs of students, but also strengthens ecological awareness and practical engagement. A diverse layout of green infrastructure facilitates the integration of educational, psychological, and cultural functions on campus [19]. Existing research has preliminarily confirmed the positive effects of plant systems on the students' psychological performance [3,20,21].

2.1.2. Assessing the Restorative Quality of Campus Spaces Using Street View Images

Quantifying the built environment of campuses alongside public perception is essential for improving the quality of street spaces. As a critical component of the campus environment, the assessment of street quality involves not only physical attributes, but also human perception [22]. Vision, as the dominant sensory modality, has been widely employed in evaluating street landscapes [23,24]. Consequently, many studies have advocated for analyzing street space quality from the perspective of visual perception.

In recent years, an increasing number of studies have applied street view imagery to evaluate spatial perception within campus environments. For instance, Meng et al. (2023) explored the relationship between visual aesthetics and physical elements in street views [25]. Qin et al. (2025) assessed the perceived quality and spatial distribution of campus walkability [26]. Ma et al. (2023) integrated street view images with the PRS-11 scale to develop a high-resolution predictive model of restorative quality [27]. Wu et al. (2025), using image-extracted features, constructed a framework for evaluating restorative benefits, identified areas for spatial optimization, and proposed corresponding improvement strategies [28].

2.1.3. Application of Image Semantic Segmentation in Landscape Feature Extraction

Deep learning has been widely applied in the fields of image segmentation and spatial perception, significantly improving the accuracy and efficiency of image interpretation [29]. It has demonstrated notable effectiveness in studies of subjective perception in urban environments [30]. Semantic segmentation techniques enable the precise delineation of urban landscape elements through pixel-level classification, leveraging neural networks to enhance feature extraction capabilities [31]. The application of this technique in landscape feature quantification has become increasingly prevalent. For example, Garré et al. (2009) employed semantic segmentation on street view images to quantify indicators such as field of view and sky visibility [32]. Xia et al. (2021) developed a sky visibility assessment method based on DeepLabV3+ [33]. Moreover, Li et al. (2023) combined UAV data with deep learning to extract multidimensional features including color, texture, and geometry [34].

However, existing studies have tended to focus primarily on either physical attributes or a single dimension of subjective perception, lacking a comprehensive analysis of multidimensional psychological indicators such as psychological restoration, emotional uplift, and social interaction. Moreover, current models often overlook visually inconspicuous elements that may nonetheless have significant psychological effects, thereby limiting the comprehensiveness of landscape evaluation.

2.2. *Theoretical Framework*

In recent years, the advancement of environmental psychology and cognitive behavioral science has brought increasing scholarly attention to the psychological impacts of campus spatial environments, positioning this topic as a critical intersection of urban planning, public health, and higher education research [35,36]. Natural elements and spatial configurations play essential roles in emotion regulation, attentional restoration, and social facilitation [37,38]. Building on classical theories, this study proposes three core psychological indicators: psychological restoration, emotional uplift, and social interaction, and constructs an evaluation framework to assess the effects of outdoor campus spaces. Figure 1 illustrates the theoretical foundations underlying the three psychological indicators.

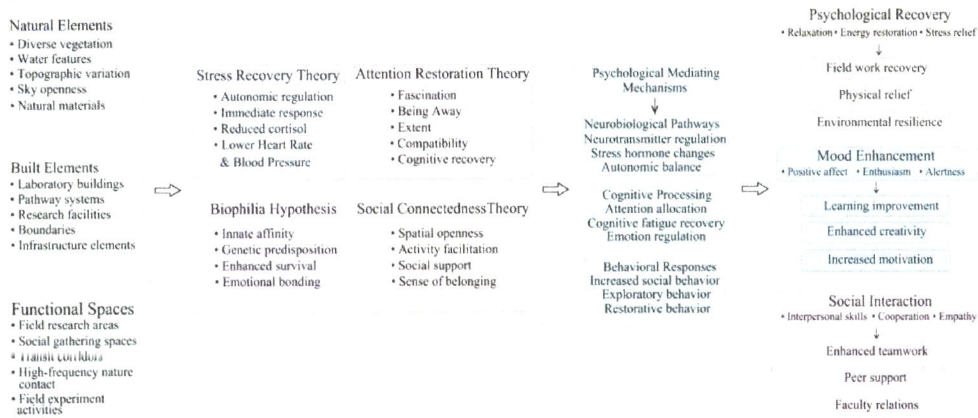

Figure 1. The theoretical foundations of this study.

The psychological restoration dimension is grounded in stress recovery theory (SRT), which posits that natural environments can reduce physiological arousal, regulate emotions, and alleviate psychological stress [38,39]. Students at forestry universities often face high cognitive loads due to demanding academic tasks and thus rely more heavily on natural settings as psychological recovery resources. Their sensitivity to restorative environmental features is correspondingly heightened [40,41].

The emotional uplift dimension is based on attention restoration theory (ART), which emphasizes the roles of "soft fascination" and "extent" in natural environments in replenishing attentional capacity, enhancing emotional states, and improving cognitive performance [37]. Studies have shown that even brief exposure to natural scenes can significantly enhance positive emotions and cognitive function [42]. Highly green and open spaces, in particular, are effective in supporting emotional recovery and relieving mental fatigue [43]. For students in forestry universities, the need for emotional regulation tends to be continuous and is especially susceptible to environmental influences.

The social interaction dimension integrates the biophilia hypothesis and social cohesion theory, highlighting how natural elements such as vegetation and water bodies can evoke feelings of safety and foster social connectedness [44,45]. Additionally, spatial openness and accessibility have a substantial influence on the frequency and quality of social interactions. Well-designed natural open spaces can facilitate communication between faculty and students, encourage peer engagement, and strengthen feelings of social belonging and support [36,46]. Evidence also suggests that group activities in natural environments enhance the students' teamwork skills and social adaptability [47]. This dimension is particularly salient for students in forestry universities, who frequently engage with nature and demonstrate strong social needs.

Collectively, these three psychological dimensions not only reveal the differentiated spatial-psychological needs of forestry university students, but also provide a robust theoretical foundation for designing targeted interventions in campus outdoor spaces.

3. Materials and Methods

This study utilized campus street view images retrieved from Baidu Maps and applied a Mask2former-based image semantic segmentation method to extract landscape features. Simultaneously, a questionnaire survey was conducted to obtain student scores on three psychological indicators: psychological restoration, emotional uplift, and social interaction. Subsequently, a decision tree algorithm integrated with the SHAP (Shapley additive explanations) mechanism was employed to identify key landscape features significantly influencing each psychological outcome and analyze their influence patterns and threshold effects.

Next, multivariate linear regression was used to uncover additional landscape features that contained limited semantic information and were potentially overlooked by interpretable machine learning models, assessing their statistical significance regarding student well-being. Finally, a heterogeneity analysis was performed by grouping participants based on high, medium, and low scores for the three psychological indicators and by applying k-means clustering to delineate functional zones within the campus. Figure 2 presents the technical framework of this study. The main research steps are detailed in Section 3.1.

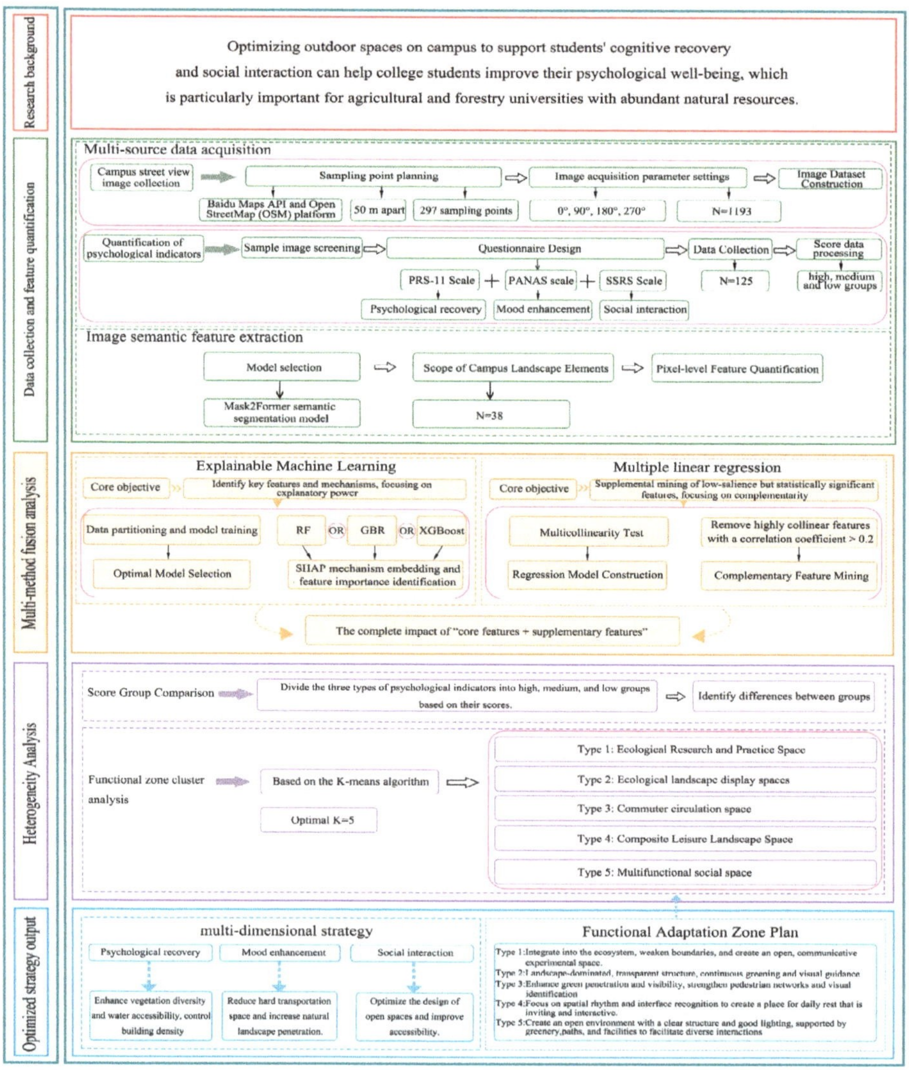

Figure 2. The technical framework of the study.

3.1. Data Collection and Feature Quantification

- **Collection of campus street view images**

Fujian Agriculture and Forestry University (FAFU), a provincially key university founded in 1936, served as the research site for this study. Its geographic coordinates can be found at https://www.openstreetmap.org/#map=16/26.08711/119.23287 (accessed on 18 July 2025). The campus spans over 2560 mu (approximately 170 hectares) and is functionally divided into five main zones: teaching, residential life, sports, faculty housing, and supporting facilities. Geographically situated between the Youxi Bridge and the Min River, the campus benefits from favorable natural and transportation conditions. Notable landscape features include the Wetland Park, the China National Botanical Garden (Fujian branch), and Guanyin Lake.

Street view images from a human-eye perspective were collected using the Baidu Maps API and OpenStreetMap (OSM). Road network data were first extracted from OSM, and a total of 297 sampling points were generated in ArcGIS at 50-m intervals. The geographic coordinates (latitude and longitude) of each point were recorded. At each sampling point, four street view images were captured facing four cardinal directions (0°, 90°, 180°, and 270°) with a pitch angle of 0°, resulting in a total of 1193 images. These images provide near-complete 360° visual coverage for each point, forming a robust dataset for subsequent semantic segmentation and psychological modeling. The process of street view data acquisition is illustrated in Figure 3.

Figure 3. The area location analysis and street view data collection.

- **Quantification of psychological indicators**

To obtain perceptual scores for the street view images across three psychological dimensions, a subjective evaluation questionnaire was developed based on three validated scales: the Perceived Restorativeness Scale (PRS-11) [48], the Positive and Negative Affect Schedule (PANAS) [49], and the Social Support Rating Scale (SSRS) [50]. An online image rating experiment was conducted accordingly. Given that this study focused on visual perception rather than traditional self-report surveys, we adapted the original scale structures by selecting 3–4 representative semantic items from each instrument. Table 1 presents examples of the semantic judgment statements that were rephrased for image-based evaluation.

Using stratified random sampling, a subset of 200 representative images was selected from the Jinshan campus street view image library of Fujian Agriculture and Forestry University. A total of 125 undergraduate students (63 females and 62 males), aged between 20 and 25, were recruited to participate in the rating task. Each image received approximately 50 independent evaluations. Participants rated each image on the three psychological indicators using a 7-point Likert scale (1 = strongly disagree, 7 = strongly agree), based on the presented semantic judgment statements. This study ensures that the image-based ratings reflect the on-site experience as accurately as possible from two aspects. First, almost all participants had visited the locations shown in the photos and were very familiar with the campus environment. Second, following the principles of environmental psychology, the street-view images were captured at eye level and under consistent lighting and weather conditions, minimizing external variations and enhancing the ecological validity of the ratings.

The final score for each image was calculated as the average rating across all participants. Based on these scores, the images were categorized into three groups: high, medium, and low for each psychological dimension including psychological restoration, emotional uplift, and social interaction. These grouped scores served as the foundation for subsequent modeling and stratified analysis.

Table 1. The rating statements and scoring criteria for the three psychological indicators in the Image-Based Semantic Perception Questionnaire.

Indicators Used	Scales Used	Items Used	7-Point Likert Scale
Psychological Restoration	PRS-11	"This image makes me feel relaxed". "I would like to stay longer in this place". "This space helps me restore my energy".	1 (Strongly Disagree)–7 (Strongly Agree)
Emotional Uplift	PANAS	"This image makes me feel pleasant". "It makes me feel energetic". "I feel uplifted when I see it".	1 (Not at all applicable)–7 (Highly applicable)
Social Interaction	SSRS	"This scene is suitable for spending time with friends". "This is a space where I would be willing to interact with others". "This place feels socially welcoming".	1 (Not suitable)–7 (Very suitable)

- **Extraction of semantic features from images**

Street view imagery, with its rich visual content and extensive spatial coverage, serves as a vital data source for perceiving and analyzing spatial environments [51]. In this study, we employed the Mask2Former model, which is built on a Transformer-based architecture, for semantic feature extraction from images [52,53]. Figure 4 illustrates the computational principle of the Mask2Former model. The model integrates a convolutional backbone with a set prediction mechanism and is designed to optimize panoptic segmentation performance in a unified manner. It enables accurate pixel-level identification of typical campus semantic elements such as vegetation, walkable areas, sky openness, and building structures.

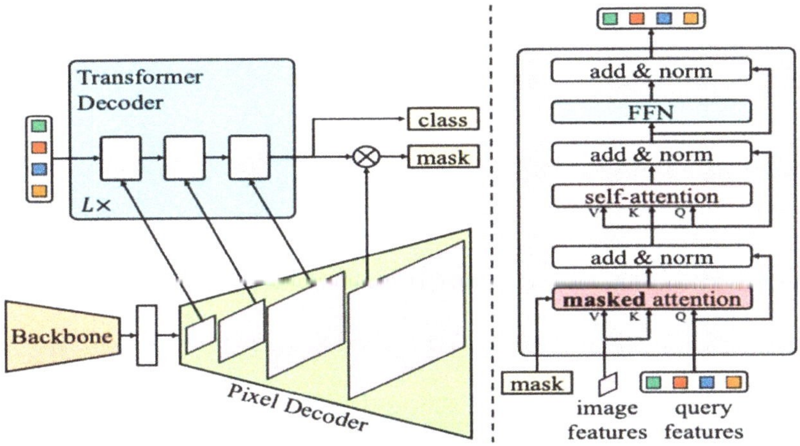

Figure 4. Computational principle of the Mask2Former model. **Note:** This figure was adapted from Reference [52].

For Mask2Former training on the Mapillary Vistas dataset [54], panoptic and semantic segmentation were trained separately with identical settings using an ImageNet-22K pre-trained backbone. Data augmentation included random scale jittering (0.5–2.0), horizontal flip, 1024 × 1024 cropping, and color perturbation to match the dataset's resolution and scene diversity [52]. A polynomial learning rate schedule ensured stable convergence (batch size 16, 300k iterations). At inference, images were resized so the long side was 2048 px. The number of queries depends on the task complexity and backbone (200 for Swin-L panoptic, otherwise 100) to balance capacity and stability.

Figure 5 illustrates the semantic segmentation results of campus panoramic images using the Mask2Former model. Using a model pretrained on the Mapillary Vistas dataset, which supports recognition of a wide range of objects, we selected 38 key semantic categories closely related to psychological perception for further analysis. These selected features reflect critical visual components hypothesized to influence students' emotional, social, and cognitive responses to the campus environment. Table 2 presents 38 semantic feature classes.

Table 2. The 38 semantic feature classes.

38 Semantic Feature Classes					
Bird	Ground animal	Curb	Fence	Guard rail	Barrier
Wall	Bike lane	Crosswalk-plain	Curb cut	Parking	Pedestrian area
Road	Sidewalk	Bridge	Building	Person	Lane marking—crosswalk
Lane marking—general	Mountain	Sand	Sky	Terrain	Vegetation
Water	Bench	Billboard	Manhole	Street light	Pole
Utility pole	Traffic light	Traffic sign (back)	Traffic sign (front)	Trash can	Bicycle
	Car			Motorcycle	

Note: The 38 semantics were trained based on the Mapillary Vistas dataset, and after obtaining the image semantic segmentation results for FAFU, semantics with no feature values were removed, leaving the final 38.

Figure 5. Semantic segmentation results of the campus panoramic images using the Mask2Former model.

3.2. Machine Learning and Multivariate Linear Regression Methods

- **Decision tree modeling with SHAP interpretation**

To evaluate the importance of landscape features on the students' psychological outcomes, this study employed three decision tree-based algorithms: random forest (RF), gradient boosting regression (GBR), and extreme gradient boosting (XGBoost) [55–57], each offering distinct advantages in nonlinear modeling and feature importance ranking. Image semantic features were used as input variables and psychological indicator scores as output variables. After dividing the dataset into training, validation, and test sets, model performance was evaluated using the mean squared error (MSE), mean absolute error (MAE), and explained variance score (EVS) on the validation set. The model with the best performance was selected for final feature importance analysis and prediction. To enhance interpretability, the optimal model was combined with the SHAP (Shapley additive explanations) mechanism [58,59], which reveals the most influential features affecting psychological scores, along with their impact directions and threshold boundaries.

In this study, SHAP analysis was performed based on the final selected model (e.g., XGBoost) trained on the training set. The "TreeExplainer" module from the Python "shap" library (version 0.47.2) was employed to compute SHAP values for each input feature. For the test set, SHAP values were calculated for every sample to quantify the local contribution of each feature to the predicted psychological scores. The mean absolute SHAP value across all samples was then used to derive the global feature importance ranking. To enhance the robustness of the results, the calculation was repeated 50 times with random resampling, and the average SHAP importance was reported.

The SHAP-based analysis of the interpretable machine learning algorithms focused only on the top six features ranked by importance, potentially overlooking semantically meaningful image features that ranked lower but may still exert significant effects on the three psychological indicators. To address this limitation, multivariate linear regression analysis was conducted further to explore the impact of additional image-based semantic features. First, a Pearson correlation matrix was used to assess multicollinearity among features. Features with high collinearity were removed while retaining as much semantic information as possible. The remaining semantic features were then used as independent variables, with the three psychological indicators as dependent variables in a multivariate linear regression model.

3.3. Heterogeneity Analysis

In addition to modeling the overall relationships between campus landscape features and psychological indicators at a forestry university, this study further addressed two key questions: (1) Which landscape features exert significant influence on high versus low psychological scores across the three psychological indicators? and (2) do such influences vary across different functional zones within the campus? To explore these questions, a heterogeneity analysis was conducted.

First, participants were grouped into high, medium, and low score categories for each psychological indicator, based on tertile (top 1/3, middle 1/3, bottom 1/3) segmentation. Separate SHAP-based decision tree analyses and multivariate regression models were applied to each subgroup. The focus was placed on identifying features that exerted significantly positive effects in the high-score groups and significantly adverse effects in the low-score groups.

Second, the study applied k-means clustering to classify the campus into five distinct functional zones. Within each zone, machine learning and multivariate regression analyses were performed separately. This functional-zone-based heterogeneity analysis provided targeted insights to support future strategies related to ecological research, leisure activities, and transportation planning within forestry university campuses.

4. Empirical Results

4.1. Optimal Decision Tree Algorithm Selection for Campus Space Optimization

In this study, three decision tree-based machine learning algorithms, random forest (RF), gradient boosting regression (GBR), and extreme gradient boosting (XGBoost), were employed to model and predict the restorative effects of campus outdoor spaces. The dataset was split into training, validation, and test sets in a 70:20:10 ratio. Model performance was evaluated using metrics such as the mean squared error (MSE), mean absolute error (MAE), and explained variance score (EVS) [60].

The strategies employed to prevent overfitting in the three decision tree algorithms were as follows: for RF, the maximum depth of individual trees was limited, the minimum number of samples per leaf node was set, and a random subset of features was introduced during node splitting. For GBR, a small learning rate was used, the number and depth of trees were restricted, and a subsampling strategy was applied. For XGBoost, both L1 and L2 regularization terms were incorporated to penalize overly complex model structures.

For each psychological indicator and campus space type, the model with the best performance on the validation set was selected as the optimal algorithm. The final selection of optimal decision tree algorithms for different tasks is presented in Appendix A Tables A1–A3. For example, on the full-sample dataset, the XGBoost model outperformed others in predicting all three psychological indicators: psychological restoration, emotional uplift, and social interaction.

4.2. Image Semantic Influence Analysis of Campus Outdoor Spaces

4.2.1. Visualization of Influential Semantic Features Across the Study Area

Figure 6 illustrates the spatial distribution of semantic features across the study area based on the full sample. The top six semantic features most influential to the three psychological indicators were selected, with duplicates removed to highlight the most representative factors.

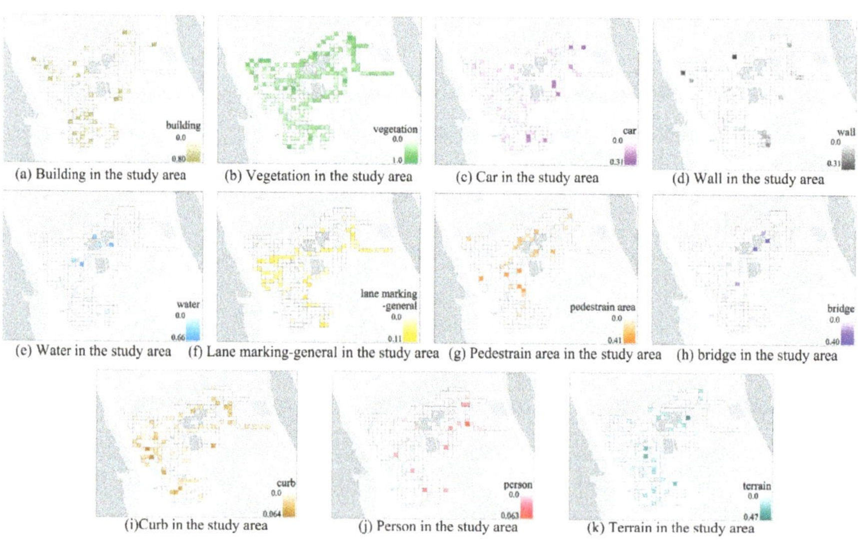

(a) Building in the study area (b) Vegetation in the study area (c) Car in the study area (d) Wall in the study area (e) Water in the study area (f) Lane marking-general in the study area (g) Pedestrain area in the study area (h) bridge in the study area (i)Curb in the study area (j) Person in the study area (k) Terrain in the study area

Figure 6. Grid-based spatial visualization of the image semantic features.

4.2.2. SHAP-Based Feature Importance Analysis

- **Key semantic image features influencing psychological restoration**

Figure 7 presents the SHAP analysis results for semantic image features on the psychological restoration indicator based on the full sample. The left panel displays the ranking of feature importance based on the absolute SHAP values, while the right panel shows the SHAP value distribution across individual samples. A notable pattern was observed: image features with higher importance tended to exhibit more polarized SHAP value distributions, whereas those with lower importance showed SHAP values more concentrated near the center.

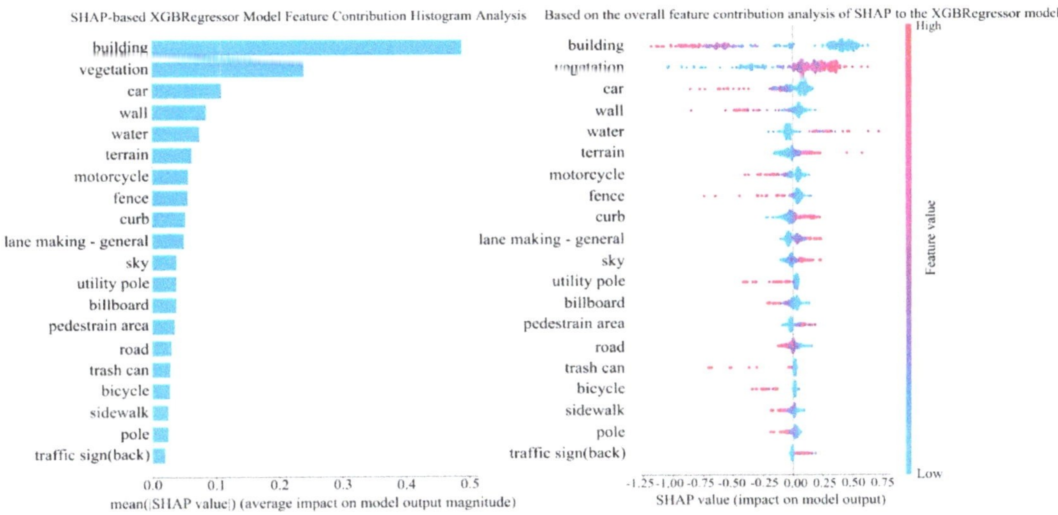

Figure 7. SHAP value analysis of the image semantic features for the psychological restoration indicator (full sample).

The SHAP analysis, utilizing the XGBRegressor model, revealed that buildings and vegetation were the two most influential semantic features in the psychological restoration of outdoor campus spaces. While buildings, though significant, had a negative impact on psychological well-being, vegetation stood out as a key positive factor. Other notable features, such as water bodies and sky visibility, also played an important role, demonstrating measurable effects on the overall quality of campus space optimization.

Figure 8 presents the influence patterns of the top six image semantic features contributing to psychological restoration, based on the full-sample SHAP analysis. As the values of these features varied, the direction and magnitude of their effects on the psychological restoration indicator also changed. Specifically, the building feature showed a positive impact when its value ranged from 0 to 0.08, but shifted to a negative influence beyond 0.08. The vegetation feature exerted an adverse effect at values below 0.3, and became a positive contributor when the value exceeded 0.3. Both car and wall features positively affected psychological restoration when their values were within 0 to 0.02, but their influence turned negative once they surpassed this threshold. The water feature consistently exerted a positive effect as long as its value was greater than zero. Finally, the terrain feature had a negative impact when its value was below 0.03, transitioning to a positive influence when it exceeded that level. These findings underscore the nonlinear and threshold-based nature of semantic landscape features in shaping students' restorative experiences in campus outdoor spaces, emphasizing their role in campus space optimization.

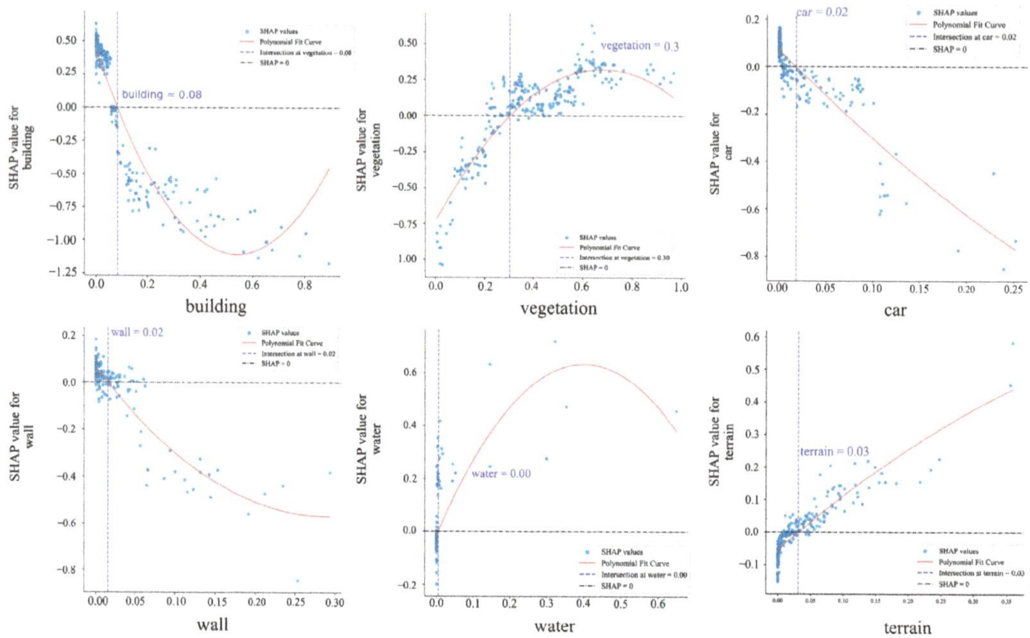

Figure 8. The influence patterns of the top-ranked semantic features contributing to psychological restoration in the full sample.

- **Key semantic image features influencing emotional uplift**

Figure 9 presents the SHAP analysis of image semantic features associated with the emotional uplift indicator based on the full-sample dataset. The results showed that hardscape elements, such as buildings, lane markings, vehicles, pedestrian areas, and bridges, exerted the most significant influence on the students' emotional uplift. Overall, these features tended to have a negative impact, indicating that environments dominated by rigid infrastructural elements are less likely to enhance the students' emotional states. These findings highlight the importance of balancing hardscape and softscape elements in campus space optimization to foster a more emotionally supportive and restorative environment.

Figure 10 illustrates the influence patterns of the top semantic features contributing to emotional enhancement in the full-sample analysis, further elucidating the mechanisms by which these key features affect the emotional enhancement scores. Building features exerted a positive effect when their values ranged from 0 to 0.09, but became detrimental beyond 0.09. The lane marking—general feature had a negative impact below 0 and a positive impact above this threshold. Car features showed a negative effect between 0 and 0.02, turning positive when exceeding 0.02. Pedestrian area features negatively influenced emotional enhancement from 0 to 0.02, but shifted to a positive effect beyond 0.02. Bridge features were negative below 0 and positive above 0. Finally, curb features negatively affected emotional enhancement when below 0.01, but became positive when exceeding this value.

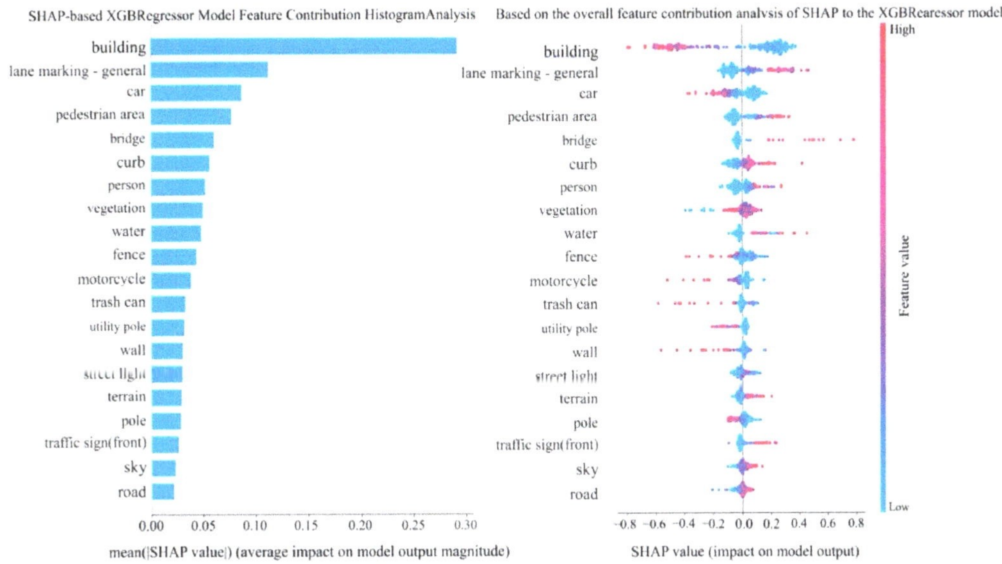

Figure 9. SHAP value analysis of the image semantic features for the emotional uplift indicator (full sample).

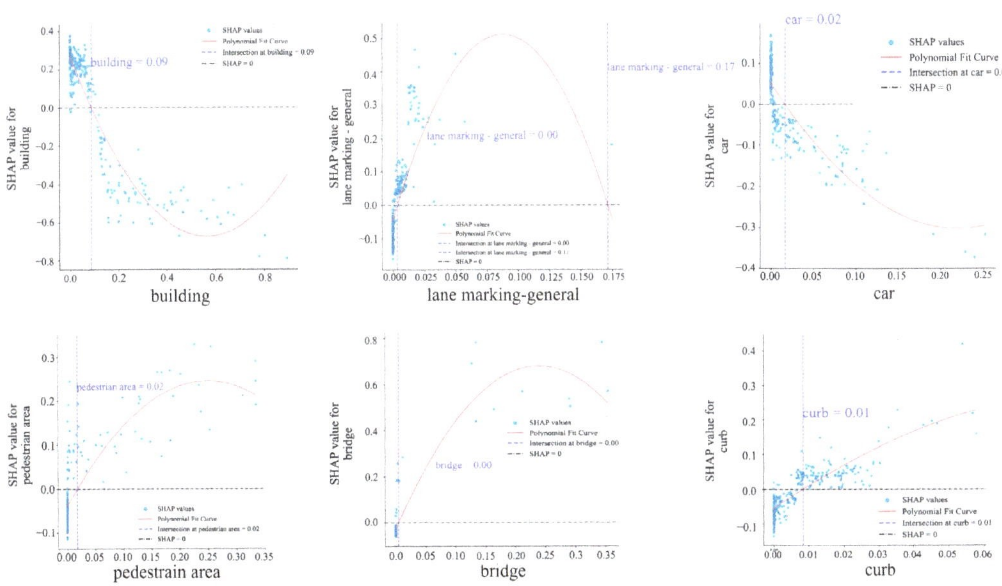

Figure 10. Influence patterns of the top-ranked semantic features contributing to emotional uplift in the full sample.

- **Key semantic image features influencing social interaction**

Figure 11 presents the SHAP analysis of image semantic features associated with the social interaction indicator based on the full-sample dataset. The results indicate that "person" was the most influential semantic feature in identifying high-social-interaction scenarios. In addition, open environmental elements such as "water" and "pedestrian area", along with structural features like "wall", "vegetation", and "sidewalk", also made notable positive contributions to the prediction of social interaction scores. In contrast, features such as "sky", "curb", and "car" demonstrated relatively limited predictive importance in the model.

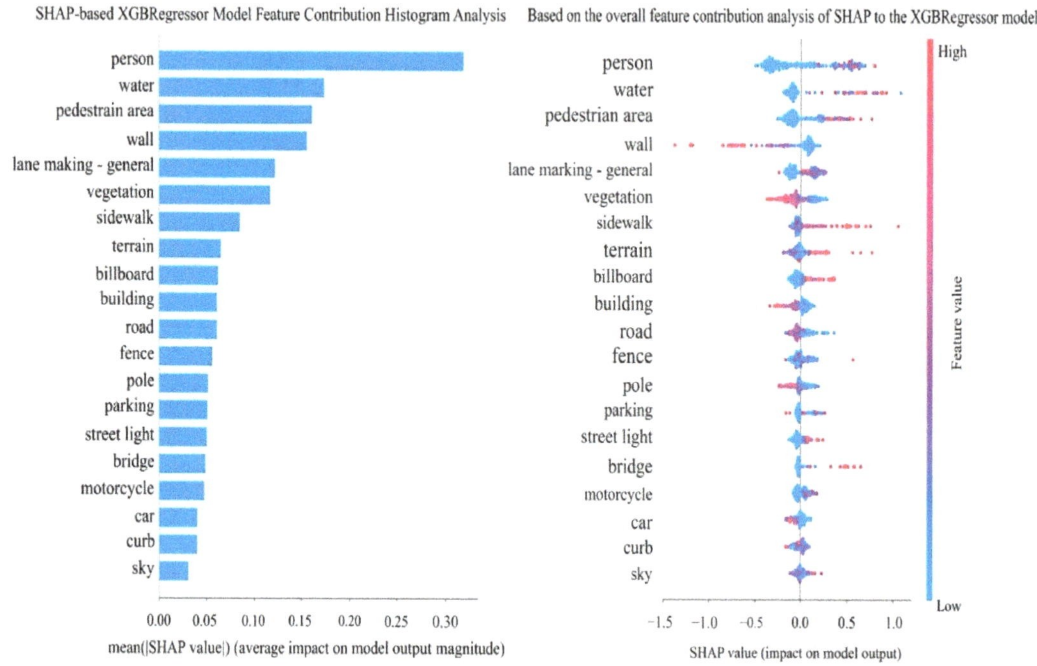

Figure 11. SHAP value analysis of the image semantic features for the social interaction indicator (full sample).

Figure 12 presents the influence patterns of the top-ranked semantic features contributing to social interaction in the full sample. A detailed analysis of the top six features ranked by SHAP values revealed distinct nonlinear influence trends. The person feature exhibited a consistently positive effect as long as its value was greater than zero. The water feature began to exert a positive influence when its value exceeded 0.0. The pedestrian area feature showed a positive effect once its value surpassed 0.01. The wall feature had a positive influence in the range of 0 to 0.01 but shifted to an adverse effect when the value exceeded 0.01. The lane marking—general feature was positively associated with social interaction when its value was greater than 0.0, but turned negative beyond 0.14. Finally, the vegetation feature positively affected social interaction within the 0–0.35 range but exhibited an adverse effect when it exceeded 0.35. Maintaining campus landscapes that have a positive impact and optimizing those that have a negative impact can foster better social interaction among students.

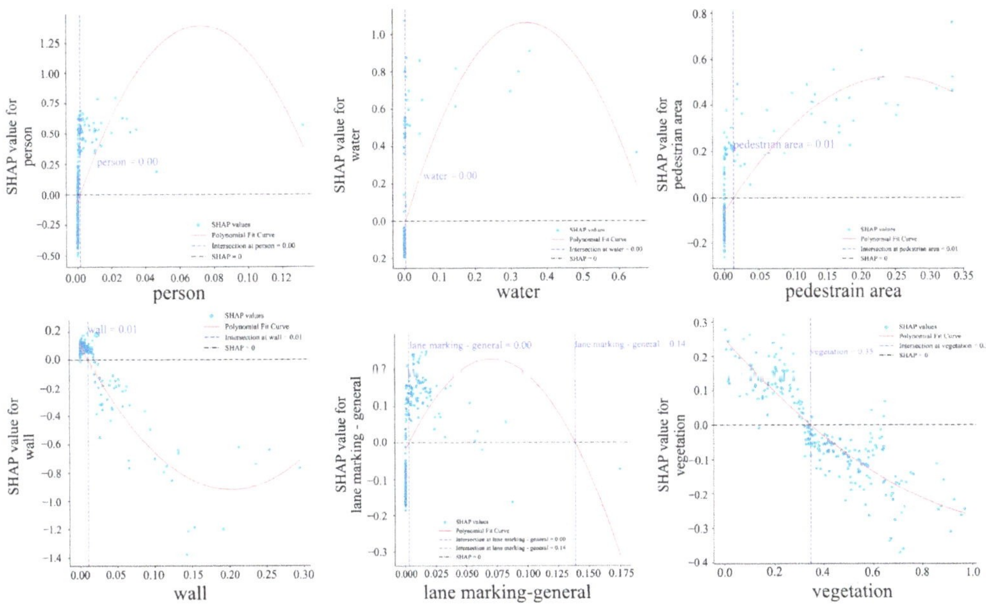

Figure 12. Influence patterns of the top-ranked semantic features contributing to social interaction in the full sample.

4.2.3. Semantic Feature Heterogeneity Analysis in Campus Outdoor Spaces Using Multivariate Linear Regression (Full Sample)

Figure 13 presents a heatmap of the Pearson correlation coefficients among the 38 semantic features extracted from campus outdoor spaces. The results showed that most features exhibited weak pairwise correlations, with the overall heatmap displaying light blue tones, indicating a high degree of independence among the image semantic features. To prevent multicollinearity in the subsequent multivariate linear regression analysis, feature pairs with Pearson correlation coefficients exceeding 0.2 were processed by preferentially removing the less important feature based on its importance ranking in the interpretable machine learning models while aiming to retain as many semantic variables as possible as explanatory factors. Some feature pairs, such as curb and lane marking—general, or mountain and utility pole, demonstrated relatively strong positive correlations. In contrast, building, sky, and vegetation showed weak negative correlations. These findings underscore the importance of carefully balancing natural and built elements in campus space optimization to foster a psychologically restorative environment for students.

The multivariate linear regression results for the psychological restoration indicator in the full sample, as presented in Table 3, revealed that several semantic features not identified as significant in the SHAP analysis exhibited statistically significant effects. For example, an increase in one unit in the fence feature was associated with a significant decrease of 3.34 units in psychological restoration scores, with the coefficient significant at the 1% level. Other features with significant adverse effects included lane marking—crosswalk (−2.92), sand (−15.78), manhole (−29.61), pole (−11.13), utility pole (−35.78), trash can (−10.04), and motorcycle (−4.98). In contrast, lane marking—general and bench demonstrated significant positive effects, with coefficients of 6.89 and 8.17, respectively, both significant at the 1% level. The adjusted R^2 of the regression model for the psychological restoration indicator was 0.43, and the F-statistic was 33.4, indicating a relatively good model fit.

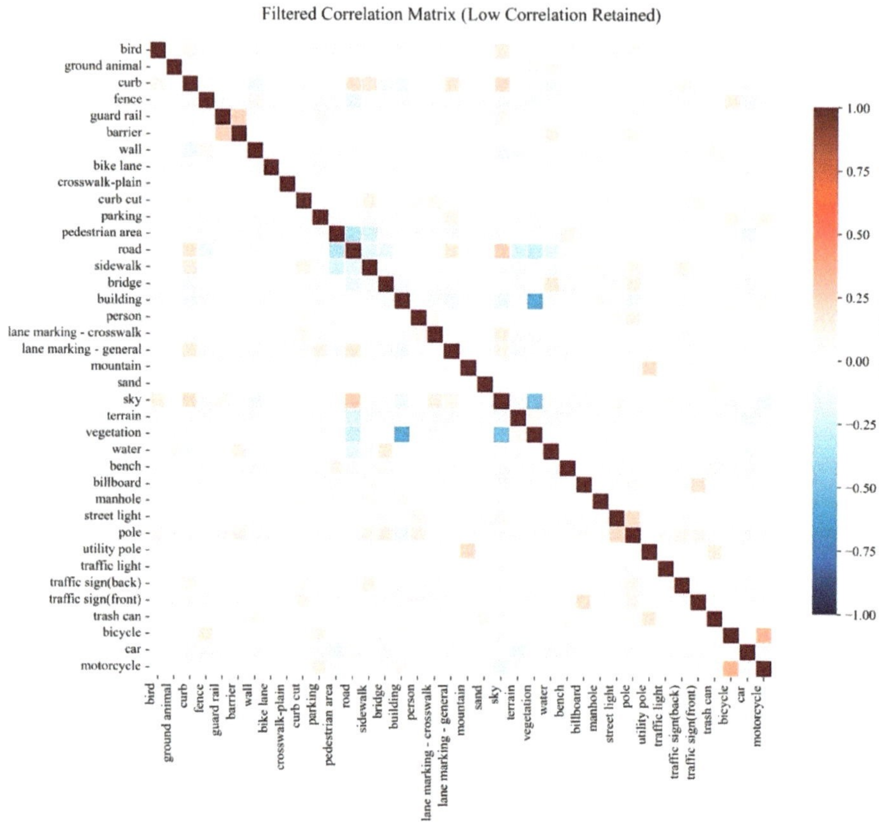

Figure 13. Heatmap of the Pearson correlation coefficients among the 38 semantic features in university outdoor spaces.

For the emotional uplift indicator, the regression results in Table 3 show that the fence feature again exerted a significant adverse effect, with a coefficient of −1.72 ($p < 0.01$). In contrast, the water feature had a significant positive effect on emotional uplift, with a coefficient of 2.08, also significant at the 1% level. The adjusted R^2 for this model was 0.24, and the F-statistic was 13.6, suggesting a moderate explanatory power. Regarding the social interaction indicator, the regression results in Table 3 revealed that the sand feature significantly decreased interaction scores, with a coefficient of −15.55 ($p < 0.01$). In contrast, seven features demonstrated significant positive associations with social interaction: curb cut (60.73), bridge (5.51), terrain (1.53), billboard (3.58), street light (97.76), traffic sign (back) (164.03), and motorcycle (5.62), all significant at the 1% level. These results further highlight the complementary value of multivariate regression in identifying features that may be underestimated in machine learning-based importance rankings.

Table 3. Multivariate linear regression results for the full sample.

	Psychological Recovery		Mood Improvement		Social Interaction
Fence	−3.34 *** (1.22×10^{-11})	Fence	−1.718 *** (0.0006)	Curb cut	60.731 *** (0.00594)
Lane marking—Crosswalk	−2.92 ** (0.0494)	Wall	−3.148 *** (0.0001)	Bridge	5.508 *** (4.67×10^{-11})

Table 3. *Cont.*

Psychological Recovery		Mood Improvement		Social Interaction	
Lane marking—General	6.89 *** (0.00792)	Bike lane	−10.7463 *** (0.0438)	Sand	−15.55 *** (0.000295)
Sand	−15.78 *** (1.8×10^{-5})	Sand	−15.9204 *** (0.0)	Terrain	1.525 *** (0.00773)
Bench	8.17 ** (0.0393)	Water	2.0821 *** (0.0006)	Billboard	3.576 ** (0.0368)
Billboard	−5.41 *** (0.000225)	Utility pole	−22.927 ** (0.0115)	Street light	97.756 ** (0.0282)
Manhole	−29.61 *** (0.00644)	Trash can	−10.2561 ** (0.0399)	Traffic sign (back)	164.028 ** (0.0362)
Pole	−11.13 ** (0.0106)				
Utility pole	−35.78 *** (6.03×10^{-5})	Motorcycle	−5.2437 *** (0.0023)	Motorcycle	5.615 *** (0.00427)
Trash can	−10.04 ** (0.0406)				
Motorcycle	−4.98 *** (0.00311)				
Adjusted R-squared	0.432	Adjusted R-squared	0.235	Adjusted R-squared	0.209
F-statistic	33.37	F-statistic	13.60	F-statistic	11.46

Note: *** indicates significance at the 1% significance level, ** indicates significance at the 5% significance level.

4.3. Heterogeneity Analysis Based on Psychological Score Groups

4.3.1. Semantic Feature Comparison Across High, Medium, and Low Perception Score Groups

One representative outdoor campus image was randomly selected from each of the high, medium, and low score groups. Figure 14 presents these images along with their corresponding semantic feature distributions, aiming to highlight the compositional differences across different perceptual score levels.

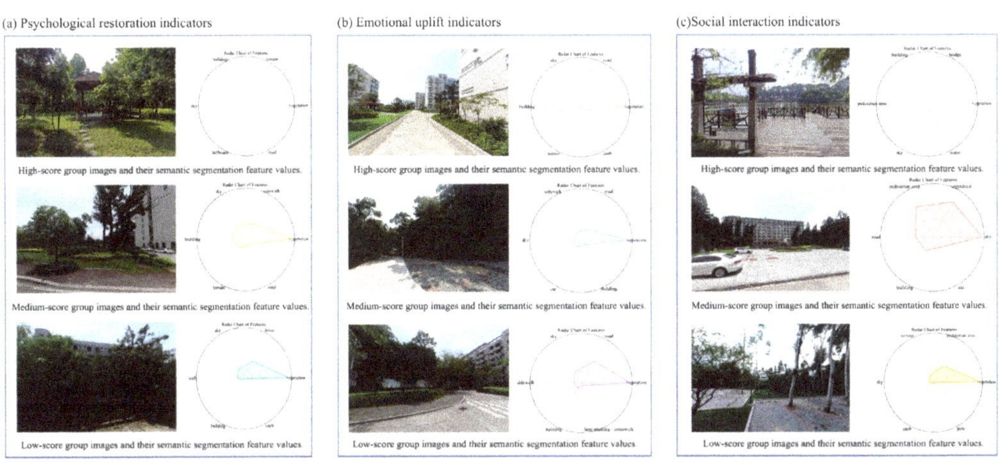

Figure 14. Semantic feature distributions across different score groups.

4.3.2. Modeling Results of Interpretable Machine Learning Across Score-Based Subgroups

- **Psychological restoration indicator**

In the SHAP value analysis for the psychological restoration indicator, vegetation and building consistently appeared among the top six features across the high-, medium-, and low-score subgroups. The billboard feature appeared in the top six only in the medium-score group; in contrast, sand, wall, and lane marking—general ranked among the top six features exclusively in the low-score group. These variations suggest that different semantic elements may be more salient in influencing psychological restoration depending on the perceived restoration level of the space.

Analysis of the influence patterns of the top-ranked semantic features contributing to psychological restoration across the high-, medium-, and low-score groups revealed that in the high-score group, features such as road (SHAP values: 0–0.07), car (0–0.01), building (0–0.04), sky (0–0.09), and water (positive SHAP values) were associated with positive effects on psychological restoration. In the medium-score group, building (0–0.07), vegetation (0.35–0.97), water (>0), road (0–0.11), billboard (>0), and sky (0–0.22) also showed positive influences. In contrast, within the low-score group, vegetation (0.0–0.21) and lane marking—general (values < 0) were associated with adverse effects on psychological restoration. These findings suggest that the impact of specific landscape features varies notably across different levels of perceived psychological restoration.

- **Emotional uplift indicator**

In the SHAP value analysis for the emotional uplift indicator, vegetation consistently appeared among the top six most important features in all three subgroups: high-, medium-, and low-score. In the high-score group, bridge, wall, and water also appeared among the top six features contributing to emotional uplift. In contrast, car and pole ranked within the top six features in the low-score group, suggesting that these elements may be more salient in lower-rated emotional environments.

In the analysis of influence patterns for the top-ranked semantic features contributing to emotional uplift across all score groups, the high-score group showed that curb (SHAP values: 0–0.01), fence (0–0.14), and water (positive SHAP values) were associated with positive effects on emotional uplift. In contrast, in the low-score group, features including vegetation (0–0.3), lane marking—general (values < 0), curb (values < 0), and car (0–0.02) showed adverse effects. These results highlight the differentiated roles that environmental features play in shaping emotional responses across perceptual levels.

- **Social interaction indicator**

In the SHAP value analysis for the social interaction indicator, the building feature consistently appeared among the top six features in all three subgroups: high-, medium-, and low-score. In the high-score group, sand, water, and pole were also among the top six features contributing to higher perceived social interaction. In contrast, in the low-score group, features such as curb, wall, sidewalk, and lane marking—general appeared in the top six, suggesting their relevance in lower-rated social interaction environments.

In the analysis of top-ranked semantic features for social interaction, the high-score group was characterized by positive associations for road (SHAP values: 0–0.08), building (0–0.11), and water (positive SHAP values), whereas the low-score group showed negative associations for curb (0–0.3), sidewalk (0–0.04), lane marking—general (negative SHAP values), car (0–0.01), and building (0–0.01). These results indicate that features beneficial to social interaction in well-designed settings may have the opposite effect in less favorable spatial configurations.

4.3.3. Multivariate Linear Regression Results Across Score-Based Subgroups

Following a similar procedure to the full-sample multivariate linear regression analysis, this section removed redundant features within each high-, medium-, and low-score group for the three psychological indicators to ensure the absence of multicollinearity among the input variables. This empirical analysis aimed to identify influential semantic features that the interpretable machine learning approach in Section 4.3.1 may have overlooked. Table 4 presents the multivariate linear regression results for the psychological restoration indicator across the three score groups. The results revealed that several semantic features not identified as necessary in the SHAP analysis nonetheless exhibited statistically significant effects, highlighting secondary yet critical visual elements that influence psychological restoration. For example, in the high-score group, each unit increase in the terrain feature was associated with a 1.05-unit increase in the psychological restoration score. Similarly, the bench feature showed a strong positive effect, with a coefficient of 6.50. Both coefficients were statistically significant. The adjusted R^2 of this regression model was 0.21, and the F-statistic was 4.94, indicating a reasonably good model fit.

Table 4. Multivariate linear regression results for the psychological restoration indicator across the high-, medium-, and low-score groups.

	High-Score Group of Psychological Recovery		Medium-Score Group of Psychological Recovery		Low-Score Group of Psychological Recovery
Parking	−3.1834 ** (0.042)	Sidewalk	0.9232 ** (0.0158)	Fence	−0.7523 ** (0.034)
				Wall	−1.3202 ** (0.027)
Terrain	1.0471 *** (0.005)	Bridge	1.5738 (0.0007)	Mountain	−2.5272 ** (0.05)
				Sand	−9.175 *** (4×10^{-5})
Bench	6.4953 *** (0.005)	Utility pole	−16.7423 ** (0.0438)	Vegetation	0.9526 *** (4.4×10^{-9})
				Car	−1.6967 ** (0.008)
Adjusted R-squared	0.211	Adjusted R-squared	0.261	Adjusted R-squared	0.168
F-statistic	4.943	F-statistic:	6.181	F-statistic	3.96

Note: *** indicates significance at the 1% significance level, ** indicates significance at the 5% significance level.

Table 5 presents the multivariate linear regression results for the emotional uplift indicator across the high-, medium-, and low-score groups. The findings show that several semantic features not identified as significant in the SHAP analysis nonetheless had statistically significant effects on emotional uplift, revealing additional factors relevant to the students' affective perceptions. For example, in the high-score group, a one-unit increase in the parking feature was associated with a 0.5-unit increase in emotional uplift. In addition, vegetation (coefficient = 0.87), water (6.89), and pole (6.04) all showed significant positive effects. The adjusted R^2 of the regression model for this group was 0.08, and the F-statistic was 2.20, indicating a good model fit.

Table 5. Multivariate linear regression results for the emotional uplift indicator across the high-, medium-, and low-score groups.

	High-Score Group of Mood Enhancement		Medium-Score Group of Mood Enhancement		Low-Score Group of Mood Enhancement
Parking	0.5031 *** (0.0006)	Sidewalk	1.2863 *** (0.0024)	Sidewalk	0.9256 ** (0.023)
Vegetation	0.8747 ** (0.0133)	Water	1.5735 (0.0017)		
Water	6.89 *** (0.00792)	Manhole	−13.6865 ** (0.0293)	Sand	−7.1171 *** (0.0001)
Pole	6.0368 ** (0.0489)	Utility pole	44.905 ** (0.0175)		
Adjusted R-squared	0.08	Adjusted R-squared	0.09	Adjusted R-squared	0.128
F-statistic	2.197	F-statistic	2.434	F-statistic	3.156

Note: *** indicates significance at the 1% significance level, ** indicates significance at the 5% significance level.

Table 6 presents the multivariate linear regression results for the social interaction indicator across the high-, medium-, and low-score groups. The analysis revealed that several semantic features not highlighted in the SHAP analysis nonetheless exhibited statistically significant effects on social interaction, uncovering additional explanatory variables. For instance, in the high-score group, a one-unit increase in the bike lane feature corresponded to a 34.58-unit increase in the social interaction score. In addition, the bridge (coefficient = 1.59) and the utility pole (19.55) also showed significant positive effects. The adjusted R^2 for this regression model was 0.18, and the F-statistic was 4.16, indicating a modest but acceptable model fit.

Table 6. Multivariate linear regression results for the social interaction indicator across the high-, medium-, and low-score groups.

	High-Score Group of Social Interaction		Medium-Score Group of Social Interaction		Low-Score Group of Social Interaction
Bike lane	34.581091 *** (0.0179)	Bridge	1.2484 ** (0.041)	Wall	−2.1237 *** (5×10^{-5})
Bridge	1.590637 *** (0.00027)	Person	10.1645 **(0.042)	Sidewalk	1.1473 *** (0.0297)
Utility pole	19.551512 *** (0.02002)	Lane marking—Crosswalk	1.7094 ** (0.041)	Mountain	−2.7529 * (0.021)
				Vegetation	−0.3282 ** (0.0125)
Adjusted R-squared	0.177	Adjusted R-squared	0.053	Adjusted R-squared	0.076
F-statistic	4.164	F-statistic	1.825	F-statistic	2.246

Note: *** indicates significance at the 1% significance level, ** indicates significance at the 5% significance level, * indicates significance at the 10% significance level.

4.4. Functional Zoning and Semantic Feature Analysis of Campus Outdoor Spaces

4.4.1. Functional Zoning of Campus Outdoor Spaces

Figure 15 illustrates the classification results of campus outdoor spaces based on k-means clustering. The inertia of clustering showed a sharp decline when the number of clusters was fewer than five, after which the rate of decrease stabilized. This pattern indicates that selecting five clusters is an appropriate choice for segmenting campus outdoor spaces [61].

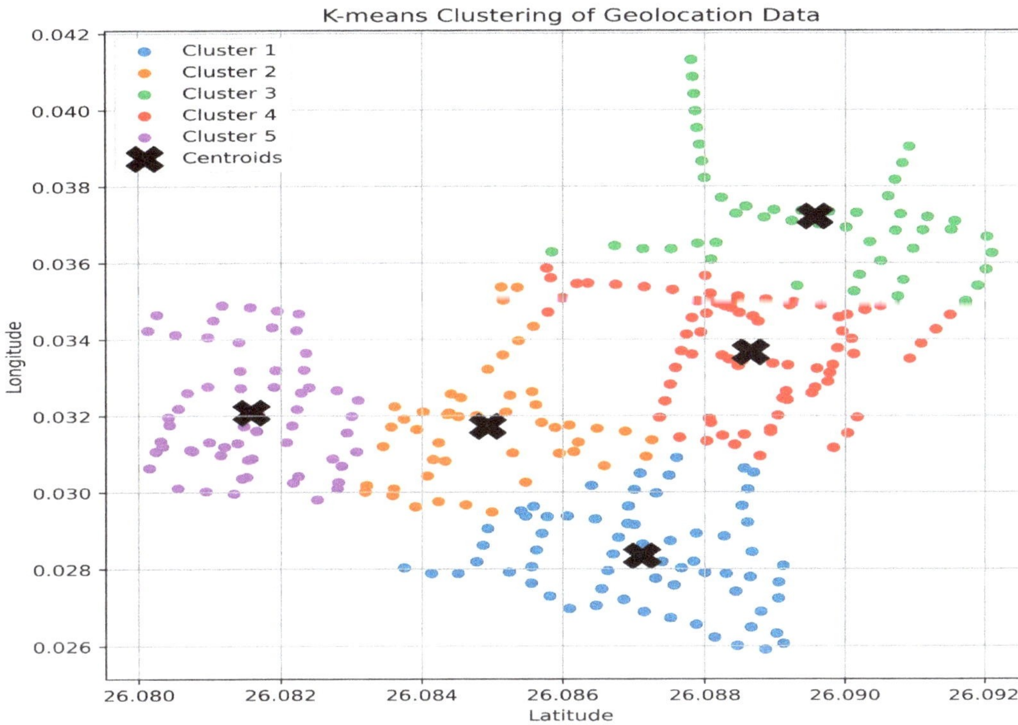

Figure 15. Classification results of the campus outdoor spaces based on k-means clustering analysis.

Based on the k-means clustering analysis, the outdoor spaces of the Fujian Agriculture and Forestry University (Jinshan Campus) were categorized into five functional zones. As shown in Figure 16, representative image semantics for each zone are illustrated. The first zone, the ***Ecological Research and Practice Zone***, is primarily located around the Forestry Building and the Sugarcane Research Center in the northern part of the campus. This area is characterized by strong academic functions, supporting activities such as plant observation, crop experimentation, and field-based practical training for agriculture and forestry disciplines. The second zone, the ***Ecological Landscape Exhibition Zone***, is centered in the Zhonghuayuan area and integrates botanical gardens, water features, and horticultural landscapes, highlighting ecological education and campus image presentation. The third zone, the ***Commuting and Circulation Zone***, runs linearly along the eastern edge of the campus, connecting teaching, residential, and entrance nodes. It is marked by high pedestrian flow and strong spatial accessibility. The fourth zone, the ***Integrated Leisure Landscape Zone***, surrounds Guanyin Lake and combines ecological, leisure, and social functions, serving as a core area for campus activities and landscape appreciation. Finally, the fifth zone, the ***Multifunctional Social Interaction Zone***, includes areas such as the Research Institute Plaza, the library forecourt, the Innovation Building, and the dormitory surroundings. This zone integrates academic exchange, recreational activities, and daily services, acting as a key node for campus social engagement.

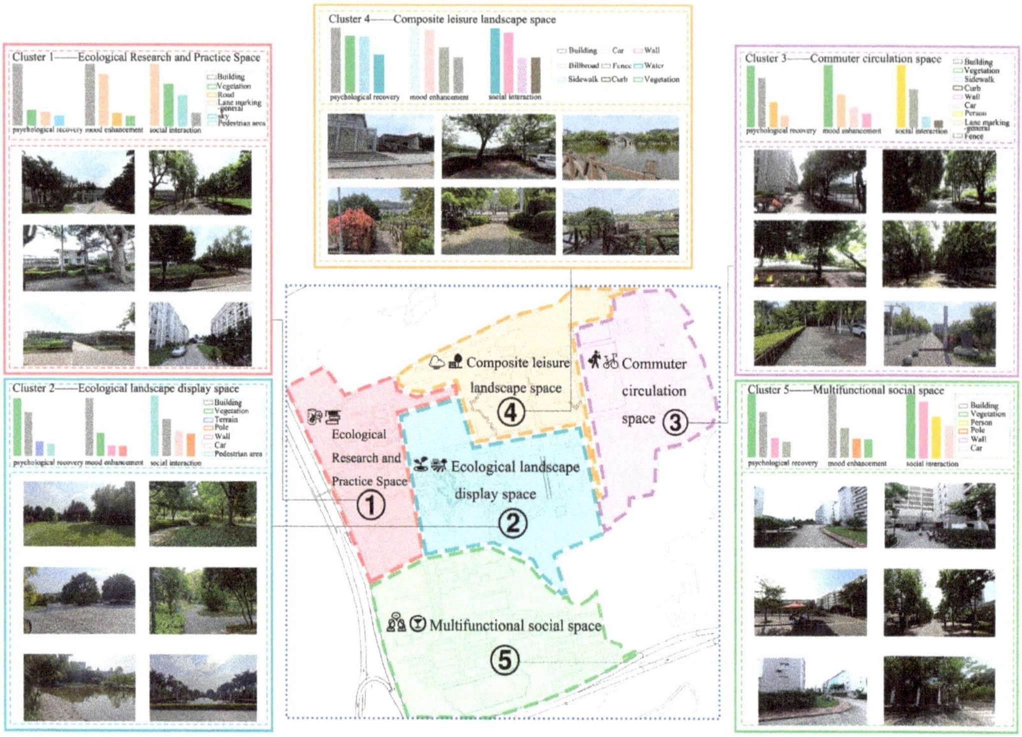

Figure 16. Semantic examples of image features across the five campus functional zones.

4.4.2. Semantic Feature Analysis of Different Campus Outdoor Spaces

- **Cluster 1: Ecological Research and Practice Zone**

Appendix A Figure A1 presents the SHAP value analysis of semantic image features for the three psychological indicators in the first cluster. At the same time, Appendix A Figure A2 illustrates the impact patterns of the top contributing features. In the ***Ecological Research and Practice Zone***, the feature building showed a consistently positive effect across all three psychological dimensions. Additionally, billboards and cars contributed positively to psychological restoration, pole enhanced emotional uplift, and both vegetation and walls facilitated social interaction. In contrast, traffic-related features such as lane marking—general, road, and pedestrian area generally exerted adverse effects. Notably, sidewalk and vegetation also showed suppressive effects on psychological restoration and emotional uplift in specific models.

- **Cluster 2: Ecological Landscape Display Spaces**

In the ***Ecological Landscape Display Spaces***, Appendix A Figure A3 presents the SHAP value analysis of semantic image features across the three psychological dimensions. At the same time, Appendix A Figure A4 illustrates the impact patterns of the top contributing features. The feature building consistently exerted a positive influence on all three psychological indicators. Specifically, bicycle contributed to psychological restoration, while the motorcycle, fence, and pole positively affected emotional uplift. Features such as car, vegetation, and a curb enhanced social interaction. Conversely, elements like vegetation and pedestrian area showed adverse effects across multiple dimensions, and features such as terrain, mountain, and wall also appeared to have adverse effects on specific psychological outcomes.

- **Cluster3: Commuting and Circulation Spaces**

In the ***Commuting and Circulation Spaces***, Appendix A Figure A5 presents the SHAP value analysis of semantic image features across the three psychological dimensions. At the same time, Appendix A Figure A6 illustrates the impact patterns of the top-ranked features. Results showed that building and pole positively contributed to psychological restoration, car and billboard enhanced emotional uplift, and wall facilitated social interaction. In contrast, multiple features, including vegetation, lane marking (general), and fences, demonstrated adverse effects across dimensions. Additionally, people and curbs tended to suppress social interaction functionality.

- **Cluster4: Composite Leisure Landscape Spaces**

In the ***Composite Leisure Landscape Spaces***, Figures A7 and A8 in the Appendix A illustrate the SHAP value analysis and the impact patterns of top-ranked semantic features across the three psychological dimensions, respectively. The results indicate that buildings and cars exhibited positive effects across multiple dimensions. Sidewalk contributed positively to psychological restoration, while motorcycle enhanced emotional uplift. Additionally, walls and billboards played supportive roles in promoting social interaction. Conversely, several natural and circulation-related elements, including vegetation, water, terrain, and specific traffic-related components, showed negative associations across dimensions, suggesting that they may diminish the restorative, emotional, or social benefits of the space.

- **Cluster5: Multi-Functional Social Interaction Spaces**

In the ***Multi-Functional Social Interaction Spaces***, Appendix A Figure A9 presents the SHAP value analysis of image semantic features under the three psychological indicators. At the same time, Appendix A Figure A10 illustrates the impact patterns of the top-ranked semantic features. Within this space category, "building" demonstrated consistently strong positive effects across all psychological outcomes. Features such as "wall", "fence", and "motorcycle" were positively associated with psychological restoration and emotional uplift, whereas "car" contributed positively to social interaction. In contrast, elements such as "vegetation", "curb", and several infrastructure-related features exhibited predominantly adverse effects, indicating a potential weakening of the space's overall psychological functionality.

4.4.3. Heterogeneity Analysis Based on Multiple Linear Regression

To further investigate the spatial heterogeneity across the five functional zones, this section employed multiple linear regression to identify semantic features that were not globally important in the SHAP analysis but demonstrated statistically significant effects within specific spatial contexts. The k-means clustering results revealed that these features exerted notable influence on the perceived psychological scores across different zones, highlighting their context-dependent value and potential for targeted spatial interventions. Appendix A Table A4 presents the regression results for the first functional zone, the ***Ecological Scientific Practice Space***. In this zone, the psychological restoration dimension was negatively affected by features such as fence, wall, sand, and utility pole. For the emotional uplift dimension, wall, sand, and billboard exhibited adverse effects, whereas water showed a significant positive contribution. In the social interaction dimension, mountain was a negative factor, while person, lane marking, crosswalk, and traffic light were positively associated with enhanced social perceptions. Notably, wall and sand consistently appeared as negative contributors across multiple psychological dimensions.

In the ***Ecological Landscape Exhibition Space***, Appendix A Table A5 presents the multiple linear regression results for the second spatial cluster. For the psychological restoration dimension, bike lane, terrain, utility pole, and bicycle exhibited significant negative associations, whereas water showed a significant positive effect. Regarding the emotional uplift dimension, bike lane was negatively associated, while terrain and traffic sign (front) exerted positive influences. In terms of social interaction, wall, terrain, traffic sign (front), and bench contributed positively, whereas bike lane, utility pole, bicycle, and curb showed adverse effects. These results highlight both supportive and inhibitory roles of various landscape elements across different psychological dimensions, suggesting the context-dependent significance of these features within the ecological landscape exhibition space.

In the ***Commuting Circulation Space***, as shown in Appendix A Table A6, the multivariate linear regression results for the third type of space indicate that the psychological restoration dimension was negatively affected by the presence of barriers, trash cans, and cars. At the same time, terrain was a positive influencing factor. In the emotional uplift dimension, manholes and cars had a negative impact, whereas lane marking—general had a positive effect. For the social interaction dimension, fence, sidewalk, person, and traffic sign (back) were positive influencing factors, while wall and utility pole had an adverse effect. The presence of cars consistently showed a negative impact on both psychological restoration and emotional uplift.

In the ***Composite Leisure Landscape Space***, the multiple linear regression results presented in Appendix A Table A7 revealed distinct semantic feature influences across the three psychological dimensions. For psychological restoration, features such as wall, billboard, utility pole, car, and motorcycle exerted significant adverse effects. In the emotional uplift dimension, wall, utility pole, and traffic sign (front) were also associated with negative impacts, whereas bird showed a positive association with emotional uplift. Regarding social interaction, wall and sand demonstrated adverse effects, while bridge and pole contributed positively. Notably, the wall feature consistently exerted a negative influence across all three psychological indicators, suggesting that it may hinder the restoration of students' psychological health in terms of restorative, emotional, and social experiences.

In the ***Multi-functional Social Interaction Space***, as presented in Appendix A Table A8, several semantic features exhibited significant associations with psychological outcomes. For the psychological restoration dimension, the presence of a manhole showed a significant adverse effect, suggesting that it may detract from the perceived restorative quality, while lane marking—general was positively associated with restoration, indicating its potential to provide spatial order or perceived navigability that enhances psychological comfort. In terms of emotional uplift, both building and water demonstrated significant adverse impacts, implying that within this multifunctional context, dense built-up elements and nearby water bodies may reduce emotional vitality, possibly due to crowding or perceived inaccessibility. For social interaction, bike lanes and street lights positively influenced perceptions of sociability, reflecting their roles in promoting accessibility and safety, which facilitate interaction. In contrast, traffic lights and ground animals were negatively associated with social interaction, potentially due to their links with traffic stress or unpredictability in the space. These findings emphasize the nuanced and context-specific effects of landscape features, highlighting the value of fine-grained spatial analysis in supporting well-being-oriented campus design.

Based on the results of the SHAP-based machine learning and multiple linear regression analyses, the functional characteristics of campus outdoor spaces can be summarized as follows. In the first clustered space, ***Ecological Research and Practice Spaces***, buildings serve as the central semantic element, comprehensively supporting psychological restoration, emotional uplift, and social interaction. This reflects a spatial character marked by structural stability and multifunctionality. Features such as billboards, cars, and poles enhance environmental legibility and emotional vitality. While vegetation and walls positively contributed to social interaction, they showed mixed effects on the psychological and emotional dimensions. Conversely, lane marking—general, sand, and walls exerted adverse effects across multiple dimensions, indicating strong traffic-oriented and spatially enclosed characteristics. Overall, this space type demonstrates a dual nature of research-focused functionality and limited social exchange.

In the second clustered space, ***Ecological Landscape Display Spaces***, buildings again functioned as a positive core element, supporting both psychological and social outcomes. This space exhibited clear structural organization and multifunctional performance. Features such as motorcycles, fences, and poles enhanced emotional vitality, while cars and interface elements promoted social interaction. However, bike lanes and bicycles were predominantly negative, suggesting that a slow-traffic system may weaken restorative and interactive experiences. The effect of terrain was differentiated across psychological dimensions, while walls and traffic signs positively supported social functionality. Overall, this space type has a dual nature, combining research-focused functionality with limited opportunities for social interaction, emphasizing the creation of a campus environment conducive to both academic and social well-being.

In the third clustered space, ***Commuting and Circulation Spaces***, buildings and poles contributed to psychological order and stability, while billboards and lane marking generally aided in emotional uplift. Cars exerted consistent adverse effects across models, indicating that vehicular interference impairs restorative potential. The role of walls in social interaction is context-dependent, while fences, sidewalks, and persons enhance social engagement. The characteristics of this space type are structural clarity and high circulation efficiency, which are conducive to promoting short-term social interactions and enhancing both the practicality and restorative potential of the campus environment.

In the fourth clustered space, ***Composite Leisure Landscape Spaces***, buildings and cars served as stabilizing elements supporting multi-dimensional functions. Sidewalks and motorcycles contributed to psychological restoration and emotional uplift, while billboards fostered social interaction. Walls consistently showed adverse effects across all three indicators, implying that spatial enclosure diminishes environmental experience. Natural and mobility-related features also tended to exert negative impacts. This space type reflects an atmosphere of openness and functional diversity, though spatial boundaries require moderate control. These insights highlight the need for a balanced approach to campus space optimization, where openness is encouraged without compromising the quality of interaction and restoration.

In the fifth clustered space, ***Multi-functional Social Interaction Spaces***, buildings remained a core component, although their effects varied slightly across dimensions. Walls, fences, and motorcycles enhanced both psychological restoration and emotional uplift, while cars, bike lanes, and street lights strengthened social interaction. On the other hand, features such as vegetation, curbs, and traffic lights showed predominantly adverse effects, suggesting that while this space type is oriented toward social engagement and structural clarity, certain elements may hinder its integrated performance.

5. Robustness Check

The first robustness check in this study was an ablation experiment that examined whether the conclusions changed when multiple linear regression was replaced with the weighted least squares (WLS) method. Compared with multiple linear regression, WLS is more appropriate when image semantic features exhibit heteroscedasticity. The WLS results indicate that the directions and relative magnitudes of the significant semantic coefficients reported in Tables 3–6 and Appendix A Tables A4–A8 remained unchanged, confirming the robustness of the conclusions. For example, in Table 2, for the full sample with the psychological restoration indicator as the dependent variable, the positive coefficients for lane marking—general (6.89) and bench (8.17) in the multiple linear regression became 7.43 and 9.52, respectively, in the WLS results. The negative coefficients for sand (−15.78), billboard (−5.41), and manhole (−29.61) changed to −15.87, −6.56, and −22.11, respectively. Similarly, in Table 4, for the high-score group of the emotional uplift indicator, the coefficients of parking, vegetation, water, and pole were 0.50, 0.87, 6.89, and 6.04, respectively, in the multiple linear regression, and 0.21, 0.96, 4.90, and 2.70, respectively, in the WLS results.

The second robustness check in this study involved replacing the image semantic segmentation method used in the main analysis. In the main text, the Mask2Former model trained on the Mapillary Vistas dataset was adopted. For this robustness test, the DeepLabv3+ model was used instead, with partial hyperparameters set as follows: AdamW optimizer, learning rate of 1×10^{-4}, weight decay of 0.05, and the same Mapillary Vistas dataset for training. During fine-tuning, the backbone's layer-wise learning rate decay was maintained at 0.95. After replacing the image segmentation method, the feature importance rankings obtained through the interpretable decision tree algorithm remained largely consistent with the main results. For example, in the low-score group of the social interaction indicator, the top six features remained unchanged, except that the feature importance of "car" surpassed that of "sidewalk." Similarly, for the psychological restoration indicator in the second cluster type, the top five features remained the same, while the previously sixth-ranked "mountain" feature was replaced by "fence."

6. Discussion

Table 7, Panel A compares the research method used in this study with those employed in previous studies. From the methodological perspective, this study integrated advanced approaches including Mask2Former-based image semantic segmentation, interpretable machine learning models combining decision tree algorithms with SHAP, and multivariate linear regression analysis. While previous research has employed subsets of these methods, most have focused primarily on developing predictive models for assessing the restorative quality of campus landscapes [9,25,28,62]. In contrast, this study emphasized identifying specific landscape features that significantly influence students' psychological restoration, emotional uplift, and social interaction. By examining both the direction and magnitude of these effects, the study aimed to provide actionable insights for landscape optimization. The integration of decision tree algorithms with SHAP enabled the interpretable identification of landscape elements most influential to the three psychological indicators. At the same time, SHAP also facilitated the detection of threshold effects in the influence of these features. Furthermore, multivariate linear regression supplemented this analysis by revealing statistically significant landscape elements that may be under-valued in feature importance rankings but still exert substantial influence. Together, these methodological innovations enhance the ability to identify key intervention points for optimizing, improving, or maintaining campus outdoor environments, particularly in forestry universities, ultimately advancing the field of campus space optimization.

Figure 17 presents a knowledge tree outlining the mainstream technological trends in this field. First, in image feature recognition, previous studies have used models such as DeepLabV3+ [33], Mask2Former [52], and object detection models like YOLO [63]. This study employed the Mask2Former model, which offers advantages over the other two models in terms of feature recognition accuracy and multitask processing capability. Second, in machine learning algorithms, the main approaches include neural network models [64] and interpretable decision tree algorithms [55–57]. This study used the latter, as it provides advantages in computational efficiency, preventing overfitting and interpretability, making it more suitable for the image feature importance analysis in this research. Third, regarding statistical analysis methods, including structural equation modeling [65], multiple linear regression [28], and weighted least squares (WLS) regression [66], this study used multiple linear regression in the main analysis and WLS for the robustness check. Overall, the research applied Mask2Former for pixel-level feature extraction, combined with SHAP for interpretable learning and multiple regression to identify feature importance, thresholds, and intergroup differences. The results were translated into heterogeneous group analysis and functional zoning optimization suggestions, providing a more integrated approach in terms of interpretability, heterogeneity characterization, and implementability.

Table 7, Panel B compares the research findings of this study with those of previous studies. From the finding perspective, in the full-sample analysis, natural elements such as vegetation, water, and terrain consistently exhibited positive effects across all three psychological dimensions: psychological restoration, emotional uplift, and social interaction. This confirms the critical role of natural environments in cognitive regulation and emotional stability [42,67]. In contrast, structural features such as walls, fences, and dense buildings generally showed nonlinear or adverse effects, underscoring the importance of careful spatial boundary design and structural scale considerations. Although these findings align with prior studies on green space in higher education environments [10,11,68–70], the present study extends this knowledge by revealing deeper heterogeneity through score-based stratification and spatial clustering. The results suggest that outdoor space optimization should be conducted through stratified interventions based on perceptual differences [71]. Furthermore, the functional zoning analysis identified five distinct campus space types, each exhibiting unique response patterns to psychological indicators. This led to the development of a multidimensional optimization pathway characterized by "natural embedding–order orientation–moderate boundary". This framework, which links semantic features with psychological perceptions while accounting for both spatial typologies and individual heterogeneity, makes a novel contribution to the literature on outdoor spaces in forestry universities, offering valuable insights for campus space optimization.

Table 7. Comparisons with previous research.

Panel A. Comparative Analysis of Research Methods				
Author (Year)	**Research Background**	**Research Method**	**Limitations**	**Supplement of This Study**
Meng et al. (2023) [25]	Campus landscape quality affects user experience; traditional methods are subjective and limited in coverage.	Collected campus street-view images; DeepLabV3+ extracted spatial element proportions; Place Pulse 2.0 calculated visual aesthetics.	Relied on cross-cultural dataset, risking adaptability bias; no validation with subjective or behavioral data.	Uses high-precision Mask2Former features, combined with multiple regression and SHAP to verify nonlinear effects and significance, improving transferability and practical value across contexts.
Zhang et al. (2023) [9]	COVID-19 lockdown made outdoor campus space the only contact with nature; need to identify emotional drivers.	Collected 7092 student facial photos and ECG (24 participants) over two weeks; FaceReader 9.0 created GIS emotion maps; extracted 12 panoramic visual indicators; regression tested spatial effects.	Single university; small sample; single HRV metric; no microclimate or multisensory control; linear models missed nonlinear thresholds.	
Wu et al. (2025) [28]	Rising mental health needs; campus public spaces lack restorative design.	PSPNet segmentation of street-view images; PRS-11 survey trained XGBoost to predict restorative scores; overlaid with sDNA walking accessibility for optimization.	Small PRS-11 sample; black-box model without interpretability; sDNA lacked real-time pedestrian flow and multimodal access.	

Table 7. *Cont.*

Panel B. Comparative Analysis of Research Findings				
Author (Year)	**Research Background**	**Main Findings**	**Limitations**	**Supplement of This Study**
Alserry et al. (2024) [10]	Universities influence student social life and well-being; outdoor spaces often neglected.	Observations, interviews, and surveys showed seating quantity, layout, and maintenance determine gathering vs. solitude; UK university rated highest for flexible furniture and spatial hierarchy; all lacked quiet focus areas.	Only 50 samples from three Egyptian universities; subjective green space quality; cross-sectional design limits causality.	Psychological restoration, emotional uplift, and social interaction respond differently to the same features; propose "nature integration–order guidance–moderate boundaries" and targeted interventions by zone; call for dynamic behavioral data to improve causal inference.
Hasan et al. (2024) [11]	Few quantitative studies on campus outdoor space; An-Najah campus offers diverse spatial structure.	High-integration and visible areas attract more people; students prefer shaded seating, using steps when lacking; deficiencies include insufficient seating, missing landmarks, and poor facilities.	Single site; limited temporal and seasonal coverage; manual observation; no control for user attributes.	
Liu et al. (2022) [69]	Depression rates higher among university students, especially in China; unclear role of green space and gender differences.	Green space improves mental health partly via academic achievement; stronger effect for males; females report higher depression.	Subjective measures; causality disorder; uncontrolled confounders.	

The method proposed in this study can be applied to other universities. By integrating online street-view image acquisition, survey-based scoring, interpretable machine learning, and econometric analysis, it is possible to identify landscape features that influence psychological health indicators among students in other institutions. However, the conclusions of this study are unique and may only be replicated in environments similar to the forestry and agricultural university examined here, particularly those with comparable landscape characteristics. As noted in Section 4.4.1, the study site features distinct ecological functional zones including an ecological research area characterized by a sugarcane research center and greenhouse cultivation, an ecological landscape zone centered on a botanical garden, and a recreational landscape zone defined by lakeside scenery. Such specific ecological and landscape zones are not present in every university, and therefore the conclusions drawn in this study are unique.

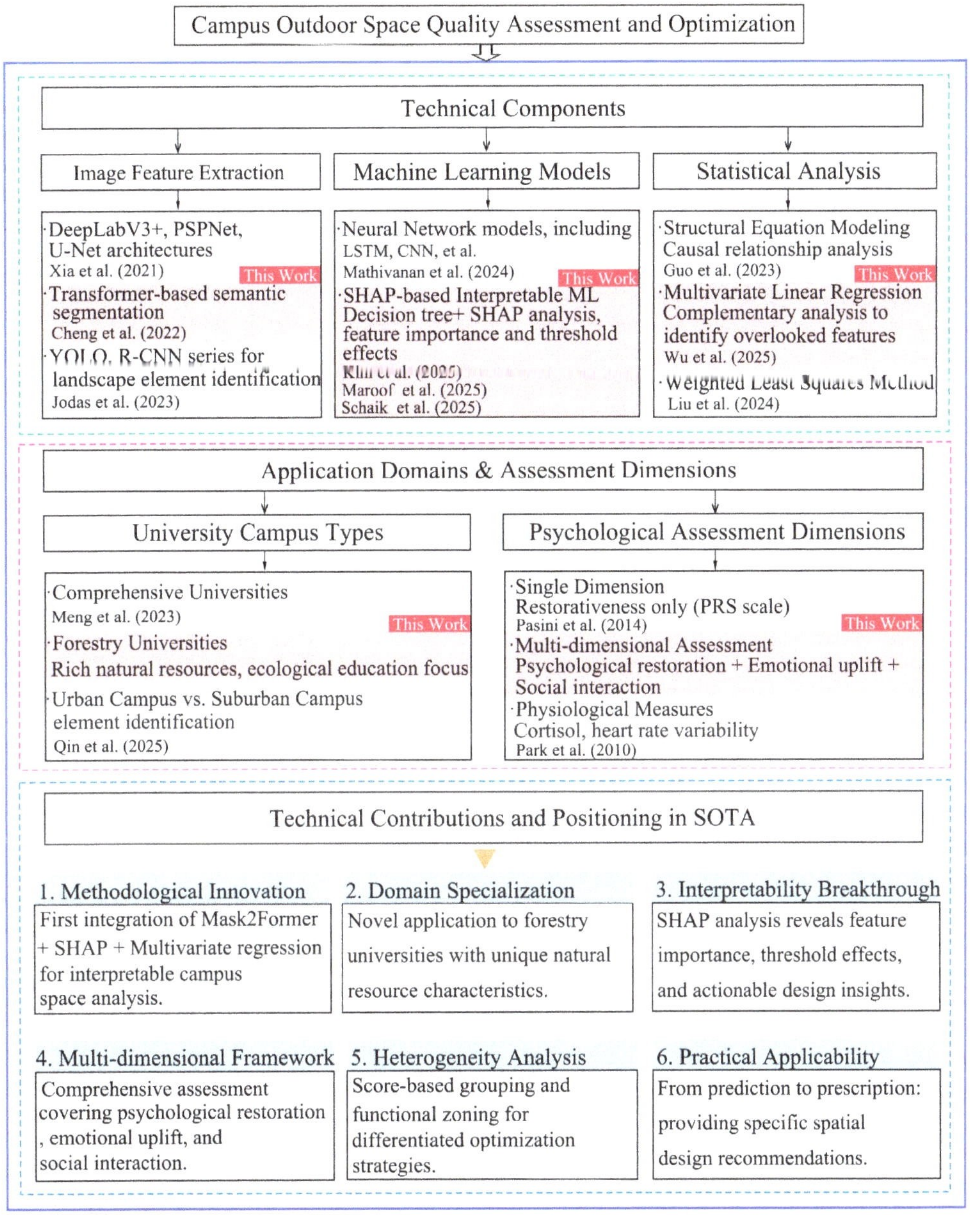

Figure 17. Knowledge tree of campus outdoor space quality assessment and optimization [25,26,28, 33,40,48,52,55–57,63–66].

7. Conclusions

This study investigated the influence of outdoor spatial features on emotional uplift, social interaction, and psychological restoration among students at a forestry university in China. Employing a comprehensive methodological framework that integrated image semantic segmentation (Mask2Former), SHAP-based interpretable machine learning, and multivariate linear regression, the study provides a data-driven understanding of how campus environments shape student well-being. Semantic features were extracted using the Mask2Former model and quantified, while psychological indicators were collected via questionnaires. SHAP analysis based on decision tree models was employed to examine the impact and threshold effects of spatial features. Subsequently, multivariate regression was applied to uncover additional features with statistical significance that were overlooked by SHAP analysis. Furthermore, psychological indicators were stratified into high-, medium-, and low-score groups to examine heterogeneity, and k-means clustering was used to analyze the functional spatial zones. The main findings are as follows:

- In the full-sample analysis, interpretable machine learning revealed that vegetation significantly promoted psychological restoration, whereas buildings generally exerted an adverse effect. Multivariate regression further highlighted that bench, water, street light, and bridge positively contributed to specific psychological indicators, while artificial features such as fence, sand, motorcycle, pole, utility pole, and trash can tend to have adverse effects. These findings suggest several design implications. First, prioritize increasing green coverage and biodiversity to enhance the natural continuity of the environment. Second, incorporate water features and terrain to support emotional regulation and foster social engagement. Finally, reduce the scale and visual dominance of built structures such as walls and fences, particularly for users with low perceptual sensitivity, to create a more balanced and restorative campus environment.
- The stratified analysis by psychological score groups revealed heterogeneity in spatial sensitivity. For psychological restoration, features such as road and terrain had positive effects in the high-score group, while sand showed adverse effects. Regarding emotional uplift, elements like the bridge and water were beneficial to specific groups, whereas the curb and sand were detrimental. For social interaction, features like road and bike lanes were positively associated, while curb and wall had negative associations. Collectively, individuals with higher psychological scores favored environments with open views, natural continuity, and walkability, while those with lower scores were more susceptible to noise, visual obstruction, and redundant infrastructure. Design strategies should thus focus on creating legible, walkable systems in academic and residential zones while incorporating water features, seating, and lighting at social nodes to enhance adaptability and user experience.
- The five spatial clusters derived from k-means clustering exhibited distinct psychological perception patterns and functional orientations. Ecological Research and Practice Spaces were centered on buildings and benefited from enhanced ecological integration and openness. Ecological Landscape Display Spaces emphasized visual appeal and interaction, requiring improved greenery continuity and sightline guidance. Commuting and Circulation Spaces focused on connectivity and order, requiring clearer pathways to facilitate movement. Composite Leisure Landscape Spaces merged multiple functions and required careful boundary management and spatial rhythm. Multi-Functional Social Interaction Spaces demanded a clear structure and strong social support elements. In conclusion, spatial optimization should adopt a context-sensitive approach, balancing ecological, mobility, and social needs to enhance psychological well-being and support campus community interaction.

In summary, the optimization of outdoor campus spaces in forestry universities should adhere to the overarching principles of "nature-led, structurally ordered, functionally integrated, and socially inclusive". This approach involves systematically strengthening the ecological foundation, strategically incorporating built elements, and creating open, connected, and accessible spatial configurations. By integrating ecological, cognitive, and social dimensions, the goal is to enhance the overall campus environment, better supporting both educational activities and daily life.

This study had several limitations that should be addressed in future research. First, the k-means clustering method may be limited by the dimensionality and coverage of the dataset, potentially leading to ambiguous classifications. Incorporating behavioral trajectory data could enhance clustering accuracy and provide more nuanced insights into spatial preferences. Second, all participants in this study were from a single university, which may limit the external validity of the findings. Future studies should broaden the sample to include students from other universities and diverse tourist groups to enhance generalizability. Third, the linear assumptions of multivariate regression may not fully capture the nonlinear relationships between spatial features and psychological indicators.

Future research should explore methods such as generalized additive models (GAMs) or structural equation modeling (SEM) to uncover deeper causal insights. Additionally, integrating multi-source dynamic data and conducting cross-regional validations with nonlinear modeling would enhance the adaptability and theoretical robustness of campus space optimization strategies.

Author Contributions: Conceptualization, X.Z.; Methodology, X.Z., Z.T. and Z.D.; Software, Z.T., S.L. and Z.D.; Writing—original draft preparation, X.Z.; Writing—review and editing, S.L. and Z.D.; Supervision, Visualization, S.L. and Z.D.; Funding acquisition, Z.T. All authors have read and agreed to the published version of the manuscript.

Data Availability Statement: The raw data supporting the conclusions of this article will be made available by the authors on request.

Funding: This research work was supported by the National Natural Science Foundation of China under grant No. 72341030, and the Special Funding Project of the China Agricultural and Forestry University Design Art Alliance under grant No. 111900050.

Conflicts of Interest: The authors declare no conflicts of interest.

Appendix A

Table A1. Performance of the decision tree algorithms on the validation set using the full sample.

Full Sample Psychological Recovery Indicators				Complete Sample Sentiment Improvement Index				Full Sample Social Interaction Indicators			
	RF	GBR	XGBOOST		RF	GBR	XGBOOST		RF	GBR	XGBOOST
MSE	0.9874	0.9677	0.9504	MSE	1.0319	1.0271	1.0125	MSE	1.2258	1.2343	1.1783
MAE	0.7771	0.7847	0.7771	MAE	0.8323	0.8383	0.8244	MAE	0.9189	0.9067	0.8766
EVS	0.4181	0.4301	0.4410	EVS	0.2570	0.2628	0.2725	EVS	0.3518	0.3468	0.3763

Note: In Tables A1–A3, RF, GBR, and XGBoost are the abbreviations for random forest, gradient boosting regression, and extreme gradient boosting, respectively. MSE, MAE, and EVS are the abbreviations for mean squared error, mean absolute error, and explained variance score, respectively.

Table A2. Performance of the decision tree algorithms across different psychological score groups.

High-Scoring Group Psychological Recovery Indicators				Middle Group Psychological Recovery Index				Low-Scoring Group Psychological Recovery Indicators			
	RF	GBR	XGBOOST		RF	GBR	XGBOOST		RF	GBR	XGBOOST
MSE	0.2278	0.2418	0.2452	MSE	0.1479	0.1436	0.1447	MSE	0.2485	0.2524	0.2810
MAE	0.3661	0.3925	0.3925	MAE	0.3096	0.3012	0.3028	MAE	0.4100	0.4094	0.4291
EVS	0.3120	0.2697	0.2596	EVS	0.3950	0.4145	0.4095	EVS	0.4178	0.4141	0.3391
High-Scoring Group Emotional Enhancement Indicators				**Emotional Uplift Index for the Middle Group**				**Low-Scoring Group Emotional Uplift Indicator**			
	RF	GBR	XGBOOST		RF	GBR	XGBOOST		RF	GBR	XGBOOST
MSE	0.2881	0.2930	0.2832	MSE	0.2231	0.2345	0.2438	MSE	0.3311	0.3286	0.3540
MAE	0.4392	0.4486	0.4446	MAE	0.3894	0.4010	0.4127	MAE	0.4512	0.4596	0.4771
EVS	0.0154	0.0029	0.0328	EVS	0.3040	0.2522	0.2343	EVS	0.0936	0.0997	0.0310

Table A2. *Cont.*

High-Scoring Social Interaction Indicators				Social Interaction Indicators for the Middle Group				Low Social Interaction Score			
	RF	GBR	XGBOOST		RF	GBR	XGBOOST		RF	GBR	XGBOOST
MSE	0.2214	0.2313	0.2345	MSE	0.2127	0.2185	0.2194	MSE	0.3482	0.3805	0.3789
MAE	0.3566	0.3782	0.3801	MAE	0.4392	0.4498	0.4512	MAE	0.4572	0.4841	0.4891
EVS	0.3188	0.2901	0.2810	EVS	0.1067	0.0820	0.0774	EVS	0.0699	0.0323	0.0121

Table A3. Performance of the decision tree algorithms across the five clustered spatial zones.

Psychological Recovery Indicators in the First Cluster Space				Emotional Uplift Index in the First Cluster Space				Social Interaction Indicators in the First Cluster Space			
	RF	GBR	XGBOOST		RF	GBR	XGBOOST		RF	GBR	XGBOOST
MSE	1.6998	1.4832	1.5388	MSE	0.7963	0.8457	0.7491	MSE	0.5517	0.5577	0.6107
MAE	0.9296	0.9014	0.9056	MAE	0.7070	0.7224	0.6758	MAE	0.5940	0.5952	0.6269
EVS	0.2624	0.3460	0.3249	EVS	0.3735	0.3215	0.3950	EVS	0.5714	0.5667	0.5271
Psychological Recovery Indicators in the Second Cluster Space				**Emotional Uplift Index for the Second Cluster Space**				**Social Interaction Indicators in the Second Cluster Space**			
	RF	GBR	XGBOOST		RF	GBR	XGBOOST		RF	GBR	XGBOOST
MSE	0.5840	0.5052	0.5200	MSE	0.8979	0.9816	0.9907	MSE	0.7278	0.7502	0.6861
MAE	0.5771	0.5339	0.5345	MAE	0.7662	0.8108	0.8225	MAE	0.6907	0.7006	0.6601
EVS	0.3362	0.4336	0.4142	EVS	0.2224	0.1461	0.6540	EVS	0.3803	0.3551	0.4008
Spatial Psychological Recovery Indicators for the Third Cluster				**Emotional Uplift Indicators in the Third Cluster Space**				**Social Interaction Indicators in the Third Cluster Space**			
	RF	GBR	XGBOOST		RF	GBR	XGBOOST		RF	GBR	XGBOOST
MSE	0.8536	0.8160	0.8264	MSE	1.8100	1.1136	2.0084	MSE	1.0347	1.0429	1.0637
MAE	0.7229	0.6859	0.6899	MAE	1.1481	1.0970	1.2264	MAE	0.7102	0.7126	0.7206
EVS	0.5772	0.6175	0.5864	EVS	0.2593	0.3033	0.1832	EVS	0.4113	0.4233	0.4094
Psychological Recovery Indicators in the Fourth Cluster Space				**Emotional Uplift Indicators in the Fourth Cluster Space**				**Social Interaction Indicators in the Fourth Cluster Space**			
	RF	GBR	XGBOOST		RF	GBR	XGBOOST		RF	GBR	XGBOOST
MSE	0.7840	0.7780	0.8003	MSE	1.0621	1.1955	1.2770	MSE	0.9014	1.0674	1.0196
MAE	0.6603	0.6613	0.6680	MAE	0.8152	0.8670	0.8859	MAE	0.7860	0.8200	0.7709
EVS	0.5468	0.5565	0.5413	EVS	0.2065	0.1152	0.0484	EVS	0.3548	0.2368	0.2694
Psychological Recovery Indicators in the Fifth Cluster Space				**Emotional Uplift Index for the Fifth Cluster Space**				**Social Interaction Indicators in the Fifth Cluster Space**			
	RF	GBR	XGBOOST		RF	GBR	XGBOOST		RF	GBR	XGBOOST
MSE	0.7541	0.8328	0.8158	MSE	0.7957	0.8381	0.8471	MSE	1.0303	0.8879	0.8781
MAE	0.7072	0.7297	0.7342	MAE	0.7114	0.7229	0.7225	MAE	0.8566	0.7599	0.7483
EVS	0.4681	0.4127	0.4227	EVS	0.2428	0.1935	0.1843	EVS	0.2732	0.3577	0.3663

Table A4. Multiple linear regression results for the first spatial cluster (Ecological Scientific Practice Space).

Psychological Recovery		Mood Improvement		Social Interaction	
Fence	−3.3268 *** (0.000552)	Wall	−5.369 *** (0.0016)	Person	73.3060 *** (0.000577)
Wall	−6.9987 *** (0.000109)	Sand	−14.2 *** (0.0002)	Lane marking—Crosswalk	6.7824 *** (0.00968)
Sand	−15.8654 *** (0.000114)	Water	4.9 ** (0.0445)	Mountain	−4.6514 ** (0.0156)
Utility pole	−52.4433 *** (0.004206)	Billboard	−4.611 ** (0.0437)	Traffic light	225.5039 ** (0.0283)

Table A4. *Cont.*

Psychological Recovery		Mood Improvement		Social Interaction	
Adjusted R-squared	0.467	Adjusted R-squared	0.284	Adjusted R-squared	0.387
F-statistic	12.7	F-statistic	7.232	F-statistic	9.427

Note: *** indicates significance at the 1% significance level, ** indicates significance at the 5% significance level.

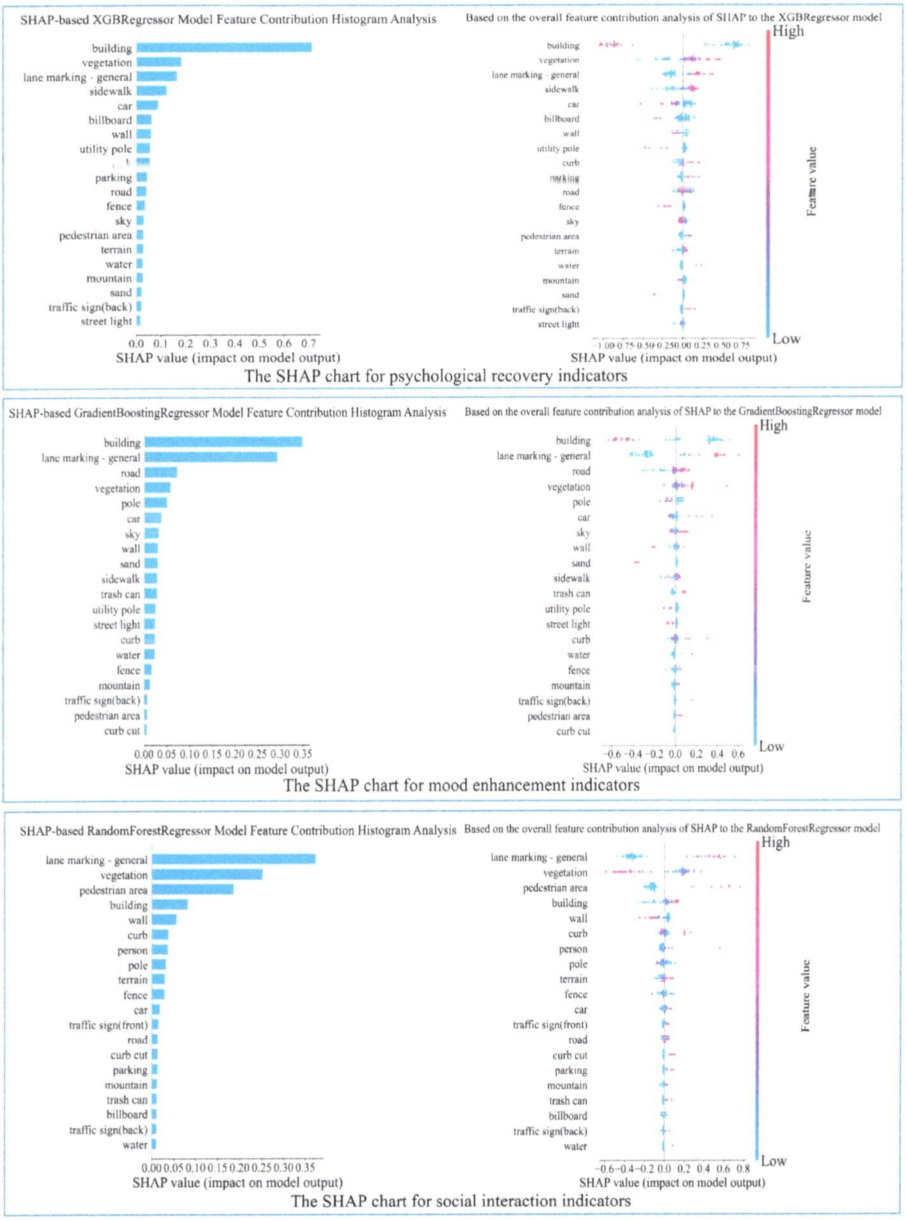

Figure A1. SHAP value analysis of the semantic image features for the three psychological dimensions in the first clustered space.

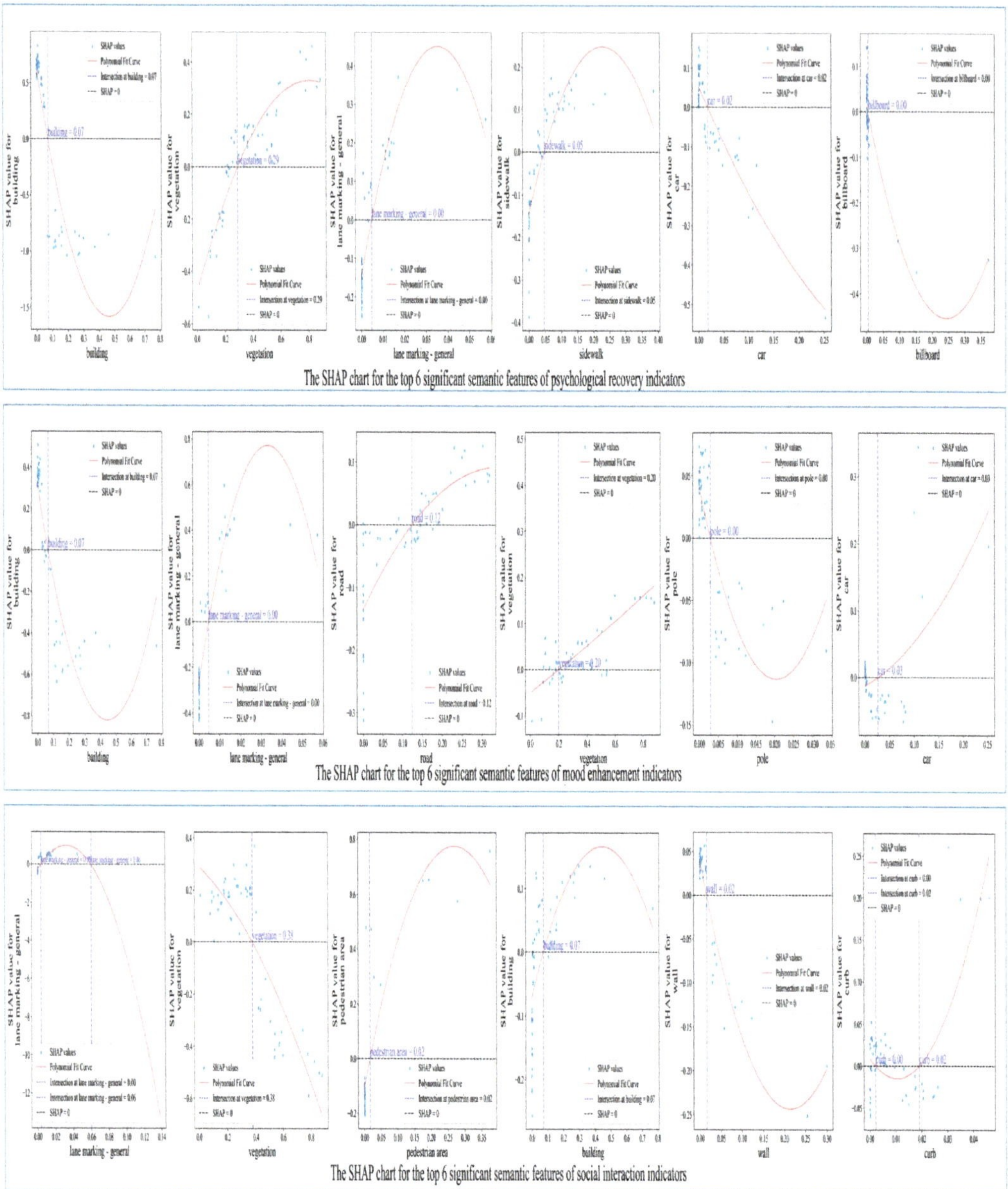

Figure A2. Impact patterns of the top-ranked semantic features for the three psychological dimensions in the first clustered space.

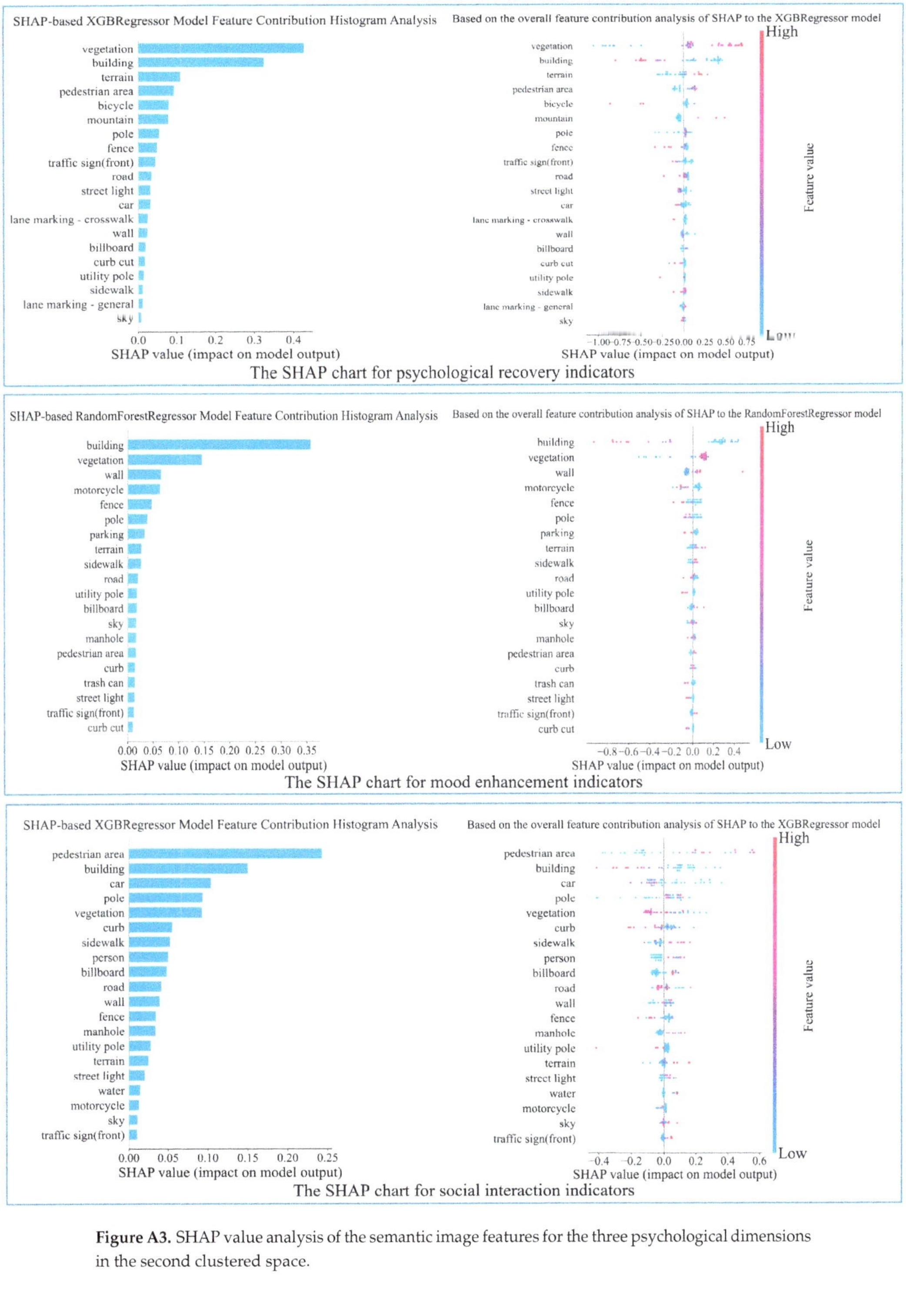

Figure A3. SHAP value analysis of the semantic image features for the three psychological dimensions in the second clustered space.

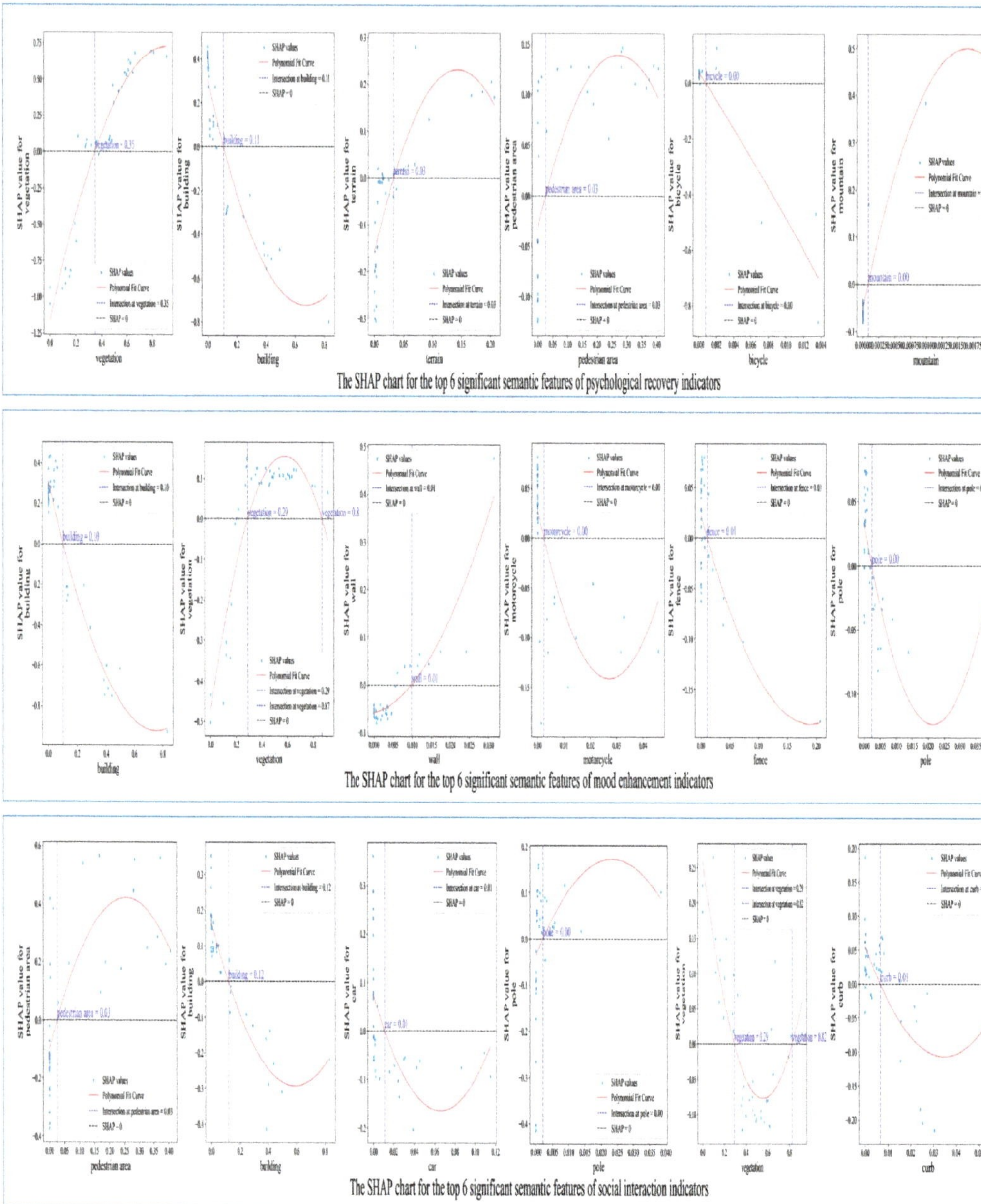

Figure A4. Impact patterns of the top-ranked semantic features for the three psychological dimensions in the second clustered space.

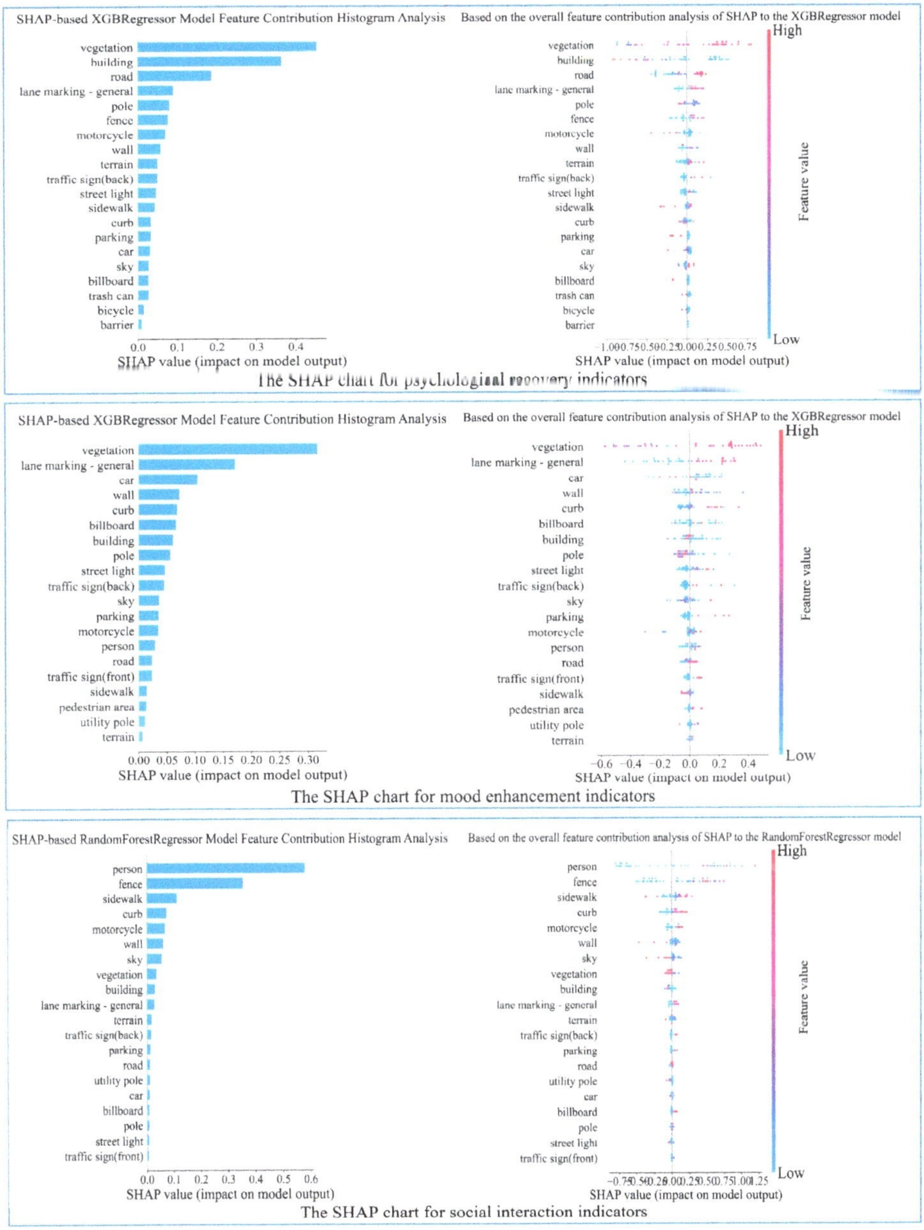

Figure A5. SHAP value analysis of the semantic image features for the three psychological dimensions in the third clustered space.

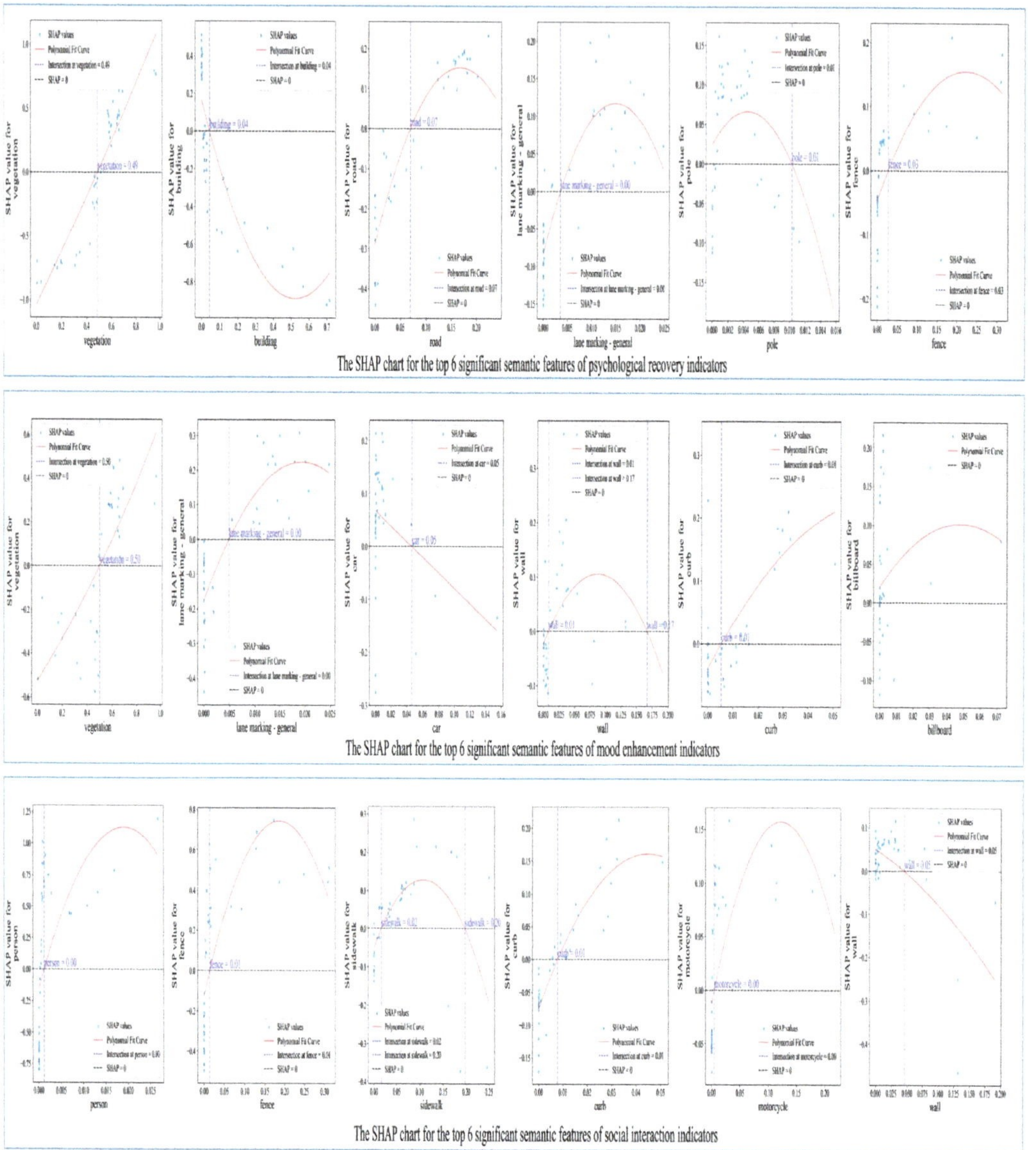

Figure A6. Impact patterns of the top-ranked semantic features for the three psychological dimensions in the third clustered space.

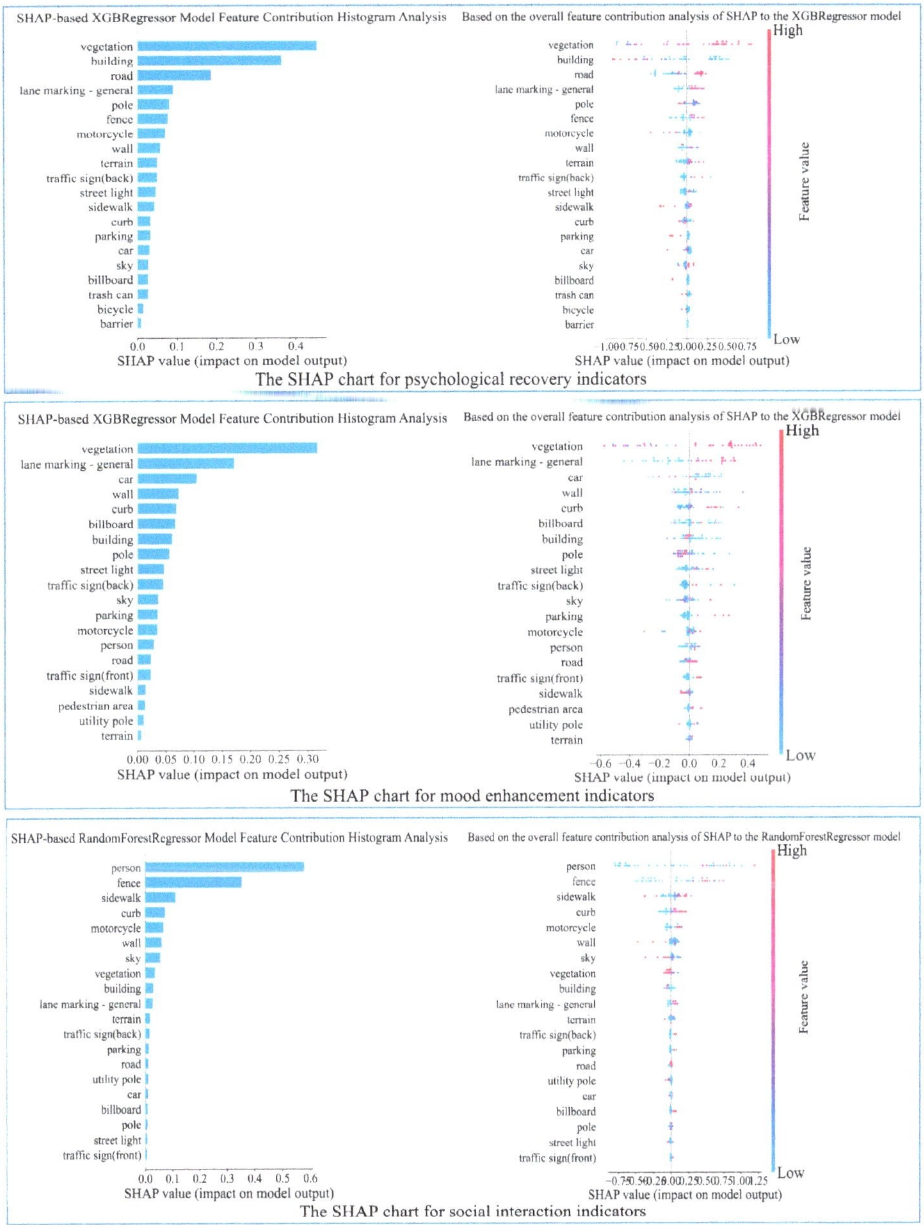

Figure A5. SHAP value analysis of the semantic image features for the three psychological dimensions in the third clustered space.

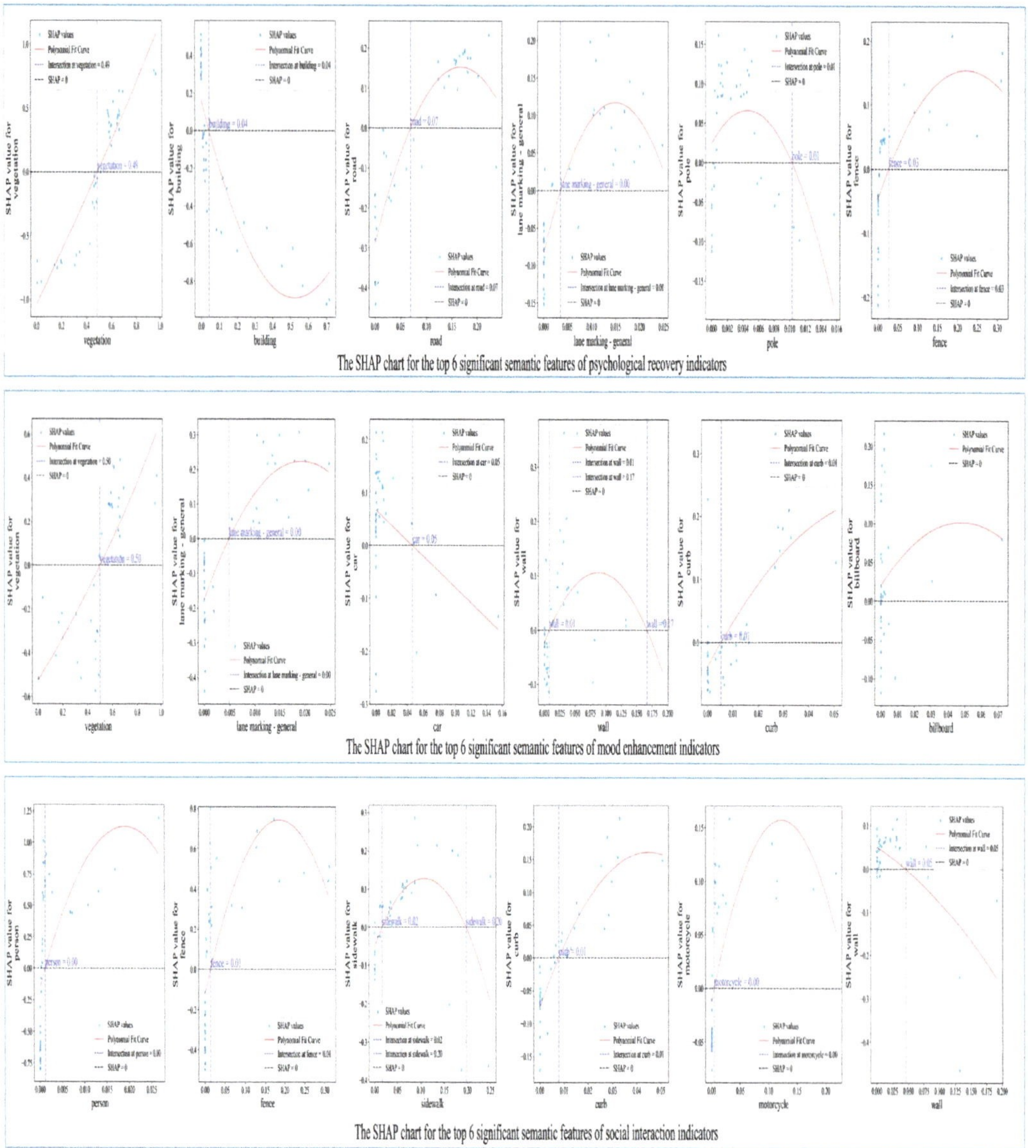

Figure A6. Impact patterns of the top-ranked semantic features for the three psychological dimensions in the third clustered space.

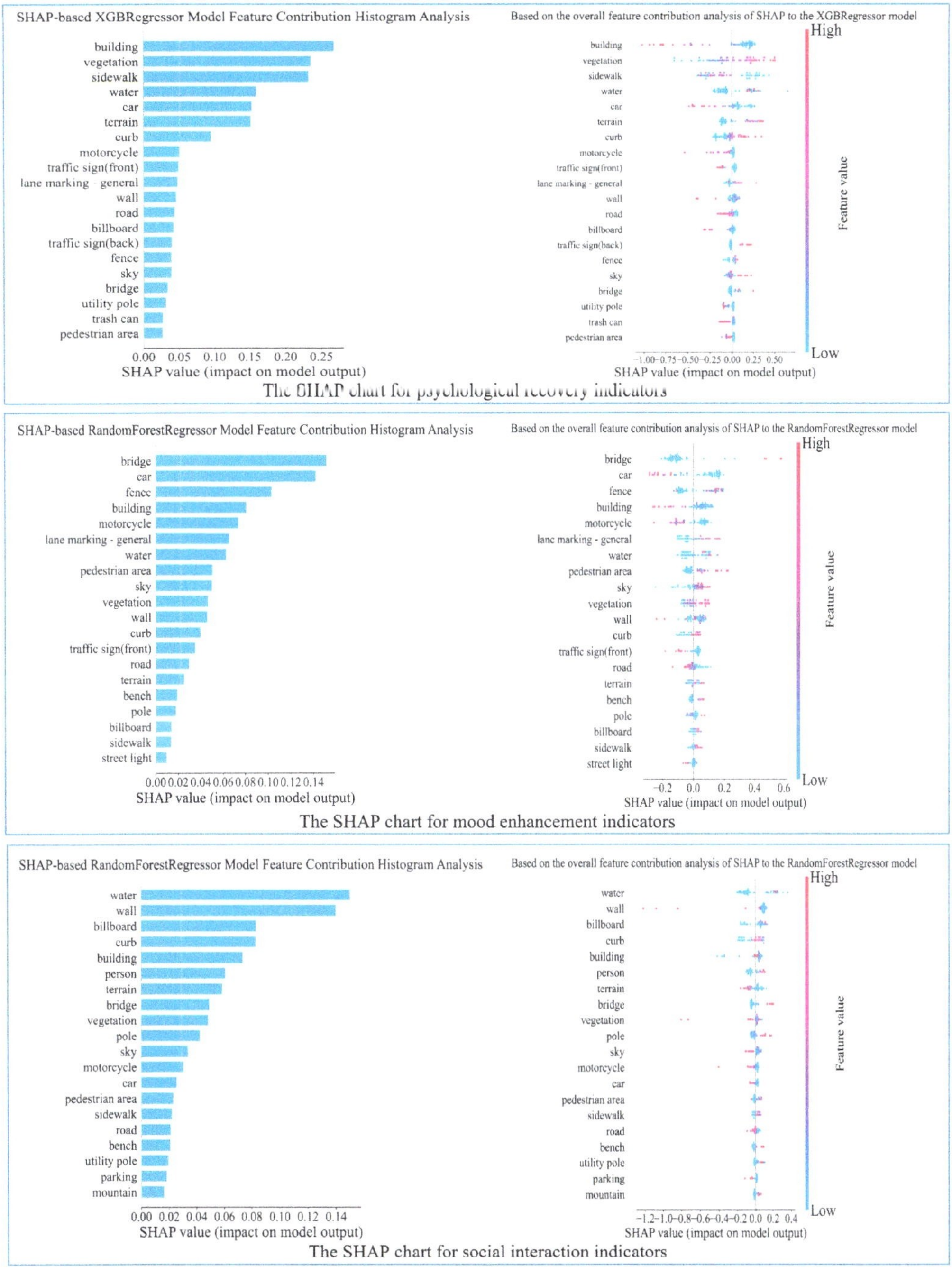

Figure A7. SHAP value analysis of the image semantic features under the three psychological indicators in the fourth clustered space.

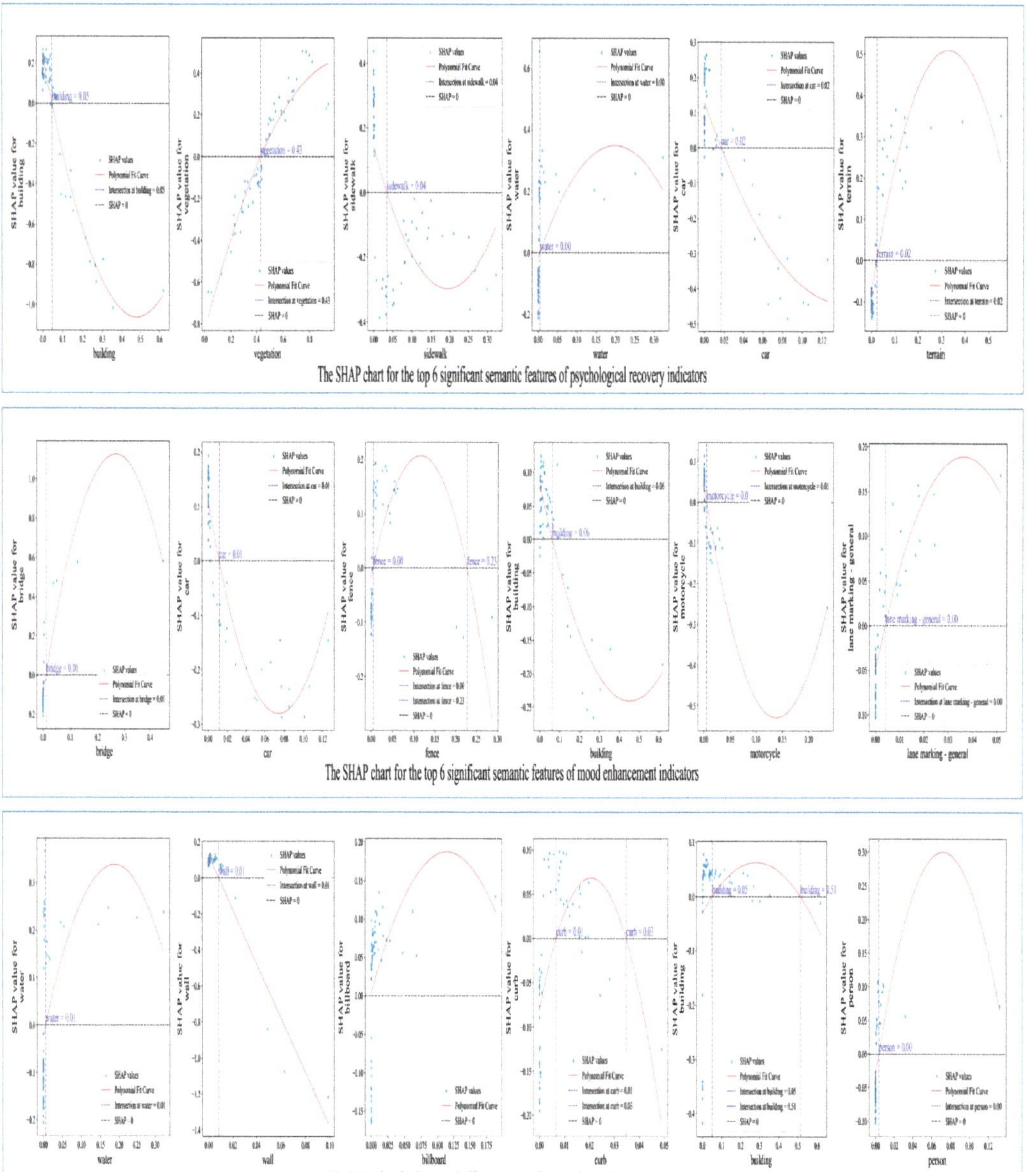

Figure A8. Impact patterns of the top-ranked semantic features for the three psychological dimensions in the fourth clustered space.

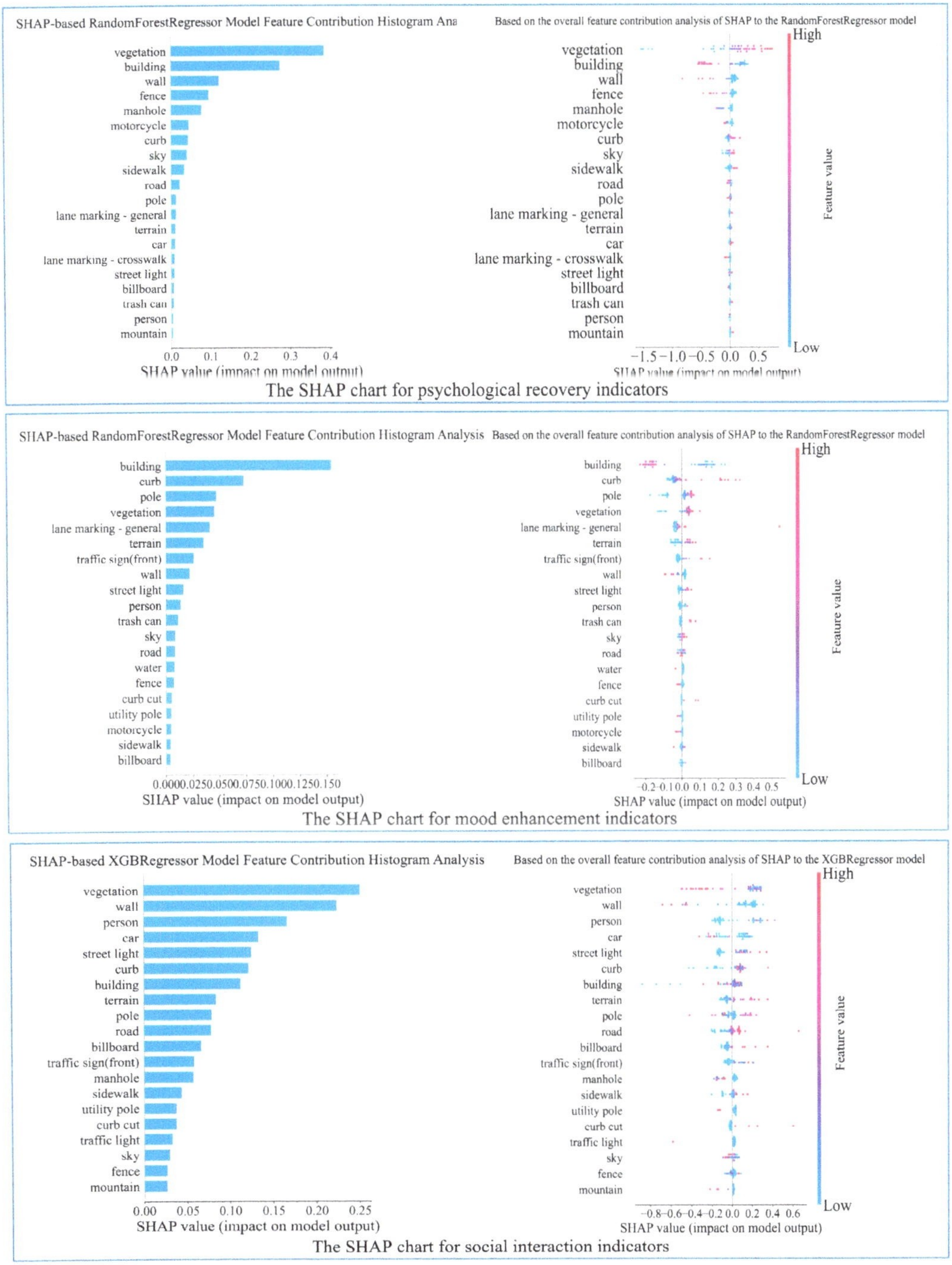

Figure A9. SHAP value analysis of the image semantic features under the three psychological indicators in the fifth clustered space.

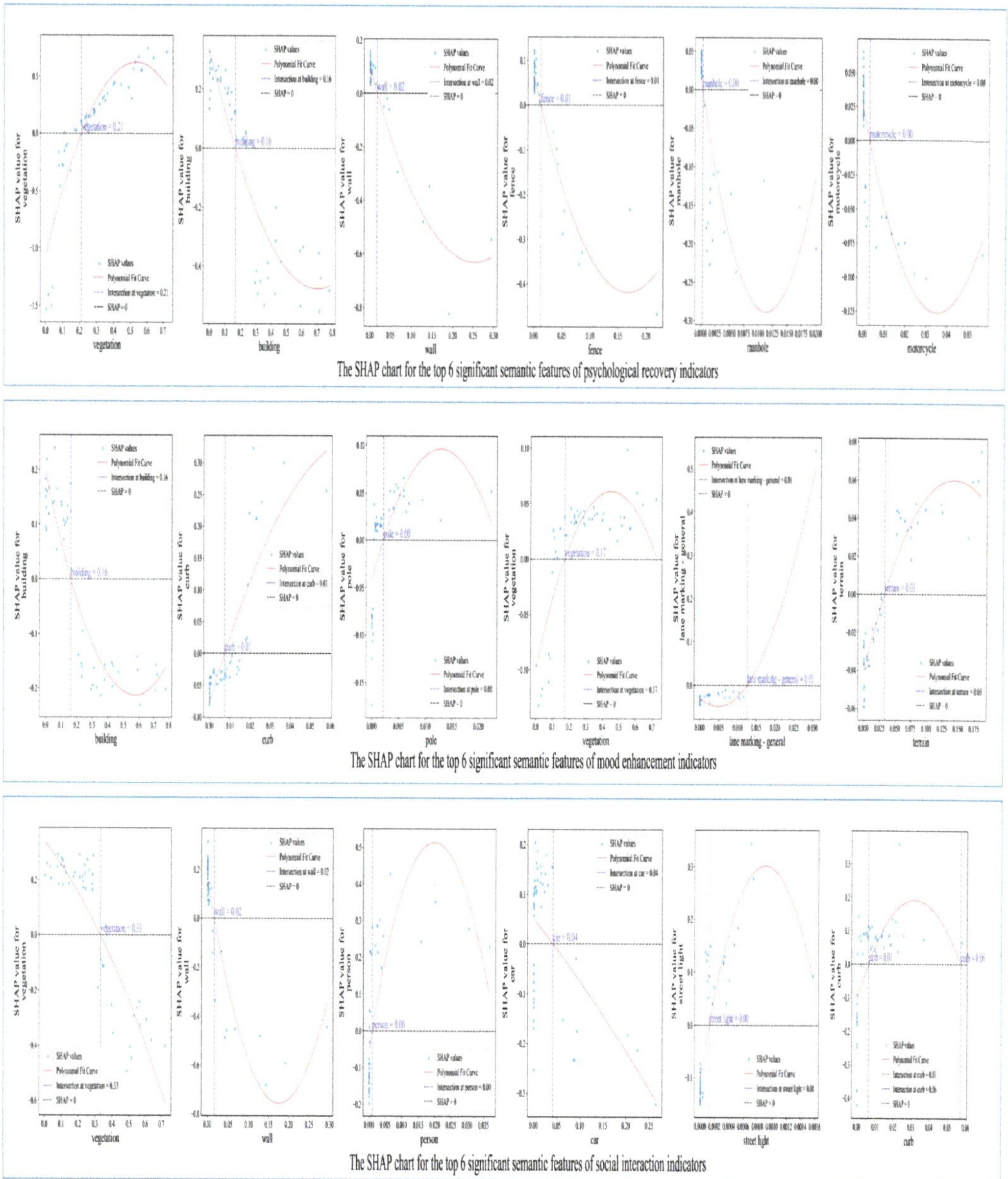

Figure A10. Influence patterns of the top-ranked semantic features under the three psychological indicators in the fifth clustered space.

Table A5. Multiple linear regression results for the second spatial cluster (Ecological Landscape Exhibition Space).

	Psychological Recovery		Mood Improvement		Social Interaction
Bike lane	−10.405 *** (0.043)	Bike lane	−13.6983 ** (0.0076135)	Wall	7.026 ** (0.027)
Terrain	3.969 *** (5.54×10^{-5})	Terrain	1.9371 *** (0.0462826)	Bike lane	−13.698 *** (0.008)
Vegetation	3.594 *** (1.13×10^{-18})	Traffic sign (front)	92.7601 *** (0.0188831)	Terrain	1.937 ** (0.046)
Water	1.788 ** (0.0496)			Utility pole	−90.562 ** (0.033)
Utility pole	−90.563 ** (0.0331)			Bicycle	−26.103 *** (0.007)
Bicycle	−26.103 *** (0.00704)			Traffic sign (front)	92.76 ** (0.019)
				Bench	137.322 ** (0.041)
				Curb	−22.034 ** (0.017)
Adjusted R-squared	0.459	Adjusted R-squared	0.296	Adjusted R-squared	0.163
F-statistic:	9.098	F-statistic:	5.006	F-statistic:	2.858

Note: *** indicates significance at the 1% significance level, ** indicates significance at the 5% significance level.

Table A6. Multiple linear regression results for the third spatial cluster (Commuting Circulation Space).

	Psychological Recovery		Mood Improvement		Social Interaction
Barrier	−167.8697 *** (0.0091)	Lane marking—General	40.2382 *** (7.94×10^{-5})	Fence	5.6208 ** (1.78×10^{-5})
Terrain	5.2990 *** (0.0009)	Manhole	−127.0590 ** (0.0117765)	Wall	−4.9034 *** (0.007509)
Trash can	−15.8654 *** (0.0373)	Car	−4.7413 ** (0.0105782)	Sidewalk	3.7862 ** (0.014283)
Car	−52.4433 *** (0.0083)			Person	47.3537 ** (6.26×10^{-5})
				Utility pole	−38.2084 *** (0.022986)
				Traffic sign (back)	425.1233 ** (0.028697)
Adjusted R-squared	0.459	Adjusted R-squared	0.223	Adjusted R-squared	0.389
F-statistic:	9.271	F-statistic:	3.790	F-statistic:	4.752

Note: *** indicates significance at the 1% significance level, ** indicates significance at the 5% significance level.

Table A7. Multiple linear regression results for the fourth clustered space (Composite Leisure Landscape Space).

	Psychological Recovery		Mood Improvement		Social Interaction
Wall	−6.741 ** (0.0047)	Bird	10898.9500 ** (0.0223)	Wall	−13.0500 *** (0.0)
Billboard	−7.363 *** (0.0155)	Wall	−7.2868 *** (0.0060)	Bridge	2.3327 *** (0.008)
Utility pole	−45.488 *** (0.0039)	Utility pole	−40.5835 ** (0.0189)	Sand	−101.3829 ** (0.0189)
Car	−10.098 ** (9.43×10^{-14})	Traffic sign (front)	−69.1259 ** (0.0292)	Pole	12.6919 ** (0.0454)

Table A7. *Cont.*

	Psychological Recovery		Mood Improvement		Social Interaction
Motorcycle	−5.972 * (0.0419)				
Adjusted R-squared	0.406	Adjusted R-squared	0.161	Adjusted R-squared	0.195
F-statistic	8.766	F-statistic	2.885	F-statistic	3.458

Note: *** indicates significance at the 1% significance level, ** indicates significance at the 5% significance level, * indicates significance at the 10% significance level.

Table A8. Multiple linear regression results for the fifth clustered space (Multi-Functional Social Interaction Space).

	Psychological Recovery		Mood Improvement		Social Interaction
Lane marking—general	17.8247 *** (0.0033)	Building	−1.2294 ** (0.0110)	Bike lane	2.96×10^{-12} ** (0.0070)
Manhole	−54.8303 *** (0.0019)	Water	−12.8585 ** (0.0256)	Street light	383.5750 *** (0.0213)
				Traffic light	−1206.2617 ** (0.0056)
				Ground animal	-1.65×10^{-11} ** (0.0341)
Adjusted R-squared	0.427	Adjusted R-squared	0.154	Adjusted R-squared	0.236
F-statistic	9.888	F-statistic	2.895	F-statistic	4.074

Note: *** indicates significance at the 1% significance level, ** indicates significance at the 5% significance level.

References

1. Hajrasouliha, A. Campus score: Measuring university campus qualities. *Landsc. Urban Plan.* **2017**, *158*, 166–176. [CrossRef]
2. Soares, I.; Weitkamp, G.; Yamu, C. Public spaces as knowledge-scapes: Understanding the relationship between the built environment and creative encounters at Dutch university campuses and science parks. *Int. J. Environ. Res. Public Health* **2020**, *17*, 7421. [CrossRef]
3. Van den Berg, A.E.; Jorgensen, A.; Wilson, E.R. Evaluating restoration in urban green spaces: Does setting type make a difference? *Landsc. Urban Plan.* **2014**, *127*, 173–181. [CrossRef]
4. Hipp, J.A.; Gulwadi, G.B.; Alves, S.; Sequeira, S. The relationship between perceived greenness and perceived restorativeness of university campuses and student-reported quality of life. *Environ. Behav.* **2016**, *48*, 1292–1308. [CrossRef]
5. Rapoport, A. *Culture, Architecture, and Design*, 1st ed.; Locke Science Publishing Company: New York, NY, USA, 2005; pp. 154–196.
6. Aydin, D.; Ter, U. Outdoor space quality: Case study of a university campus plaza. *Archnet-IJAR Int. J. Archit. Res.* **2008**, *2*, 189–203.
7. Abu-Ghazzeh, T.M. Communicating behavioral research to campus design: Factors affecting the perception and use of outdoor spaces at the University of Jordan. *Environ. Behav.* **1999**, *31*, 764–804. [CrossRef]
8. Kaplan, R.; Kaplan, S.; Ryan, R.L. *People Places: Design Guidelines for Urban Open Space*, 1st ed.; John Wiley & Sons: New York, NY, USA, 1997; pp. 154–196.
9. Zhang, J.; Li, Y. The impact of campus outdoor space features on students' emotions based on the emotion map. *Int. J. Environ. Res. Public Health* **2023**, *20*, 4277. [CrossRef]
10. Alserry, S.T.; ElGohary, A.S. The impact of campus sustainable outdoor spaces quality on students' social needs, behavior, and satisfaction. *Int. J. Eng. Mod. Sci.* **2024**, *3*, 24014.
11. Hasan, N.; Hijazi, I.; Enab, D.; Qanazi, S.; Shahrour, I.; Al-Qadi, H.; Qamhieh, K. Influence of campus outdoor spaces on students' behavior: Enhancing social interaction and learning at An-Najah University. An-Najah Univ. *J. Res. A (Nat. Sci.)* **2024**, *39*, 125–136. [CrossRef]
12. Kim, W. Effects of students' perceived safety of public outdoor environment on academic achievement at university campus. *Archit. Res.* **2015**, *17*, 13–20. [CrossRef]
13. Peker, E.; Ataöv, A. Exploring how campus open space design influences students' learning experiences. *Landsc. Res.* **2020**, *45*, 310–326. [CrossRef]

14. Coulson, J.; Roberts, P.; Taylor, I. *University Planning and Architecture: The Search for Perfection*, 1st ed.; Routledge: London, UK, 2015; pp. 154–196.
15. Williams, D.R.; Patterson, M.E. *Place, Leisure, and Well-Being, in Sense of Place, Health and Quality of Life*; Routledge: London, UK, 2016; pp. 125–140.
16. Dolan, P.; Metcalfe, R. The relationship between innovation and subjective well-being. *Res. Policy* **2012**, *41*, 1489–1498. [CrossRef]
17. Sheldon, E.; Simmonds-Buckley, M.; Bone, C.; Mascarenhas, T.; Chan, N.; Wincott, M.; Gleeson, H.; Sow, K.; Hind, D.; Barkham, M. Prevalence and risk factors for mental health problems in university undergraduate students: A systematic review with meta-analysis. *J. Affect. Disord.* **2021**, *287*, 282–292. [CrossRef]
18. Gulwadi, G.B.; Mishchenko, E.D.; Hallowell, G.; Alves, S.; Kennedy, M. The restorative potential of a university campus: Objective greenness and student perceptions in Turkey and the United States. *Landsc. Urban Plan.* **2019**, *187*, 36–46. [CrossRef]
19. Speake, J.; Edmondson, S.; Nawaz, H. Everyday encounters with nature: Students' perceptions and use of university campus green spaces. *Hum. Geogr.-J. Stud. Res. Hum. Geogr.* **2013**, *7*, 21–31. [CrossRef]
20. Guo, L.N.; Zhao, R.L.; Ren, A.H.; Niu, L.-X.; Zhang, Y.-L. Stress recovery of campus street trees as visual stimuli on graduate students in autumn. *Int. J. Environ. Res. Public Health* **2020**, *17*, 148. [CrossRef] [PubMed]
21. Hami, A.; Abdi, B. Students' landscaping preferences for open spaces for their campus environment. *Indoor Built Environ.* **2021**, *30*, 87–98. [CrossRef]
22. Lee, J.; Kim, D.; Park, J. A machine learning and computer vision study of the environmental characteristics of streetscapes that affect pedestrian satisfaction. *Sustainability* **2022**, *14*, 5730. [CrossRef]
23. Zhou, H.; He, S.; Cai, Y.; Wang, M.; Su, S. Social inequalities in neighborhood visual walkability: Using street view imagery and deep learning technologies to facilitate healthy city planning. *Sustain. Cities Soc.* **2019**, *50*, 101605. [CrossRef]
24. Zhao, Y.; Liu, J.H.; Zheng, Y.L. Preservation and renewal: A study on visual evaluation of urban historical and cultural street landscape in Quanzhou. *Sustainability* **2022**, *14*, 8775. [CrossRef]
25. Meng, Y.; Li, Q.; Ji, X.; Yu, Y.; Yue, D.; Gan, M.; Wang, S.; Niu, J.; Fukuda, H. Research on campus space features and visual quality based on street view images: A case study on the Chongshan campus of Liaoning University. *Buildings* **2023**, *13*, 1332. [CrossRef]
26. Qin, Y.; Wu, X.; Yu, T.; Jiang, S.; Zhu, B. Enhancing student-centered walking environments on university campuses through street view imagery and machine learning. *PLoS ONE* **2025**, *20*, e0321028. [CrossRef]
27. Ma, H.; Xu, Q.; Zhang, Y. High or low? Exploring the restorative effects of visual levels on campus spaces using machine learning and street view imagery. *Urban For. Urban Green.* **2023**, *88*, 128087. [CrossRef]
28. Wu, T.; Lin, D.; Chen, Y.; Wu, J. Integrating street view images, deep learning, and sDNA for evaluating university campus outdoor public spaces: A focus on restorative benefits and accessibility. *Land* **2025**, *14*, 610. [CrossRef]
29. Zhang, J.; Hu, A. Analyzing green view index and green view index best path using Google street view and deep learning. *J. Comput. Des. Eng.* **2022**, *9*, 2010–2023. [CrossRef]
30. Wei, J.; Yue, W.; Li, M.; Gao, J. Mapping Human Perception of Urban Landscape from Street-View Images: A Deep-Learning Approach. *Int. J. Appl. Earth Obs. Geoinf.* **2022**, *112*, 102886. [CrossRef]
31. He, Y.Y. Overview of image semantic segmentation based on deep learning. *Chang. Inf. Commun.* **2023**, *36*, 77–79.
32. Garré, S.; Meeus, S.; Gulinck, H. The dual role of roads in the visual landscape: A case-study in the area around Mechelen (Belgium). *Landsc. Urban Plan.* **2009**, *92*, 125–135. [CrossRef]
33. Xia, Y.; Yabuki, N.; Fukuda, T. Sky view factor estimation from street view images based on semantic segmentation. *Urban Clim.* **2021**, *40*, 100999. [CrossRef]
34. Li, P.; Khan, J. Feature extraction and analysis of landscape imaging using drones and machine vision. *Soft Comput.* **2023**, *27*, 18529–18547. [CrossRef]
35. Koning, M.; Kim, J.H.; Saeidi-Rizi, F.; Durst, N. The role of the campus outdoor environment on university student mental health. *J. People Plants Environ.* **2022**, *25*, 263–272. [CrossRef]
36. Pretty, J.; Peacock, J.; Sellens, M.; Griffin, M. The mental and physical health outcomes of green exercise. *Int. J. Environ. Health Res.* **2005**, *15*, 319–337. [CrossRef]
37. Kaplan, R.; Kaplan, S. *The Experience of Nature: A Psychological Perspective*, 1st ed.; Cambridge University Press: Cambridge, UK, 1989; pp. 154–196.
38. Ulrich, R.S. Aesthetic and Affective Response to Natural Environment. In *Behavior and the Natural Environment*; Altman, I., Wohlwill, J.F., Eds.; Springer: Boston, MA, USA, 1983; pp. 85–125.
39. Ulrich, R.S.; Simons, R.F.; Losito, B.D.; Fiorito, E.; Miles, M.A.; Zelson, M. Stress recovery during exposure to natural and urban environments. *J. Environ. Psychol.* **1991**, *11*, 201–230. [CrossRef]
40. Park, B.J.; Tsunetsugu, Y.; Kasetani, T.; Miyazaki, Y. The physiological effects of Shinrin-yoku (taking in the forest atmosphere or forest bathing): Evidence from field experiments in 24 forests across Japan. *Environ. Health Prev. Med.* **2010**, *15*, 18–26. [CrossRef]
41. Antonelli, M.; Barbieri, G.; Donelli, D. Effects of forest bathing (shinrin-yoku) on levels of cortisol as a stress biomarker: A systematic review and meta-analysis. *Int. J. Biometeorol.* **2019**, *63*, 1117–1134. [CrossRef] [PubMed]

42. Berman, M.G.; Jonides, J.; Kaplan, S. The cognitive benefits of interacting with nature. *Psychol. Sci.* **2008**, *19*, 1207–1212. [CrossRef]
43. Tyrväinen, L.; Ojala, A.; Korpela, K.; Lanki, T.; Tsunetsugu, Y.; Kagawa, T. The influence of urban green environments on stress relief measures: A field experiment. *J. Environ. Psychol.* **2014**, *38*, 1–9. [CrossRef]
44. Wilson, E.O. *Biophilia*, 1st ed.; Harvard University Press: Cambridge, MA, USA, 1986; pp. 154–196.
45. Fischer, C.S. Widespread Likings: The Biophilia Hypothesis. *Science* **1994**, *263*, 1161–1162. [CrossRef]
46. Kweon, B.S.; Sullivan, W.C.; Wiley, A.R. Green common spaces and the social integration of inner-city older adults. *Environ. Behav.* **1998**, *30*, 832–858. [CrossRef]
47. Hattie, J.; Marsh, H.W.; Neill, J.T.; Richards, G.E. Adventure education and Outward Bound: Out-of-class experiences that make a lasting difference. *Rev. Educ. Res.* **1997**, *67*, 43–87. [CrossRef]
48. Pasini, M.; Berto, R.; Brondino, M.; Hall, R.; Ortner, C. How to Measure the Restorative Quality of Environments: The PRS-11. *Procedia Soc. Behav. Sci.* **2014**, *159*, 293–297. [CrossRef]
49. Watson, D.; Clark, L.A.; Tellegen, A. Development and validation of brief measures of positive and negative affect: The PANAS scales. *J. Pers. Soc. Psychol.* **1988**, *54*, 1063. [CrossRef]
50. Xiao, S. Theoretical basis and research application of social support rating scale. *J. Clin. Psychiatry* **1994**, *4*, 98.
51. Biljecki, F.; Ito, K. Street view imagery in urban analytics and GIS: A review. *Landsc. Urban Plan.* **2021**, *215*, 104217. [CrossRef]
52. Cheng, B.; Misra, I.; Schwing, A.G.; Kirillov, A.; Girdhar, R. Masked-attention mask transformer for universal image segmentation. In Proceedings of the 2022 IEEE/CVF Conference on Computer Vision and Pattern Recognition (CVPR), New Orleans, LA, USA, 18–24 June 2022; pp. 1290–1299.
53. Zhou, B.; Zhao, H.; Puig, X.; Xiao, T.; Fidler, S.; Barriuso, A.; Torralba, A. Semantic understanding of scenes through the ADE20K dataset. *Int. J. Comput. Vis.* **2019**, *127*, 302–321. [CrossRef]
54. Neuhold, G.; Ollmann, T.; Rota Bulo, S.; Kontschieder, P. The mapillary vistas dataset for semantic understanding of street scenes. In Proceedings of the IEEE International Conference on Computer Vision, Venice, Italy, 22–29 October 2017; pp. 4990–4999.
55. Kim, J.; Lee, D. Comparative Study on Hyperparameter Tuning for Predicting Concrete Compressive Strength. *Buildings* **2025**, *15*, 2173. [CrossRef]
56. Maroof, A.; Karballaeezadeh, N.; Mohammadzadeh, S.D.; Mirzaei, H. Prediction of the wetting-induced compression of collapsible soils using ensemble machine learning. *Indian Geotech. J.* **2025**. [CrossRef]
57. Schaik, P.v.; Clements, H.; Karayaneva, Y.; Imani, E.; Knowles, M.; Vall, N.; Cotton, M. Using Machine Learning to Model the Acceptance of Domestic Low-Carbon Technologies. *Sustainability* **2025**, *17*, 6668. [CrossRef]
58. Lundberg, S.M.; Lee, S.-I. A Unified Approach to Interpreting Model Predictions. In Proceedings of the Advances in Neural Information Processing Systems, Long Beach, CA, USA, 4–9 December 2017; Volume 30.
59. Lundberg, S.M.; Erion, G.; Chen, H.; DeGrave, A.; Prutkin, J.M.; Nair, B.; Katz, R.; Himmelfarb, J.; Bansal, N.; Lee, S.-I. From local explanations to global understanding with explainable AI for trees. *Nat. Mach. Intell.* **2020**, *2*, 56–67. [CrossRef]
60. Chai, T.; Draxler, R.R. Root mean square error (RMSE) or mean absolute error (MAE)?—Arguments against avoiding RMSE in the literature. *Geosci. Model Dev.* **2014**, *7*, 1247–1250. [CrossRef]
61. MacQueen, J. Some Methods for Classification and Analysis of Multivariate Observations. In Proceedings of the Berkeley Symposium on Mathematical Statistics and Probability, Berkeley, CA, USA, 21 June–18 July 1965; pp. 281–297.
62. Chen, S.; Chen, Z.; Hong, J.; Zhuang, X.; Su, C.; Ding, Z. Exploring the Relationship Between Audio-Visual Perception in Fuzhou Universities and College Students' Attention Restoration Quality Using Machine Learning. *Front. Psychol.* **2023**, *16*, 1572426. [CrossRef]
63. Jodas, D.S.; Pierezan, J.; Lima, M.; Nakamura, R.Y.M.; Papa, J.P. Detection of Trees on Street-View Images Using a Convolutional Neural Network. *Int. J. Neural Syst.* **2022**, *32*, 2150042. [CrossRef] [PubMed]
64. Mathivanan, S.K.; Rajadurai, H.; Cho, J.; Kim, M.; Gwak, J.; Rho, S. A Multi-Modal Geospatial–Temporal LSTM Based Deep Learning Framework for Predictive Modeling of Urban Mobility Patterns. *Sci. Rep.* **2024**, *14*, 31579.
65. Guo, W.; Wen, H.; Liu, X. Research on the psychologically restorative effects of campus common spaces from the perspective of health. *Front. Public Health* **2023**, *11*, 1131180. [CrossRef]
66. Liu, K.; Zhang, L.; Tsou, S.; Wang, L.; Hu, Y.; Yang, K. Exploring the complex association between urban built environment, sociodemographic characteristics and crime: Evidence from Washington, DC. *Land* **2024**, *13*, 1886. [CrossRef]
67. Bratman, G.N.; Hamilton, J.P.; Hahn, K.S.; Daily, G.C.; Gross, J.J. Nature Experience Reduces Rumination and Subgenual Prefrontal Cortex Activation. *Proc. Natl. Acad. Sci. USA* **2015**, *112*, 8567–8572. [CrossRef] [PubMed]
68. Foellmer, J.; Kistemann, T.; Anthonj, C. Academic greenspace and well-being—Can campus landscape be therapeutic? Evidence from a German university. *Wellbeing Space Soc.* **2021**, *2*, 100003. [CrossRef]
69. Liu, W.; Sun, N.; Guo, J.; Zheng, Z. Campus green spaces, academic achievement and mental health of college students. *Int. J. Environ. Res. Public Health* **2022**, *19*, 8618. [CrossRef]

70. Liprini, R.M. *Students' Perceptions of Green Space on a University Campus: An Attention Restoration Theory Perspective*, 1st ed.; University of Pretoria: Pretoria, South Africa, 2014; pp. 154–196.
71. Bonell, C.; Parry, W.; Wells, H.; Jamal, F.; Fletcher, A.; Harden, A.; Thomas, J.; Campbell, R.; Petticrew, M.; Murphy, S.; et al. The effects of the school environment on student health: A systematic review of multi-level studies. *Health Place* **2013**, *21*, 180–191. [CrossRef]

buildings

MDPI

Article

Research on Park Perception and Understanding Methods Based on Multimodal Text–Image Data and Bidirectional Attention Mechanism

Kangen Chen [1,*], Xiuhong Lin [1,2,*], Tao Xia [3] and Rushan Bai [1]

[1] Faculty of Innovation and Design, City University of Macau, Macau 999078, China; u22092120379@cityu.edu.mo
[2] School of Urban Architecture, Guangzhou Huali College, Guangzhou 511325, China
[3] Zhuhai Dechuang Construction Engineering Consulting Limited Company, Zhuhai 519000, China; so19929013@126.com
* Correspondence: u22092120427@cityu.edu.mo (K.C.); u22092120351@cityu.edu.mo (X.L.)

Abstract: Parks are an important component of urban ecosystems, yet traditional research often relies on single-modal data, such as text or images alone, making it difficult to comprehensively and accurately capture the complex emotional experiences of visitors and their relationships with the environment. This study proposes a park perception and understanding model based on multimodal text–image data and a bidirectional attention mechanism. By integrating text and image data, the model incorporates a bidirectional encoder representations from transformers (BERT)-based text feature extraction module, a Swin Transformer-based image feature extraction module, and a bidirectional cross-attention fusion module, enabling a more precise assessment of visitors' emotional experiences in parks. Experimental results show that compared to traditional methods such as residual network (ResNet), recurrent neural network (RNN), and long short-term memory (LSTM), the proposed model achieves significant advantages across multiple evaluation metrics, including mean squared error (MSE), mean absolute error (MAE), root mean squared error (RMSE), and the coefficient of determination (R^2). Furthermore, using the SHapley Additive exPlanations (SHAP) method, this study identified the key factors influencing visitors' emotional experiences, such as "water", "green", and "sky", providing a scientific basis for park management and optimization.

Keywords: deep learning; park environment perception; multimodality; sentiment analysis; attention mechanism

1. Introduction

As an important component of urban ecosystems, parks serve key functions in daily leisure, recreation, and social interaction, fostering a connection with nature [1–8]. With the acceleration of urbanization and the improvement of people's demand for quality of life, parks not only need to satisfy the basic ecological functions but also need to provide quality emotional experiences [9], which poses new challenges to the design and management of parks. Studies have shown that visitors' emotional experience affects their satisfaction, willingness to visit, and health and well-being in parks [1,10]. Therefore, it is very important to accurately assess the emotional tendencies of park visitors to improve the quality of park services and social values.

Public perception of a park environment is a complex cognitive process that is influenced by a combination of factors such as natural landscape, artificial facilities, social

https://doi.org/10.3390/buildings15091552

interaction, and air quality [11–14]. Measuring environmental perception is challenging because of its subjective and intangible nature [15,16]. Traditional assessment methods mainly include questionnaire surveys [17,18], interviews [19,20], field observations [21,22], and case studies [23–25]. These conventional methods have obvious shortcomings, such as consuming more time and manpower, having a single source of data, and being susceptible to subjective factors [26,27], which make it difficult to meet the needs of large-scale, real-time, and objective park emotion assessment. Therefore, how to scientifically assess the emotional experience of park visitors with the help of modern computing methods in park operation and management has become an important issue to be solved.

With advancements in computer technology, deep learning has gradually become an important tool for studying the perception of the park environment [16,28,29]. More accurate analysis and automated assessment of affective tendencies are gradually becoming a research hotspot. However, most existing research on park perception relies on single-modal data, such as text or images. Studies based on textual data analyze textual content, such as social media posts, user comments, etc., to infer the public's emotional disposition towards the environment [30–34]. Models such as RNN, BERT, and LSTM have been widely used in recent years in textual sentiment disposition analysis [30,32,35]. Studies based on image data utilize the visual features of environmental photographs to assess the potential emotional impact of the environment on the public [36–38]. Models such as CNN, R-CNN, and ViT (Vision Transformer) are able to detect the effect of images on people's emotional tendencies [39,40]. Notably, in recent years, Transformer models have demonstrated strong capabilities in handling sentiment analysis and complex data patterns [41,42]. However, single-modal data can only represent human perception from a particular perspective, which not only makes it difficult to complete a full-picture description of complex emotional experiences independently but also fails to accurately reflect the emotional fluctuations of tourists in the changing park environment [43,44].

As a cutting-edge technology in the field of deep learning, multimodal learning methods are able to provide more comprehensive and accurate results by fusing data from different sources and forms [43–46]. In the research field of park environment perception, multimodal deep learning is still in the exploratory stage. Studies from Yang et al. and Liu et al. suggest that integrating text and image data not only enhances the overall understanding of spatial environments by the model but also improves the accuracy of emotional tendency prediction [27,28]. However, despite the potential of multimodal learning methods to integrate text and image data, existing research still faces two major challenges. First, effectively fusing text and image features remains a critical issue. Text and images belong to fundamentally different modalities [47], with distinct feature spaces and representational structures. Developing models that seamlessly integrate these two types of information is a key challenge in multimodal learning. Second, capturing the intricate interactions between text and images presents another major challenge. Traditional multimodal fusion techniques mostly rely on simple feature concatenation or weighted averaging [48], which limits their ability to fully uncover deep connections between textual and visual data.

To overcome these challenges, this study proposes a park perception and understanding method based on multimodal text–image data and a bidirectional attention mechanism. Specifically, we developed a cross-modal framework that integrates a bidirectional encoder representations from transformers (BERT)-based text feature extraction module, a Swin Transformer-based image feature extraction module, and a bidirectional cross-attention fusion module. Through the deep fusion of multimodal information, the framework enhances the model's ability to capture visitors' emotional experiences. This technical framework not only addresses the limitations of unimodal assessment methods but also overcomes the technical obstacle of simple multimodal fusion that makes it difficult to capture deep semantic associations across modalities. The method is able to accurately assess the emotional tendencies of visitors while capturing textual information and image features of the park environment, which enhances the comprehensiveness of urban public space assessment. This data-driven analysis framework not only provides actionable recommendations for urban planning but is also versatile enough to be extended to a wide range of urban public spaces.

The contributions of this paper are as follows:

1. The BERT model effectively captures semantic information in text using a bidirectional encoder, whereas the Swin Transformer efficiently extracts visual features from images through a local window self-attention mechanism. This study proposed a park perception and understanding method based on multimodal text–image data to effectively fuse text and image features and capture their interactions.
2. To better integrate BERT and Swin Transformer features, a bidirectional cross-attention mechanism that enables mutual reinforcement of text and image information is introduced in order to enhance the accuracy and robustness of sentiment analysis.
3. In addition, a method that enables an interpretability analysis of the model results is employed, focusing on the significance of key terms, which allowed us to identify key factors of the park environment that influence visitors' emotional experience, for optimization and providing suggestions for park management.

2. Materials and Methods

2.1. Overall Model Architecture

Parks provide important spaces for leisure and recreation. Their environmental quality directly influences visitors' emotional experiences. However, fully capturing these emotional experiences by relying only on single-source data is difficult. Therefore, we propose a cross-modal framework that integrates textual descriptions and corresponding image data, to more accurately assess visitors' emotional tendencies.

The overall model architecture consisting of a text feature extraction, an image feature extraction, and a multimodal cross-fusion module is depicted in Figure 1. First, the input layer receives textual descriptions and associated image data. The text data are processed by a BERT-based text feature extraction module, which captures deep semantic information within the text. For image processing, the Swin Transformer effectively reduces computational complexity and enhances the modeling of long-range dependencies by segmenting the image into multiple local windows and applying self-attention computations within each window. Finally, the bidirectional cross-attention fusion module enhances the interaction between text and image features through text-to-image and image-to-text cross-attention computations. This ensures that the generated comprehensive feature representation includes key information from both modalities, accurately reflecting the emotional sentiment associated with the park.

2.2. BERT-Based Text Feature Extraction Module

BERT is a language model based on the Transformer architecture. It employs bidirectional encoders to capture deep semantic information in text and conducts unsupervised learning on large-scale text corpora [49]. Unlike traditional unidirectional language models, BERT simultaneously considers both preceding and succeeding contextual information of a word to capture not only the intrinsic meaning of words but also their relationships with other words, resulting in a more comprehensive and precise text representation. In this study, we utilized a pre-trained BERT model to extract deep semantic information from park-related text. The model was fine-tuned on our dataset to enhance its understanding of sentiment recognition in park perception tasks.

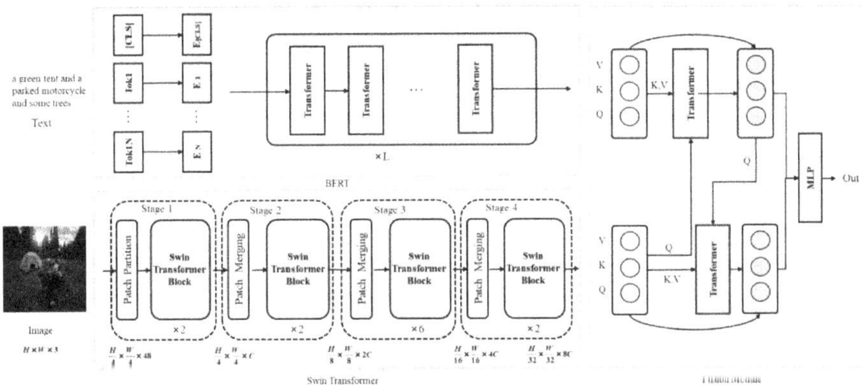

Figure 1. Overall model architecture.

BERT consists of multilayer bidirectional Transformer encoders, with each encoder layer comprising a multi-head self-attention mechanism and a feed-forward neural network (FFN). BERT first converts the input text into a corresponding word embedding representation. Given a text T, BERT generates a series H of context-dependent embedding vectors with dimensions $[L, d]$, where L represents the text length and d is the hidden layer dimension. Since the Transformer lacks recurrent or convolutional structures, explicit positional information is added to capture the sequential order of words. Positional embeddings are fixed mappings that assign different vectors to words based on their position in the sentence.

The self-attention mechanism captures dependencies between different positions by applying three independent linear transformations to generate the query, key, and value matrices. The scaled dot-product attention score is computed as follows:

$$\text{Attention}(Q, K, V) = \text{Softmax}\left(\frac{QK^T}{\sqrt{d_k}}\right)V \tag{1}$$

where Q, K, and V are the query, key, and value matrices, respectively, and d_k is the dimension of the key vector.

To capture information from various subspaces, BERT employs a multi-head self-attention mechanism. Assuming that there are h attention heads, the output of the i-th head is given as follows:

$$\text{Head}_i = \text{Attention}\left(QW_i^Q, KW_i^K, VW_i^V\right) \tag{2}$$

where W_i^Q, W_i^K, and W_i^V are the projection matrices of the query, key, and value of the i-th attention head, respectively.

The outputs of all attention heads are concatenated and transformed linearly:

$$\text{MultiHead}(Q, K, V) = [\text{Head}_1; \text{Head}_2; \ldots; \text{Head}_h]W_O \tag{3}$$

where W_O is the output projection matrix.

Each Transformer encoder layer also includes an FFN, which is applied independently to the output at each position:

$$\text{FFN}(x) = \text{ReLU}(xW_1 + b_1)W_2 + b_2 \tag{4}$$

where x is the data input to the FFN; W and b are the learnable parameters in FFN; and the rectified linear unit (ReLU) is the activation function.

To prevent vanishing gradients and accelerate training, BERT employs residual connections and layer normalization after each layer. Figure 2 illustrates the architecture of the BERT-based text feature extraction module and its attention mechanism.

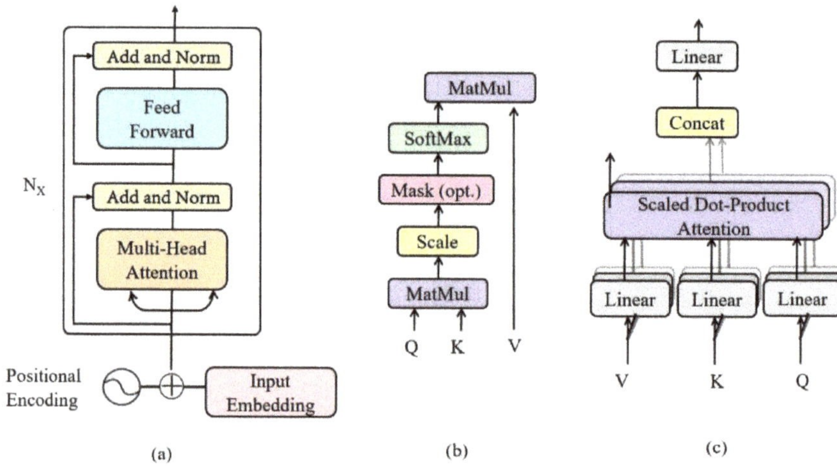

Figure 2. BERT-based text feature extraction module and structure of attention mechanism: (**a**) attention mechanism; (**b**) substructures of attention mechanism; (**c**) multi-head attention.

To mitigate overfitting, a specific fine-tuning strategy was employed where only the parameters of BERT's final layer were fine-tuned, while the weights of the remaining layers were frozen and not updated during the training process. During training, the maximum sequence length was set to 512.

2.3. Swin Transformer-Based Image Perception Module

The Swin Transformer is a vision model based on the Transformer architecture that integrates the concept of local connectivity of the traditional convolutional neural network (CNN) with the self-attention mechanism of the Transformer, enabling the capture of contextual information in images at multiple scales [50]. Traditional Transformers compute self-attention across the entire image, leading to high computational complexity, especially for high-resolution images. The Swin Transformer partitions the image into multiple non-overlapping windows and performs self-attention calculations within each window. Through the "shifted window" mechanism, it efficiently captures both local and global self-attention.

As shown in Figure 3, the input image is first divided into non-overlapping windows, and self-attention is computed within each window. In subsequent layers, the window configuration shifts to facilitate cross-window information exchange, enhancing the model's ability to capture long-range dependencies while maintaining computational efficiency. The Swin Transformer consists of four hierarchical stages, each comprising several Transformer blocks. As the stages progress, the spatial resolution of the feature map decreases while the number of channels increases accordingly.

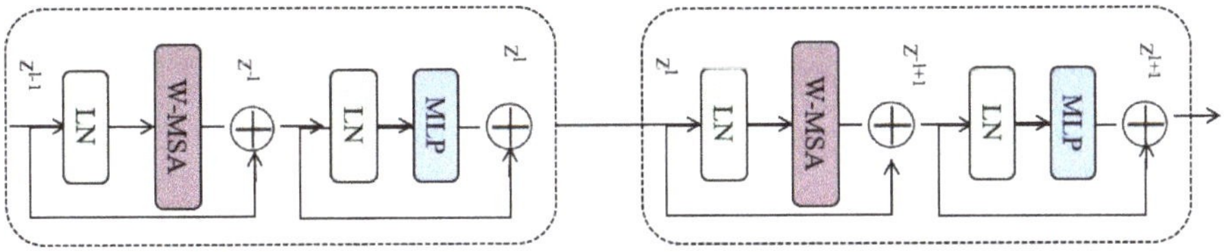

Figure 3. Swin-transformer structure.

2.4. Bidirectional Cross-Attention Fusion Module

In this study, a bidirectional cross-attention fusion module was designed to enhance the interaction and integration of text and image features, thereby improving the accuracy of park emotion analysis. Text features are extracted using the pre-trained language model BERT, and visual features are obtained from image data via the Swin Transformer. To achieve efficient feature fusion, we employed a cross-attention mechanism to process cross-modal information.

The text-to-image cross-attention computation maps key information from features onto image features, capturing the associations between textual descriptions and visual content. Next, the image-to-text cross-attention computation maps key information from image features back onto text features, further reinforcing their mutual interactions. These two stages not only enable deep interaction between text and image features but also preserve the integrity of the original features. Moreover, residual connections enhance the expressiveness of the model.

Finally, the text and image features processed through both cross-attention computations are concatenated to form a comprehensive feature representation. This representation integrates information from both modalities, providing a more complete and accurate reflection of the emotional information embedded in the park environment.

2.5. Data and Evaluation Metrics

In this study, the COCO Caption dataset (https://cocodataset.org) was the primary data source (accessed on 3 February 2025). This dataset is widely used in image recognition and natural language processing tasks. It contains a rich collection of images and corresponding textual descriptions. The COCO Caption dataset has 82,783 samples in the training set; 40,504 samples in the validation set; and 40,775 samples in the test set. Each image is associated with 5 to 7 sentences of descriptive text. This dataset not only spans a broad range of scenarios but also provides detailed annotations, making it highly suitable for cross-modal analysis and research [51]. To focus on the emotional perception of the park environment, we extracted a subset from the COCO dataset comprising image–text pairs related to parks. By extracting samples containing keywords such as "park" and "visitor" and filtering them, we ultimately obtained 30,682 valid samples. We preprocessed and filtered these data, to ensure that the model focuses on learning features relevant to the park environment. For the emotion tendency annotation, we first initially categorize the text data according to "0" (negative), "1" (neutral) or "2" (positive). Then, the text was rated by each of the five experts on its emotional tendency. Ultimately, we determine the sentiment propensity score for each comment by calculating the average of multiple expert ratings to obtain a more precise continuum value to ensure the scientific validity of the data. This approach not only considers sentiment categorization but also quantifies the intensity of sentiment, providing more detailed sentiment analysis results.

As shown in Figure 4, a word cloud analysis was conducted on the extracted park-related texts. This analysis helps visualize key phrases and high-frequency words associated with parks, providing an important reference for subsequent sentiment analysis. The word cloud highlights keywords such as "trees", "play equipment", and "roads", which are closely related to the park environment, indicating that the data selection process is both effective and representative.

Figure 4. Word cloud statistics for data.

To comprehensively assess the model performance, we employed multiple evaluation metrics to measure the differences between predicted results and actual values. Mean squared error (MSE), mean absolute error (MAE), root mean squared error (RMSE), and the coefficient of determination (R^2) were used as evaluation metrics. The formulas are as follows:

$$\text{MSE} = \frac{1}{n}\sum_{i=1}^{n}(c - \hat{y}_i)^2 \tag{5}$$

$$\text{MAE} = \frac{1}{n}\sum_{i=1}^{n}|y_i - \hat{y}_i| \tag{6}$$

$$\text{RMSE} = \sqrt{\frac{1}{n}\sum_{i=1}^{n}(y_i - \hat{y}_i)^2} \tag{7}$$

$$R^2 = 1 - \frac{\sum_{i-1}^{n}(y_i - \hat{y}_i)^2}{\sum_{i-1}^{n}(y_i - \overline{y})^2} \tag{8}$$

where y_i is the actual value; $\hat{y}_i$ is the predicted value; $\overline{y}$ is the mean of the actual values; and n is the sample size.

3. Results

In our experiments, we trained and evaluated the model using filtered data to verify its performance. The model was trained on a device equipped with a 4090 GPU graphics card, utilizing the PyTorch framework. During training, the learning rate was set to 0.0001; the batch size was 96; and a total of 4000 iterations were performed. This configuration ensured that the model could efficiently learn key features from the data within a reasonable timeframe, thus improving its predictive accuracy.

3.1. Experimental Results and Analysis

Figure 5 shows how the loss values and validation errors change during the iterative training process. As the number of training steps increased, the loss value steadily decreased. This continuous reduction in training loss indicates that the model gradually optimizes its parameters to better capture the emotional features in the input data. Meanwhile, the validation error on the validation dataset also showed a steady decline and eventually stabilized. This further demonstrates that the model not only fit the training data well but also has a strong generalization ability for new data to some extent.

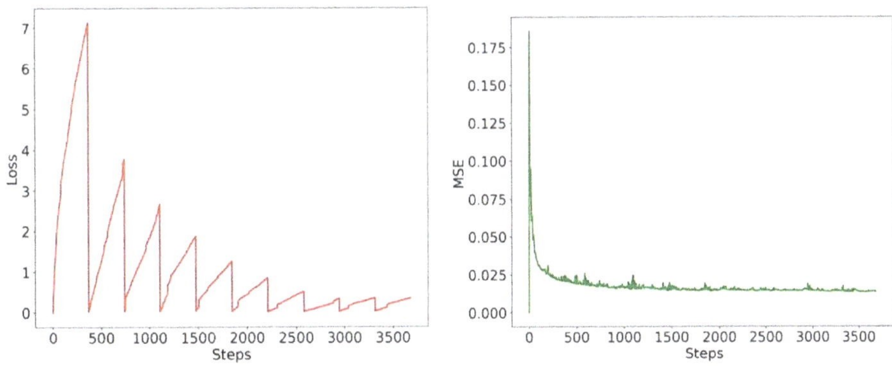

Figure 5. Training loss and validation error curves for different iteration steps: (**a**) training loss curve; (**b**) validation error curve.

To validate the effectiveness of the model, we compared it with models used in previous urban perception studies. The residual network (ResNet), as a deep CNN, excels in processing image data [52]. The recurrent neural network (RNN) was used by Hong et al. for park environment perception analysis [53], and long short-term memory (LSTM) was employed by Ren et al. in their study of urban park satisfaction [54]. As shown in Table 1, significant differences exist in the performance of the various models in the multimodal park affective tendency prediction task. ResNet has relatively higher error metrics in this task, with an R^2 value of 0.7992, indicating limitations in fitting complex relationships. In contrast, traditional sequence models such as RNN and LSTM perform better. The error metrics for RNN, including MSE, MAE, and RMSE, are 0.0278, 0.1325, and 0.1764, respectively, outperforming ResNet. Moreover, the R^2 value of RNN is 0.8174. These results demonstrated its advantage in handling time series. LSTM further improved RNN's performance. Its error metrics of MSE, MAE, and RMSE are at 0.0199, 0.1163, and 0.1345, respectively, and its R^2 value is 0.94.

Table 1. Comparison of different models on MSE, MAE, RMSE, and R2 metrics.

Method	MSE	MAE	RMSE	R^2
ResNet	0.0352	0.1578	0.2216	0.7992
RNN	0.0278	0.1325	0.1764	0.8174
LSTM	0.0199	0.1163	0.1345	0.9424
Our method	0.0109	0.0781	0.1044	0.9819

The proposed model outperformed all other models across all evaluation metrics, with an MSE of 0.0109, an MAE of 0.0781, an RMSE of 0.1044, and an R^2 value of 0.9819 (Table 1). Our model not only fits the training data more accurately but also excels in generalization. It effectively captures the complex relationships between text and images, thereby more precisely predicting emotional tendencies. The results show that by integrating multimodal information and introducing the bidirectional cross-attention mechanism, our model significantly enhances prediction performance.

In order to assess the consistency of the model predictions with subjective user experience, a small sample of manual evaluations was conducted. We first randomly selected 50 samples from the test set, each consisting of an image and its corresponding textual description. Next, we invited 50 users to participate in the assessment process, and all participants went through a short training to understand the process and criteria of the assessment. During the evaluation process, each user will view 50 samples. Subsequently, the user needs to evaluate the prediction results of the four models (our model, ResNet, RNN, LSTM) for each sample, categorized as "Agree" and "Disagree". "Agree" indicates that the user believes that the model's predictions are consistent with their emotional experience. "Disagree" indicates that the user believes that the model's predictions are inconsistent with their emotional experience. Each choice will be recorded on a rating scale, which results in an evaluation that includes each user, each sample, and each model. As shown in Table 2, in the manual evaluation, our model has a higher percentage of agreement than the other three models among all users. The evaluation indicates that our model achieves better consistency in the subjective user experience, proving the validity of the model.

Table 2. Comparison of the consistency of users' subjective experiences.

Method	Satisfaction Rate of Manual Assessment of Prediction Results
ResNet	0.74
RNN	0.82
LSTM	0.86
Our method	0.92

3.2. Theme Word Analysis

As shown in Figure 6, a heat map visualization of the input text was generated for the model using the SHapley Additive exPlanations (SHAP) method. SHAP is a game theory-based approach that explains the output of a complex model by calculating the contribution of each feature to the prediction [55]. The SHAP value measures the marginal contribution of a feature across all possible combinations of features, thus providing a global and local framework for interpretation. This paper reveals which keywords have the most significant impact on predicting the emotional tendencies of park visitors using SHAP, helping intuitively understand the logic behind the model's decisions. Figure 6 illustrates four specific input text cases. Red represents positive sentiment polarity, while blue represents negative sentiment polarity. The darker the color depth of the text vocabulary, the more critical the park visitor's emotional tendency embodied by the word.

Figure 7 is the analysis of term importance, which compares the overall word frequency in the entire dataset with the word frequency related to specific sentiment polarity themes. The horizontal axis represents the frequency of occurrence of words. The vertical axis lists the different words. Light blue represents the overall word frequency of the term in the entire dataset, while red represents the frequency of the term related to specific sentiment polarity themes.

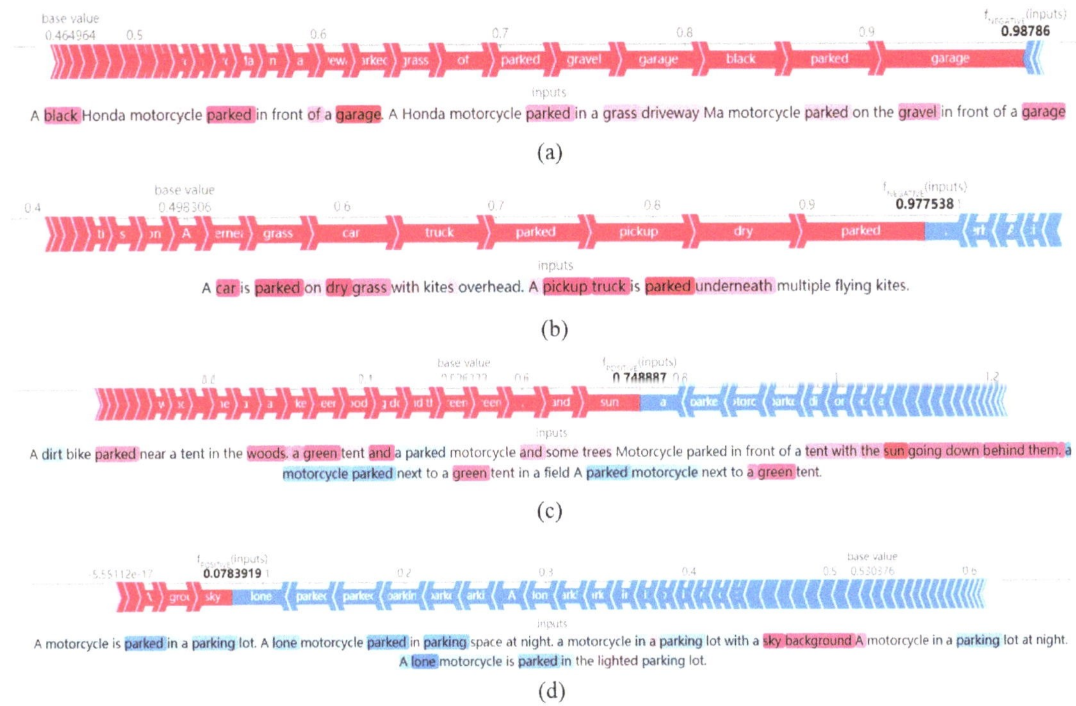

Figure 6. SHAP-based heat map analysis of theme words: (**a**) Example 1; (**b**) Example 2; (**c**) Example 3; (**d**) Example 4. Red represents positive sentiment polarity, while blue represents negative sentiment polarity. The darker the color depth of the text vocabulary, the more critical the park visitor's emotional tendency embodied by the word.

As shown in Figures 6 and 7, the theme words "water", "driveway", "sky", "green", and "near" show relatively high heat values in the input text, indicating that these words have a significant effect on the model's prediction of park visitors' emotional tendencies. For instance, the high heat values of "water" and "green" reflect the importance of natural landscapes to visitors' emotional experiences, whereas the high heat values of "driveway" and "near" suggest the influence of external environmental factors on visitor satisfaction. In addition, red words such as "grass" and "car" indicate positive affective tendencies towards the park. Blue words such as "dirt" and "lone" indicate negative affective tendencies towards the park. This analysis not only helps identify key emotional drivers but also provides a scientific basis for subsequent park management and optimization. By focusing on these high-impact keywords, park managers can make targeted improvements to the park's overall environmental quality, thus enhancing visitors' emotional experiences and satisfaction.

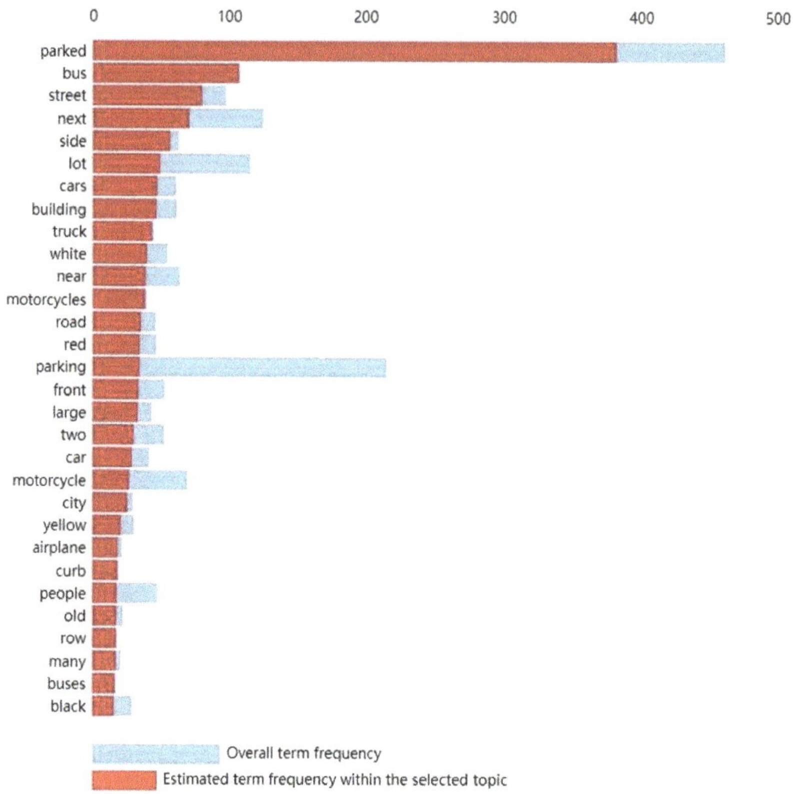

Figure 7. Theme word importance analysis. Light blue represents the overall word frequency of the word across the dataset. The red color represents the number of times the word has a word frequency related to a specific emotional tendency theme.

3.3. Saliency Analysis

Saliency analysis plays a key role in image understanding, especially when combining text and image data. Saliency refers to certain areas of an image or specific words in a text having a particularly significant impact on the overall emotional experience. By identifying these salient elements, the emotional tendencies of visitors and their interaction with the environment can be captured more accurately.

As shown in Figure 8, salient region analysis helps us identify the parts of the image that attract the most attention from visitors, such as kites, animals, and landscapes in the park. These regions are often closely related to visitors' emotional experiences, and an in-depth analysis of these regions can reveal visitors' points of focus and preferences in the park.

Figure 8. Analysis of salient regions in the image. The red areas in the figure represent the regions that attract the most attention.

4. Discussion

4.1. Methodological Contributions

The proposed park perception understanding method, which is based on multimodal text–image data and the bidirectional attention mechanism, performed well in experiments. Compared with traditional methods such as ResNet, RNN, and LSTM, our method excels in several evaluation metrics, including MSE, MAE, RMSE, and R^2. This advantage is mainly attributed to the model's ability to effectively integrate text and image features, thereby capturing the complex interactions between them through the bidirectional cross-attention mechanism. The design of the bidirectional attention mechanism allows textual and image information to mutually enhance each other, generating more comprehensive and representative feature representations. The introduction of the attention mechanism effectively handles the multimodal fusion of textual and visual features [56]. This approach demonstrates a strong adaptability for processing multimodal data.

4.2. Theoretical Reflections Based on Phenomenological and Neuroscientific Perspectives

In the study of public space perception, theories from phenomenology and neuroscience provide a deeper conceptual foundation for our research. Phenomenology emphasizes that humans perceive the world through multiple senses, and these perceptions are closely linked to emotions [57]. Our model comprehensively evaluates emotional experiences in parks by integrating both textual and visual sensory information. Neuroscience theories suggest that the integration of emotional cues is primarily driven by multisensory processes, where the brain synthesizes information from multiple senses [58]. Our model is capable of simultaneously analyzing both textual and visual information, similar to how the brain processes these types of data. Furthermore, research has shown that factors such as sound and odor can also influence an individual's emotional state [59]. Future research will further explore the incorporation of multisensory factors into park perception models.

4.3. Key Factors Influencing Visitors' Emotional Experiences

Thematic word analysis and significance analysis demonstrated the key elements that may influence the affective tendencies of park visitors. The theme words "water", "driveway", "sky", and "green" have a significant impact on the model's prediction of park visitors' affective tendencies. Managers can prioritize improving water systems and vegetation coverage based on the significant influence of "water" and "green". Popular urban green spaces need to incorporate favorable natural elements and facilities that meet user needs [60]. The type of park elements can influence the importance of visitors' perceptual indicators. "Water" and "green" are both among the more significant perceptual indicators in the park environment [61]. Meanwhile, the faunal elements identified by the saliency analysis suggest that urban parks may need to focus on enhancing biodiversity. Urban green spaces with rich biodiversity can improve the attention levels of respondents after they are in close contact with such environments [62]. Managers and researchers should monitor the dynamic changes in visitors' perceived satisfaction with the environment and make timely adjustments to the perceived elements as required [63,64]. As a theme word, "sky" has a high level of heat in the input text. It has not received as much attention as "water" and "green" in previous park perception factors. Understanding that "sky" may be an important perceptual factor for park visitors, park managers, and designers can optimize park activity areas and landscapes for weather and seasonal changes to enhance the emotional experience of visitors. The practical application of the above analysis results may be influenced by the representativeness of the data and the accuracy of model interpretation. Their reliability can be further validated through field research in the future.

4.4. Practical Applications of the Model

The model can provide decision support for future smart park planning. Designers can use the model for emotional simulation assessment and feature sensitivity analysis. During the park design phase, designers can evaluate the emotional effects of different scenarios by inputting renderings and associated text of park design options and allowing the model to predict the emotional responses of potential visitors. Additionally, designers can identify key environmental elements that affect visitors' emotions through theme word analysis based on the SHAP methodology and then reinforce important positive elements in the design.

The model can be used as a real-time monitoring system to dynamically assess the park environment and visitor experience, providing a feedback system for city managers. The model is integrated with a Geographic Information System (GIS) to generate an "emotion map" using pictures and comments uploaded by visitors, which visualizes the distribution of emotions in different areas of the park. Managers can use the model to identify emotional hotspots and cold zones, providing a basis for resource allocation in the park. In addition, park environments may change seasonally, producing different emotional effects on visitors. The model is able to analyze this seasonal variation and assist managers in developing seasonal park environment optimization strategies.

4.5. Limitations and Future Work

The model has high computational complexity, particularly that the bidirectional cross-attention mechanism and windowing computations of the Swin Transformer require high-performance hardware (e.g., 4090 GPUs). This may limit its deployment in resource-constrained environments. Future research may consider further exploration around model lightweight, efficient computation, and hardware co-optimization.

The experiment utilized a park subset of the COCO Caption dataset. Although it provides high-quality images and corresponding descriptive texts, these contents were not generated by real users in natural interaction scenarios. As a result, these data may lack the nuance, naturalness, and complexity of sentiment expression found in real social media data. This discrepancy implies that although our model performs well on the COCO Caption dataset, its generalization ability may be somewhat compromised when applied directly to analyze real social media graphic data, especially in terms of capturing nuanced sentiment or understanding specific cultural contexts. Future research will further explore the effectiveness of the model in real-world social media data assessment tasks to improve the reliability and usefulness of its application.

The applicability of the model in different types of park environments has to be evaluated and validated. Urban parks contain a variety of types, such as urban pocket parks, forest parks, coastal parks, and community green spaces. Different types of parks exhibit variations in spatial scale, functional positioning, and landscape features. Visitors' emotional responses may vary depending on the type, scale, or function of the park. For example, people may be more concerned with natural elements such as vegetation and landscape in forest parks [65]. At the same time, users in urban pocket parks may be more concerned with the facilities and social interactions within the site [66]. The ability of the model to generalize for different types of parks requires more testing and validation in the future.

The research model focuses on experimental validation based on the COCO Caption dataset, with no case validation studies or consideration of the effects of differences in geographic patterns, environmental factors, and socio-cultural backgrounds. Differences in the above factors may also affect the adaptability of the model. Therefore, future research will further assess the generalizability and validity of the model by incorporating diverse case studies covering different park geographic, environmental, and cultural contexts.

In our approach, we used expert ratings for the emotional tendency labeling of texts. In the emotion classification system, Plutchik's wheel subdivided emotions into a variety of basic emotions, and by refining the emotion labels, it is possible to capture the diversity of emotions in a text [67]. The valence–arousal model, on the other hand, is able to quantify the positive and negative affectivity as well as the degree of activation, which can enrich the dimensionality of affect analysis [68]. In future research, we plan to further explore park visitors' emotion assessment studies by refining emotion labels and emotion intensity in conjunction with the emotion classification system.

As part of this work, we analyzed the contribution of the input text to the prediction results through the SHAP tool, which enhances the interpretability of the text part. At the same time, we investigated the distribution of visual elements through the significance analysis in the experiment and explored the visitors' points of interest in the visual elements of the park, which further improved the interpretability of the model. Future research could incorporate statistical data and utilize visualization tools from machine learning models such as XGBoost (XGB) and LightGBM (LGB) to further explore the impact of various features, such as the number of park visitors, types and quantities of infrastructure, and other numerical characteristics, thereby providing more comprehensive interpretability analysis.

5. Conclusions

To address the challenges of multimodal feature fusion and cross-modal interaction, this study developed an innovative cross-modal framework that effectively integrates textual and image data using a bidirectional attention mechanism to improve the understanding and prediction of visitors' emotional tendencies in a park environment. The main conclusions are as follows:

1. The combination of BERT-based text feature extraction and Swin Transformer-based image feature extraction, along with the bidirectional cross-attention mechanism, enables the model to capture the deep interaction between text and images, thereby improving the accuracy of emotional tendency prediction.
2. Compared with traditional methods such as ResNet, RNN, and LSTM, our method excels in several evaluation metrics, including MSE, MAE, RMSE, and R^2. This confirms the advantages of the model in handling complex sentiment analysis tasks.
3. The theme word analysis conducted using the SHAP method identified key factors, such as "water", "green", and "sky", that significantly impact visitors' emotional experiences. These findings not only contribute to a deeper understanding of the emotional drivers of visitors but also provide park managers with concrete directions for enhancing the park environment.

This study has limitations in terms of model computational complexity, data characteristics, scenario applicability, and case validation. First, the computational complexity of the model is high and requires high-performance hardware support, limiting its ability to be deployed in resource-constrained environments. Second, the COCO Caption dataset used in the study, while containing high-quality image–text pairs, may lack the nuance and sentiment expression found in real social media data, which may affect the generalization ability of the model. Further, the validation of the model's applicability in different types of park environments has not yet been undertaken, and differences in park types may influence visitors' affective responses. Finally, the current study does not take into account differences in geographical patterns, environmental factors, and socio-cultural contexts, and there is a need to assess the generalizability of the model in more real-world cases.

Future research can be further developed in terms of model weights, the application of real social media data, different park types, and geo-cultural adaptations. Specifically, future work will explore how to make the model run efficiently in environments with limited computational resources and validate the model's effectiveness in real-world social media data, as well as conduct in-depth tests of the effects of differences in park type and cultural context. In addition, the study plans to further improve the emotion assessment and interpretability of the model by refining the emotion labels, introducing an emotion classification system, and combining other interpretability methods, among other things.

In summary, this study offers an efficient solution for park emotion perception and understanding through an innovative multimodal framework and bidirectional attention mechanism. This technical framework not only addresses the limitations of unimodal assessment methods but also overcomes the technical obstacle of simple multimodal fusion that makes it difficult to capture deep semantic associations across modalities. The method is able to accurately assess the emotional tendencies of visitors while capturing textual information and image features of the park environment, which enhances the comprehensiveness of urban public space assessment. With future technological optimizations and expanded applications, this method is expected to play an even greater role in environmental management and design optimization.

Author Contributions: Conceptualization, data curation, methodology, software, validation, original draft, visualization, writing—review and editing, K.C.; conceptualization, methodology, visualization, writing—review and editing, X.L.; conceptualization, methodology, software, validation, T.X.; data curation, methodology, visualization, R.B. All authors have read and agreed to the published version of the manuscript.

Funding: This research was funded by the Guangdong Province Philosophy and Social Sciences 2023 Annual Discipline Co-construction Project (Grant No. GD23XSH15).

Data Availability Statement: The original contributions presented in this study are included in the article. Further inquiries can be directed to the corresponding author.

Acknowledgments: We thank Editage (https://www.editage.cn) for their linguistic assistance during the preparation of this manuscript (accessed on 21 March 2025).

Conflicts of Interest: Author Tao Xia from company Zhuhai Dechuang Construction Engineering Consulting Limited Company. The remaining authors declare that the research was conducted in the absence of any commercial or finaical relationships that could be construed as a potential conflicts of interest.

Abbreviations

The following abbreviations are used in this manuscript:

BERT	Bidirectional Encoder Representations from Transformers
CNN	Convolutional Neural Network
FFN	Feed-Forward Neural Network
LSTM	Long Short-Term Memory
MAE	Mean Absolute Error
MSE	Mean Squared Error
ReLU	Rectified Linear Unit
RMSE	Root Mean Squared Error
RNN	Recurrent Neural Network
SHAP	SHapley Additive exPlanations

References

1. Kong, L.; Liu, Z.; Pan, X.; Wang, Y.; Guo, X.; Wu, J. How Do Different Types and Landscape Attributes of Urban Parks Affect Visitors' Positive Emotions? *Landsc. Urban Plan.* **2022**, *226*, 104482. [CrossRef]
2. Kabisch, N.; Kraemer, R.; Masztalerz, O.; Hemmerling, J.; Püffel, C.; Haase, D. Impact of Summer Heat on Urban Park Visitation, Perceived Health and Ecosystem Service Appreciation. *Urban For. Urban Green.* **2021**, *60*, 127058. [CrossRef]
3. Bedla, D.; Halecki, W. The Value of River Valleys for Restoring Landscape Features and the Continuity of Urban Ecosystem Functions—A Review. *Ecol. Indic.* **2021**, *129*, 107871. [CrossRef]
4. Wang, Y.; Chen, F. Research on Environmental Behavior of Urban Parks in the North of China during Cold Weather—Nankai Park as a Case Study. *Buildings* **2024**, *14*, 2742. [CrossRef]
5. Wu, Y.; Zhou, W.; Zhang, H.; Liu, Q.; Yan, Z.; Lan, S. Relationships between Green Space Perceptions, Green Space Use, and the Multidimensional Health of Older People: A Case Study of Fuzhou, China. *Buildings* **2024**, *14*, 1544. [CrossRef]
6. Fagerholm, N.; Eilola, S.; Arki, V. Outdoor Recreation and Nature's Contribution to Well-Being in a Pandemic Situation—Case Turku, Finland. *Urban For. Urban Green.* **2021**, *64*, 127257. [CrossRef]
7. Liu, S.; Wang, X. Reexamine the Value of Urban Pocket Parks under the Impact of the COVID-19. *Urban For. Urban Green.* **2021**, *64*, 127294. [CrossRef]
8. Puhakka, R. University Students' Participation in Outdoor Recreation and the Perceived Well-Being Effects of Nature. *J. Outdoor Recreat. Tour.* **2021**, *36*, 100425. [CrossRef]
9. Wei, H.; Hauer, R.J.; Sun, Y.; Meng, L.; Guo, P. Emotional Perceptions of People Exposed to Green and Blue Spaces in Forest Parks of Cities at Rapid Urbanization Regions of East China. *Urban For. Urban Green.* **2022**, *78*, 127772. [CrossRef]
10. Cheng, Y.; Zhang, J.; Wei, W.; Zhao, B. Effects of Urban Parks on Residents' Expressed Happiness before and during the COVID-19 Pandemic. *Landsc. Urban Plan.* **2021**, *212*, 104118. [CrossRef]
11. Chen, X.; Kang, J. Natural Sounds Can Encourage Social Interactions in Urban Parks. *Landsc. Urban Plan.* **2023**, *239*, 104870. [CrossRef]
12. Bi, X.; Gan, X.; Jiang, Z.; Li, Z.; Li, J. How Do Landscape Patterns in Urban Parks Affect Multiple Cultural Ecosystem Services Perceived by Residents? *Sci. Total Environ.* **2024**, *946*, 174255. [CrossRef] [PubMed]
13. Himschoot, E.A.; Crump, M.C.; Buckley, S.; Cai, C.; Lawson, S.; White, J.; Beeco, A.; Taff, B.D.; Newman, P. Feelings of Safety for Visitors Recreating Outdoors at Night in Different Artificial Lighting Conditions. *J. Environ. Psychol.* **2024**, *97*, 102374. [CrossRef]
14. Xie, C.; Zhao, M.; Li, Y.; Tang, T.; Meng, Z.; Ding, Y. Evaluating the Effectiveness of Environmental Interpretation in National Parks Based on Visitors' Spatiotemporal Behavior and Emotional Experience: A Case Study of Pudacuo National Park, China. *Sustainability* **2023**, *15*, 8027. [CrossRef]

15. Tang, Z.; Zhao, Y.; Fu, M.; Wang, Y.; Xue, J. Which Factors Influence Public Perceptions of Urban Attractions?—A Comparative Study. *Ecol. Indic.* **2023**, *154*, 110541. [CrossRef]
16. Huai, S.; Van De Voorde, T. Which Environmental Features Contribute to Positive and Negative Perceptions of Urban Parks? A Cross-Cultural Comparison Using Online Reviews and Natural Language Processing Methods. *Landsc. Urban Plan.* **2022**, *218*, 104307. [CrossRef]
17. Yao, W.; Yun, J.; Zhang, Y.; Meng, T.; Mu, Z. Usage Behavior and Health Benefit Perception of Youth in Urban Parks: A Case Study from Qingdao, China. *Front. Public Health* **2022**, *10*, 923671. [CrossRef]
18. Subiza-Pérez, M.; Hauru, K.; Korpela, K.; Haapala, A.; Lehvävirta, S. Perceived Environmental Aesthetic Qualities Scale (PEAQS)—A Self-Report Tool for the Evaluation of Green-Blue Spaces. *Urban For. Urban Green.* **2019**, *43*, 126383. [CrossRef]
19. Rivera, E.; Timperio, A.; Loh, V.H.Y.; Deforche, B.; Veitch, J. Critical Factors Influencing Adolescents' Active and Social Park Use: A Qualitative Study Using Walk-along Interviews. *Urban For. Urban Green.* **2021**, *58*, 126948. [CrossRef]
20. Yoon, J.I.; Lim, S.; Kim, M.-L.; Joo, J. The Relationship between Perceived Restorativeness and Place Attachment for Hikers at Jeju Gotjawal Provincial Park in South Korea: The Moderating Effect of Environmental Sensitivity. *Front. Psychol.* **2023**, *14*, 1201112. [CrossRef]
21. Mak, B.K.L.; Jim, C.Y. Contributions of Human and Environmental Factors to Concerns of Personal Safety and Crime in Urban Parks. *Secur. J.* **2022**, *35*, 263–293. [CrossRef]
22. Chitra, B.; Jain, M.; Chundelli, F.A. Understanding the Soundscape Environment of an Urban Park through Landscape Elements. *Environ. Technol. Innov.* **2020**, *19*, 100998. [CrossRef]
23. Powers, S.L.; Mowen, A.J.; Webster, N. Development and Validation of a Scale Measuring Public Perceptions of Racial Environmental Justice in Parks. *J. Leis. Res.* **2024**, *55*, 1–24. [CrossRef]
24. Zhao, J.; Abdul Aziz, F.; Song, M.; Zhang, H.; Ujang, N.; Xiao, Y.; Cheng, Z. Evaluating Visitor Usage and Safety Perception Experiences in National Forest Parks. *Land* **2024**, *13*, 1341. [CrossRef]
25. Gkoltsiou, A.; Paraskevopoulou, A. Landscape Character Assessment, Perception Surveys of Stakeholders and SWOT Analysis: A Holistic Approach to Historical Public Park Management. *J. Outdoor Recreat. Tour.* **2021**, *35*, 100418. [CrossRef]
26. Gosal, A.S.; Geijzendorffer, I.R.; Václavík, T.; Poulin, B.; Ziv, G. Using Social Media, Machine Learning and Natural Language Processing to Map Multiple Recreational Beneficiaries. *Ecosyst. Serv.* **2019**, *38*, 100958. [CrossRef]
27. Liu, W.; Hu, X.; Song, Z.; Yuan, X. Identifying the Integrated Visual Characteristics of Greenway Landscape: A Focus on Human Perception. *Sustain. Cities Soc.* **2023**, *99*, 104937. [CrossRef]
28. Yang, C.; Zhang, Y. Public Emotions and Visual Perception of the East Coast Park in Singapore: A Deep Learning Method Using Social Media Data. *Urban For. Urban Green.* **2024**, *94*, 128285. [CrossRef]
29. Zhao, X.; Huang, H.; Lin, G.; Lu, Y. Exploring Temporal and Spatial Patterns and Nonlinear Driving Mechanism of Park Perceptions: A Multi-Source Big Data Study. *Sustain. Cities Soc.* **2025**, *119*, 106083. [CrossRef]
30. He, H.; Sun, R.; Li, J.; Li, W. Urban Landscape and Climate Affect Residents' Sentiments Based on Big Data. *Appl. Geogr.* **2023**, *152*, 102902. [CrossRef]
31. Wang, Z.; Miao, Y.; Xu, M.; Zhu, Z.; Qureshi, S.; Chang, Q. Revealing the Differences of Urban Parks' Services to Human Wellbeing Based upon Social Media Data. *Urban For. Urban Green.* **2021**, *63*, 127233. [CrossRef]
32. Li, J.; Fu, J.; Gao, J.; Zhou, R.; Wang, K.; Zhou, K. Effects of the Spatial Patterns of Urban Parks on Public Satisfaction: Evidence from Shanghai, China. *Landsc. Ecol.* **2023**, *38*, 1265–1277. [CrossRef] [PubMed]
33. Shang, Z.; Cheng, K.; Jian, Y.; Wang, Z. Comparison and Applicability Study of Analysis Methods for Social Media Text Data: Taking Perception of Urban Parks in Beijing as an Example. *Landsc. Archit. Front.* **2023**, *11*, 8. [CrossRef]
34. Huai, S.; Liu, S.; Zheng, T.; Van De Voorde, T. Are Social Media Data and Survey Data Consistent in Measuring Park Visitation, Park Satisfaction, and Their Influencing Factors? A Case Study in Shanghai. *Urban For. Urban Green.* **2023**, *81*, 127869. [CrossRef]
35. Kanahuati-Ceballos, M.; Valdivia, L.J. Detection of Depressive Comments on Social Media Using RNN, LSTM, and Random Forest: Comparison and Optimization. *Soc. Netw. Anal. Min.* **2024**, *14*, 44. [CrossRef]
36. Luo, J.; Zhao, T.; Cao, L.; Biljecki, F. Water View Imagery: Perception and Evaluation of Urban Waterscapes Worldwide. *Ecol. Indic.* **2022**, *145*, 109615. [CrossRef]
37. Yang, C.; Liu, T.; Zhang, S. Using Flickr Data to Understand Image of Urban Public Spaces with a Deep Learning Model: A Case Study of the Haihe River in Tianjin. *IJGI* **2022**, *11*, 497. [CrossRef]
38. Zhang, K.; Chen, Y.; Li, C. Discovering the Tourists' Behaviors and Perceptions in a Tourism Destination by Analyzing Photos' Visual Content with a Computer Deep Learning Model: The Case of Beijing. *Tour. Manag.* **2019**, *75*, 595–608. [CrossRef]
39. Pereira, R.; Mendes, C.; Ribeiro, J.; Ribeiro, R.; Miragaia, R.; Rodrigues, N.; Costa, N.; Pereira, A. Systematic Review of Emotion Detection with Computer Vision and Deep Learning. *Sensors* **2024**, *24*, 3484. [CrossRef]
40. Chutia, T.; Baruah, N. A Review on Emotion Detection by Using Deep Learning Techniques. *Artif. Intell. Rev.* **2024**, *57*, 203. [CrossRef]

41. Naseem, U.; Razzak, I.; Musial, K.; Imran, M. Transformer Based Deep Intelligent Contextual Embedding for Twitter Sentiment Analysis. *Future Gener. Comput. Syst.* **2020**, *113*, 58–69. [CrossRef]
42. Hittawe, M.M.; Harrou, F.; Sun, Y.; Knio, O. Stacked Transformer Models for Enhanced Wind Speed Prediction in the Red Sea. In Proceedings of the 2024 IEEE 22nd International Conference on Industrial Informatics (INDIN), Beijing, China, 18–20 August 2024; pp. 1–7. [CrossRef]
43. Filali, H.; Riffi, J.; Boulealam, C.; Mahraz, M.A.; Tairi, H. Multimodal Emotional Classification Based on Meaningful Learning. *BDCC* **2022**, *6*, 95. [CrossRef]
44. Wang, Z.; Zhou, X.; Wang, W.; Liang, C. Emotion Recognition Using Multimodal Deep Learning in Multiple Psychophysiological Signals and Video. *Int. J. Mach. Learn. Cyber.* **2020**, *11*, 923–934. [CrossRef]
45. Gandhi, A.; Adhvaryu, K.; Poria, S.; Cambria, E.; Hussain, A. Multimodal Sentiment Analysis: A Systematic Review of History, Datasets, Multimodal Fusion Methods, Applications, Challenges and Future Directions. *Inf. Fusion.* **2023**, *91*, 424–444. [CrossRef]
46. Zhang, S.; Yang, Y.; Chen, C.; Zhang, X.; Leng, Q.; Zhao, X. Deep Learning-Based Multimodal Emotion Recognition from Audio, Visual, and Text Modalities: A Systematic Review of Recent Advancements and Future Prospects. *Expert Syst. Appl.* **2024**, *237*, 121692. [CrossRef]
47. Wang, Q.; Xu, H.; Wen, Z.; Liang, B.; Yang, M.; Qin, B.; Xu, R. Image-to-Text Conversion and Aspect-Oriented Filtration for Multimodal Aspect-Based Sentiment Analysis. *IEEE Trans. Affect. Comput.* **2024**, *15*, 1264–1278. [CrossRef]
48. Jiao, T.; Guo, C.; Feng, X.; Chen, Y.; Song, J. A Comprehensive Survey on Deep Learning Multi-Modal Fusion: Methods, Technologies and Applications. *CMC* **2024**, *80*, 1–35. [CrossRef]
49. Devlin, J.; Chang, M.-W.; Lee, K.; Toutanova, K. BERT: Pre-Training of Deep Bidirectional Transformers for Language Understanding. In Proceedings of the 2019 Association for Computational Linguistics North American Chapter Conference: Human Language Technologies (NAACL HLT 2019), Minneapolis, Minnesota, 2–7 June 2019; Volume 1, pp. 4171–4186.
50. Liu, Z.; Lin, Y.; Cao, Y.; Hu, H.; Wei, Y.; Zhang, Z.; Lin, S.; Guo, B. Swin Transformer: Hierarchical Vision Transformer Using Shifted Windows. In Proceedings of the 2021 IEEE/CVF International Conference on Computer Vision (ICCV), Montreal, QC, Canada, 10–17 October 2021; pp. 9992–10002. [CrossRef]
51. Li, K.; Zhang, Y.; Li, K.; Li, Y.; Fu, Y. Image-Text Embedding Learning via Visual and Textual Semantic Reasoning. *IEEE Trans. Pattern Anal. Mach. Intell.* **2023**, *45*, 641–656. [CrossRef]
52. Stateczny, A.; Praveena, H.D.; Krishnappa, R.H.; Chythanya, K.R.; Babysarojam, B.B. Optimized Deep Learning Model for Flood Detection Using Satellite Images. *Remote Sens.* **2023**, *15*, 5037. [CrossRef]
53. Hong, X.; Huang, Z.; Wang, G.; Liu, J. Long-Term Perceptual Soundscape Modeling of Urban Parks: A Case Study of Three Urban Parks in Vancouver, Canada. *Landsc. Archit.* **2022**, *29*, 86–91. [CrossRef]
54. Ren, W.; Zhan, K.; Chen, Z.; Hong, X.-C. Research on Landscape Perception of Urban Parks Based on User-Generated Data. *Buildings* **2024**, *14*, 2776. [CrossRef]
55. Lundberg, S.M.; Lee, S.-I. A Unified Approach to Interpreting Model Predictions. In Proceedings of the Advances in Neural Information Processing Systems 30 (NIPS 2017), Long Beach, CA, USA, 4–9 December 2017; Volume 30.
56. Lu, S.; Liu, M.; Yin, L.; Yin, Z.; Liu, X.; Zheng, W. The Multi-Modal Fusion in Visual Question Answering: A Review of Attention Mechanisms. *PeerJ Comput. Sci.* **2023**, *9*, e1400. [CrossRef] [PubMed]
57. Lewicka, M. Place Attachment: How Far Have We Come in the Last 40 Years? *J. Environ. Psychol.* **2011**, *31*, 207–230. [CrossRef]
58. Choo, C.M.; Bai, S.; Privitera, A.J.; Chen, S.-H.A. Brain Imaging Studies of Multisensory Integration in Emotion Perception: A Scoping Review. *Neurosci. Biobehav. Rev.* **2025**, *172*, 106118. [CrossRef] [PubMed]
59. Rodriguez, M.; Kross, E. Sensory Emotion Regulation. *Trends Cogn. Sci.* **2023**, *27*, 379–390. [CrossRef]
60. Xu, Z.; Georgiadis, T.; Cremonini, L.; Marini, S.; Toselli, S. The Perceptions and Attitudes of Residents Towards Urban Green Spaces in Emilia-Romagna (Italy)—A Case Study. *Land* **2024**, *14*, 13. [CrossRef]
61. Zheng, T.; Yan, Y.; Lu, H.; Pan, Q.; Zhu, J.; Wang, C.; Zhang, W.; Rong, Y.; Zhan, Y. Visitors' Perception Based on Five Physical Senses on Ecosystem Services of Urban Parks from the Perspective of Landsenses Ecology. *Int. J. Sustain. Dev. World Ecol.* **2020**, *27*, 214–223. [CrossRef]
62. Douglas, J.W.A.; Evans, K.L. An Experimental Test of the Impact of Avian Diversity on Attentional Benefits and Enjoyment of People Experiencing Urban Green-space. *People Nat.* **2022**, *4*, 243–259. [CrossRef]
63. Wang, H.; Zhu, L.; Zhao, C.; Zheng, S. Urban Ecological Risk Assessment Management Platform. *Int. J. Sustain. Dev. World Ecol.* **2018**, *25*, 477–482. [CrossRef]
64. Tang, L.; Wang, L.; Li, Q.; Zhao, J. A Framework Designation for the Assessment of Urban Ecological Risks. *Int. J. Sustain. Dev. World Ecol.* **2018**, *25*, 387–395. [CrossRef]
65. Chen, J.; Konijnendijk Van Den Bosch, C.C.; Lin, C.; Liu, F.; Huang, Y.; Huang, Q.; Wang, M.; Zhou, Q.; Dong, J. Effects of Personality, Health and Mood on Satisfaction and Quality Perception of Urban Mountain Parks. *Urban For. Urban Green.* **2021**, *63*, 127210. [CrossRef]

66. Dong, J.; Guo, R.; Guo, F.; Guo, X.; Zhang, Z. Pocket Parks-a Systematic Literature Review. *Environ. Res. Lett.* **2023**, *18*, 083003. [CrossRef]
67. Wang, X.; Tang, L.R.; Kim, E. More than Words: Do Emotional Content and Linguistic Style Matching Matter on Restaurant Review Helpfulness? *Int. J. Hosp. Manag.* **2019**, *77*, 438–447. [CrossRef]
68. Wu, J.-L.; Chung, W.-Y. Sentiment-Based Masked Language Modeling for Improving Sentence-Level Valence–Arousal Prediction. *Appl. Intell.* **2022**, *52*, 16353–16369. [CrossRef]

buildings

MDPI

Review

Machine Learning for Resilient and Sustainable Cities: A Bibliometric Analysis of Smart Urban Technologies

Bin Luan [1,2] and Xinqun Feng [1,*]

[1] College of Fashion and Design, Donghua University, Shanghai 200051, China; luanbin@ldy.edu.rs

[2] School of Architecture and Landscape Architecture, Shandong University of Art & Design, Jinan 250014, China

* Correspondence: fengxinqun@ldy.edu.rs; Tel.: +86-13901700591

Abstract: With the acceleration of urbanization, the construction of smart cities has become a global focal point, with machine learning technology playing a crucial role in this process. This study aims to conduct a bibliometric analysis of the published research in the fields of smart cities and machine learning, using visualization techniques to reveal the spatiotemporal distribution patterns, research hotspots, and collaborative network structures. The goal is to provide systematic references for academic research and technological innovation in related fields. The results indicate that the development of this field exhibits distinct phases and regional characteristics. From a temporal perspective, research has undergone three stages: initial development, rapid growth, and stable consolidation, with the period from 2017 to 2021 marking a critical phase of rapid expansion. In terms of spatial distribution, countries such as China and the United States are at the forefront of this field, whereas regions like Africa and South America have a relatively low research output due to constraints in research resources and technological infrastructure. A hotspot analysis revealed that research topics are increasingly diverse and dynamically evolving. Issues such as data privacy, cybersecurity, sustainable development, and intelligent transportation have gradually become focal points, reflecting the dual demand of smart city development for technological innovation and green growth. Furthermore, collaboration network analysis indicates that international academic cooperation is becoming increasingly close, with research institutions in China, the United States, and Europe playing a central role in the global collaboration system, thereby promoting technology sharing and interdisciplinary integration. Through a systematic bibliometric analysis, this study identifies key application directions and future development trends in the research on smart cities and machine learning, providing valuable insights for academic research and technological advancements in related fields.

Keywords: machine learning; smart cities; bibliometrics; Scimago; VOSviewer

1. Introduction

In recent years, the process of global urbanization has been accelerating. According to data from the United Nations, by 2050, it is expected that 68% of the world's population will live in cities [1]. This trend has led to a dramatic increase in the complexity of urban structures, manifesting in multiple issues such as traffic congestion [2], environmental pollution [3], energy shortages [4], and increased public safety risks [5], all of which severely affect the quality of life for urban residents and the sustainable development of cities [6]. These challenges not only impact the quality of life for city dwellers but also impose new demands on urban managers [7]. To effectively address these issues, countries around the world are advancing smart city construction, aiming to achieve more efficient and sustainable urban management through technological innovation [8]. In the process of building smart cities, emerging technologies such as the Internet of Things (IoT) [9], big data [10], cloud computing [11], and 5G have played important roles [12,13]. However, with the explosive growth of urban data, establishing how to extract valuable information from vast amounts of data and apply it to practical urban management decisions has become a major challenge in the construction of smart cities.

Machine learning, as one of the core technologies of artificial intelligence, is playing an increasingly important role in the construction of smart cities [14]. It is a method that learns and optimizes algorithmic models through data, enabling the automatic discovery of hidden patterns and trends from vast amounts of urban data, which can then be used to predict future behaviors and trends [15]. Compared to traditional urban management and service methods, machine learning-based smart city applications offer advantages such as real-time responsiveness, dynamic adaptability, and personalization, allowing for more efficient, precise, and flexible responses to the complexity and uncertainty of urban environments [16]. Many developed countries and regions are actively promoting the application of machine learning across various domains of smart cities, achieving remarkable results. For instance, New York City in the United States uses machine learning to analyze traffic flow data and optimize traffic signal timing, reducing travel time by 10% [17]. Singapore employs computer vision and deep learning technologies to enable the real-time monitoring and analysis of urban environments, enhancing the efficiency and accuracy of city management [18]. In 2016, the Dubai government, in collaboration with the Dubai Future Foundation and Smart Dubai, launched the Dubai Blockchain Strategy, aiming to widely adopt blockchain technology in government operations to improve efficiency, transparency, and security. Tokyo, Japan, utilizes machine learning algorithms to process massive energy consumption data, achieving the precise prediction and optimized allocation of urban energy demand, resulting in a 20% reduction in energy consumption [19]. These case studies not only demonstrate the potential of machine learning in smart cities but also reflect the urgent global demand for intelligent management solutions in the process of urbanization.

Meanwhile, Beatley proposed the concept of "biophilic cities", emphasizing that cities should closely integrate nature with urban life through design, policies, and community engagement [20]. In recent years, the field of urban planning and construction has gradually placed greater emphasis on the concept of "Biophilic Cities", which involves deeply integrating natural elements into urban spaces to improve residents' health, enhance social resilience, and promote urban sustainability [21]. Some scholars study how to achieve the goals of biophilic cities through green infrastructure, ecological design, and community participation. The development of smart cities is not isolated from the concept of biophilic cities; rather, the two can form a complementary relationship. Machine learning technologies can play a critical role in this integration process. Machine learning algorithms can optimize the distribution of urban green spaces and increase vegetation coverage, thereby improving air quality and regulating urban microclimates. Ecological monitoring systems based on sensor networks and computer vision technologies can track and analyze urban biodiversity in real time, providing essential support for formulating scientifically sound ecological conservation policies. Additionally, intelligent water resource management systems can utilize machine learning to predict precipitation patterns and optimize rainwater collection and reuse strategies, thereby enhancing urban ecological adaptability. Through these technological means, smart city construction can not only improve urban operational efficiency but also strengthen urban resilience—the ability of cities to cope with natural disasters, environmental changes, and socio-economic shocks. The rise of urban resilience can enhance residents' subjective well-being by increasing their income or consumption, improving their mental and physical health, and boosting their social trust or social integration [22].

However, despite the enormous potential of machine learning in smart city applications, the challenges it faces cannot be ignored. First, smart city systems are typically highly heterogeneous and dynamic, involving multi-source, multi-modal data from various fields, making it a key challenge to handle the complexity and uncertainty of such data [23]. Second, data privacy and security issues are particularly prominent in the context of smart cities, especially when it comes to public safety and personal privacy [24]. Ensuring the security and legal compliance of data has become an important consideration in the application of technology. In addition, the issues of explainability and fairness in machine learning models have also posed new requirements for the construction of smart cities [25]. Urban management decisions often involve multiple stakeholders, and the black-box nature of some models may lead to non-transparent and unfair decision making, further exacerbating social conflicts [26]. Therefore, how to effectively apply machine learning technology in the complex environment of smart cities while ensuring its transparency, fairness, and security is a key challenge that academia and industry urgently need to resolve.

Given that the application of machine learning in smart cities has achieved preliminary results and demonstrated significant potential in urban management, public service optimization, and resource allocation, it is necessary to systematically review and analyze the current state of research in this field. However, existing studies often focus on specific application scenarios for the optimization of individual technologies, lacking a comprehensive examination of overall development trends. Additionally, discussions on issues such as model transparency, data security, fairness assurance, and cross-domain collaborative applications remain insufficient. Therefore, to systematically review the research progress on machine learning in smart cities, identify the developmental trajectory of key technologies and emerging critical issues, and reveal possible future research directions, conducting bibliometric analyses is particularly important. A bibliometric analysis, based on large-scale academic research data, can quantitatively uncover the research scale, thematic evolution, core research forces, and frontier trends in this field, providing a scientific foundation for subsequent studies and valuable references for policymakers and practitioners.

Therefore, this study focuses on the application prospects of machine learning technology in smart cities, highlighting the theoretical and practical significance of bibliometric research. The field is currently experiencing rapid growth, but the highly fragmented knowledge system and its interdisciplinary nature have led to three significant challenges. First, a vast number of research findings are dispersed across multiple disciplines, such as computer science, urban planning, and public administration, lacking the systematic integration of core issues, evolutionary pathways, and paradigm shifts. Second, the dynamic adaptation mechanism between the rapid iteration of technology and the evolving demands of urban governance remains unclear, resulting in a significant knowledge gap between fundamental theoretical research and technological applications. Third, the global distribution of research efforts is uneven, and the efficiency of constructing cross-institutional and cross-regional collaborative innovation networks needs to be improved. Using bibliometric methods, this study aims to construct a knowledge map to reveal the migration patterns of research hotspots, identify key academic communities and their collaboration models, and provide empirical support for overcoming technological application bottlenecks. This approach not only helps optimize resource allocation and facilitates interdisciplinary knowledge integration but also provides a scientific foundation for policymakers to establish technology risk assessment frameworks and promote a data-driven paradigm shift in urban governance. Ultimately, it contributes to the realization of sustainable smart city development goals. This research perspective offers important insights for both academic inquiry and practical applications, fostering the responsible use of machine learning in smart cities and cultivating a more intelligent, equitable, and trustworthy human–machine collaboration ecosystem.

The remaining sections of this study are as follows: Section 2 is on the research methodology and data sources; Section 3 is on the temporal and spatial analyses; Section 4 is on the hotspot analyses; Section 5 is on the collaborative analyses; Section 6 is on the results and the discussion; and Section 7 is on the conclusions, which contain the limitations and outlook.

2. Research Methodology and Data Sources

2.1. Data-Screening Process

The first step of this study involved selecting reference journal articles from the Web of Science (WoS) database to create a unified analytical database. The WoS is widely recognized as a primary source of authoritative and representative citation data, with its indexed journal articles representing high-quality achievements in international social science research [27]. Additionally, it enables more comprehensive citation analyses. The data for this study were sourced from five sub-databases within the Web of Science Core Collection: the Science Citation Index Expanded (SCI-EXPANDED), the Social Sciences Citation Index (SSCI), the Conference Proceedings Citation Index—Science (CPCI-S), Current Chemical Reactions (CCR-EXPANDED), and the Index Chemicus (IC). The search query was set to "Topic = ('machine learning') AND Topic = ('smart city')", with the retrieval period extending until 31 December 2024. A total of 2326 published records were generated and saved in the Plain Text file format.

In the second step, this study strictly adhered to the standardized framework of the PRISMA statement (Preferred Reporting Items for Systematic Reviews and Meta-Analyses) and follows the process guidelines of PRISMA-ScR (an extension for Scoping Reviews) to ensure transparency and reproducibility in the literature-screening process. Given that this study aims to systematically map the knowledge evolution and structure of the interdisciplinary field of "machine learning + smart cities", the scoping review method was employed to conduct a comprehensive scan of the existing literature, rather than focusing on specific intervention effects, as in a systematic review. Specifically, starting from the initial 2326 records retrieved from the WoS database, automated deduplication was performed using the EndNote X9 reference management software. Subsequently, two researchers independently conducted a double-blind screening of the titles and abstracts, excluding non-research studies, such as book reviews and conference announcements. Discrepancies were resolved through group discussions, resulting in a final analytical sample of 828 valid articles. The entire screening decision process was visualized using a PRISMA flow diagram (Figure 1), strictly following the PRISMA workflow, which includes three stages: Identification, screening, and inclusion, to enhance the systematicity and reliability of the study.

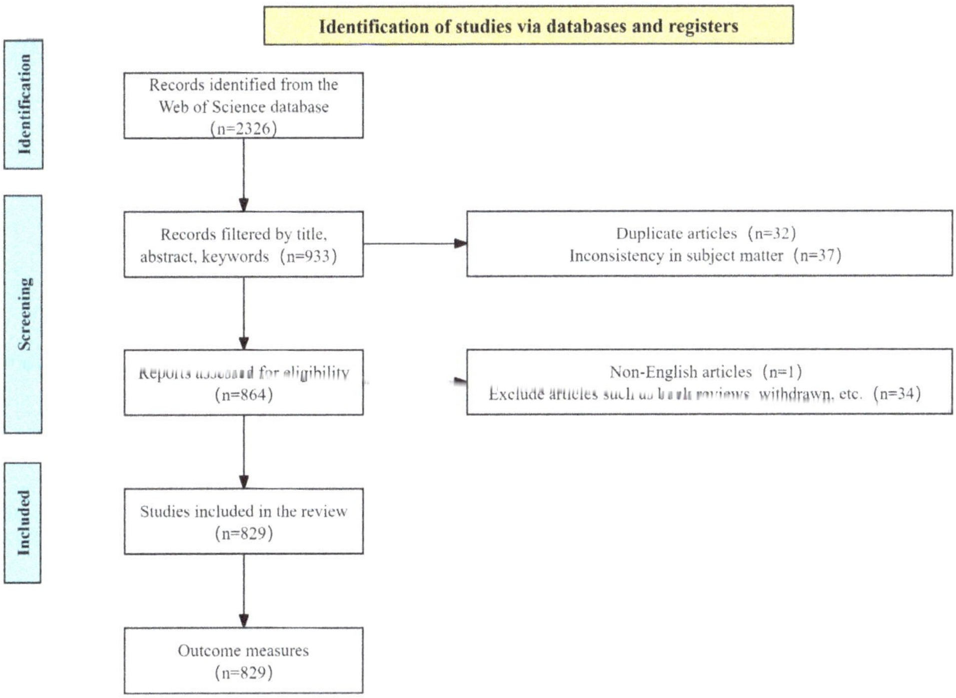

Figure 1. PRISMA flowchart.

2.2. Inclusion and Exclusion Criteria

This section helps readers reproduce the selection process and explains why only a limited number of papers met the selection criteria. Additionally, the following inclusion and exclusion criteria were used to filter the relevant papers:

- Inclusion criteria

Research topic: the studies must involve both machine learning and smart cities, with a direct connection between the two.

Source of articles: the papers must be published in one of the five sub-databases of the Web of Science (WoS) Core Collection, including SCI-EXPANDED, SSCI, CPCI-S, CCR-EXPANDED, and IC.

Language: to ensure accessibility and comprehension, only the papers written in English were included.

Full-text access: only the articles with full-text availability were included to allow for an in-depth analysis.

Publication date: the papers must have been published before 31 December 2024, to maintain the timeliness of the study.

- Exclusion criteria

Research topic: the studies that focused solely on either machine learning or smart cities, without addressing their intersection, were excluded.

Not peer-reviewed: the non-scientific publications, such as grey publications and opinion papers, which are commonly found in non-peer-reviewed sources, were excluded.

Language: the papers that were not published in English were excluded to ensure international readability and comparability.

Duplicate studies: the redundant studies were excluded to avoid data duplication.

Inaccessible papers: the papers that could not be fully accessed were not considered for the final selection.

After applying the above screening criteria, a total of 829 high-quality research papers were included, providing a solid data foundation for the subsequent analysis.

2.3. Research Methods

The Scimago Graphica 1.0.49 method is a commonly used scientific evaluation approach in bibliometric analyses, primarily relying on the Scimago Journal and Country Rank (SJR) and the Scopus database. It systematically analyzes academic publications, journals' impact, and research trends by integrating multiple bibliometric indicators. This method is widely applied in scientometrics and knowledge mapping research, effectively revealing the academic influence, research hotspots, and developmental trends of a particular field. In the research domain of smart cities and machine learning, a bibliometric analysis using the Scimago method can help researchers identify high-impact journals, core research institutions, key research topics, and their evolutionary trends, thereby providing data support and scientific evidence for future studies. Furthermore, this method can also uncover international collaboration patterns and the research contributions of different countries and institutions, offering insights that can be utilized for the coordinated development of global smart city construction and intelligent technology applications.

CiteSpace 6.3.R1 is a widely used software tool for scientometric and visual analyses, developed by Chaomei Chen. Based on co-citation analysis, it visually presents the structure, evolution, and dynamics of knowledge domains [28]. This study employed CiteSpace to create knowledge maps, identify research frontiers and breakthroughs in the field, and reveal the evolutionary trajectory of research topics. Through features such as time zone views, cluster views, and timeline views in CiteSpace, the developmental context and key nodes of research topics can be intuitively displayed, aiding in understanding the global structure and trends of machine learning applications in smart cities.

VOSviewer 1.6.20 is a text mining and visual analysis software developed by Nees Jan van Eck and Ludo Waltman, utilizing VOS (Visualization of Similarities) technology to generate network structure diagrams based on similarity calculations between nodes [29]. This study used VOSviewer for a keyword co-occurrence analysis and a cluster analysis to identify the semantic structure and knowledge base of research topics, revealing the connections and distinctions between different themes. Through features such as density visualization and overlay visualization in VOSviewer, the distribution characteristics and evolutionary trends of machine learning applications in smart cities can be systematically and intuitively revealed.

Bibliometrix is a bibliometric analysis toolkit written in R, developed by Aria and Cuccurullo. It provides a series of functions and commands for the descriptive analysis, collaboration network analysis, and thematic co-occurrence analysis of research data [30]. This study employed Bibliometrix for an author collaboration network analysis, revealing the distribution of research strengths and collaboration patterns in the field and identifying core authors and research institutions. Through features such as collaboration network maps and country collaboration maps in Bibliometrix, the collaborative relationships and centrality among authors and institutions in the field of machine learning applications in smart cities can be clearly displayed.

In summary, to comprehensively analyze the research status and trends of machine learning applications in smart cities, this study adopted a quantitative bibliometric analysis method, integrating tools, such as Scimago, CiteSpace, VOSviewer, and Bibliometrix, to conduct systematic research from multiple dimensions. The specific research process is illustrated in Figure 2. First, the time-slicing technique of Scimago was employed to observe the temporal evolution trends of research topics, and Bibliometrix was used to identify high-impact journals, authors, and institutions, thereby clarifying the core literature and foundational research in the field. Second, VOSviewer and CiteSpace were utilized for study clustering and a burst analysis to identify research frontiers and trend shifts at different stages. Third, Scimago was used to map collaboration networks among authors, institutions, and countries, analyzing academic collaboration patterns. Finally, the research results from various dimensions were synthesized to summarize the knowledge system of the field, predict future development trends, and explore emerging technological directions.

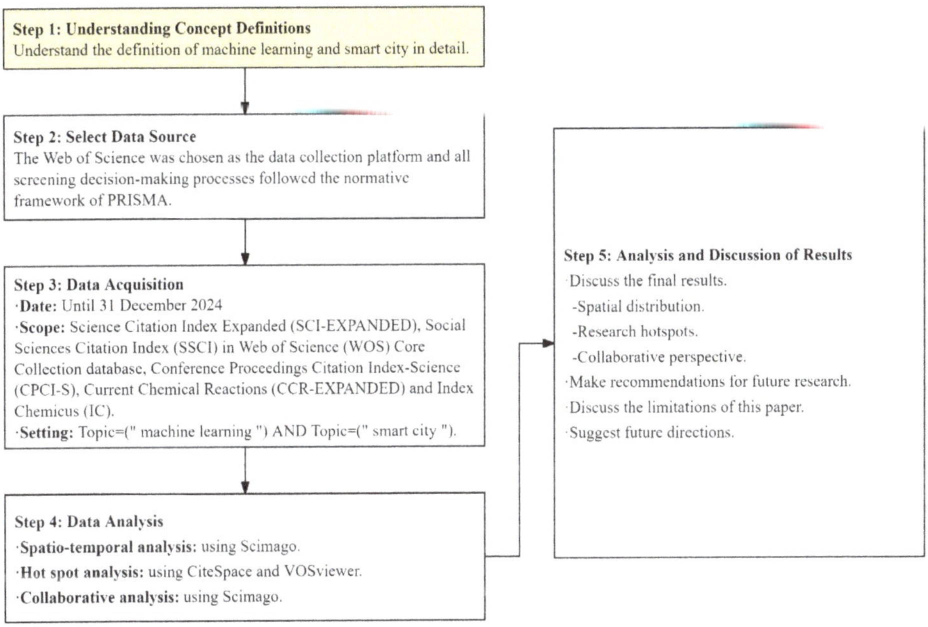

Figure 2. Research flowchart.

3. Analysis of the Overall Research Situation

3.1. Analysis of Publication Quantity and Trends

As shown in Figure 2, based on the trend in the publication volume from 2006 to 2024, the research process can be divided into three distinct periods, each characterized by an increase in the output and notable indicators.

- Phase 1: initial development stage (2014–2017).

During this phase, the number of publications was relatively low and grew slowly, with annual publications consistently remaining below 50. The number increased gradually from 2 in 2014 to 24 in 2017. This indicates that the application of machine learning in smart cities was still in the exploratory stage, with limited attention from the academic community. Research primarily focused on theoretical discussions and preliminary technical attempts, with few practical applications, and it had not yet garnered widespread attention. In the early 2010s, machine learning technologies were not yet mature, and their applications in smart cities remained experimental, with research mainly concentrating on exploring fundamental algorithms. Additionally, the governmental and corporate interest in smart city development had not yet peaked, resulting in limited investment and policy support.

- Phase 2: rapid growth stage (2017–2021).

Starting in 2017, the number of publications entered a period of rapid growth, reaching a peak of 147 in 2022. With advancements in deep learning, the capabilities of machine learning in urban data analyses significantly improved, driving explosive growth in smart city-related research during this phase. Research directions gradually expanded to include key areas such as traffic optimization [31], energy management [32], environmental monitoring [33] and urban planning [34]. Concurrently, governments worldwide introduced smart city development strategies, such as the EU's "Digital Europe Programme" and China's "Guidelines for New Smart City Construction", coupled with increased investment from governments and enterprises, further propelling research in this field.

- Phase 3: stabilization stage (2022–present).

After 2022, the number of publications fluctuated but remained at a relatively high level. Following the peak of 147 in 2022, the number slightly decreased to 127 in 2023 but rebounded to 136 in 2024. This suggests that research in this field matured. With the maturation of foundational technologies, the research focus gradually shifted from theoretical exploration to the optimization of specific applications, such as intelligent traffic management, urban energy scheduling, and smart security systems. At the same time, there was an in-depth exploration of issues such as technological ethics, data privacy, and model interpretability [25].

Looking at the trend in cumulative postings in Figure 3, research was limited before 2014 but entered a period of rapid development after 2014, peaking in 2022. Overall, research on machine learning in smart cities underwent a process of initial exploration $\rightarrow$ rapid growth $\rightarrow$ stable development. In the future, this field may further deepen in areas such as intelligent governance, cross-domain integration, and sustainable urban development, aiming to enhance the intelligence level and practical application value of smart cities.

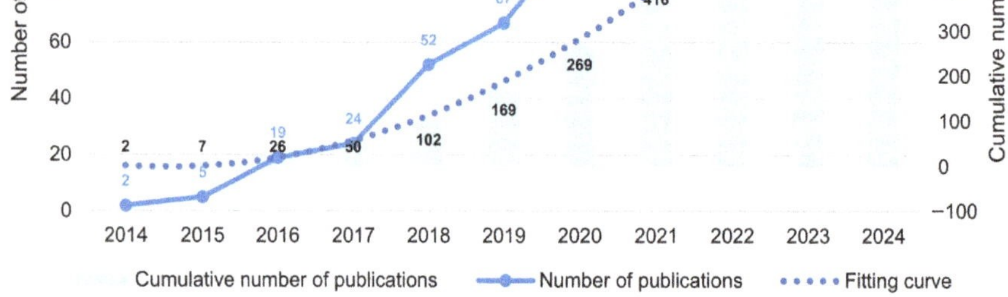

Figure 3. Trends in the issuance of communications by year (created by the author, based on 829 publications).

3.2. High-Yield Country Information

Figure 4 illustrates the global distribution of scientific research achievements related to machine learning and smart cities. The number of publications from each country is represented by varying shades of blue, with darker colors indicating a higher output of related research articles. As shown in Figure 3, China and the United States are the two countries with the highest number of publications globally, with the darkest colors, indicating their dominant positions in research on machine learning and smart cities. This dominance may stem from the long-term, large-scale funding support provided by the Chinese and U.S. governments for artificial intelligence (AI) and smart city research. For instance, the U.S. National Science Foundation (NSF) and China's National Natural Science Foundation (NSFC) have established numerous specialized funding programs. Additionally, both countries possess world-leading cloud computing platforms (e.g., Alibaba Cloud, AWS) and smart city pilot projects, while many developing countries still lack high-performance computing centers, extensive data collection systems, and infrastructure such as smart transportation and power grids, which hinder related research. Furthermore, several European countries (e.g., Germany, the United Kingdom, France) also stand out, with darker shades indicating a high scientific output.

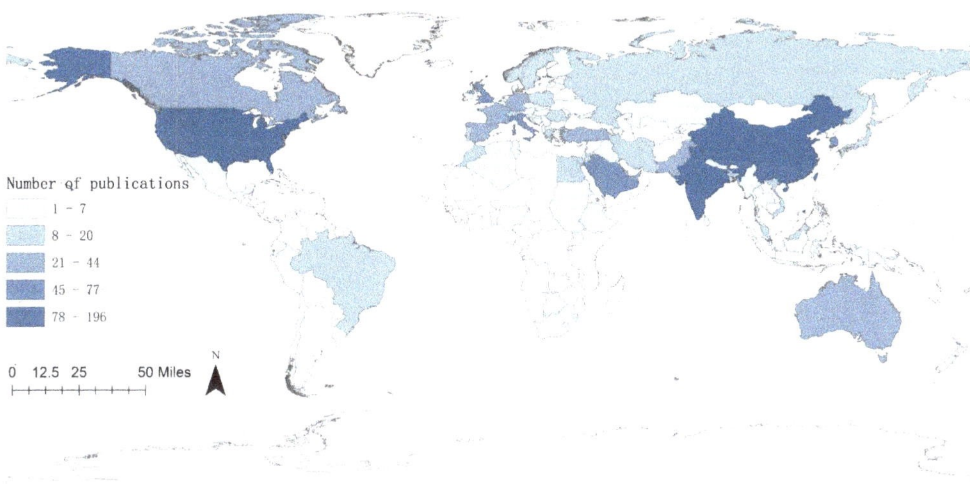

Figure 4. Volume of communications by country (created by the author, based on 829 publications).

However, the lighter colors in South America, Africa, and parts of Asia suggest relatively lower research outputs in these regions. This disparity is not merely a matter of research interest or capability but stems from multiple systemic barriers. First, funding gaps are a critical factor. Developing and underdeveloped countries often face severe shortages in research funding, limiting their investments in high-tech fields [35]. Second, infrastructure limitations pose another significant obstacle. Many African and South American countries lack advanced research facilities, high-speed internet, and a stable electricity supply, all of which are essential for conducting research into machine learning and smart cities. Moreover, the United States boasts world-leading universities and research institutions that attract scholars globally, while China has strengthened talent recruitment in recent years through policies such as the "Thousand Talents Plan". In contrast, higher education systems in Africa and South America have fewer resources for scientific training, and many talented scholars migrate to Europe and the United States due to better salaries or research conditions, further weakening local research capabilities [36].

This map provides a global perspective, highlighting the disparities in AI and smart city development across different regions. Overall, developed countries, particularly those with advanced economies and technological capabilities, are significantly ahead in this field. As technological innovation plays an increasingly vital role in driving economic and social development, strengthening international cooperation and promoting the sharing of knowledge and experience will help narrow the technological gap between nations.

3.3. *High-Yield Journal Information*

As shown in Figure 4, *IEEE Access* ranks first, with 136 publications, making it one of the most influential journals in this field and highlighting its significant role in research related to smart cities and machine learning. *Sensors* follows closely, with 114 publications, indicating that sensor technology also holds a critical position in smart city research, likely due to its close connection with technologies such as the Internet of Things (IoT). The journal *Sustainability* has 69 publications, demonstrating that research in this field is closely tied to sustainable development. The construction of smart cities involves not only technology but also emphasizes the sustainable use of the environment and resources. *Sustainable Cities and Society* (60 publications) and *Applied Sciences-Basel* (56 publications) reflect the broad links between smart city development and society, as well as scientific applications, indicating a trend toward interdisciplinary research. The *IEEE Internet of Things Journal* also published 56 articles, underscoring the importance of the IoT in smart cities, as IoT technology is one of the core technologies in this domain. *Electronics* (39 publications) and *Energies* (25 publications) reflect the research interest in and applications of electronic technology and energy systems in smart cities. Additionally, *Future Generation Computer Systems-The International Journal* and *Multimedia Tools and Applications* have 25 and 19 publications, respectively, indicating the potential applications of future computing systems and multimedia tools in smart cities.

As shown in Figure 5, *IEEE Access* ranks first, with 45 publications, making it one of the most influential journals in this field, indicating its important position in research related to smart cities and machine learning [37]. *Sensors* follows closely, with 44 publications, demonstrating the significant role of sensor technology in smart cities, likely closely related to technologies such as the Internet of Things (IoT) [38]. The journal *Sustainability* has 29 publications, indicating that research in this field is closely linked to sustainable development, emphasizing that the construction of smart cities involves not only technology but also the sustainable use of the environment and resources [39]. The *IEEE Internet of Things Journal* (24 publications) [40] and *Applied Sciences-Basel* (22 publications) [41] show the broad connections between smart city construction and social and scientific applications, indicating a trend of multidisciplinary research. *Sustainable Cities and Society* also published 17 articles [42], underscoring the importance of the IoT in smart cities, with IoT technology being one of the core technologies of smart cities. *Electronics* (nine publications) [43], *Energies* (nine publications) [44], and *Sustainable Energy Technologies and Assessments* (nine publications) [45] reflect the applications of and research interest in electronics technology and energy systems in smart cities. Additionally, the publication volume of *CMC-Computers, Materials & Continua* (seven publications) [46] reflects the application of computer science and materials science in this field.

Overall, the distribution of these journals reflects the interdisciplinary nature of smart city research, encompassing multiple fields, such as the Internet of Things (IoT), machine learning, sustainable development, and energy management, showcasing the research landscape of smart cities as a comprehensive innovation product. Research trends indicate that technological exploration in smart cities is not limited to engineering and computer science but is also closely integrated with fields such as social sciences and environmental sciences. In the future, with the continuous advancement of the IoT, next-generation computing systems, and multimedia tools, the depth and breadth of smart city research will further expand, offering more possibilities for the intelligent and sustainable development of cities.

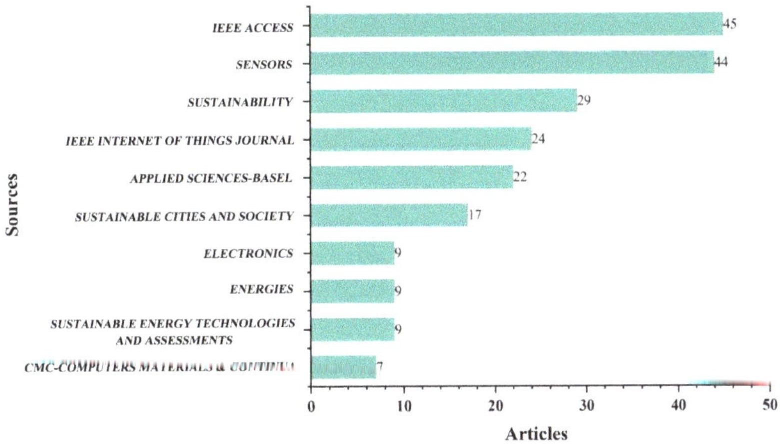

Figure 5. Top 10 journals in terms of publications (created by the author, based on 829 publications).

Figure 6 shows the top ten most frequently cited journals in the field of machine learning and smart cities. From the figure, it is evident that *IEEE Access* is the most cited journal, leading significantly with 1158 local citations. This large number of citations is likely due to its broad coverage and the high number of articles it publishes. Following this is *Sensors-Basel*, with 608 citations, indicating its important position in research within this field. The third-ranked, the *IEEE Internet of Things Journal*, has significantly fewer citations, with only 541. Additionally, *Sustainable Cities and Society* (390 citations), *Arxiv* (383 citations), and *Future Generation Computer Systems* (363 citations) are also on the list. Although their citation counts are not as high as the top journals, their content still holds considerable influence in the fields of smart cities and machine learning. From the distribution of citations, it can be seen that research outcomes in the field of machine learning and smart cities are concentrated in a few high-impact journals, reflecting the concentration and authority of research hotspots. Moreover, the research in this field is concentrated in a few high-impact journals, demonstrating the dissemination and cross-disciplinary influence of research outcomes across different academic fields.

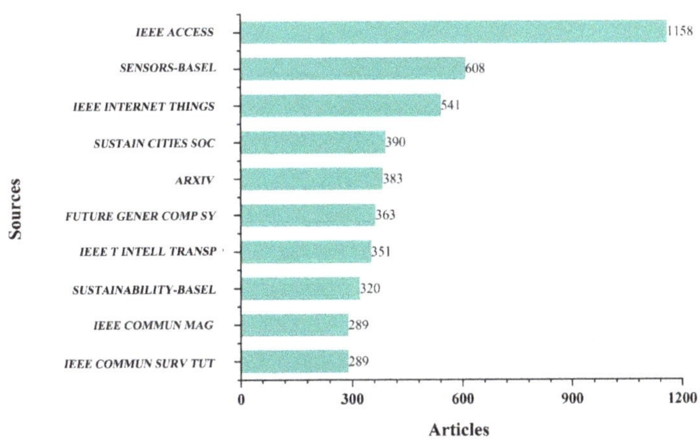

Figure 6. Top 10 cited journals (created by the author, based on 829 publications).

3.4. High-Yield Author Information

Figure 7 shows the top 10 authors with the most publications in the field of machine learning and smart cities, along with their specific number of publications. From the figure, it is evident that NESI P ranks first, with 10 articles. NESI P's research mainly focuses on data management and optimization in smart cities, including the intelligentization of urban infrastructure, traffic flow optimization, and energy management based on machine learning [47–49]. His research outcomes have played an important role in enhancing the operational efficiency of smart cities and promoting sustainable urban development. Following closely are KHAN MA and KIM K, both of whom have published eight articles. KHAN MA mainly focuses on the development of intelligent systems in smart cities, such as intelligent transportation, environmental monitoring, and urban safety systems, aiming to enhance urban governance through deep learning and data analysis [50–52]. KIM K focuses on the integration of big data analysis and smart city applications, using machine learning techniques to optimize urban resource scheduling and improve the quality of public services [53]. ABBAS S and KUMAR A have each published six articles. ABBAS S primarily researches the application of the Internet of Things (IoT) and machine learning in smart cities, including smart sensor networks, real-time data processing, and urban monitoring systems [54–57]. KUMAR A has contributed to network security and privacy protection in smart cities, focusing on the application of machine learning methods to ensure data security and prevent network threats [58,59]. Additionally, BELLINI P [60], BOUKERCHE A [61], CHEN X [62], and GUPTA S [63] have each published five articles. Their research covers several key areas of smart cities, such as the optimization of intelligent transportation systems, urban health monitoring, large-scale data processing, and AI-driven urban service innovation.

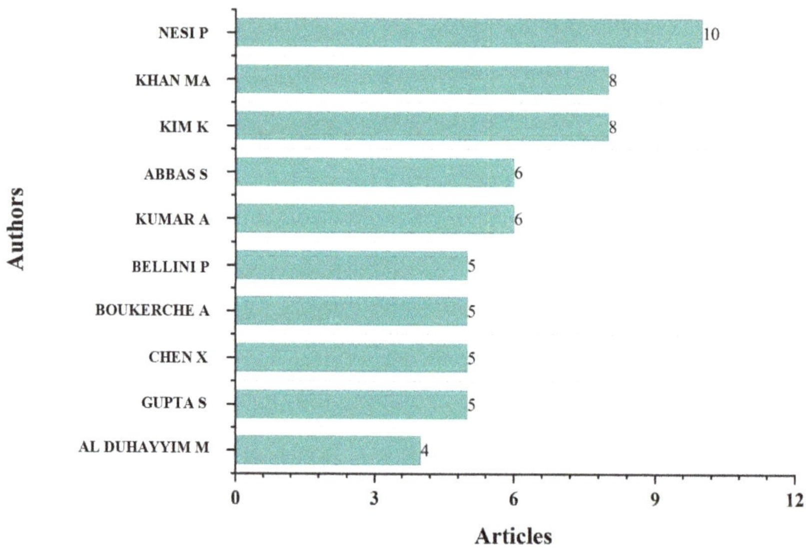

Figure 7. Top 10 authors in terms of publications (created by the author, based on 829 publications).

The research of these prolific scholars is concentrated on different aspects of smart city development, particularly in the application of machine learning. Their contributions provide a solid theoretical foundation and technical support for the development of core technologies in smart cities. Smart cities involve the real-time processing and analysis of large volumes of heterogeneous data, spanning areas such as transportation, energy, environment, and public services. The application of machine learning has significantly enhanced the intelligence of urban management. The studies conducted by these scholars not only address practical issues in smart cities, such as traffic congestion [64], energy consumption optimization [65], and environmental monitoring [66], but also offer innovative ideas for building smarter and more efficient cities in the future [67].

From Figure 8, it can be observed that there are significant differences in the scientific research output across different countries in this field, indicating a certain degree of imbalance. China leads by a large margin in the research output in this area, with its single-country publications (SCPs) far exceeding those of other countries, demonstrating the strong capability of Chinese scholars to conduct independent research domestically. At the same time, China has also engaged in a certain level of international collaboration (MCP), but compared to its substantial SCP volume, the proportion of international collaboration is relatively low. Following closely are India and the United States, whose research outputs are roughly comparable, though they still lag significantly behind China. India has a higher SCP count, indicating strong domestic independent research capabilities, while the United States has a relatively higher proportion of international collaboration, reflecting its more open academic cooperation model.

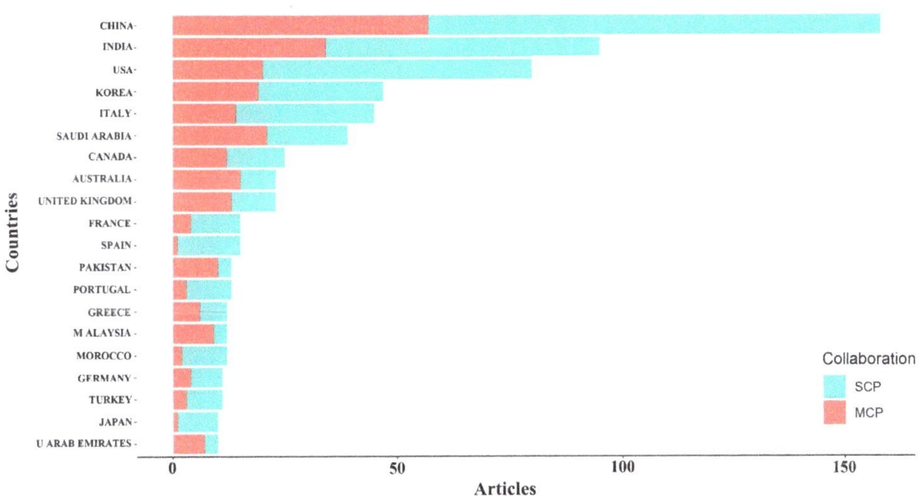

Figure 8. National distribution of corresponding authors (created by the author, based on 829 publications).

From a regional perspective, East Asian countries (such as China and South Korea) play a significant role in research in this field, with China's dominance being particularly notable. Meanwhile, Western European countries (such as Italy, the United Kingdom, France, and Spain) and North American countries (such as the United States and Canada) also contribute significantly to the research output, with higher MCP proportions, indicating their active participation in international collaboration. Additionally, countries in the Middle East and South Asia (such as Saudi Arabia, Pakistan, and the United Arab Emirates) have shown a certain level of research activity in this field, reflecting a growing regional interest in related studies. Other countries (such as Malaysia, Morocco, Greece, and Turkey) have relatively lower research outputs, highlighting the regional disparities in research resources and development in this field.

Overall, China has the highest proportion of single-country research, while countries such as the United States, Canada, and the United Kingdom exhibit higher proportions of international collaboration. This distribution pattern reflects the regional development characteristics of machine learning and smart city research, while also revealing the importance and diversity of international scientific collaboration.

3.5. High-Yield Institutions Information

As shown in Figure 9, in terms of publication volume, King Saud University ranks first, with 17 publications. It is followed by the National Institute of Technology (NIT System) and Wuhan University, both with 15 publications each. The Egyptian Knowledge Bank (EKB), the Indian Institute of Technology System (IIT System), King Abdulaziz University, and the Vellore Institute of Technology (VIT) have each contributed 13 papers, demonstrating notable performance. Additionally, the University of Florence and the University of Ottawa are tied, with 12 publications each, while the Chinese Academy of Sciences follows closely, with 11 publications.

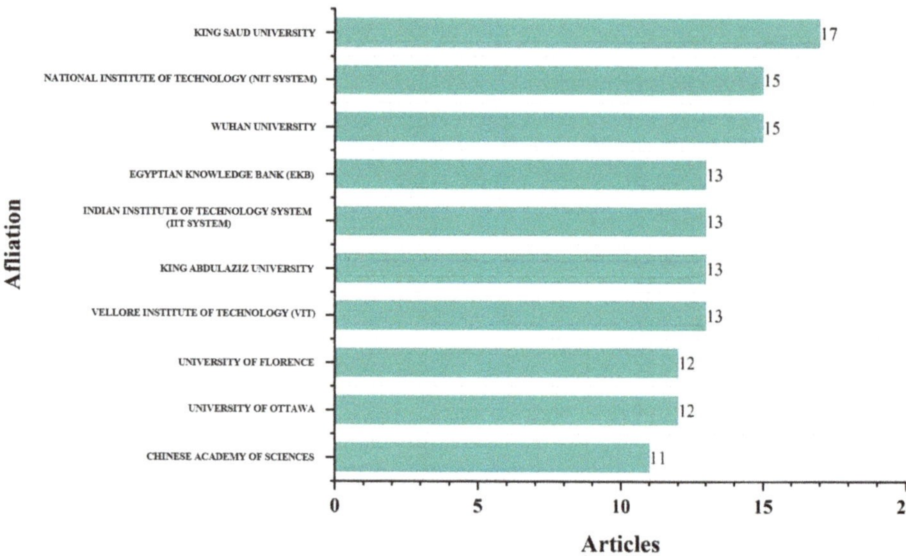

Figure 9. Top 10 institutions in terms of publications (created by the author, based on 829 publications).

From an institutional perspective, universities and research institutions in Saudi Arabia, India, Egypt, and China exhibit strong research capabilities in this field. Saudi Arabia's King Saud University and King Abdulaziz University are particularly active in research on smart cities and artificial intelligence, likely due to government investments in scientific research and the emphasis on smart city development. India's NIT System, IIT System, and VIT also show high publication volumes, reflecting the country's growing research focus on smart cities and technological innovation. Egypt's Egyptian Knowledge Bank (EKB), as a key academic resource platform, plays a significant role in advancing research on smart cities and artificial intelligence. Meanwhile, the Chinese Academy of Sciences and Wuhan University, as leading research institutions in China, have also demonstrated considerable influence in the fields of smart cities and machine learning.

Overall, smart cities and artificial intelligence technologies are attracting attention from universities and research institutions worldwide. Research institutions in countries such as Saudi Arabia, India, Egypt, and China have shown outstanding performance in this field, driving advancements in related technologies.

4. Analysis of Research Hotspots and Frontiers

4.1. Keyword Co-Occurrence Analysis

Figure 10 illustrates the clustering of keywords in the fields of machine learning and smart cities, reflecting the interconnections between different concepts and thematic focuses in the research. The red cluster revolves around "smart city" and is closely associated with keywords such as the IoT, neural network, deep learning, and intelligent transportation system. The research focuses on using technologies like deep learning and neural networks to optimize data processing, pattern recognition, and predictive analysis in smart cities [68,69]. It also involves applications such as traffic flow optimization, route planning, and smart parking to improve urban transportation efficiency [70]. The green cluster mainly focuses on artificial intelligence, data mining, and data analysis, and also involves keywords like renewable energy, sentiment analysis, and energy efficiency. Machine learning and data mining are widely applied in smart cities, including social media analysis, urban sentiment analysis, and energy optimization. The blue cluster involves keywords like cybersecurity, information security, digital twin, and challenges, reflecting that with the extensive application of IoT devices [71], information security and cybersecurity have become key research directions. This involves issues like data privacy and network attack defense, and it highlights the challenges faced in smart city construction, such as data privacy, infrastructure security, and social acceptance [72]. The yellow cluster revolves around internet-of-things, edge computing, wireless networks, and 5G, indicating that the high-speed, low-latency characteristics of 5G technology, combined with the local data processing capabilities of edge computing, provide more efficient computing and communication methods for smart cities. The purple cluster covers keywords like big data analytics, smart grid, reinforcement learning, and artificial neural network [73]. By optimizing energy consumption through machine learning, the efficiency of power systems can be enhanced, such as with smart meters and demand response.

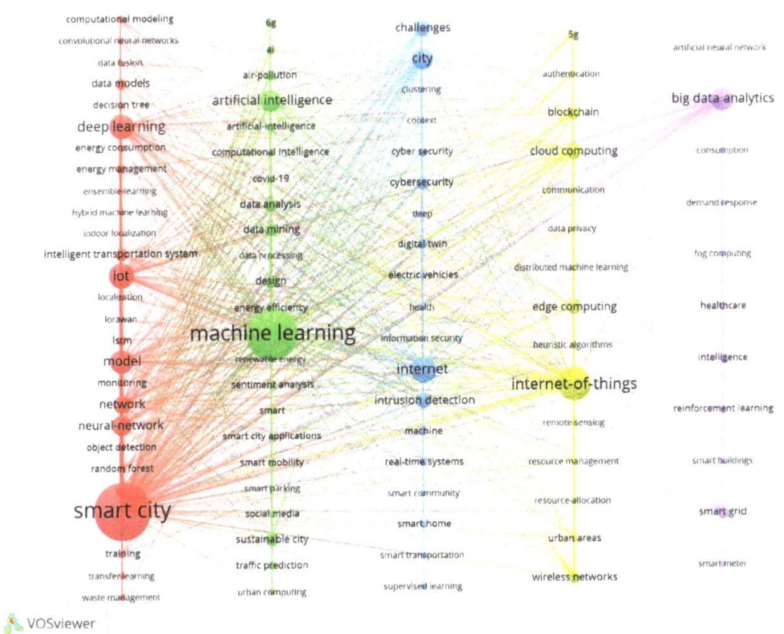

Figure 10. Keyword clustering diagram (created by the author, based on 829 publications).

Keywords are interconnected through lines, showing their interdependence and co-occurrence in research. From the figure, it can be seen that machine learning is the core concept of multiple clusters and is closely linked with keywords like big data analytics, cloud computing, and internet-of-things, indicating that big data analysis in smart cities relies on cloud computing and AI technologies for processing. In addition, 5G and edge computing show that smart cities are evolving toward more efficient computing and communication methods. These technological advancements provide more efficient data processing and communication means for smart cities, promoting intelligent urban management and improving residents' quality of life.

4.2. Keyword Emergence Analysis

Figure 11 presents the "Top 15 Keywords with the Strongest Citation Bursts", which refer to the keywords that experienced a sharp increase in citations during specific time periods. This analysis is significant for revealing dynamic changes and emerging hotspots in the research field. The "Year" of a keyword indicates its first appearance, "Strength" quantifies the intensity of the citation burst, and "Begin" and "End" mark the duration of the burst. The red bars represent the active periods of the keywords.

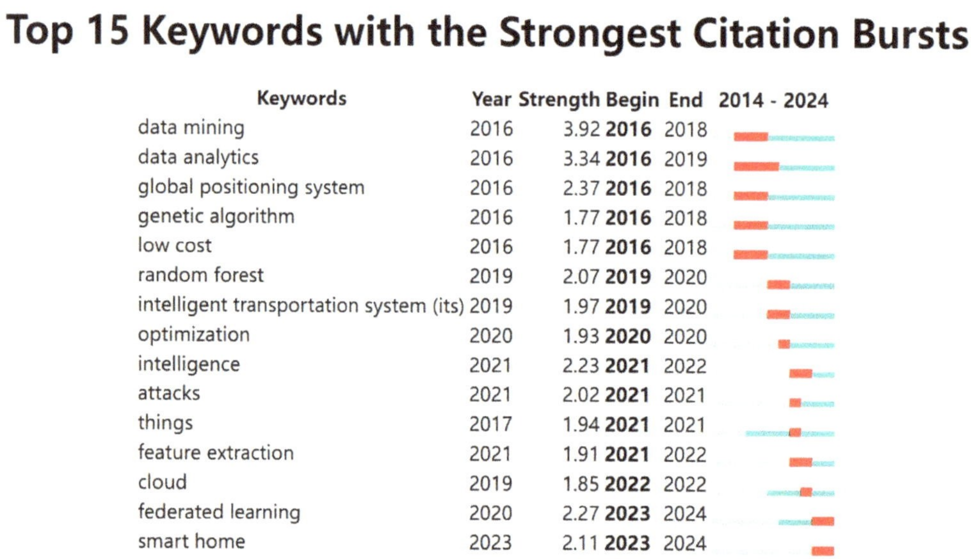

Figure 11. Keyword surfacing chart (created by the author, based on 829 publications).

The year 2016 marked an important milestone, with keywords such as "data mining" (strength 3.92), "data analytics" (strength 3.34), "global positioning system" (strength 2.37), "genetic algorithm" (strength 1.77), and "low cost" (strength 1.77) emerging and forming bursts that lasted until 2018–2019. These keywords are closely related to research areas such as big data, machine learning, positioning technologies, and intelligent algorithms, reflecting the academic community's strong focus on data mining, analytics, and low-cost optimization technologies at the time. From 2019 to 2020, the burst keywords "random forest" (strength 2.07), "intelligent transportation system (ITS)" (strength 1.97), and "optimization" (strength 1.93) indicated a shift in the research focus toward intelligent transportation systems and optimization algorithms. This trend reflects the widespread application of artificial intelligence and data-driven technologies in smart transportation, such as intelligent scheduling, route optimization, and autonomous driving. In 2021, the keywords "intelligence" (strength 2.23), "attacks" (strength 2.02), and "things" (strength 1.94) highlighted the growing attention to topics such as artificial intelligence, cybersecurity attacks, and the Internet of Things (IoT). This is likely linked to the rise of AI safety concerns and IoT security threats, reflecting the academic community's emphasis on the security of intelligent systems. From 2022 to 2023, the keywords "feature extraction" (strength 1.91), "cloud" (strength 1.85), "federated learning" (strength 2.27), and "smart home" (strength 2.11) indicated a shift in research hotspots toward feature extraction, cloud computing, federated learning, and smart home technologies. Notably, federated learning emerged as a burst keyword in 2023 and continues into 2024, demonstrating increasing attention to privacy-preserving distributed machine learning techniques.

This trend illustrates the evolution of the research field from foundational technologies in data and algorithms to the security of intelligent systems, and then to broader smart applications, such as smart homes and cloud computing. It reflects the continuous advancement of technology and the expansion of application scenarios.

4.3. Keyword Time Zone Analysis

Figure 12 presents a keyword co-occurrence time zone map in the fields of machine learning and smart cities. The size of the nodes reflects the frequency of the keywords in the literature, while the color gradient from blue to yellow indicates the temporal evolution, spanning from around 2020 to 2022. As shown in the figure, "smart city", "machine learning", and "internet of things" are core concepts in the research, with larger and densely connected nodes, highlighting their importance in smart city technology applications. From a temporal perspective, the gradual transition of some keywords from blue to green and yellow reflects the evolution of research hotspots in recent years. For example, keywords such as "cybersecurity" and "data privacy" exhibit colors leaning toward green and yellow, indicating that data security and privacy protection have gained increasing attention as smart cities develop [42]. Given that smart cities involve the collection, storage, and analysis of vast amounts of data, ensuring data security has become a key research focus. Additionally, technologies such as "cloud computing", "edge computing", and "blockchain" also hold significant positions in smart city research [74], with node colors showing their rising research prominence during 2021–2022. In particular, cloud computing and edge computing play critical roles in the intelligent infrastructure of smart cities, such as in data storage, real-time computing, and resource optimization.

Notably, keywords such as "5G", "smart grid", and "renewable energy" exhibit node colors closer to yellow, indicating increased research activity in these areas in recent years [75]. The widespread adoption of 5G technology supports efficient communication in smart cities, while advancements in smart grids and renewable energy reflect progress in sustainability and energy management within smart cities [76]. In artificial intelligence (AI)-related research, the appearance of keywords such as "deep learning", "LSTM", and "ensemble learning" demonstrates the growing application of deep learning methods in smart cities, including intelligent transportation, air pollution prediction, and urban planning.

Overall, in the construction of smart cities, machine learning, as a core technology, is being widely applied across various subfields, such as cloud computing, the Internet of Things, data privacy, and cybersecurity. Over time, the co-occurrence relationships among these keywords have become increasingly close, indicating their synergistic roles within the overall framework of smart cities, thereby driving continuous development in this field.

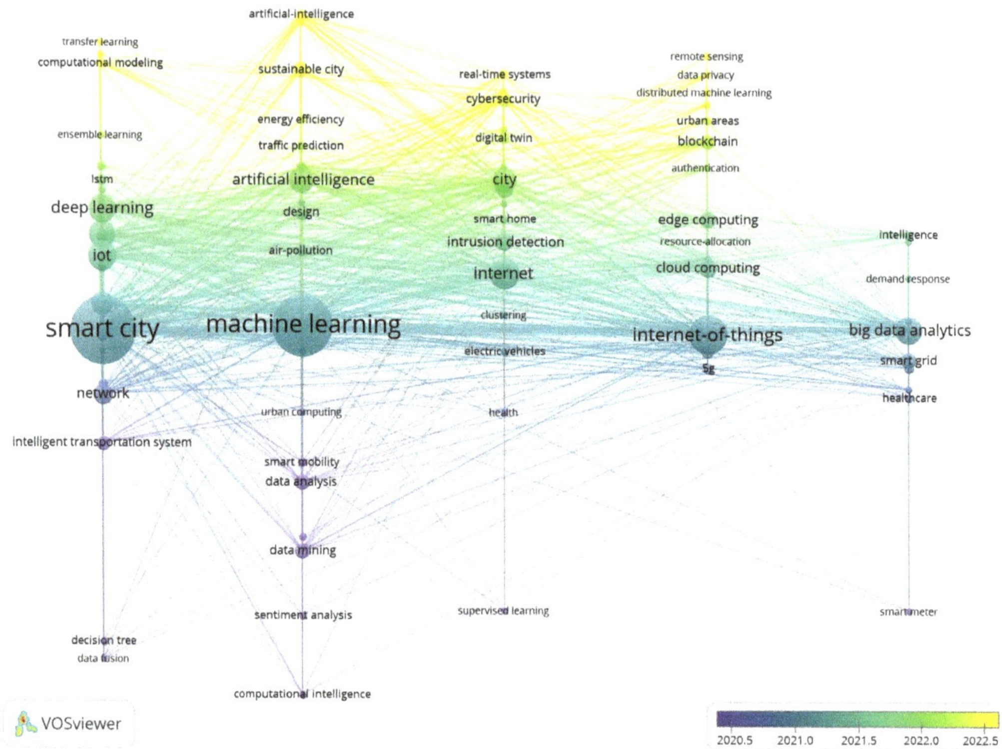

Figure 12. Keyword time zone map (created by the author, based on 829 publications).

Figure 13 uses different colors to represent distinct research themes, ranging from #0 to #9, which include "digital twin", "electric vehicles", "internet of things", "big data", "smart grid", "neural networks", "attacks", "edge computing", "support vector machines", and "5G". The size of each keyword's bubble indicates the activity level of the research theme at a specific time point, while darker colors signify greater influence or citation concentration in the field.

First, digital twin is one of the most prominent research topics, spanning the entire timeline and showing rapid growth in recent years. Digital twin technology, through the synchronous mapping of virtual models with the physical world, provides robust data support for smart cities, industrial manufacturing, and infrastructure management. With advancements in artificial intelligence and IoT technologies, digital twin applications have become increasingly widespread in areas such as smart cities, energy management, and intelligent transportation [77]. Second, IoT technology plays a critical role in smart city development by enhancing urban management efficiency through sensor networks, data collection, and intelligent analysis [58]. As seen in the figure, IoT-related keywords, such as "data analytics", "intelligent transportation systems", and "deep learning", are closely linked, reflecting the mainstream trend of data-driven smart cities. Third, research on smart grids and electric vehicles is highly active, particularly in areas such as energy consumption and resource management. As the world transitions toward sustainable energy, smart grid technologies, integrated with big data, AI, and IoT, optimize power distribution and improve energy efficiency [78]. Fourth, machine learning techniques, such as neural networks and support vector machines, play significant roles in smart cities, anomaly detection, security protection, and intelligent decision-making [79]. Fifth, with the increasing digitalization of smart cities, cybersecurity has become a key research focus [80]. Keywords such as "attacks", "privacy", and "intrusion detection" reflect growing concerns about cybersecurity, especially in IoT and smart infrastructure applications. Meanwhile, edge computing, as a critical technology for optimizing data processing and reducing latency, is also widely applied in smart cities, 5G networks, and intelligent manufacturing.

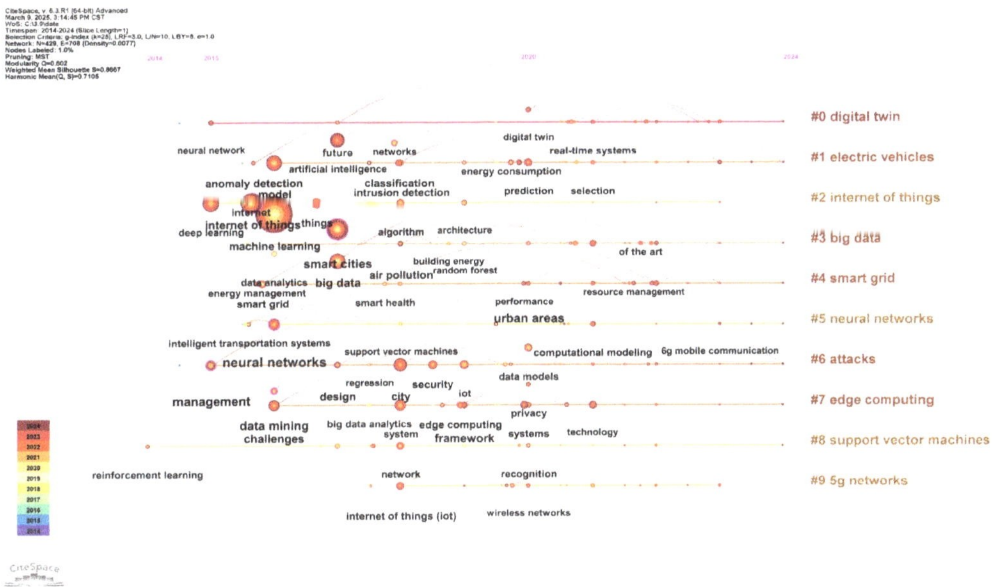

Figure 13. Keywords timeline mapping (created by the author, based on 829 publications).

Research in the smart city domain is evolving toward data-driven and intelligent decision-making. With the continuous maturation of technologies such as deep learning, edge computing, and digital twins, smart city development is gradually transitioning from theoretical exploration to practical applications, aiming to address real-world challenges in energy management, environmental optimization, and security protection. In the future, advancements in 5G networks and 6G mobile communication will further drive the upgrade of smart city network infrastructure, enhancing data transmission efficiency and intelligent capabilities.

5. Analysis of Cooperation Networks

5.1. Core Author Collaborative Network

Using the Scimago software, the literature database in the field of "machine learning and smart cities" was analyzed, and an author collaboration network diagram was generated. Figure 14 presents the collaboration network among different scholars in the form of a chord diagram, with the colors distinguishing various research groups and the thickness of the connecting lines representing the strength of collaboration (weight), reflecting the research hotspots and collaboration trends in this field.

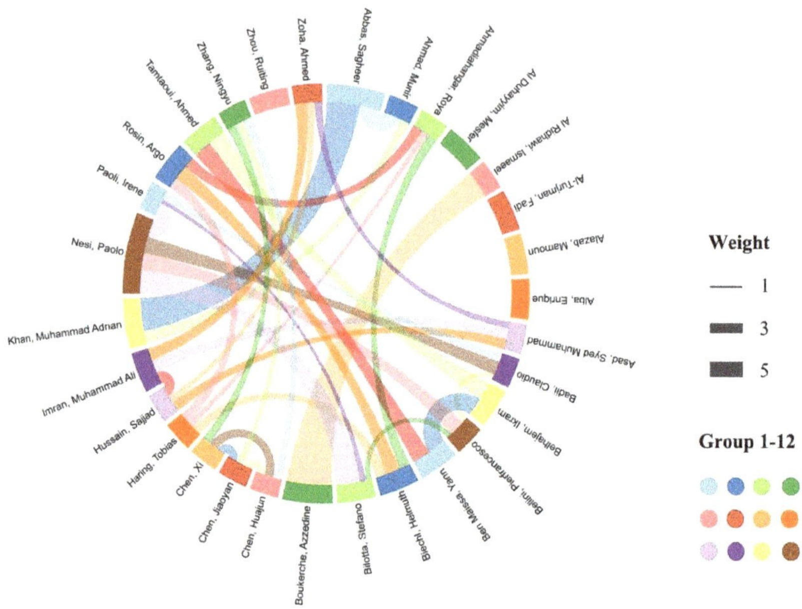

Figure 14. Cluster diagram of author cooperation (created by the author, based on 829 publications).

From Figure 14, it can be observed that researchers such as Paolo Nesi, Muhammad Adnan Khan, Muhammad Ali Imran, and Azzedine Boukerche occupy central positions in the collaboration network. Their research teams maintain close collaborations with multiple scholars, forming concentrated collaboration clusters [47,49,60,81,82]. Paolo Nesi's collaboration network is extensive, involving multiple research groups, and he has significant research influence in areas such as intelligent transportation and urban big data analytics. Muhammad Adnan Khan and Muhammad Ali Imran also play important roles in this field, with their collaboration clusters being tightly knit and focused on network optimization, smart city communication architectures, and IoT-related research [57,83]. Azzedine Boukerche and his team concentrate on research in security, privacy protection, and intelligent computing infrastructure in smart cities, collaborating with numerous scholars [84]. Additionally, scholars such as Zohra Ahmed and Sergeher Abbas form a relatively independent collaboration network, focusing on specific applications in smart cities, such as intelligent healthcare or smart energy management [85]. Chen Xi, Chen Huijun, and Chen Jiayuan exhibit strong collaborative relationships, with in-depth cooperation in areas such as edge computing and data mining for smart cities [86,87].

Overall, academic collaboration in this research field exhibits a multi-centric characteristic, with several core scholars and their teams forming dense collaboration networks, while some relatively independent research groups also exist. This indicates that research in smart cities and machine learning not only relies on international cross-team collaboration but also includes scholars focusing on specific research directions. In the future, this field could further strengthen cross-team and interdisciplinary collaboration to more comprehensively advance smart city technologies, achieving the deeper integration of data intelligence, the IoT, edge computing, and other technologies, thereby enhancing the overall intelligence level of smart cities.

5.2. Institutional Cooperation Network

From Figure 15, it can be observed that the institutional collaboration network in the current research field exhibits a high degree of complexity and an internationalized trend of cooperation. The connections between core institutions are tight, forming multiple academic collaboration hubs, reflecting the close cooperation patterns among institutions from different regions and types. The structure of this network reveals several prominent central nodes, which represent the major research institutions in the field and play a key role in global academic exchange and collaboration.

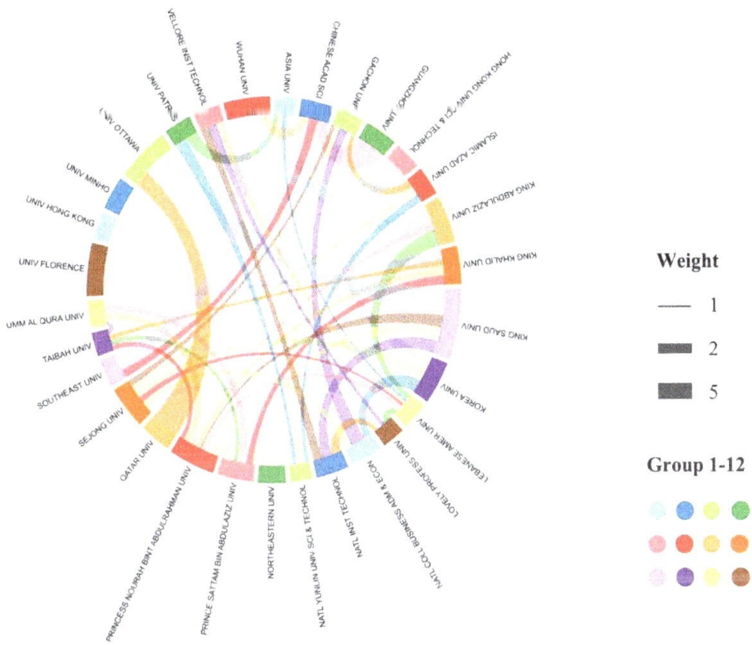

Figure 15. Chart of cooperation of issuing bodies (created by the author, based on 829 publications).

The Chinese Academy of Sciences occupies a central position in the collaboration network, indicating its significant influence in global academic research. It maintains extensive collaborations with multiple international institutions, such as Wuhan University, Shanghai Jiao Tong University, and Korea University, forming a relatively tight collaboration network. This reflects the trend of international cooperation among Chinese research institutions in related technological fields and underscores China's important position in the global landscape of scientific and technological innovation. Additionally, King Saud University and King Abdulaziz University, as representative research institutions in the Middle East, also exhibit strong connectivity in the collaboration network. They have formed regional collaboration clusters with multiple international universities, such as Qatar University, Taibah University, and Umm Al Qura University. The Middle East's investment in research areas such as smart cities, artificial intelligence, and related fields is continuously growing, and its academic influence is expanding through strengthened international cooperation.

Notably, the Hong Kong University of Science and Technology and the University of Hong Kong, as representative institutions in Hong Kong, China, maintain collaborative ties with multiple international institutions. This demonstrates the bridging role of Hong Kong universities in the global academic network, facilitating academic exchanges between Asia and regions such as Europe, America, and the Middle East, while also promoting the cross-border flow of knowledge. In terms of collaboration intensity, the connections between some institutions are thicker, indicating closer cooperation in this field. For example, the connection between the Chinese Academy of Sciences and Wuhan University is strong, reflecting deep collaboration among domestic research institutions. Similarly, the collaboration between King Saud University and King Abdulaziz University is relatively tight, indicating that research institutions within the same country or region often form stable collaboration networks.

Overall, Figure 15 highlights the evident trends of the internationalization and diversification of collaboration in the fields of machine learning and smart cities. By establishing cross-border and cross-institutional cooperation hubs, it has facilitated the global sharing and dissemination of academic research and technological innovation. In the future, by further strengthening cross-border and cross-institutional collaboration, the field of smart cities is expected to achieve more efficient technology sharing and academic exchange, providing innovative momentum for the intelligent and sustainable development of cities worldwide.

5.3. International Cooperative Research

Figure 16 illustrates the collaborative relationships among multiple countries, presented in the form of an arc network that reflects the intensity of cooperation. The lines of different colors in the figure represent distinct collaborative groups, while the thickness of the lines indicates the strength of collaboration. Table 1 displays the top ten countries ranked by centrality scores. As shown in Figure 16, the collaboration exhibits a clear global characteristic, involving major regions such as Asia, Europe, North America, and Oceania, and forming a transcontinental collaboration network. Countries such as China, the USA, and several European nations (e.g., Germany, France, and the UK) play central roles in the collaboration network. The dense and widely radiating collaboration lines of these countries indicate their strong international influence in research or technological cooperation and their leading role in promoting global collaboration. The close collaborative relationships among European countries (e.g., Germany, France, the UK, Spain, and Sweden) reflect the tight connections in academic research or technological fields within Europe. This phenomenon may be related to the European Union's internal research collaboration frameworks, such as the Horizon Europe program, which has facilitated cooperation among member states.

Table 1. Top 10 countries in centrality score (created by the author, based on 829 publications).

National	Centrality Score
China	0.31
United Kingdom	0.24
India	0.2
South Korea	0.12
United Arab Emirates	0.12
USA	0.11
Saudi Arabia	0.1
Australia	0.1
Italy	0.09
Germany	0.09

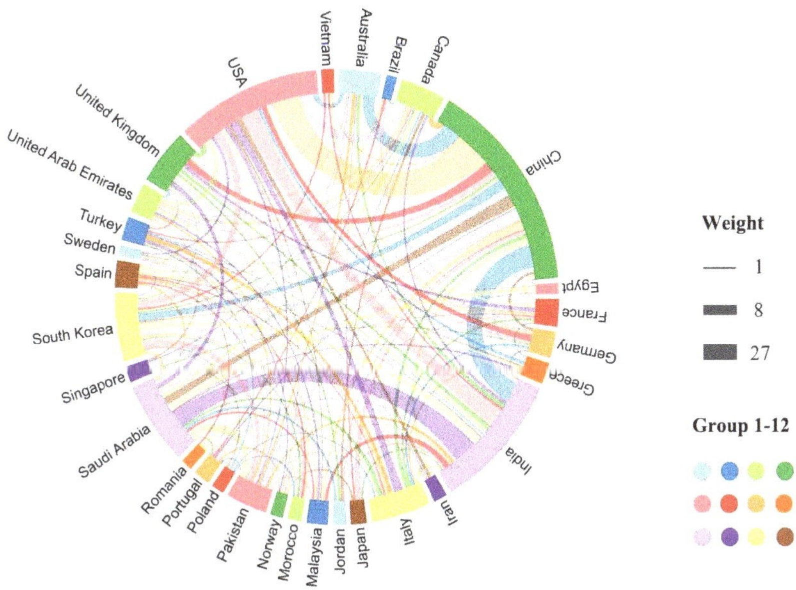

Figure 16. Map of cooperation networks in the issuing countries (created by the author, based on 829 publications).

Collaboration in Asia is particularly active, with strong connections among countries such as China (0.31), India (0.2), Japan, and South Korea (0.12). Among these, China serves as the collaborative hub in Asia, maintaining close ties with multiple Asian countries (e.g., Malaysia, Pakistan, and Saudi Arabia (0.1)), demonstrating its growing technological capabilities and international influence in this field. Additionally, the USA (0.11), as a key player in global scientific research and technological collaboration, maintains strong connections with numerous countries. Italy (0.09) and Australia (0.1) also hold significant positions in the collaboration network, particularly in their notable collaborations with the USA, Europe, and Asia. This may be attributed to the high degree of internationalization in higher education and technological research and development in these countries.

The figure reveals the patterns of global scientific research or technological collaboration, showcasing a network centered around China, the USA, and several European countries, while also highlighting the tight intra-regional cooperation and transcontinental collaboration trends. This globalized collaboration model not only promotes knowledge sharing but also drives technological advancements and industrial development. In the future, with the deepening of technological innovation and the strengthening of international cooperation, this collaboration network is likely to expand and deepen further.

6. Discussion and Research Needs

Research on machine learning in the field of smart cities has made significant progress in recent years, with expanding research hotspots and increasing international collaboration. However, addressing issues such as fairness, privacy and security, and technological sustainability in urban digital transformation remains a critical challenge for both academia and industry. In the future, the deep integration of machine learning and smart cities will further drive innovation in urban governance models, enhance the ability to optimize resource allocation, and provide global urban residents with a more intelligent, efficient, and sustainable urban living experience, as illustrated in Figure 17.

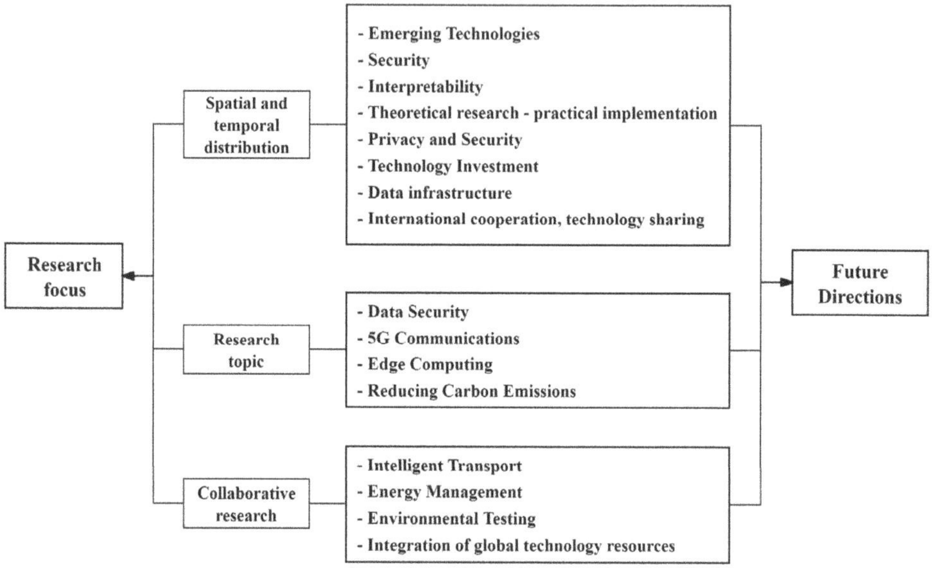

Figure 17. Research priorities and future directions.

6.1. Spatial and Temporal Distribution

The phased and regional characteristics of research in this field are influenced by multiple factors. From a temporal perspective, the evolution of research is driven not only by technological advancements but also by policy support, industrial application demands, and the improvement of data infrastructure. During the initial development stage (2014–2017), research was primarily constrained by computational capabilities and algorithm development, focusing on theoretical exploration and preliminary applications. The rapid growth stage (2017–2021) benefited from breakthroughs in deep learning technologies, the widespread adoption of Internet of Things (IoT) devices, and the promotion of smart city construction policies in various countries, leading to the expansion of research hotspots into multiple areas of urban governance. Entering the stable stage (2022 to the present), the research focus has gradually shifted from technological breakthroughs to application optimization, interdisciplinary integration, and ethical and privacy concerns, reflecting the maturation of smart city construction and the increasing trend of practice-driven academic research.

From a spatial perspective, the imbalance in research capabilities is primarily related to economic development levels, government investment, the distribution of scientific resources, and international collaboration networks. Countries such as China, the United States, and India have high research outputs in the fields of smart cities and machine learning, benefiting from rapid urbanization, strong technological industrial foundations, and government policy support. European countries exhibit high activity in international collaborations, reflecting the openness of their research systems and their capacity for transnational cooperation. In contrast, regions such as South America and Africa have lower research outputs, mainly due to limited funding, inadequate infrastructure, and a shortage of high-end technical talent.

Future research needs to place a greater emphasis on practicality and inclusivity. While maintaining technological innovation, attention should also be given to lowering the barriers to technology application and promoting the equitable sharing of smart city construction achievements. At the same time, interdisciplinary and cross-domain research collaborations will become crucial pathways for advancing this field. It is recommended to strengthen the construction of international collaboration networks, promote balanced global development in smart city research, and provide more diverse solutions to address urbanization challenges.

6.2. Hotspot Analyses

From the hotspot analyses, it is evident that the research fields of machine learning and smart cities exhibit a multi-thematic and dynamically evolving characteristic. This trend is primarily driven by factors such as technological advancements, the demand for urban digital transformation, and challenges related to data security. The red cluster revolves around technologies like smart cities, the Internet of Things (IoT), and deep learning, reflecting core application scenarios such as intelligent traffic optimization. This highlights the critical demand for smart cities to enhance urban operational efficiency and residents' quality of life. The green cluster focuses on artificial intelligence, data mining, and energy efficiency, indicating that the widespread application needs—from social media analysis to urban energy management—are driving the popularization of data-driven decision-making. The blue cluster emphasizes cybersecurity and digital twins, underscoring the urgency of data privacy protection and the security of critical infrastructure, especially as urban systems increasingly rely on interconnected technologies. The yellow cluster integrates new infrastructures, such as 5G and edge computing, providing efficient computational and communication support for smart cities, enabling distributed intelligent computing and real-time data processing. The purple cluster focuses on big data analytics and smart grids, showcasing the potential of machine learning in energy optimization and sustainable development.

From a temporal perspective, research hotspots have undergone significant evolution. In the early stages (2016–2019), the focus was primarily on foundational technologies, such as data mining, analysis, and optimization algorithms, laying the theoretical groundwork for subsequent applications. From 2019 to 2020, research began shifting toward specific application scenarios, like intelligent transportation systems, marking the transition of smart city technologies from laboratories to real-world environments. Starting in 2021, issues such as AI security and IoT security gained more attention, reflecting the growing complexity of smart city systems and the increasing scale of data, making security a critical concern. Recently (2022–2023), emerging hotspots include feature extraction, cloud computing, federated learning, and smart homes. The rise of privacy-preserving technologies like federated learning indicates that, in the context of increasingly widespread data sharing, achieving efficient intelligent computing while ensuring data security has become one of the core challenges in research.

In the future, smart city research needs to further focus on interdisciplinary integration, data security and privacy protection, the optimization of intelligent infrastructure, sustainable development and carbon neutrality, human–machine collaboration, and explainability. It is recommended to strengthen industry–academia–research collaboration, promote the deep integration of technological innovation and practical applications, and drive the high-quality development of smart city construction. At the same time, it is essential to enhance ethical norms and standard systems to ensure that the direction of technological development aligns with societal values.

6.3. Analysis of Cooperation

The global collaboration network in the field of smart cities and machine learning is becoming increasingly close. This trend is not only due to the complexity and interdisciplinary nature of the technology itself but also driven by support from national policies, growing industry demands, and the need for data resource sharing. From the perspective of scholars' collaboration networks, research teams involving key scholars, such as Paolo Nesi, Muhammad Adnan Khan, Muhammad Ali Imran, and Azzedine Boukerche, continue to deepen their cooperation in critical subfields of smart cities, including intelligent transportation, network optimization, security and privacy, and the Internet of Things. This close academic connection helps drive technological innovation and application implementation. Meanwhile, from the perspective of institutional collaboration networks, institutions like the Chinese Academy of Sciences and King Saud University play a leading role in international cooperation. In particular, the Chinese Academy of Sciences, leveraging its extensive international cooperation network, has established solid collaborative relationships with both domestic and international universities and research institutions, making it an important hub for research in smart cities and machine learning.

The formation of this global scientific cooperation model is partly due to the fact that the construction of smart cities involves multiple technical fields, such as data analysis, artificial intelligence, the Internet of Things, and 5G communication. No single institution or country can independently achieve breakthroughs in the entire technological chain, thus making transnational cooperation an essential path for technological innovation. On the other hand, the growing demand for smart city applications from governments and enterprises also prompts research institutions to strengthen international cooperation to share research outcomes, optimize resource allocation, and accelerate technology transformation. Additionally, with the development of artificial intelligence technology bringing challenges in data privacy and security, the international academic community is increasingly focusing on issues of security, ethics, and regulations. This also encourages scholars from different countries to enhance cooperation and jointly explore feasible technical standards and policy frameworks on a global scale.

Future research should focus on establishing multi-level international cooperation platforms to promote knowledge sharing and technology transfer; enhancing exchanges among young scholars to cultivate cross-cultural research capabilities; promoting the deep integration of industry, academia, and research to accelerate the transformation of scientific achievements; improving the evaluation mechanisms for international cooperation projects to increase cooperation efficiency; and exploring new cooperation models in the "Internet+" era to overcome geographical limitations. By driving continuous innovation in the fields of smart cities and machine learning and promoting international cooperation to deeper levels, collaborative research is expected to achieve groundbreaking progress in the intelligent transformation of cities, providing strong support for sustainable urban development worldwide.

7. Conclusions

This study employs bibliometric methods to systematically analyze research on machine learning in the field of smart cities, revealing the spatial and temporal distribution characteristics, research hotspots, and cooperation network structures in this area. The results show that research on smart cities and machine learning exhibits distinct phases and regional characteristics, continuously evolving with technological advancements and application demands. The study clarifies the developmental stages and regional distribution characteristics of the smart cities and machine learning field, aiding researchers in understanding the evolutionary trajectory and global landscape of this domain, especially highlighting the period from 2017 to 2021 as a critical phase of rapid growth. This provides a temporal reference for future research directions. Additionally, by analyzing research hotspots, this paper highlights the importance of issues such as data privacy, network security, sustainable development, and intelligent transportation, reflecting the dual demand for technological innovation and green development in smart city construction. This offers priority directions for policymakers and technology developers. Furthermore, the cooperation network analysis indicates that international academic collaboration is becoming increasingly close, with research institutions in China, the United States, and Europe playing central roles in the global cooperation system. This drives technology sharing and interdisciplinary integration, providing practical guidance for fostering collaborative innovation in global smart city development. This study not only provides a systematic analytical framework for academic research and technological innovation in the field of smart cities and machine learning but also offers a scientific basis for optimizing future research directions and rationalizing resource allocation, possessing significant theoretical and practical value.

However, this study has several major limitations. Firstly, in terms of the research methodology, relying solely on the Web of Science database for publication retrieval may result in the omission of some important studies. Specifically, some high-quality conference papers or relevant research included in other databases might not have been analyzed. Secondly, the selection of only studies published in English introduces a language bias, potentially overlooking significant research contributions from non-English academic communities, especially from countries and regions with innovative practices in smart city development. To mitigate these limitations, future research could consider expanding to other major research databases and attempt to include multilingual publications to obtain a more comprehensive research perspective. Additionally, smart cities are a multidisciplinary field involving computer science, urban planning, sociology, environmental science, and more. However, this study focuses more on machine learning technology itself, with less exploration of its interdisciplinary applications and impacts, which may lead to an incomplete understanding of the overall development of smart cities. Future research should further integrate multidisciplinary perspectives, examining the intersectional impacts of machine learning on the social, economic, cultural, and policy aspects within smart cities. For example, it could investigate how machine learning technology promotes social inclusiveness and equitable resource distribution in smart cities. With the rapid development of technologies such as generative artificial intelligence, edge computing, and blockchain, future research directions could focus on the potential applications of these emerging technologies in smart cities. Furthermore, future studies should strengthen the focus on the practices of smart cities in developing countries and non-English-speaking regions to produce more globally representative research findings.

Overall, the research on machine learning in the field of smart cities has entered a mature stage, gradually deepening towards multidisciplinary integration, technology security, and sustainable development. In the future, ensuring technological innovation while protecting data privacy, enhancing technological inclusiveness, and promoting sustainable development will become important research topics in this field. By strengthening international cooperation, promoting technology sharing, and advancing policy formulation, the construction of smart cities is expected to achieve more intelligent, efficient, and sustainable development, creating a more livable future for urban residents worldwide.

Author Contributions: Conceptualization, B.L.; methodology, B.L.; software, B.L.; validation, B.L.; formal analysis, B.L.; investigation, B.L.; resources, B.L.; data curation, B.L.; writing—original draft preparation, B.L.; writing—review and editing, X.F.; visualization, B.L.; supervision, X.F.; project administration, X.F.; funding acquisition, X.F. All authors have read and agreed to the published version of the manuscript.

Funding: This research received no external funding.

Data Availability Statement: Not applicable.

Acknowledgments: This study is grateful to all the authors who participated in writing and collating the data.

Conflicts of Interest: The authors declare no conflicts of interest.

References

1. Birch, E.L.; Wachter, S.M. (Eds.) Chapter 1. World Urbanization: The Critical Issue of the Twenty-First Century. In *Global Urbanization*; University of Pennsylvania Press: Philadelphia, PA, USA, 2011; pp. 3–23. [CrossRef]
2. Han, F.; Xie, R.; Lai, M. Traffic density, congestion externalities, and urbanization in China. *Spat. Econ. Anal.* **2018**, *13*, 400–421. [CrossRef]
3. Zhang, Y.; Cai, Q. Impact Mechanism of New Urbanization on Environmental Pollution: Empirical Analysis Based on Spatial Panel Model. *Front. Public Health* **2022**, *10*, 928100. [CrossRef] [PubMed]
4. Wang, Q.; Su, M.; Li, R.; Ponce, P. The effects of energy prices, urbanization and economic growth on energy consumption per capita in 186 countries. *J. Clean. Prod.* **2019**, *225*, 1017–1032. [CrossRef]
5. Berlotti, M.; Di Grande, S.; Cavalieri, S. Proposal of a Machine Learning Approach for Traffic Flow Prediction. *Sensors* **2024**, *24*, 2348. [CrossRef]
6. Zhang, X.Q. The trends, promises and challenges of urbanisation in the world. *Habitat Int.* **2016**, *54*, 241–252. [CrossRef]
7. Madlener, R.; Sunak, Y. Impacts of urbanization on urban structures and energy demand: What can we learn for urban energy planning and urbanization management? *Sustain. Cities Soc.* **2011**, *1*, 45–53. [CrossRef]
8. Javed, A.R.; Shahzad, F.; ur Rehman, S.; Zikria, Y.B.; Razzak, I.; Jalil, Z.; Xu, G. Future smart cities: Requirements, emerging technologies, applications, challenges, and future aspects. *Cities* **2022**, *129*, 103794. [CrossRef]
9. Xia, X.; Ji, S.; Vijayakumar, P.; Shen, J.; Rodrigues, J.J.P.C. An efficient anonymous authentication and key agreement scheme with privacy-preserving for smart cities. *Int. J. Distrib. Sens. Netw.* **2021**, *17*, 15501477211026804. [CrossRef]
10. Abdalla, A.A.; Abdalla, Y.A.; Haddad, A.M.; Bhavani, G.; Zabalawi, E. Connections between Big Data and Smart Cities from the Supply Chain Perspective: Understanding the Impact of Big Data. *Sustainability* **2022**, *14*, 16161. [CrossRef]
11. Massobrio, R.; Nesmachnow, S.; Tchernykh, A.; Avetisyan, A.; Radchenko, G. Towards a Cloud Computing Paradigm for Big Data Analysis in Smart Cities. *Program. Comput. Softw.* **2018**, *44*, 181–189. [CrossRef]
12. Jiang, D. The construction of smart city information system based on the Internet of Things and cloud computing. *Comput. Commun.* **2020**, *150*, 158–166. [CrossRef]
13. Han, T.; Ge, X.; Wang, L.; Kwak, K.S.; Han, Y.; Liu, X. 5G Converged Cell-Less Communications in Smart Cities. *IEEE Commun. Mag.* **2017**, *55*, 44–50. [CrossRef]
14. Allam, Z.; Dhunny, Z.A. On big data, artificial intelligence and smart cities. *Cities* **2019**, *89*, 80–91. [CrossRef]
15. Sarker, I.H. Machine Learning: Algorithms, Real-World Applications and Research Directions. *SN Comput. Sci.* **2021**, *2*, 160. [CrossRef]
16. Sharma, A.; Jain, P. Adaptability of IoT and Cloud for Enabling the Smart City: Applications and Challenges. In *Handbook of Research on Network-Enabled IoT Applications for Smart City Services*; IGI Global Scientific Publishing: Hershey, PA, USA, 2023; pp. 54–74. [CrossRef]
17. Ghanim, M.S.; Abu-Lebdeh, G. Real-Time Dynamic Transit Signal Priority Optimization for Coordinated Traffic Networks Using Genetic Algorithms and Artificial Neural Networks. *J. Intell. Transp. Syst.* **2015**, *19*, 327–338. [CrossRef]
18. Kumar, K.; Poonam; Kaur, R.; Rani, V. Different Methodologies for Smart City and Urbanization System. In *5G Enabled Technology for Smart City and Urbanization System*; Chapman and Hall/CRC: Boca Raton, FL, USA, 2024.
19. Qolomany, B.; Al-Fuqaha, A.; Gupta, A.; Benhaddou, D.; Alwajidi, S.; Qadir, J.; Fong, A.C. Leveraging Machine Learning and Big Data for Smart Buildings: A Comprehensive Survey. *IEEE Access* **2019**, *7*, 90316–90356. [CrossRef]
20. Beatley, T.; Newman, P. Biophilic Cities Are Sustainable, Resilient Cities. *Sustainability* **2013**, *5*, 3328–3345. [CrossRef]
21. Beatley, T. *Handbook of Biophilic City Planning and Design*; Island Press/Center for Resource Economics: Washington, DC, USA, 2016. [CrossRef]
22. Liao, L.; Du, M.; Huang, J. The Effect of Urban Resilience on Residents' Subjective Happiness: Evidence from China. *Land* **2022**, *11*, 1896. [CrossRef]
23. Liu, B.; Li, Q.; Zheng, Z.; Huang, Y.; Deng, S.; Huang, Q.; Liu, W. A Review of Multi-Source Data Fusion and Analysis Algorithms in Smart City Construction: Facilitating Real Estate Management and Urban Optimization. *Algorithms* **2025**, *18*, 30. [CrossRef]
24. Braun, T.; Fung, B.C.M.; Iqbal, F.; Shah, B. Security and privacy challenges in smart cities. *Sustain. Cities Soc.* **2018**, *39*, 499–507. [CrossRef]
25. Ahmad, K.; Maabreh, M.; Ghaly, M.; Khan, K.; Qadir, J.; Al-Fuqaha, A. Developing Future Human-Centered Smart Cities: Critical Analysis of Smart City Security, Interpretability, and Ethical Challenges. *arXiv* **2021**, arXiv:2012.09110. [CrossRef]

26. Kinney, M.; Anastasiadou, M.; Naranjo-Zolotov, M.; Santos, V. Expectation management in AI: A framework for understanding stakeholder trust and acceptance of artificial intelligence systems. *Heliyon* **2024**, *10*, e28562. [CrossRef] [PubMed]
27. Mongeon, P.; Paul-Hus, A. The journal coverage of Web of Science and Scopus: A comparative analysis. *Scientometrics* **2016**, *106*, 213–228. [CrossRef]
28. Chen, C. CiteSpace II: Detecting and visualizing emerging trends and transient patterns in scientific literature. *J. Am. Soc. Inf. Sci. Technol.* **2006**, *57*, 359–377. [CrossRef]
29. van Eck, N.; Waltman, L. Software survey: VOSviewer, a computer program for bibliometric mapping. *Scientometrics* **2009**, *84*, 523–538. [CrossRef]
30. Bakır, M.; Özdemir, E.; Akan, Ş.; Atalık, Ö. A bibliometric analysis of airport service quality. *J. Air Transp. Manag.* **2022**, *104*, 102273. [CrossRef]
31. Bigler, L.; Gopalakrishnan, V. Artificial Intelligence for Infrastructure Monitoring and Traffic Optimization. In *Artificial Intelligence for Autonomous Networks*; Chapman and Hall/CRC: Boca Raton, FL, USA, 2018.
32. Ahmad, T.; Madonski, R.; Zhang, D.; Huang, C.; Mujeeb, A. Data-driven probabilistic machine learning in sustainable smart energy/smart energy systems: Key developments, challenges, and future research opportunities in the context of smart grid paradigm. *Renew. Sustain. Energy Rev.* **2022**, *160*, 112128. [CrossRef]
33. Hui, C.X.; Dan, G.; Alamri, S.; Toghraie, D. Greening smart cities: An investigation of the integration of urban natural resources and smart city technologies for promoting environmental sustainability. *Sustain. Cities Soc.* **2023**, *99*, 104985. [CrossRef]
34. Abid, M.T.; Aljarrah, N.; Shraa, T.; Alghananim, H.M. Forecasting and managing urban futures: Machine learning models and optimization of urban expansion. *Asian J. Civ. Eng.* **2024**, *25*, 4673–4682. [CrossRef]
35. Ciocca, D.R.; Delgado, G. The reality of scientific research in Latin America; an insider's perspective. *Cell Stress Chaperones* **2017**, *22*, 847–852. [CrossRef]
36. Docquier, F.; Rapoport, H. Globalization, Brain Drain, and Development. *J. Econ. Lit.* **2012**, *50*, 681–730. [CrossRef]
37. Ajao, L.A.; Apeh, S.T. Blockchain Integration with Machine Learning for Securing Fog Computing Vulnerability in Smart City Sustainability. In Proceedings of the 2023 1st International Conference on Advanced Innovations in Smart Cities (ICAISC), Jeddah, Saudi Arabia, 23–25 January 2023; pp. 1–6. [CrossRef]
38. Hancke, G.P.; Silva, B.D.C.e.; Hancke, G.P., Jr. The Role of Advanced Sensing in Smart Cities. *Sensors* **2013**, *13*, 393–425. [CrossRef]
39. Zhao, L.; Tang, Z.; Zou, X. Mapping the Knowledge Domain of Smart-City Research: A Bibliometric and Scientometric Analysis. *Sustainability* **2019**, *11*, 6648. [CrossRef]
40. Ud Din, I.; Almogren, A.; Rodrigues, J.J.P.C. AIoT Integration in Autonomous Vehicles: Enhancing Road Cooperation and Traffic Management. *IEEE Internet Things J.* **2024**, *11*, 35942–35949. [CrossRef]
41. Alnaser, A.A.; Maxi, M.; Elmousalami, H. AI-Powered Digital Twins and Internet of Things for Smart Cities and Sustainable Building Environment. *Appl. Sci.* **2024**, *14*, 12056. [CrossRef]
42. Bibri, S.E. The IoT for smart sustainable cities of the future: An analytical framework for sensor-based big data applications for environmental sustainability. *Sustain. Cities Soc.* **2018**, *38*, 230–253. [CrossRef]
43. Xu, Z.; Abawajy, J. Smart Electronics, Energy, and IoT Infrastructures for Smart Cities. *Electronics* **2025**, *14*, 407. [CrossRef]
44. Mishra, P.; Singh, G. Energy Management Systems in Sustainable Smart Cities Based on the Internet of Energy: A Technical Review. *Energies* **2023**, *16*, 6903. [CrossRef]
45. Zhang, X.; Manogaran, G.; Muthu, B. IoT enabled integrated system for green energy into smart cities. *Sustain. ENERGY Technol. Assess.* **2021**, *46*, 101208. [CrossRef]
46. Houichi, M.; Jaidi, F.; Bouhoula, A. Cyber Security within Smart Cities: A Comprehensive Study and a Novel Intrusion Detection-Based Approach. *CMC-Comput. Mater. Contin.* **2024**, *81*, 393–441. [CrossRef]
47. Nesi, P.; Paolucci, M.; Zamperlin, P. Snap4City: A big data platform for smart cities. *Contesti Città Territ. Progetti* **2020**, *1*, 90–107. [CrossRef]
48. Nesi, P. *Smart City verso i Digital Twins*; ITA: Rome, Italy, 2023.
49. Nesi, P.; Po, L.; Viqueira, J.R.R.; Trillo-Lado, R. An Integrated Smart City Platform. In *Semantic Keyword-Based Search on Structured Data Sources*; Szymański, J., Velegrakis, Y., Eds.; Springer International Publishing: Cham, Switzerland, 2018; pp. 171–176. [CrossRef]
50. Khan, M.; Babar, M.; Ahmed, S.H.; Shah, S.C.; Han, K. Smart city designing and planning based on big data analytics. *Sustain. Cities Soc.* **2017**, *35*, 271–279. [CrossRef]
51. Khan, M.A.; Siddiqui, M.S.; Rahmani, M.K.I.; Husain, S. Investigation of Big Data Analytics for Sustainable Smart City Development: An Emerging Country. *IEEE Access* **2022**, *10*, 16028–16036. [CrossRef]
52. Shah, S.S.; Ali, M.; Malik, A.W.; Khan, M.A.; Ravana, S.D. vFog: A Vehicle-Assisted Computing Framework for Delay-Sensitive Applications in Smart Cities. *IEEE Access* **2019**, *7*, 34900–34909. [CrossRef]
53. Kim, K.; Jung, J.-K.; Choi, J.Y. Impact of the Smart City Industry on the Korean National Economy: Input-Output Analysis. *Sustainability* **2016**, *8*, 649. [CrossRef]

54. Saleem, M.; Abbas, S.; Ghazal, T.M.; Adnan Khan, M.; Sahawneh, N.; Ahmad, M. Smart cities: Fusion-based intelligent traffic congestion control system for vehicular networks using machine learning techniques. *Egypt. Inform. J.* **2022**, *23*, 417–426. [CrossRef]
55. Rahman, A.; Abbas, S.; Gollapalli, M.; Ahmed, R.; Aftab, S.; Ahmad, M.; Khan, M.A.; Mosavi, A. Rainfall Prediction System Using Machine Learning Fusion for Smart Cities. *Sensors* **2022**, *22*, 3504. [CrossRef]
56. Abbas, S.; Khan, M.A.; Falcon-Morales, L.E.; Rehman, A.; Saeed, Y.; Zareei, M.; Zeb, A.; Mohamed, E.M. Modeling, Simulation and Optimization of Power Plant Energy Sustainability for IoT Enabled Smart Cities Empowered With Deep Extreme Learning Machine. *IEEE Access* **2020**, *8*, 39982–39997. [CrossRef]
57. Abbas, S.; Khan, M.A.; Athar, A.; Shan, S.A.; Saeed, A.; Alyas, T. Enabling Smart City With Intelligent Congestion Control Using Hops With a Hybrid Computational Approach. *Comput. J.* **2022**, *65*, 484–494. [CrossRef]
58. Syed, A.S.; Sierra-Sosa, D.; Kumar, A.; Elmaghraby, A. IoT in Smart Cities: A Survey of Technologies, Practices and Challenges. *Smart Cities* **2021**, *4*, 429–475. [CrossRef]
59. Syed, A.S.; Sierra-Sosa, D.; Kumar, A.; Elmaghraby, A. Making Cities Smarter—Optimization Problems for the IoT Enabled Smart City Development: A Mapping of Applications, Objectives, Constraints. *Sensors* **2022**, *22*, 4380. [CrossRef] [PubMed]
60. Bellini, P.; Nesi, P.; Pantaleo, G. IoT-Enabled Smart Cities: A Review of Concepts, Frameworks and Key Technologies. *Appl. Sci.* **2022**, *12*, 1607. [CrossRef]
61. Boukerche, A.; Coutinho, R.W.L. Smart Disaster Detection and Response System for Smart Cities. In Proceedings of the 2018 IEEE Symposium on Computers and Communications (ISCC), Natal, Brazil, 25–28 June 2018; pp. 01102–01107. [CrossRef]
62. Chen, X.; Deng, Y.; Ding, H.; Qu, G.; Zhang, H.; Li, P.; Fang, Y. Vehicle as a Service (VaaS): Leverage Vehicles to Build Service Networks and Capabilities for Smart Cities. *IEEE Commun. Surv. Tutor.* **2024**, *26*, 2048–2081. [CrossRef]
63. Jain, S.; Gupta, S.; Sreelakshmi, K.K.; Rodrigues, J.J.P.C. Fog computing in enabling 5G-driven emerging technologies for development of sustainable smart city infrastructures. *Clust. Comput.* **2022**, *25*, 1111–1154. [CrossRef]
64. Ei Leen, M.W.; Jafry, N.H.A.; Salleh, N.M.; Hwang, H.; Jalil, N.A. Mitigating Traffic Congestion in Smart and Sustainable Cities Using Machine Learning: A Review. In *Computational Science and Its Applications—ICCSA 2023*; Gervasi, O., Murgante, B., Taniar, D., Apduhan, B.O., Braga, A.C., Garau, C., Stratigea, A., Eds.; Springer Nature: Cham, Switzerland, 2023; pp. 321–331. [CrossRef]
65. Bibri, S.E.; Krogstie, J. Environmentally data-driven smart sustainable cities: Applied innovative solutions for energy efficiency, pollution reduction, and urban metabolism. *Energy Inform.* **2020**, *3*, 29. [CrossRef]
66. Silva, B.N.; Khan, M.; Han, K. Towards sustainable smart cities: A review of trends, architectures, components, and open challenges in smart cities. *Sustain. Cities Soc.* **2018**, *38*, 697–713. [CrossRef]
67. Bibri, S.E.; Huang, J.; Jagatheesaperumal, S.K.; Krogstie, J. The synergistic interplay of artificial intelligence and digital twin in environmentally planning sustainable smart cities: A comprehensive systematic review. *Environ. Sci. Ecotechnol.* **2024**, *20*, 100433. [CrossRef]
68. Satpathy, I.; Nayak, A.; Jain, V.; Padmadas, S.S. Applying Data Into Action: AI-Powered Solutions for Mitigating Climate Change and Fostering Sustainable Future. In *AI Applications for Clean Energy and Sustainability*; IGI Global Scientific Publishing: Hershey, PA, USA, 2024; pp. 50–74. [CrossRef]
69. Shahbazi, Z.; Shahbazi, Z.; Nowaczyk, S. Enhancing Air Quality Forecasting Using Machine Learning Techniques. *IEEE Access* **2024**, *12*, 197290–197299. [CrossRef]
70. Ullah, U.; Usama, M.; Muhammad, Z.; Akbar, A.; Latif, S.; Ullah, R. Intelligent transportation channels for smart cities. In *Artificial Intelligence for Intelligent Systems*; CRC Press: Boca Raton, FL, USA, 2024.
71. Songhorabadi, M.; Rahimi, M.; Farid, A.M.M.; Kashani, M.H. Fog Computing Approaches in Smart Cities: A State-of-the-Art Review. *arXiv* **2020**, arXiv:2011.14732. [CrossRef]
72. Barbierato, E.; Gatti, A. Decoding Urban Intelligence: Clustering and Feature Importance in Smart Cities. *Future Internet* **2024**, *16*, 362. [CrossRef]
73. Wright, J.; Weber, V.; Walton, G.F. Identifying potential emerging human rights implications in Chinese smart cities via machine-learning aided patent analysis. *Internet Policy Rev.* **2023**, *12*, 3. [CrossRef]
74. Sivagurunathan, S.; Sebastian, A.; Prathapchandran, K. Internet of Things for Developing Smart Sustainable Cities (SSC): A Security Perspective. In *Connectivity Frameworks for Smart Devices: The Internet of Things from a Distributed Computing Perspective*; Mahmood, Z., Ed.; Springer International Publishing: Cham, Switzerland, 2016; pp. 307–331. [CrossRef]
75. Qu, F. How Can Smart Buildings Enabled by IoT Technology Accelerate Sustainable Green City Development: Case Studies in China. *Eur. Acad. J.—II CHINA* **2024**, *1*. Available online: https://eaj.ebujournals.lu/index.php/EAJ_II/article/view/105 (accessed on 18 March 2025).
76. Pandey, R.; Rajput, I.S.; Tyagi, S.; Koranga, M.; Sharma, P.K. Enhancing Smart City Resilience: Integrating Cloud Computing and 5G for Secure and Sustainable Urban Environments. In *Secure and Intelligent IoT-Enabled Smart Cities*; IGI Global Scientific Publishing: Hershey, PA, USA, 2024; pp. 77–90. [CrossRef]

77. Jafari, M.; Kavousi-Fard, A.; Chen, T.; Karimi, M. A Review on Digital Twin Technology in Smart Grid, Transportation System and Smart City: Challenges and Future. *IEEE Access* **2023**, *11*, 17471–17484. [CrossRef]
78. Ranjan, R.; Firoz, C.M.; Rajendran, L.P. Sustainable Development Goals. In *The Palgrave Encyclopedia of Urban and Regional Futures*; Brears, R.C., Ed.; Springer International Publishing: Cham, Switzerland, 2022; pp. 1757–1770. [CrossRef]
79. Ullah, Z.; Al-Turjman, F.; Mostarda, L.; Gagliardi, R. Applications of Artificial Intelligence and Machine learning in smart cities. *Comput. Commun.* **2020**, *154*, 313–323. [CrossRef]
80. Ma, C. Smart city and cyber-security; technologies used, leading challenges and future recommendations. *Energy Rep.* **2021**, *7*, 7999–8012. [CrossRef]
81. Bellini, P.; Bologna, D.; Nesi, P.; Pantaleo, G. A Unified Knowledge Model for Managing Smart City/IoT Platform Entities for Multitenant Scenarios. *Smart Cities* **2024**, *7*, 2339–2365. [CrossRef]
82. Adreani, L.; Bellini, P.; Bilotta, S.; Bologna, D.; Collini, E.; Fanfani, M.; Nesi, P. Smart City Scenario Editor for General What-If Analysis. *Sensors* **2024**, *24*, 2225. [CrossRef]
83. Hasan, M.K.; Khan, M.A.; Issa, G.F.; Atta, A.; Akram, A.S.; Hassan, M. Smart Waste Management and Classification System for Smart Cities using Deep Learning. In Proceedings of the 2022 International Conference on Business Analytics for Technology and Security (ICBATS), Dubai, United Arab Emirates, 16–17 February 2022; pp. 1–7. [CrossRef]
84. Habibzadeh, H.; Kaptan, C.; Soyata, T.; Kantarci, B.; Boukerche, A. Smart City System Design: A Comprehensive Study of the Application and Data Planes. *ACM Comput. Surv.* **2019**, *52*, 41:1–41:38. [CrossRef]
85. Karas, I.R.; Ben Ahmed, M.; Boudhir, A.A.; Ane, B.K. The sixth international conference on smart city applications: Preface. *Int. Arch. Photogramm. Remote Sens. Spat. Inf. Sci.* **2021**, *XLVI-4-W5-2021*, 1–6. [CrossRef]
86. Chen, J.; Yi, C.; Okegbile, S.D.; Cai, J.; Shen, X. Networking Architecture and Key Supporting Technologies for Human Digital Twin in Personalized Healthcare: A Comprehensive Survey. *IEEE Commun. Surv. Tutor.* **2024**, *26*, 706–746. [CrossRef]
87. Chen, X.; Duan, Z.; Li, Y.; Wang, M.; Zhu, H. Coordination control and start-up strategy in flexible DC urban distribution networks for smart cities. In Proceedings of the 2017 IEEE Conference on Energy Internet and Energy System Integration (EI2), Beijing, China, 26–28 November 2017; pp. 1–5. [CrossRef]

buildings

MDPI

Article

Study on the Influence Mechanism of Machine-Learning-Based Built Environment on Urban Vitality in Macau Peninsula

Chen Pan [1,2], Jiaming Guo [1,*,†], Haibo Li [1,*,†], Jiawei Wu [1], Nengjie Qiu [1] and Shengzhen Wu [3]

[1] Architecture and Civil Engineering Institute, Guangdong University of Petrochemical Technology, Maoming 525000, China; jourdan29@gdupt.edu.cn (C.P.); wayne@gdupt.edu.cn (J.W.); qnj906@gdupt.edu.cn (N.Q.)

[2] Faculty of Innovation and Design, City University of Macau, Macau 999078, China

[3] College of Arts and Design, Jimei University, Xiamen 361021, China; 2361000218@jmu.edu.cn

* Correspondence: gjm11@gdupt.edu.cn (J.G.); hbl@gdupt.edu.cn (H.L.)

† These authors contributed equally to this work.

Abstract: Clarifying the mechanisms by which the micro-scale built environment influences urban vitality is an important scientific challenge, to guide precise urban planning in the context of urban renewal. In this study, we quantify the intensity of human activities through Baidu heat maps, analyze social interaction patterns using social media check-in data, and integrate built environment elements such as road network topology, 3D building morphology, and the spatial distribution of points of interest (POIs). A machine learning technique combining a real-encoded Accelerated Genetic Algorithm-Projective Pathfinding Model (RAGA-PPM) and Shapley Additive Projection for Interpretability (SHAP) for Interpretability Analysis (IPA) was used to investigate the nonlinear mechanisms of 17 factors affecting urban vitality in Macau Peninsula, China. Firstly, the explanatory power of the built environment for comprehensive vitality was significantly better than the other dimensions. Two factors, population vitality and microblogging check-in vitality, contributed the most to the composite vitality value. Secondly, road network density was the most important built environment factor affecting urban vitality in Macau Peninsula (SHAP = 0.025). Finally, the impacts of built environment factors on urban vitality showed nonlinearities, and the threshold effects of the core factors (road network density, spatial fractal dimension, and openness to the sky) showed a consistent neighborhood-level pattern. This study establishes a framework for micro-vitality mechanisms in high-density cities, addressing the limitations of traditional methods in modeling complex nonlinear relationships. The methodological integration of RAGA-PPM and SHAP advances the innovative paradigm of applying interpretable machine learning to the study of urban form.

Keywords: built environment; urban vitality; projection pursuit; genetic algorithm; Macau Peninsula

1. Introduction

With the acceleration of global urbanization, the relationship between urban vitality and the built environment has increasingly become a crucial research topic in the fields of urban planning and management. Urban vitality generally refers to the comprehensive performance of a city in economic, social, cultural, and environmental aspects, directly influencing residents' quality of life, social interactions, and economic development levels [1]. The built environment, as a core component supporting urban functions, encompasses multiple dimensions such as urban spatial structure, transportation facilities, green space distribution, and building density. Its quality and layout play a vital role in the generation and development of urban vitality [2]. Traditional research has often simplified the relationship between urban vitality and the built environment into linear correlations, employing statistical methods like regression analysis for exploration [3]. However, the relationship between urban vitality and the built environment is inherently nonlinear, influenced by the intertwining and interaction of various factors, and this interaction is highly complex [4]. Existing linear models fail to effectively capture this complex nonlinearity, leading to certain limitations in the prediction and optimization of urban vitality. In the context of sustainable urban development, there is an urgent need for more refined research on how to balance high-density development with ecological preservation, optimize transportation networks to reduce carbon emissions, and enhance environmental comfort. Although existing studies have recognized the nonlinear association between the built environment and urban vitality, there are still significant gaps in the analysis of specific nonlinear mechanisms of action. Complex relationships such as the inverse inhibitory characteristics of built environment elements for urban vitality in the threshold effect, the differential response mechanism of the same element for vitality under spatial heterogeneity, as well as the delayed and cumulative impacts of environmental changes on vitality in the dynamic time lag effect have not yet been adequately quantified and deciphered [5]. Traditional linear models have difficulty capturing multidimensional nonlinear features, due to the restriction of fixed coefficient assumptions, leading to insufficient guidance on key issues such as synergistic paths for high-density development and ecological protection, marginal benefit thresholds for transportation network optimization, and scale effects of green infrastructure implementation. Therefore, exploring a modeling method that can systematically identify and reveal the multiple types of nonlinear relationships between the built environment and urban vitality is an important direction, to promote the progress of urban research theory and practice.

In recent years, with the rapid development of artificial intelligence and big data technologies, modeling methods based on complex systems have gradually become important tools in urban research [6,7]. In particular, the Real-coded Adaptive Genetic Algorithm (RAGA) can achieve global optimization in complex, multidimensional, and multivariate data, while the Probabilistic Potential Model (PPM) reveals deep-seated correlations among various factors in urban systems through latent factor modeling [8]. Although existing research has made some progress in theoretical construction and model application, there has been a lack of exploration into urban vitality measurement based on the RAGA-PPM model [9]. The integration of machine learning algorithms with Shapley Additive Explanations (SHAP) provides intuitive interpretability for complex nonlinear models. By quantifying the contribution of each feature to the model's output, it can reveal how built environment elements nonlinearly influence urban vitality, further enhancing the model's transparency and practicality [10].

In this study, we aim to construct a hybrid computational framework capable of characterizing the nonlinear dynamics of the built environment and urban vitality by coupling the global optimization capability of the real-encoded adaptive genetic algorithm (RAGA) with the potential factor inference advantages of the probabilistic potential model (PPM). On this basis, SHAP is introduced to decouple the differentiated paths of the contribution of multi-dimensional built environment elements to vitality generation and their interaction effects. Based on multi-source high-precision urban data, a cross-scale empirical study was carried out to quantitatively verify the superiority of the model for vitality prediction accuracy, generation of spatial intervention strategies, and resolution of nonlinear relationships, forming a complete closed loop of "theory construction–mechanism analysis–decision-making support", and providing a combination of machine learning and explanatory planning and decision-making capabilities for the optimization of the built environment in the context of a smart city. We provide a methodological tool for the optimization of built environment in the context of smart cities, which combines machine learning, sophistication, and interpretability of planning and decision-making.

2. Literature Review

The term "vitality" in Chinese was originally used to describe and express the state of activity of living organisms. Later, it was understood as the vigorous vitality of living organisms or the activity of individual organisms in their actions, communication and expression, and thinking. The term "vitality" was initially mainly applied in the field of physics. With the research and exploration of the natural world, it further was expanded to the field of biology. Later, it was introduced into various scientific research fields and has also developed different connotations in various research processes within different disciplinary systems. Ultimately, the term "vitality" has a rich meaning that transcends its original meaning, and there are certain differences in its representation and interpretation in different fields and specific research issues. The concept of "vitality" in this study is mainly in reference to cities and urban areas. It mainly represents the "vitality" within the city itself or the urban block space. The definition of "urban vitality" begins with understanding the nature of a city's continuous operation and regards the overall framework of the city as a complex organic life or physical system. Urban vitality, as a crucial indicator for assessing urban health and sustainable development, has long attracted attention from the academic community. At the macro level, urban statistical data, interviews, and questionnaires have been utilized as data sources [11,12]. However, these data sources are often plagued by issues such as insufficient timeliness, limited sample sizes, and being time-consuming and labor-intensive, making them difficult to apply to fine-grained perspectives. At the meson and micro levels, Jacob's explanatory indicator framework, which approaches from various dimensions, has been widely accepted [13–15]. Scholars have explored urban vitality from aspects such as urban morphology, design, land use, population density, mobility, and transportation accessibility [16–21]. Most research on urban vitality has focused on the exploration of indicators and their impact levels, with relatively limited studies on the formation mechanisms of vitality itself and comprehensive evaluations. Furthermore, more granular indicators have also not been fully developed [22,23].

2.1. Research on the Relationship Between Urban Vitality and Built Environment

Early studies on urban vitality primarily focused on the role of urban economic development, social structure optimization, and infrastructure improvement in promoting urban vitality. Scholars have extensively investigated the impact of built environment variables such as traffic density [1], green space [14], public facility provision [20], and building density [24] on urban vitality through regression analysis and spatial econometric models [24]. In her seminal work *"The Death and Life of Great American Cities"*, Jacobs defined urban vitality as a self-regenerating force capable of nurturing and promoting diverse pedestrian activities and the connections between neighborhood and street-level activities [25].

With the rapid advancement of information technology, a new data environment centered around big data and open data has gradually emerged. The availability of diverse data sources, including GPS data, mobile phone signaling, social media check-ins, and street view images, has provided richer possibilities for measuring urban vitality, thereby expanding the depth and breadth of built environment research [26–32]. Building upon this foundation, studies conducted at various urban scales—from administrative districts [30,33,34] to neighborhoods [12,35], streets [36–38], and public spaces [39,40]—have consistently demonstrated the significant role of the built environment in fostering urban vitality.

2.2. *Application of Nonlinear Models and Complex Systems Approache*

Existing studies have mainly used linear regression models, spatial measurement models, and geographically weighted regression models to analyze the effects of built environment factors on urban vitality and their significance [20,31,33,34,36,39–41]. Although these methods can verify the statistical significance among variables, their linear assumptions have difficultly capturing the nonlinear nature of multi-factor interactions within urban systems. In recent years, scholars have gradually introduced complex modeling techniques to overcome this limitation. For example, the adaptive genetic algorithm (RAGA) optimizes the global solution of a high-dimensional nonlinear problem by simulating biological evolutionary mechanisms (e.g., selection, crossover, and mutation), which significantly improves the efficiency of parameter finding [42]. Its advantage lies in its ability to process multi-objective optimization tasks in parallel and adaptively adjust the search interval to approximate the optimal solution [43]. However, the RAGA is sensitive to the initial parameter settings, and the computational complexity increases exponentially with the increase in the dimensionality of the variables, which may lead to a decrease in the convergence speed, or falling into a local optimum [42,44]. In addition, its "black-box" nature limits the interpretability of the model results, making it difficult to directly guide the planning practice. Gradient boosted decision tree (GBDT) is an iterative approach that builds weak classifiers and weighted combinations by integrating learning strategies, and it has been shown to perform well in identifying the threshold effects and interactions of built environments for urban vitality [35,44–47]. The GBDT method is particularly suitable for multi-source heterogeneous urban big data analysis, as it can deal with multiple covariations between features under loose assumptions on data distribution. However, its limitations are equally significant: on the one hand, the model tends to over-rely on high-frequency features and may ignore low-frequency but high-impact variables; on the other hand, the complexity of the tree structure undermines the interpretability of the results, and although methods such as SHAP (Shapley Additive Properties for Interpretation) have been used for post hoc interpretations, they are computationally costly and rely on sample independence assumptions [48]. In addition, GBDT is more sensitive to outliers and noisy data, which may affect the robustness of threshold determination. Future research could compensate for the shortcomings of a single algorithm through a hybrid modeling framework. The RAGA could be combined with local search algorithms to balance global exploration and local exploitation capabilities; or an attention mechanism could be introduced to improve the feature weight allocation logic of GBDT and enhance the ability to capture sparse features. Meanwhile, the integration path between Explainable ML and Spatially Explicit Modeling should be further explored to reveal the spatial heterogeneity and spillover effect of built environment elements, to provide improved operational theoretical support for refined urban planning.

In summary, significant progress has been made in the current literature regarding the exploration of nonlinear relationships between urban vitality and the built environment. However, there remains considerable room for improvement regarding the complexity of model construction and the breadth and depth of multifactor analysis in existing studies. Most research still relies on linear assumptions or simplified variable settings, failing to comprehensively capture the complex interactive effects between the built environment and urban vitality. Future research should focus on developing more sophisticated and refined modeling methods, while emphasizing the integration of interdisciplinary theories, particularly the incorporation of sociological and behavioral economic theories. This will not only enhance the explanatory power of existing models but also provide crucial support for advancing innovation in urban planning theory and practice.

3. Study Area and Data Sources

3.1. Study Area

As one of the three core areas of the Macau Special Administrative Region, the Macau Peninsula is located on the west bank of the Pearl River Delta (latitude 22°11′ to 22°13′ N, longitude 113°32′ to 113°35′ E), adjacent to the Gongbei Border Crossing of Zhuhai City in the north, and connected to the islands of Taipa and Coloane in the south by crossing the Sai Van Bridge, making it the political, economic, and cultural hub of Macau (Figure 1). According to the 2024 data of the Statistics and Census Bureau of the Macau SAR Government [47], the region has a land area of 9.1 km^2 and a resident population of about 452,700, with a population density of 49,747 inhabitants/km^2, which is among the most densely populated regions in the world. The topographic features are dominated by hilly terrain, with Pine Hill as the highest point (93 m above sea level), and artificial land areas such as the New Border Crossing and Nam Van Lake formed along the coast through reclamation projects, constituting a spatial superposition of the characteristics of a historical urban area and a modern new district. The Macau Peninsula is divided into 159 neighborhoods, among which the Historic District (which concentrates 22 historical building clusters such as the Ruins of St. Paul's, the A-Ma Temple, the Senado Square, etc.) demonstrates the unique urban texture of Sino-Portuguese cultural fusion.

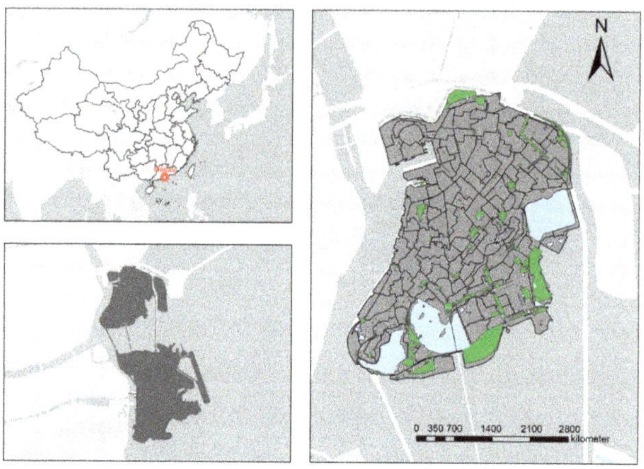

Figure 1. Map of the study area.

As an important node of the Guangdong–Hong Kong–Macau Greater Bay Area, the Macau Peninsula has close transportation and economic links with cities such as Zhuhai and Hong Kong through the Hong Kong–Zhuhai–Macau Bridge and the Guangzhou–Zhuhai Intercity Railway. The urban space has undergone a transformation from a single-core agglomeration in the historical city center to a multi-core expansion in the peripheral new reclamation area, forming a "dual-center" structure: first, a historical, cultural, and tourism center with the Senado Square as the core, concentrating 80% of the world cultural heritage and traditional commercial districts, and facing contradictions between heritage preservation and high-density development; and second, a modern business and entertainment center in the Outer Harbor New Reclamation Area. The second area is the modern business and entertainment center of the Outer Harbor New Reclamation Area, which contains integrated resorts such as MGM MACAU and Wynn Palace, but suffers from an imbalance in daytime and nighttime vitality due to its single function. The transportation network in the area is dense but limited by the narrow street pattern of the historical city center, where walking and public transportation are the main modes of travel. The Macau Peninsula, with its high-density urbanization, cultural diversity, and spatial compactness, has become a typical sample for exploring the non-linear interaction mechanisms between the built environment and urban vitality, and its experience has important theoretical value and practical significance for the sustainable regeneration and vitality enhancement of historic cities around the world.

3.2. Variables and Data

3.2.1. Vitality Measurement

The measurement of urban vitality is conducted from three perspectives: human activity, online interaction, and built environment. Among these, human activity serves as a crucial indicator for assessing urban vitality, with the distribution and movement patterns of people being key drivers of regional prosperity [48]. Baidu's heatmap population data provide real-time population density and activity information, revealing areas of high human concentration and dynamic population movements, thereby offering scientific support for urban planning and management. As behavioral records from a mainstream social media platform, Weibo check-in data reflect peak demand and distribution characteristics of users of public facilities and services [46,49–52], unveiling the activity patterns and preferences of both residents and tourists, which provide valuable references for urban operations and decision-making.

3.2.2. Measurement of the Built Environment

In this study, we adopted the "5D" built environment indicator system proposed by Ewing et al. [2], which includes density, diversity, design, destination accessibility, and distance to transit. Building upon previously established indicator systems [33,35,45], and considering data availability and indicator suitability, we selected 15 indicators to measure the built environment across seven dimensions: urban vitality, spatial form, functional characteristics, socio-economic environment, psychological perception, transportation accessibility, and street quality. Given that geographical elements exhibit spatial dependency effects, empirical models using machine learning algorithms are unable to capture the dependency between spatial units or measure the direct effects and spatial spillover effects between variables [53]. The specific calculation methods for each indicator are detailed in Table 1.

Table 1. Variable definitions and descriptive statistics.

Variable	Detection Metric	Calculation Formula	Formula Description
Urban Vitality	Population Density Weibo Check-in Density	$R = \frac{H}{A}$	R represents the density value of a specific neighborhood unit; H denotes the kernel density value of a particular neighborhood unit; A is the area of a given neighborhood unit.
Spatial Configuration	Spatial Compactness	$c = \frac{\sqrt[2]{\pi^{*}\text{Aera}}}{l}$	C represents the compactness of a block, which is determined by the area of the land parcel within the block unit and the lengths of the boundaries of the parcel.
	Fractal Dimension	$D = \frac{2\ln(\frac{h}{4})}{\ln(A)}$	D value typically ranges between 1.0 and 2.0, with a higher D value indicating greater complexity in the spatial configuration of land parcels within the region.

Table 1. *Cont.*

Variable	Detection Metric	Calculation Formula	Formula Description
Functional Attribute	Functional Density	$fd_i = \frac{N_i}{S_i}$	fd_i represents the total number of Points of Interest (POI) within the i-th urban block, S_i refers to the area of the i-th category of urban blocks.
	Functional Mix	$I = -\sum_{i=1}^{n} (p_i * \text{Inp}_i)$	I represents the functional mix of the block, "i" denotes the number of POIs within the i-th block unit, and "Pi" signifies the ratio of the i-th category of POIs to the total number of POIs within the block unit.
Socio-Economic Environment	Cultural Facility Density Public Facility Density	$D = \frac{N}{A}$	D represents the ratio of the total number of cultural facilities, public facilities, and commercial facilities within a specific region to the total area of that region.
Psychological Perception	Building Density	$P = \frac{\sum_{i=1}^{n} m_i}{s}$	P represents the building density, i denotes the identification number of different buildings, m_i refers to the base area of the building with the i-th identification number, and s stands for the usable area.
	Hydrophilic index	$WI = K - NER\_DIST$	WI denotes the hydrophilicity index of spatial blocks, while K is a constant term representing the search radius value for neighborhood analysis.
Transportation Accessibility	Road Network Densit Bus Stop Density	$D = \frac{N}{A}$	D represents the ratio of the total number of commercial facilities and bus stops to the total area within the study region.
Street Quality	Green View Index	$GVI_i = P_{Vegetation}$	The visibility of vegetation as perceived by the human eye was measured to assess the level of greenery on urban streets.
	sky view factor	$SVF_i = P_{sky}$	The term refers to the proportion of the sky visible within the field of view at a specific location, typically expressed as a value ranging from 0 to 1.
	Enclosure	$Enclosure_i = P_{Building} + P_{Wall} + P_{Fence} + P_{Pole} + P_{Light} + P_{Sign} + P_{Vegetation}$	The sense of enclosure measures the extent to which an individual feels surrounded by the surrounding environment within a space.
	Walkability	$Walkability_i = \frac{P_{Fence} + P_{Sidewalk}}{P_{road}}$	Refers to the level of pedestrian-friendliness of a street or urban area.

3.3. Data Sources and Preprocessing

The data were categorized into two main dimensions: comprehensive urban vitality and built environment. The datasets utilized primarily consisted of fundamental geographic data and open-source data obtained from the Internet, as detailed in Table 2.

Table 2. Data sources and acquisition timeline.

Data Type	Data Name	Data Source	Time	Preprocessing
Fundamental Geospatial Data	Macau Road Network Macau Architectural Structures Hydrological Data	Tianditu (https://www.tianditu.gov.cn/ accessed on 3 April 2024)	2024	The data were uniformly projected onto the WGS84 coordinate system to classify road and building categories, and metrics such as road density, building area, and water system area were subsequently calculated.
Open-source Web Data	Baidu Heatmap Population Data	Baidu Map Smart Eye Population Big Data Platform (https://huiyan.baidu.com/ accessed on 3 April 2024)	2024	The Kernel Density analysis tool within ArcGIS was employed to convert the data into a continuous raster density map.
	Macau Housing Price Data	Estate Information Network of Macau (https://www.malimalihome.net/ accessed on 3 April 2024)	2023	The core fields, including unit price, total price, gross floor area, and property type, were collected to calculate key metrics such as price per square meter and average housing prices by region.

Table 2. *Cont.*

Data Type	Data Name	Data Source	Time	Preprocessing
Open-source Web Data	Points of Interest (POI) Data	GaoDe Map	2024	The dataset underwent comprehensive data cleaning, including the handling of missing values and outliers. Classification and standardization processes were implemented to ensure field uniformity across all 45,708 records.
	Weibo Check-in Data	Sina Weibo (https://weibo.com/ accessed on 3 April 2024)	2020	The data cleaning process involved handling missing values and outliers, followed by categorizing the data based on check-in locations. The total number of Weibo check-ins processed was 145,370.
	Street View Imagery	Baidu Street View Map Open Platform (https://lbs.baidu.com/ accessed on 3 April 2024)	2023	Sampling points were generated at 100 m intervals based on road network data. Utilizing the Baidu API endpoint with a pitch angle of 0 degrees, images were captured at four different azimuth angles: 0°, 90°, 180°, and 270°. A total of 10,442 images were collected.

3.3.1. Basic Geographic Data

The basic geographic data included the administrative division data of Macau Peninsula, road network data, water system data, building data, and road network data, all of which came from the Tianditu.

3.3.2. Open-Source Data

(1) Baidu Heat Population Data

Baidu thermal population data were obtained based on the Baidu Map Wise Eye demographic geographic big data platform, whose location information covers almost every aspect of daily life. Compared with "check-in" or census data, this heat map dataset is almost unbiased [32], can provide almost real-time demographic data, and can well demonstrate the direction of urban population movement and spatial aggregation status. The platform includes functions within a certain spatial range to provide online data analysis and visualization services, including statistical and dynamic visualization information such as regional population, occupancy and residence, passenger flow, facilities, brands, ODs, and heat maps. Baidu Map Wise Eye population location data are mainly derived from calling Baidu Map Positioning SDK terminal positioning data. It has been demonstrated that crowd movement conforms to similar distribution and evolution laws based on Baidu heat map data from Monday–Friday (weekdays) and Saturday and Sunday (weekends) [48]. In this paper, the statistical information of the heat map in Baidu Map Wise Eye urban population data platform was used to obtain the data of 24 moments on May 15 (Monday) and May 20 (Saturday), 2024 as the basic data.

(2) Macau Property Price Data

Macau house price data were obtained based on the preferred real estate information platform of Macau (Malimali Home), which integrates real-time transaction information of residential and commercial properties in the Macau Peninsula, covering core fields such as the unit price of the property, the total price, the floor area, the type of property ownership, and the time of the transaction. Compared with the government's periodic statistical reports, such market transaction data can capture the spatial and temporal characteristics of house prices more dynamically and avoid the lagging bias of traditional statistics [47]. The platform provides historical transaction records and regional price indices through a standardized data interface, and its data cover more than 90% of the real estate agencies in the Macau Peninsula. It has been demonstrated that such data have high reliability and validity in parsing the spatial differentiation of urban settlements and the distribution of commercial vitality [51]. In this study, we extracted transaction data for the whole year of 2023, cleaned the outliers (excluding records with unit prices exceeding three times the standard deviation of the mean or with missing key fields), and calculated the average house price, price gradient, and the proportion of the distribution of the property types in each neighborhood, in order to quantify the mechanism by which the built-up environment affects the spatial differentiation of house prices.

(3) POI data

POI data, as a kind of point data representing geographic entities, include attribute information, such as name, type, location, etc., which is mainly used for density calculation and functional area diversity calculation. According to the Urban Land Use Classification and Planning and Construction Land Use Standards [49] and the actual situation of Nanjing, POIs are classified into seven major categories, which are residential land, commercial land, industrial land, public service land, science, education and culture land, green space and square land, and transportation facility land. In this paper, the POI data points of MSAR in 2024 obtained from Gaode Map totaled 145,370.

(4) Weibo check-in data

The Weibo check-in function allows users to record their whereabouts in real time and show their life status and activity experience. Users can express their favorites, and support or comment on a certain place or activity, by checking in, and at the same time, they can also browse the check-in information of others to understand their activity trajectories and locations. By analyzing Weibo check-in data, we could understand the distribution of people flow, heat index, and activity density of different locations in the city, revealing the population flow patterns and social activity hotspots in the city. In this paper, a total of 145,370 Weibo check-ins of Macau SAR in 2020 were obtained from Sina Weibo.

(5) Street View Image

Street View image data were acquired based on the Baidu Street View map open platform, which covers urban street space through high-resolution panoramic photography technology and can objectively present the material form and spatial enclosure characteristics of street façades. Compared with traditional field research, this dataset has the advantages of standardized collection process and simultaneous acquisition from multiple viewpoints, which can effectively avoid the subjective bias of human observation [50]. The platform provides API interface services to support researchers in batch acquiring street view images of specified locations based on geographic coordinates or road network nodes, and it allows customization of parameters such as shooting angle, elevation angle, and image resolution. It has been shown that streetscape images have significant application value in the fields of street greening-rate measurement [51], building interface continuity assessment [52], and walking environment quality quantification [46]. In this study, based on the road network data of Macau Peninsula, sampling points were set up at 100 m intervals, and street view images were obtained from four horizontal viewpoints, (0° pitch angle 0°) at 0° (due north), 90° (due east), 180° (due south), and 270° (due west), of each point location through the Baidu Street View API, and 10,442 effective street view images were finally collected for the year 2023.

4. Research Methodology

This study employed the Macau Peninsula as a case study, dividing the research area into 159 urban blocks based on road network data. A comprehensive urban vitality assessment framework was constructed, encompassing three dimensions: human activity, network interaction, and built environment. The projection pursuit model optimized by the accelerated genetic algorithm (RAGA-PPM) was utilized to reduce the dimensionality of block indicators, extract optimal projection directions, and calculate both single-dimensional and comprehensive vitality values across various aspects. Subsequently, vitality centers were identified and compared with results from the entropy weight method (EWM) to validate the effectiveness of RAGA-PPM. Furthermore, the study investigated the nonlinear impact mechanisms of the built environment on urban vitality by integrating machine learning with Shapley additive explanations (SHAP), providing scientific support for urban planning. The technical framework is illustrated in Figure 2.

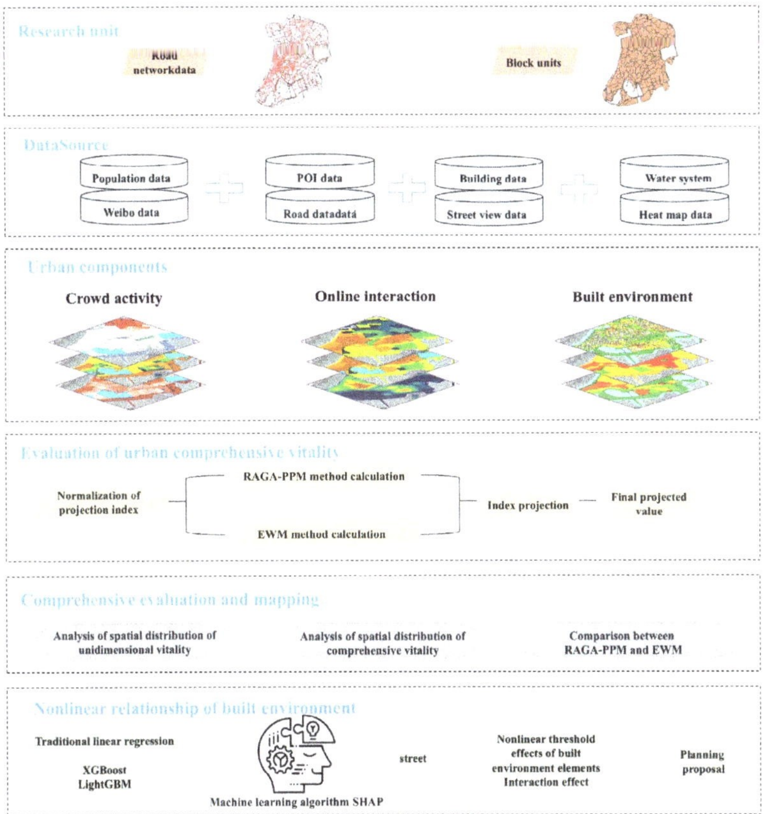

Figure 2. Technical roadmap.

4.1. Calculation of Projection Pursuit Model Based on Real-Coded Genetic Algorithm

4.1.1. Projection Pursuit Model (PPM)

In urban vitality assessment, traditional methods typically rely on expert scoring and multi-indicator comprehensive evaluation, such as factor analysis, the entropy [54] weight method, and TOPSIS method. Although these methods have achieved certain results, they are limited by strong subjectivity, fewer dimensions, and difficulty in handling high-dimensional nonlinear data. Given that urban vitality involves multiple high-dimensional and nonlinear dimensions, 15 indicators were selected for calculation. The Projection Pursuit Model (PPM) [55] can effectively identify the optimal projection direction in high-dimensional data. This method reveals the structural characteristics of data by mapping high-dimensional data into a low-dimensional space [26]. The specific steps are as follows:

(1) Since the data sources of projection indicators have different numerical types and dimensions, in order to eliminate the dimensions and unify the range of indicator changes, the following formula is used for standardization; for defined positive indicators, as shown in Formula (1), and for negative indicators, as shown in Formula (2):

$$X(i,j) = \frac{x^*(i,j) - x_{min}(j)}{x_{max}(j) - x_{min}(j)} \tag{1}$$

$$X(i,j) = \frac{x_{min}(j) - x^*(i,j)}{x_{max}(j) - x_{min}(j)} \tag{2}$$

In the given formula, $x^*(i,j)$ represents the value of the j indicator for the i sample. Additionally, $x_{max}(j)$ and $x_{min}(j)$ correspond to the minimum and maximum values, respectively, of the j evaluation indicator within the i sample.

(2) Construct the projection index function $z(i)$, project x(i,j) onto the projection direction a, let a= (a (1), a (2), . . ., a (j)), where a is the unit length vector, and the projection value z(i) is

$$z(i) = \sum_{j=1}^{p} (a_j \times x_{ij}) \tag{3}$$

where $z(i)$ is the projection vector of the i th sample, x_{ij} is the i th value in the j th index, and a_j is the projection direction of the projection vector $z(i)$.

(3) Construct the projection objective function $\rho(a)$ to identify the structural combination features of the data within the multidimensional indicators. During the comprehensive projection process, the projection value $Z(i)$ should maximize the extraction of variation information from x(i,j), ensuring that the standard deviation S_z of $Z(i)$ is as large as possible. Simultaneously, the local density D_z of the projection values $Z(i)$ should be maximized. Based on these principles, the projection objective function can be formulated as ρ.

$$\rho(a) = S_z D_z \tag{4}$$

$$S_z = \sqrt{\frac{\sum_{i=1}^{p} (Z(i) - Z^*)^2}{n-1}} \tag{5}$$

$$D_z = \sum_{i=1}^{n} \sum_{j=1}^{n} (R - r_{ij}) u(R - r_{ij}) \tag{6}$$

where S_z is the standard deviation of Z; D_z is the local density of Z; Z^* is the mean of $Z(i)(i = 1{\sim}n)$; r_{ij} is the window radius of the local density, which is related to the search radius of the projection point; r_{ij} refers to the difference between $Z(i)$ and $Z(j)$; $u(R - r_{ij})$istheunitstepfunction. Its function value is 1 when $R - r_{ij} \geq 0$, and its function value is 0 when $R - r_{ij} \geq 0$.

(4) Optimizing the projection objective function to determine the optimal projection direction, different projection directions reflect various structural characteristics of the data. The optimal projection direction is the one that maximizes the likelihood of revealing specific structural features of high-dimensional data. To estimate the optimal projection direction, the projection objective function is maximized, ensuring that the most informative structural properties are exposed.

$$\text{Max} = (\rho(a)) = S_z D_z \tag{7}$$

$$s^{*}t\sum_{j=i}^{p} a^{2}(j) = 1, -1 \leq a(j) \leq 1 \tag{8}$$

(5) Substituting the optimal projection direction into Equation (3), the projection values for each sample can be obtained, enabling a unified evaluation of the sample set. Given that Equations (7) and (8) represent a complex nonlinear optimization problem and involve parameter settings, conventional optimization methods may struggle to provide an effective solution. Therefore, a genetic algorithm (GA), a general global optimization approach that incorporates the principles of biological selection and intra-population chromosome information exchange, is introduced to address the high-dimensional global optimization problem.

4.1.2. Accelerated Genetic Algorithm Based on Real-Coded Encoding (RAGA)

In this study, an improved genetic algorithm—the Accelerated Genetic Algorithm based on Real-Coded Encoding (RAGA)—was employed to optimize the Projection Pursuit Model (PPM) and address challenges in the solution process. RAGA integrates the global search capability of genetic algorithms with the continuous optimization advantages of real-coded encoding, significantly enhancing optimization performance [41–44]. By simulating the biological evolution process, this algorithm executes selection, crossover, and mutation operations in parallel, expanding the search space and increasing the likelihood of identifying the global optimal solution. The adaptive mechanism of RAGA progressively narrows the search range of optimization variables, thereby improving the solution accuracy. As the number of iterations increases, the precision of the solution continues to improve [42].

4.1.3. Projection Pursuit Model Optimized by Accelerated Genetic Algorithm with Real-Coded Encoding (RAGA-PPM)

The Projection Pursuit Model (PPM) effectively handles complex nonlinear relationships, while the Real-coded Adaptive Genetic Algorithm (RAGA) enhances PPM's capability in processing high-dimensional nonlinear data through global search and adaptive adjustment. Compared to traditional methods, RAGA autonomously optimizes parameters, minimizing human intervention, improving the efficiency and accuracy of the evaluation process, and mitigating the risk of local optima. Notably, RAGA has demonstrated outstanding performance in multimodal function optimization. However, the model has certain limitations, such as a high computational complexity and sensitivity to parameter settings. By optimizing the calculation of 17 indicator factors, this study provides a more accurate representation of urban vitality, as illustrated in Figure 3.

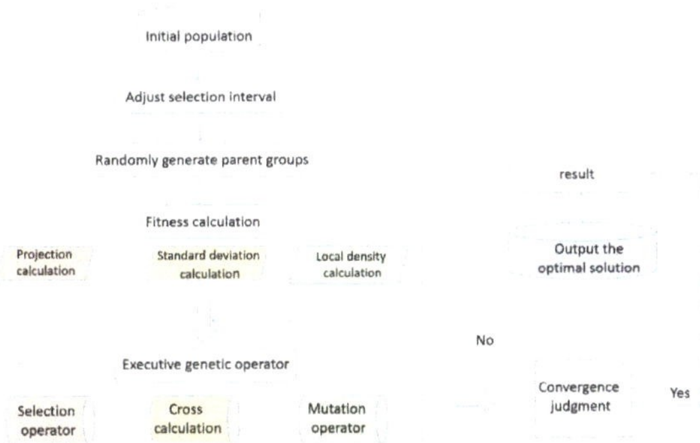

Figure 3. Solution Process of the RAGA-PPM Model.

4.2. Machine Learning Models and SHAP Interpretability Method

Both XGBoost [34] and LightGBM [35] are based on the Gradient Boosting Decision Tree (GBDT) algorithm, which iteratively adds new trees to optimize the model. Although the two algorithms differ in specific details and optimization strategies, they share a similar fundamental formulation. Their objective functions consist of two main components: a loss function and a regularization term.

$$\text{Objective} = \sum_{i=1}^{n} L(y_i, \check{y}_i) + \sum_{k=1}^{k} \Omega(f_k) \tag{9}$$

The loss function is defined as the squared error loss, which measures the squared error between the model predicted value y $\check{y}_i$ and the true value y_i.

$$L(y_i, \check{y}_i) = (y_i, \check{y}_i)^2 \tag{10}$$

The regularization term controls the complexity of the model and prevents overfitting.

$$\Omega(f_k) = \gamma M + \frac{1}{2}\lambda\sum_{j=1}^{M} \omega_j^2 \tag{11}$$

where $\Omega(f_k)$ is the regularization term of the k th leaf in the tree; γ is the penalty coefficient of the number of leaf nodes, used to limit the complexity of the model and reduce overfitting; M is the number of leaf nodes; λ is the penalty coefficient of the leaf node fraction, used to control the L2 regularization strength of the leaf node fraction in the model; and ω_j is the fraction of the j th leaf node.

For the t th tree, the objective function is

$$\text{Objective} = \sum_{i=1}^{n} \left\{ y_i - \left[\check{y}_i^{(t-1)} + f_t(x_i) \right] \right\}^2 + \gamma M + \frac{1}{2}\lambda\sum_{j=1}^{M} \omega_j^2 \tag{12}$$

where $\check{y}_i^{(t-1)}$ is the predicted value of the $t-1$ tree for the i th sample, n is the number of training samples, and ft(xi) is the predicted value of the t th tree for the sample xi.

The final prediction model is obtained by summing up of the iteratively generated trees:

$$\check{y}_i = \sum_{t-1}^{T} f_t(x_i) \tag{13}$$

where $\check{y}_i$ is the predicted value of the i th sample, which is the predicted vitality of the block sample; T is the total number of trees; and $f_t(x_i)$ is the predicted value of the t th tree for x_i.

The SHAP interpretability method [36] aims to quantify the contribution of each feature to the model's prediction, expressed as the SHAP value. This method considers all possible feature combinations and calculates the average contribution of each feature across all possible combinations. The calculation formula is as follows:

$$\text{SHAP}_{ij} = \sum_{S \subseteq \{x1,x2,\cdots,xn\} \setminus \{xj\}} \frac{|S|!(n-|S|-1)!}{n} \times [f(S \cup \{xj\}) - f(S)] \tag{14}$$

5. Results and Analysis

The projection pursuit model optimized by the accelerated genetic algorithm based on real-number encoding (RAGA-PPM) was applied to calculate the comprehensive vitality of the streets in the Macau Peninsula. A database consisting of 15 indicators was used as the foundation, and multiple experiments were conducted to determine the optimal projection direction when $\alpha = 0.05$. The resulting best projection direction was found to be a = (−0.038, 0.002, 0.003, −0.056, −0.056, −0.097). The values of each component represent the contribution of the corresponding indicators to the overall evaluation goal. The order of the indicators represented by each component was as follows: population density, Weibo check-in density, spatial compactness, functional density, and functional mix. It is evident that the largest coefficients in the projection direction correspond to population density and Weibo check-in density. These two factors contribute the most to the comprehensive vitality value, indicating that they have the greatest impact on the overall vitality of urban neighborhoods. For visualization, the vitality was classified into five levels using the natural breaks classification method, arranged from high to low vitality: high-vitality areas, relatively high-vitality areas, moderate-vitality areas, relatively low-vitality areas, and low-vitality areas.

5.1. Spatial Distribution Analysis of Unidimensional Activity

The distribution of the vitality of people's activities in the Macau Peninsula presents a significant spatial characteristic of "high value at the core and low value at the periphery", which reflects a high degree of agglomeration in the functional layout and spatial utilization of the city, as shown in Figure 4a. High-vitality areas are mainly concentrated in the integrated commercial and entertainment functional areas centered on the Grand Lisboa Hotel, Wynn Macau, and MGM MACAU, and at the same time cover important historical and cultural landmarks such as the Senado Square, the Ruins of St. Paul's, and Our Lady of the Rosary Church. The Central Vitality Zone is distributed in Hac Sa Wan and New Boundary Control Point, showing the transitional characteristics of function and vitality. Hac Sa Wan is predominantly residential in function but has the advantage of strong spatial connectivity due to its proximity to the core business district. Low-vitality areas are mainly located in traditional residential districts such as Chopstick Key and Taishan. These areas are single-function, mainly residential, and lack large public spaces and highly attractive commercial facilities, resulting in a low frequency of crowd activity.

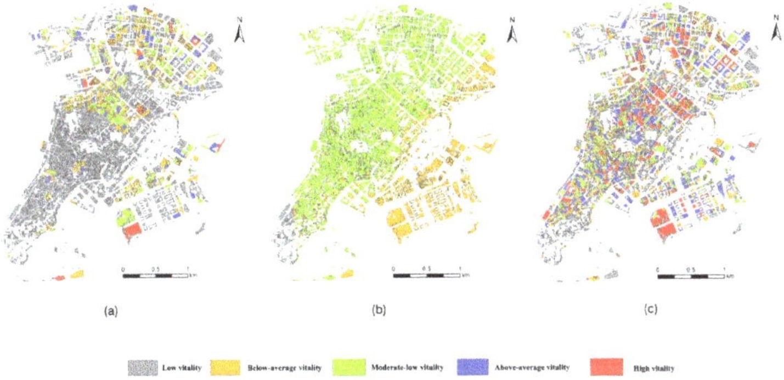

Figure 4. Activity diagram of one-dimensional aspect. (**a**) Vitality of crowd activities; (**b**) network interaction vitality; (**c**) built environment vitality.

The spatial pattern of network interaction vitality in the Macau Peninsula generally showed "predominantly medium-vitality and scattered low-vitality", as shown in Figure 4b. Medium-vitality areas are mainly concentrated in functionally diversified areas such as the New Boundary Crossing Area and Hac Sa Wan. As a combination of business and life service functions, the New Port area has created active online participation through digital activities such as check-ins and reviews, while the Hac-Sa-Huan area, although mainly residential, shows a certain degree of online interaction activity due to its proximity to the core business district and good transportation conditions. Areas with low online interaction vitality are mainly located in traditional residential areas such as Chopsticks

Key, Tsing Chau, and Ha Wan. These districts are significantly less active, due to their single function and lack of large public spaces and commercial facilities.

The vibrancy of the built environment of the Macau Peninsula shows a clear hierarchical distribution, closely related to the functional layout of the area, the land use pattern, and the transportation conditions, as shown in Figure 4c. Highly vibrant areas are mainly concentrated around the Outer Harbour, Nam Van, Senado Square, and Ruins of St. Paul's. These areas are characterized by commercial, tourism, and cultural functions. These areas attract many people due to the high integration of commercial, tourism, and cultural functions. With Hotel Lisboa and Wynn Macau as the core, these landmark districts have become the main driving centers of the vibrancy of Macau's built environment by virtue of their high-density building layouts, diversified forms of land use, and well-developed transportation networks. The central vitality areas are distributed at the periphery of the core area, mainly covering the New Border Crossing and Hac Sa Wan areas. These areas are mainly residential, with certain commercial and living service functions, moderate building density, and convenient transportation conditions. Low-vitality areas are concentrated in the urban fringe areas such as Chopstick Key, Taishan, and Xia Huan.

In terms of unidimensional vitality, high-vitality areas in the Macau Peninsula are primarily concentrated around locations such as the New Lisboa, Ruins of St. Paul's, and A-Ma Temple. These areas exhibit significant advantages in terms of human activity, network interaction, and built environment. In contrast, low-vitality areas are primarily found in the northern region, mainly due to the lack of diverse functions and insufficient infrastructure. This distribution of vitality provides important data support for optimizing urban functions and adjusting the spatial structure of Macau.

5.2. Comprehensive Vitality Space Analysis

The comprehensive vitality evaluation results based on the RAGA-PPM algorithm indicate that the urban spatial structure of Macau exhibits significant hierarchical and central aggregation characteristics in Figure 5. The values for the spatial distribution pattern of vitality show a gradual transition from high-vitality areas in the core region to low-vitality areas at the periphery, forming a distinct central–peripheral hierarchical structure. The core area, characterized by its highly concentrated commercial activities, convenient transportation network, and high-quality spatial environment, emerged as the most vibrant zone. High- and relatively high-vitality areas on the Macau Peninsula are primarily concentrated in the urban center, which includes landmarks such as the Ruins of St. Paul's, Senado Square, and the areas surrounding the A-Ma Temple and Nam Van. These regions, where the city's historical, cultural, and commercial resources converge, exhibit a prosperous commercial atmosphere and dense pedestrian traffic, thus serving as the main sources of urban vitality. Additionally, areas such as Praça de Amizade, Wynn Macau, Fisherman's Wharf, Casino Lisboa, and the New Porto Area, owing to their rich entertainment, accommodation, and dining functions, also constitute relatively high-vitality zones. In contrast, low-vitality areas are mainly found in peripheral regions such as the Lower Ring, Ka I, and Taipa, where high building density, complex road networks, and a lack of diversified functions result in relatively lower vitality levels. Compared with previous studies, this research identified more potential vitality centers. Key vitality regions, such as Hác Sá, Ka Ho, Senado Square, New Porto Area, and Nam Van, were recognized as the main vitality hubs, forming the core vitality zone of the Macau Peninsula. These areas not only concentrate rich historical and cultural resources, but also feature bustling commercial districts and diverse transportation networks, making them the largest gathering points for crowds and economic activities. Specifically, notable vitality centers in various districts include the Vista Bay (1) and Hai Ming Ju (2) in the

Our Lady of Fátima Parish, the Macau German School (3), the Post Office Building (4) in the Senado Square area, Casino Lisboa (5) and the Xian Xinghai Memorial (6) in the New Porto Area, Wynn Macau (7) and the Gung Ding Theatre (8) in Praça de Amizade, the Port Authority (9) in the Lower Ring, Ka I Square (10), the New Era Commercial Center (11) in Nam Van, and the Haihui School (12) in the Border Gate area. These vitality centers reflect the close relationship between the spatial distribution and functional layout of Macau's urban vitality, highlighting the intensity of regional commerce and population flows.

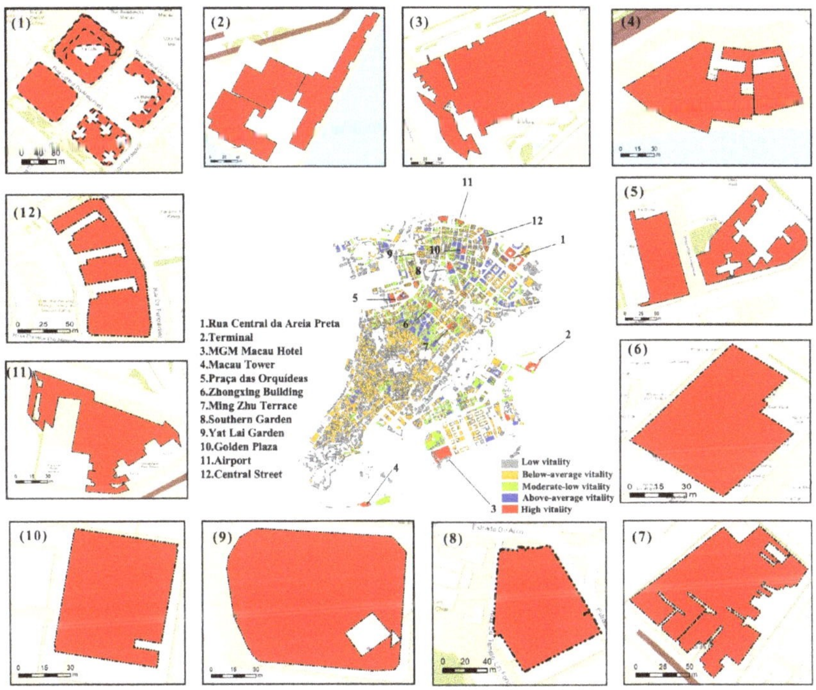

Figure 5. Comprehensive activity and activity center map based on RAGA-PPM.

The comprehensive vitality evaluation results based on the EWM entropy weight method indicate that the spatial distribution of urban vitality in the Macau Peninsula presents a hierarchical pattern of "core-functional nodes-periphery" in Figure 6. The core area of the Gongbei Port exhibits the highest vitality level, due to its high commercial density, convenient transportation, and strong population mobility. The functional node areas, such as the MGM Macau Hotel District and the harbor area, show concentrated vitality, due to their unique service functions and high foot traffic, though their radiation effect is limited. The peripheral areas, such as Heshawan and Haoshi, primarily serve ecological and residential functions, displaying relatively weaker vitality but with significant potential for development. Overall, the urban vitality distribution shows a trend of multi-core development, with the core areas maintaining a dominant advantage.

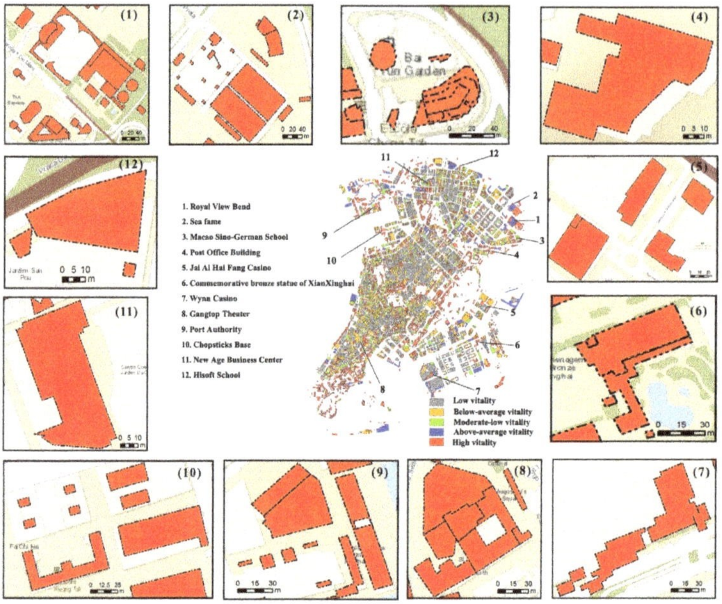

Figure 6. Comprehensive vitality and vitality center map based on EWM.

5.3. Analysis of Spatial Differences in Vitality Using RAGA-PPM and EWM

A comparative analysis of the spatial distribution of urban vitality in Macau using the RAGA-PPM and EWM revealed significant differences in vitality assessment between the two methods. The RAGA-PPM approach primarily relies on a genetic algorithm to optimize the weights of multivariable data. Its results predominantly reflect the vitality distribution patterns of urban functional zones, emphasizing the vitality contributions from functional nodes and ecological areas. In contrast, the EWM method uses objective weighting based on information entropy, focusing on the spatial distribution characteristics of regional vitality under a comprehensive set of indicators. It placed greater emphasis on the performance of core areas and functional complexity, as illustrated in Figure 7.

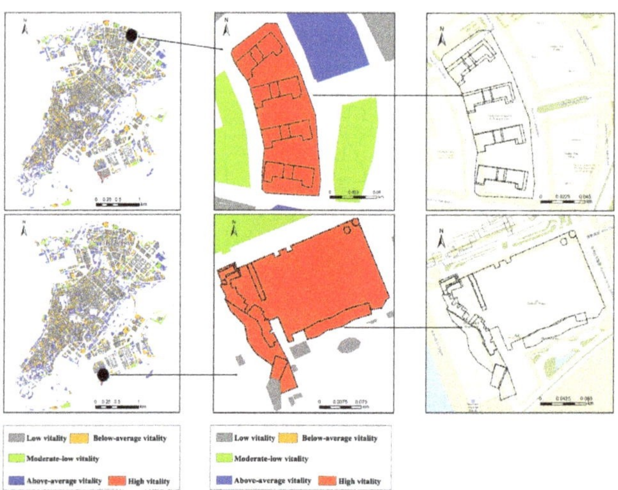

Figure 7. Differences in activity levels between RAGA-PPM and EWM.

The differences between the two methods In vitality evaluation are also reflected In the level of spatial scale detail. RAGA-PPM provides a more refined description of vitality distribution in small-scale functional areas, effectively identifying small commercial and public spaces around hotel districts. In contrast, EWM focuses more on the urban vitality patterns at a macro scale, making it suitable for holistic planning analysis.

5.4. The Impact of the Built Environment on Urban Vitality

In this study, an urban vitality prediction model was constructed based on a Bayesian optimization framework and the LightGBM algorithm, with 14 built environment elements of 159 neighborhoods as feature variables, systematically balancing the model accuracy and generalization ability in Table 3. A hierarchical random partitioning strategy was used to construct a training set (80%) and independent test set (10%), and the optimal hyperparameter combinations were determined by Bayesian optimization (5-fold cross-validation, 25 iterations): learning rate (0.262), number of trees (364), maximal depth (9), and feature/test sub-sampling rate (0.860/0.823). The early stopping method (terminate without improvement in 10 consecutive rounds of validation set) was enabled in training, the best model converged in the 30th round and the independent test set showed significant performance (RMSE = 0.082, MAE = 0.065, R^2 = 0.555), which confirmed its effective capture of nonlinear relationships for urban vitality, without overfitting. The method was further combined with SHAP interpretability analysis to reveal the threshold effect and nonlinear influence mechanism of key elements such as BC, BFD, FD, etc., which provided a quantitative decision-making basis for the optimization of the built environment. The method achieves a synergistic breakthrough in accuracy, efficiency, and mechanism analysis for multi-source urban data modeling through an efficient search of hyperparameters, an anti-overfitting strategy, and interpretability enhancement.

Table 3. The fitting accuracy of the indicators.

Dataset	MAE	RMSE	R^2
Training Set Results	0.0256	0.0375	0.8170
Test Set Results	0.0650	0.0817	0.5548

5.4.1. Relative Importance of Built Environment Elements

This study revealed the hierarchical characteristics of the impact of built environment elements on vitality in Macau's districts through SHAP value analysis (Figure 8). Panel (a) of Figure 8 shows the contribution and importance ranking of the 14 built environment elements, while panel (b) presents the specific influence patterns of each element on vitality prediction through SHAP value distribution. In the feature importance ranking, the top six built environment elements by relative importance were road network density (0.025), spatial fractal dimension (0.020), sky openness (0.015), public facility density (0.012), functional mix (0.010), and building density (0.008). Road network density, spatial fractal dimension, and sky openness were identified as key factors influencing district vitality. Road network density enhances traffic connectivity and accessibility, facilitating economic activities and the walking experience, thereby attracting more pedestrian flow. The spatial fractal dimension strengthens spatial diversity and social interaction through complex spatial structures, improving environmental attractiveness. Sky openness, by offering a sense of open space and environmental comfort, promotes outdoor activities and the gathering of crowds. These three factors work synergistically to improve traffic convenience, spatial diversity, and environmental quality, significantly enhancing the vitality and attractiveness of a district. Public facility density, functional mix, and building density are secondary key factors influencing district vitality. The high rankings of sky openness and green visibility indicate that the micro-scale street design plays a crucial role in the overall vitality of the district. In comparison, factors such as enclosure and green visibility have a relatively smaller impact, yet still influence district vitality to some extent. A greater sense of enclosure may enhance walking comfort, while districts with higher green visibility are likely to attract more outdoor activities. Walkability, waterfront index, and functional density have a relatively minor impact on vitality, possibly due to the specific environmental characteristics of the study area.

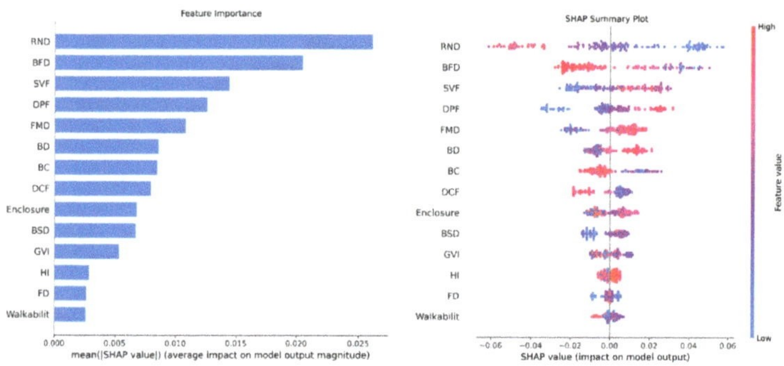

Figure 8. The relative importance of built environment elements on vitality in the Macau Peninsula districts.

5.4.2. Nonlinear Relationship Between Built Environment Elements and Urban Vitality

To further investigate the localized nonlinear relationship between urban vitality and the built environment, we focused on the top three most globally significant built environment elements. Figure 9 illustrates the nonlinear relationship between these top three elements at the block level and their contributions to predicting vitality, represented by SHAP values. The color legend indicates the value of the interacting element that had the strongest influence on each sample. The impact of road network density on urban vitality exhibited a complex nonlinear relationship. When the road network density was less than 0.2, the SHAP values were relatively high. As the density increased, the SHAP values decreased rapidly and reached negative values within the range of road network density from 0.2 to 0.4. This suggests that overly dense road networks may reduce block vitality, possibly due to street fragmentation or excessive road occupation of public spaces. However, when the road network density exceeded 0.6, the SHAP values stabilized, indicating that the marginal effect of extremely high road network density on vitality had diminished. The SHAP values for spatial fractal dimension generally showed a downward trend, particularly when the spatial fractal dimension exceeded 0.8, where the SHAP values decreased sharply. This suggests that excessively high spatial fractal dimensions may negatively impact urban vitality. Such an effect could be attributed to overly complex spatial forms, which may hinder block vitality due to factors like visual obstruction or decreased accessibility. Sky openness demonstrated a positive relationship with urban vitality, especially when the sky openness increased from 0.2 to 0.8, during which the SHAP values steadily rose. This suggests that appropriately increasing sky openness could help enhance block vitality. However, when sky openness exceeded 0.8, a decrease in SHAP values for some points was observed, possibly reflecting the negative impact of excessively open environments on block vitality.

In summary, urban vitality is influenced by the complex and nonlinear effects of built environment factors. A moderate spatial fractal dimension, an appropriate road network density, and a balanced sky openness may be more conducive to enhancing vitality. Future planning and design should consider the interactions between different elements, in order to optimize urban spatial forms.

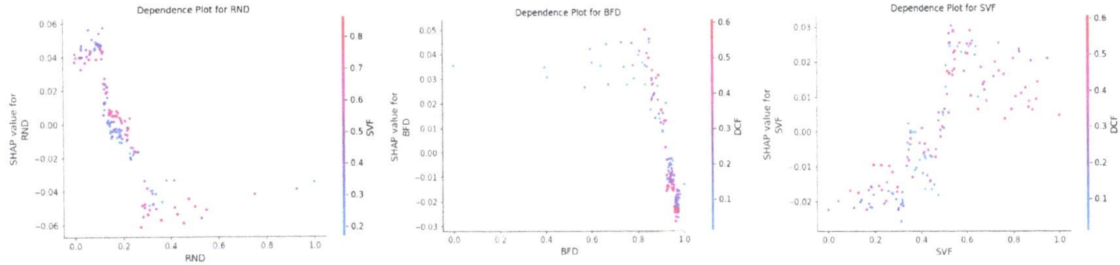

Figure 9. Nonlinear relationship between built environment elements and vitality in the Macau Peninsula district.

6. Conclusions and Discussion

6.1. Conclusions

This paper proposes a new framework for evaluating urban vitality that integrates crowd activities, network interactions, and the built environment. By introducing the RAGA-PPM method, it provides a quantitative assessment of the overall vitality of the Macau Peninsula. Unlike traditional methods that introduce biases due to subjective weighting, this approach identifies the contribution of each indicator to overall vitality based on the inherent structural characteristics of the data, resulting in a more objective and accurate identification of vitality outcomes. Using machine learning algorithms combined with the SHAP interpretability method, this study explored the nonlinear impacts and interactive effects of the built environment on urban vitality in the urban areas of the Macau Peninsula. The findings were as follows:

(1) There are certain spatial distribution differences in the urban vitality across the three dimensions—crowd activity, network interaction, and the built environment—on the Macau Peninsula. However, a common feature across all dimensions is that they exhibit high values at the center, with urban vitality values gradually decreasing toward the periphery. The vitality of crowd activities is concentrated around tourism, commerce, entertainment, and cultural functions, with a high value at the center and secondary vitality clusters in each district. The vitality of network interactions is centered around the New Port and Hac-Sa Bay, presenting a spatial pattern of "moderate vitality as the dominant feature, with scattered areas of low vitality". The spatial distribution of built environment vitality shows clear hierarchical patterns, closely related to the regional functional layout, land use patterns, and transportation conditions, with high-vitality areas primarily concentrated around the Outer Harbor, Nanhai, Senado Square, and the area surrounding the Ruins of St. Paul's.

(2) The spatial distribution of overall vitality on the Macau Peninsula is like that of the individual dimensions. Specifically, the largest vitality center is located around the historic district, with vitality decreasing as it moves outward, and the lower vitality areas are distributed at the urban periphery. The RAGA-PPM method used in this study provided a more accurate identification of vitality areas, revealing more potential vitality centers. Each district contains a vitality center, with the historic district—comprising the Ruins of St. Paul's, New Road, A-Ma Temple, and Senado Square—serving as the main vitality hub for the entire Macau Peninsula. The areas around New Road and Nanhai exhibited the highest vitality, while the areas around Lower Harbor and Taishan showed the lowest vitality. The spatial distribution of

urban vitality on the Macau Peninsula is influenced by various factors, showing heterogeneity in distribution.

(3) This study reveals that, at the neighborhood level, the six built environment factors most significantly influencing urban vitality in the Macau Peninsula are, in order of importance, street network density, spatial fractal dimension, sky openness, public facility density, functional mix, and building density. These factors were ranked based on their relative significance to urban vitality and form a key set of built environment indicators that impact the urban vitality of the Macau Peninsula.

(4) The built environment on the Macau Peninsula has a nonlinear effect on urban vitality. Regarding the key factors that influence the vitality of streets and blocks, road network density, spatial fractal dimension, and sky openness had similar threshold values, with all demonstrating a positive impact on vitality when the value was below 0.2.

6.2. Discussion

Exploring the similarities and differences in the spatial distribution of vitality across different dimensions and the overall vitality of the central urban area of the Macau Peninsula can contribute to refining the urban comprehensive vitality assessment system and offer targeted strategies for urban governance. The urban vitality of the Macau Peninsula is predominantly concentrated in the historic districts and commercial centers, while areas with lower vitality are mainly located at the outskirts of the city. Therefore, greater emphasis should be placed on fostering and developing a polycentric structure, to prevent excessive resource concentration in a single center. By clearly defining the developmental direction of each area, identifying distinct regional growth paths, and optimizing the layout of public service facilities, commercial outlets, and cultural and entertainment amenities, the vitality of each district can be enhanced, ultimately achieving a more balanced distribution of urban vitality. Additionally, lower vitality areas are primarily situated at the urban periphery. It is necessary to further strengthen the radiating role of transportation hubs and implement specific development policies for low-vitality regions. Improving the quality and coverage of public infrastructure and transportation services will facilitate better connections between peripheral and central areas. By leveraging the flow of resources and population, both within and outside the city, industries can be guided toward development, employment opportunities increased, and living conditions improved, thus gradually elevating the vitality levels of these areas. Furthermore, the historic districts play a central role in the vitality of the Macau Peninsula. In the planning and construction process, it is essential to protect and utilize the rich historical and cultural resources, while preserving the historical features and cultural characteristics of these areas. Through appropriate updates and renovations, the livability and attractiveness of the old districts can be enhanced. Lastly, the distribution of urban spatial vitality in the Macau Peninsula is influenced by multiple factors and exhibits significant heterogeneity. Smart city development should be promoted, utilizing big data, the Internet of Things, and other technologies for refined management and the establishment of innovative platforms. Real-time monitoring and analysis of dynamic changes in urban vitality can provide scientific evidence for urban planning and decision-making. Based on the development status and potential of different areas, targeted policies and measures should be formulated to improve urban governance levels and efficiency.

Based on the traditional Geographic Information System (GIS) and general map tools (Google Maps), this study overcomes the limitations of existing tools in the analysis of complex urban systems through multidimensional data fusion, nonlinear modeling approaches, and interpretability enhancement techniques in the following three respects. First, traditional GIS and map tools mainly rely on static spatial data (road networks, building contours) and basic statistical functions, while this study constructed a more comprehensive framework for urban vitality assessment by fusing heterogeneous big data from multiple sources. Secondly, compared with the linear regression or spatial autocorrelation models relied on by traditional GIS, this study combined machine learning algorithms and optimization methods to achieve an accurate resolution of complex nonlinear relationships. This paper constructed a systematic and refined set of independent variables using multi-source data science, analyzing the distribution of urban vitality in the Macau Peninsula at the micro-scale of street block units. By integrating machine learning algorithms and the SHAP interpretability method, it explored the nonlinear effects of built environment factors on urban vitality. This approach provided new insights into the mechanisms of how the built environment influences urban vitality, offering practical and actionable guidance for more detailed urban planning. It can contribute to the coordinated design of street and block planning, strengthening the connection and interaction between the urban skeleton and its "muscles". Although this study has made progress in methodological innovation and data analysis, it still has some limitations. The study mainly focused on the measurement of dynamic population density, and future research could incorporate Jiang Difei's theory to comprehensively assess vitality from economic, social, and cultural perspectives. Furthermore, this study investigated the spatial heterogeneity of the built environment's impact on urban vitality at the block level; however, future studies could explore temporal heterogeneity to deepen the understanding of the dynamic mechanisms of urban vitality. Lastly, the conclusions drawn from the Macau Peninsula may need further validation in other regions. Future research could expand data sources and incorporate more dimensions, such as economic, social, and cultural data, to build a more comprehensive urban vitality evaluation system. Additionally, time-series data could be integrated to investigate the dynamic changes in urban vitality, providing more accurate decision-making support for urban planning.

Author Contributions: C.P., Conceptualization; Supervision; Validation; Writing—original draft; Writing—review and editing. J.G., Writing—original draft; Validation; Investigation. H.L., Writing—original draft; Data curation; Resources; Validation. J.W., Writing—original draft; Investigation; Software. N.Q., Writing—review and editing; Visualization. S.W., Writing—review and editing; Software. All authors have read and agreed to the published version of the manuscript.

Funding: This research was funded by (1) Guangdong Philosophy and Social Science Planning Project (GD24XSH06), (2) Guangdong Philosophy and Social Science Planning Project (GD24XYS018), (3) 2023 Maoming Philosophy and Social Science Planning Project (2023YB18), (4) Projects of Talents Recruitment of GDUPT (2023rcyj2015), (5) Projects of PhDs' Start-up Research of GDUPT (2023bsqd1008) (2022bsqd2004), and (6) Science and Technology Programme of Maoming of Guangdong Province of China (2023398), (2023411), (2024055), (2024061).

Data Availability Statement: The datasets used and/or analyzed during the current study are available from the corresponding author on reasonable request.

Conflicts of Interest: The authors declared no potential conflicts of interest with respect to the research, authorship, and/or publication of this article.

References

1. Ye, Y.; Zhuang, Y.; Zhang, L.Z. Morphological Study of Vitality Creation in Urban Design: Based on Quantitative Analysis of Urban Spatial Morphology and Resident Activity Verification. *Int. J. Urban Plan.* **2016**, *31*, 26–33.
2. Ewing, R.; Cervero, R. Travel and the Built Environment: A Synthesis. *Transp. Res. Rec. J. Transp. Res. Board* **2001**, *1780*, 87–114. [CrossRef]
3. Montgomery, J. Making a City: Urbanity, Vitality, and Urban Design. *J. Urban Des.* **1998**, *3*, 93–116. [CrossRef]
4. Wang, Z.M.; Liu, Y.F.; Luo, X. Nonlinear Relationship Btween Urban Vitality and Built Environment Based on Multi-source Data: A Case Study of the Main Urban Area of Wuhan on Weekends. *Prog. Geogr.* **2023**, *42*, 716–729. [CrossRef]

5. Wu, W.; Ma, Z.; Guo, J.; Zhao, K. Nonlinear effect of built environment on street vitality: A multi-source big data analysis based on XGBoost model. *Chin. Landsc. Archit.* **2022**, *38*, 82–87.
6. Hong, Y.M.; Wang, S.Y. Big Data, Machine Learning, and Statistics: Challenges and Opportunities. *J. Econom.* **2021**, *1*, 17–35.
7. Feng, Z.M.; Zheng, H.X.; Liu, B.Q. Comprehensive Evaluation of Agricultural Water Resource Utilization Efficiency Based on Genetic Projection Pursuit Model. *Trans. Chin. Soc. Agric. Eng.* **2005**, *21*, 66–70.
8. Jin, J.L.; Yang, X.H.; Ding, J. An Accelerated Genetic Algorithm Based on Real Coding. *J. Sichuan Univ. (Eng. Sci. Ed.)* **2000**, *32*, 20–24.
9. Long, Y.; Huang, C.C. Does Block Size Matter? The Impact of Urban Design on Economic Vitality in Chinese Cities. *Environ. Plan. B Urban Anal. City Sci.* **2019**, *46*, 406–422. [CrossRef]
10. Yang, D.F.; Wang, X.M.; Han, R.N. Nonlinear Effects and Interaction of Built Environment on Street Vitality: A Case Study of Shenyang. *Urban Plan. J.* **2023**, *05*, 93–102.
11. Sung, H.G.; Go, D.H.; Choi, C.G. Evidence of Jacobs' Street Life in Great Seoul: Identifying the Association Between Physical Environment and Walking Activity on Streets. *Cities* **2013**, *35*, 164–173. [CrossRef]
12. Wu, J.Y.; Ta, N.; Song, Y.; Lin, J.; Chai, Y. Urban Form Breeds Neighborhood Vibrancy: A Case Study Using a GPS-Based Activity Survey in Suburban Beijing. *Cities* **2018**, *74*, 100–108. [CrossRef]
13. Delclòs-Alió, X.; Gutiérrez, A.; Miralles-Guasch, C. The Urban Vitality Conditions of Jane Jacobs in Barcelona: Residential and Smartphone-Based Tracking Measurements of the Built Environment in a Mediterranean Metropolis. *Cities* **2019**, *86*, 220–228. [CrossRef]
14. Jiang, D.F. *Urban Morphology and Vitality Theory*; Southeast University Press: Nanjing, China, 2007.
15. Tu, W.; Zhu, T.T.; Xia, J.Z.; Zhou, Y.; Lai, Y.; Jiang, J.; Li, Q. Portraying the Spatial Dynamics of Urban Vibrancy Using Multisource Urban Big Data. *Comput. Environ. Urban Syst.* **2020**, *80*, 101428. [CrossRef]
16. He, Q.S.; He, W.S.; Song, Y.; Wu, J.Y.; Yin, C.H.; Mou, Y.C. The Impact of Urban Growth Patterns on Urban Vitality in Newly Built-up Areas Based on Association Rules Analysis Using Geographical Big Data. *Land Use Policy* **2018**, *78*, 726–738. [CrossRef]
17. Zhang, A.Q.; Li, W.F.; Wu, J.Y.; Lin, J.; Chu, J.; Xia, C. How Can the Urban Landscape Affect Urban Vitality at the Street Block Level? A Case Study of 15 Metropolises in China. *Environ. Plan. B Urban Anal. City Sci.* **2021**, *48*, 1245–1262. [CrossRef]
18. Gómez-Varo, I.; Delclòs-Alió, X.; Miralles-Guasch, C. Jane Jacobs Reloaded: A Contemporary Operationalization of Urban Vitality in a District in Barcelona. *Cities* **2022**, *123*, 103565. [CrossRef]
19. Hajrasouliha, A.; Yin, L. The Impact of Street Network Connectivity on Pedestrian Volume. *Urban Stud.* **2015**, *52*, 2483–2497. [CrossRef]
20. Lan, F.; Gong, X.Y.; Da, H.L.; Wen, H. How Do Population Inflow and Social Infrastructure Affect Urban Vitality? Evidence from 35 Large- and Medium-Sized Cities in China. *Cities* **2020**, *100*, 102454. [CrossRef]
21. Li, X.; Li, Y.; Jia, T.; Zhou, L.; Hijazi, I.H. The Six Dimensions of Built Environment on Urban Vitality: Fusion Evidence from Multisource Data. *Cities* **2022**, *121*, 103482. [CrossRef]
22. Yue, W.Z.; Chen, Y.; Thy, P.T.M.; Fan, P.; Liu, Y.; Zhang, W. Identifying Urban Vitality in Metropolitan Areas of Developing Countries from a Comparative Perspective: Ho Chi Minh City versus Shanghai. *Sustain. Cities Soc.* **2021**, *65*, 102609. [CrossRef]
23. Zhang, Z.R.; Liu, J.P.; Wang, C.Y.; Zhao, Y.; Zhao, X.; Li, P.; Sha, D. A Spatial Projection Pursuit Model for Identifying Comprehensive Urban Vitality on Blocks Using Multisource Geospatial Data. *Sustain. Cities Soc.* **2024**, *100*, 104998. [CrossRef]
24. Zhang, Y. Study on the vitality measurement and influence mechanism of urban blocks—A case study of Wuhan main city. Diss. Master's Thesis, Wuhan University, Wuhan, China, 2019.
25. Jacobs, J. *The Death and Life of Great American Cities*; Random House: New York, NY, USA, 1961.
26. Ye, Y.; Van Nes, A. Quantitative Tools in Urban Morphology: Combining Space Syntax, Space Matrix, and Mixed-Use Index in a GIS Framework. *Urban Morphol.* **2014**, *18*, 97–118. [CrossRef]
27. Cao, Z.M.; Zhen, F.; Li, Z.X. Study on Urban Temporal Vitality Patterns and Influencing Factors Based on Mobile Phone Signaling Data: A Case Study of the Central Urban Area of Nanjing. *Hum. Geogr.* **2022**, *37*, 109–117.
28. Liu, S.J.; Zhang, L.; Long, Y. Urban Vitality Area Identification and Pattern Analysis from the Perspective of Time and Space Fusion. *Sustainability* **2019**, *11*, 4032. [CrossRef]
29. Wu, G.S.; Dang, Y.T.; Zhao, K. Study on Urban Vitality Spatial Characteristics Based on Multi-dimensional Perception. *J. Earth Inf. Sci.* **2022**, *24*, 1867–1882.
30. Wang, B.; Zhen, F.; Zhang, H. Study on the Temporal and Spatial Dynamics of Urban Activities Based on Check-in Data and Area Division. *Geogr. Sci.* **2015**, *35*, 151–160.
31. Liu, Y.S.; Zhao, P.J.; Liang, J.S. Study on Urban Vitality Based on Location-Based Service Data: A Case Study of the Sixth Ring Area in Beijing. *Reg. Res. Dev.* **2018**, *37*, 64–69.
32. Wu, Z.Q.; Ye, Z.N. Study on Urban Spatial Structure Based on Baidu Map Heat Maps: A Case Study of the Central Urban Area of Shanghai. *Urban Plan.* **2016**, *40*, 33–40.

33. Jia, J.Y.; Song, J.F. Research on the Relationship Between Urban Vitality and Built Environment "3D" Characteristics: A Case Study of Wuhan. *Mod. Urban Res.* **2020**, *35*, 59–66.
34. Wang, N.; Wu, J.S.; Li, S. Research on the Spatial Characteristics of Urban Vitality and the Impact Mechanisms of the Built Environment Based on Multisource Data: A Case Study of Shenzhen. *Trop. Geogr.* **2021**, *41*, 1280–1291.
35. Long, Y.; Zhou, Y. Quantitative Evaluation and Influencing Factors Analysis of Street Vitality: A Case Study of Chengdu. *New Archit.* **2016**, 52–57.
36. Niu, X.Y.; Wu, G.S.; Li, M. Study on the Impact of Built Environment on Street Vitality and Its Temporal and Spatial Characteristics Based on LBS Location Data. *Int. J. Urban Plan.* **2019**, *34*, 28–37. [CrossRef]
37. Zhang, L.M.; Zhang, R.X.; Yin, B. The Impact of Built Environment on Pedestrian Activities in Historical Areas. *Alex. Eng. J.* **2021**, *60*, 285–300. [CrossRef]
38. Zou, S.C.; Zhang, S.Q.; Zhen, F. Study on Community Daily Activity Space Measurement and Vitality Influencing Factors Based on Residents' Spatiotemporal Behavior: A Case Study of Shazhou and Nanyuan Streets in Nanjing. *Prog. Geogr.* **2021**, *40*, 580–596. [CrossRef]
39. Luo, S.Z.; Zhen, F. Study on Urban Public Space Vitality Evaluation Method Based on Mobile Phone Data: A Case Study of Parks in Nanjing. *Geogr. Res.* **2019**, *38*, 1594–1608.
40. Liu, S.; Lai, S.Q. Study on Influencing Factors of Urban Public Space Vitality Based on Multisource Data: A Case Study of the Huangpu River Waterfront Area in Shanghai. *Landsc. Archit.* **2021**, *28*, 75–81.
41. Shi, Y.; Li, T.T.; Yang, J.Y. Study on Urban Waterfront Space Vitality Based on Mobile Phone Signaling Data: A Case Study of Jinji Lake in Suzhou. *Landsc. Archit.* **2021**, *28*, 31–38.
42. Wang, M.W.; Jin, J.L.; Li, L. Application of Accelerated Genetic Algorithm-Based Projection Pursuit in Liquefaction Potential Evaluation of Sandy Soil. *Rock Soil Mech.* **2004**, *23*, 631–634.
43. Fu, Q.; Fu, H.; Wang, L.K. Application of Accelerated Genetic Algorithm-Based Projection Pursuit Model in Water Quality Evaluation. *Geogr. Sci.* **2003**, *23*, 236–239.
44. Ta, N.; Zeng, Y.T.; Zhu, Q.Y.; Wu, J.Y. Relationship Between Built Environment and Urban Vitality in Central Shanghai Based on Big Data Analysis. *Geogr. Sci.* **2020**, *40*, 60–68.
45. Wang, C.G.; Wang, B.; Wang, Q.Z. Nonlinear Relationship and Threshold Effect Between Urban Vitality and Built Environment: A Case Study of Guangzhou's Central Urban Area. *Prog. Geogr.* **2023**, *42*, 79–88. [CrossRef]
46. Han, Y.; Qin, C.P.; Xiao, L.Z.; Ye, Y. The Nonlinear Relationships Between Built Environment Features and Urban Street Vitality: A Data-Driven Exploration. *Environ. Plan. B Urban Anal. City Sci.* **2024**, *51*, 195–215. [CrossRef]
47. Macau Statistics and Census Service. Available online: https://www.dsec.gov.mo/zh-MO/Statistic/Database (accessed on 3 April 2024).
48. Fang, L.; Huang, J.L.; Zhang, Z.Y.; Nitivattananon, V. Data-Driven Framework for Delineating Urban Population Dynamic Patterns: A Case Study on Xiamen Island, China. *Sustain. Cities Soc.* **2020**, *62*, 102365. [CrossRef]
49. Huang, B.; Zhou, Y.L.; Li, Z.G.; Song, Y.; Cai, J.; Tu, W. Evaluating and Characterizing Urban Vibrancy Using Spatial Big Data: Shanghai as a Case Study. *Environ. Plan. B Urban Anal. City Sci.* **2020**, *47*, 1543–1559. [CrossRef]
50. Ding, J.; Luo, L.; Shen, X.; Xu, Y. Influence of built environ-ment and user experience on the waterfront vitality of his-torical urban areas: A case study of the Qinhuai River in Nanjing, China. *Front. Archit. Res.* **2023**, *12*, 820–836. [CrossRef]
51. Yu, L.; Sheng, Y.H.; Liu, X.Y. Research on urban vitality evaluation method based on RAGA-PPM model: A case study of central urban area in Nanjing. *J. Geo-Inf. Sci.* **2024**, *26*, 2567–2582.
52. Cao, Q.F.; Jiang, Q.X.; Zheng, L.L. Improving the compactibility of Gardeniae Fructus extract by HPMC based on systematic optimization with QbD concept and FAHP-entropy weight method. *Chin. Tradit. Herb. Drugs* **2024**, *55*, 5448–5458.
53. Liu, Y.; Wang, K.L.; Xing, X.Y. Spatial Effects in Geographic Analysis. *Acta Geogr. Sin.* **2023**, *78*, 517–531.
54. Shannon, C.E. A mathematical theory of communication. *Bell Syst. Tech. J.* **1948**, *27*, 379–423. [CrossRef]
55. Friedman, J.H.; Tukey, J.W. A projection pursuit algorithm for exploratory data analysis. *IEEE Trans. Comput.* **1974**, *23*, 881–890. [CrossRef]

buildings

MDPI

Article

Toward Symmetry in Accessible Restrooms Design: Integrating KE, RST, and SVM for Optimized Emotional-Functional Alignment

Zimo Chen [1,†], Jingwen Tian [2,*,†], Hongtao Zhou [2,3,*] and Duan Wu [2]

1 School of Design Art and Media, Nanjing University of Science and Technology, Nanjing 210094, China; czm010704@njust.edu.cn
2 College of Design and Innovation, Tongji University, Shanghai 200092, China; wuduan@tongji.edu.cn
3 Faculty of Arts and Humanities, University of Macau, Macao 999078, China
* Correspondence: 2311608@tongji.edu.cn (J.T.); zhouhongtao@tongji.edu.cn (H.Z.)
† These authors contributed equally to this work.

Abstract: Accessible restrooms must reconcile code-based functionality with the affective expectations of disabled users. This study develops an integrated Kansei Engineering (KE)–Rough Set Theory (RST)–Support Vector Machine (SVM) workflow that converts user emotions into verifiable design guidelines. Surveys and semi-structured interviews with 50 disabled participants produced nine Kansei words; factor analysis extracted three principal emotional factors—tidiness, utility and care—capturing 75.8% of total variance. The morphological decomposition of 60 restroom samples yielded 41 design attributes, from which RST attribute reduction isolated six critical features. An SVR model with a radial-basis kernel, trained on 90% of the data and validated on the remaining 10%, achieved R^2 = 0.931 and RMSE = 0.085. The exhaustive prediction of 15,750 feasible design combinations pinpointed an optimal configuration; follow-up user testing confirmed the improvement in satisfaction (mean 5.1 on a seven-point scale). The KE–RST–SVM workflow thus offers a reproducible, data-driven path for harmonizing emotional and functional objectives in inclusive restroom design, and can be extended to other barrier-free facilities.

Keywords: Kansei engineering; rough set theory; support vector machine; emotional-functional alignment; data-driven

1. Introduction

Ensuring the accessibility of public facilities for individuals with disabilities is both a fundamental technical requirement and a reflection of social inclusion. Accessible design, an essential element in built environment planning, seeks to create barrier-free spaces that support equal participation for diverse users, such as individuals with disabilities [1]. Within high-traffic public areas, the availability of properly designed accessible restrooms is a key determinant of mobility and life satisfaction for this group. While many countries have implemented standards for accessible design, these guidelines often fail to address the real needs and emotional experiences of disabled users, leading to facilities that meet technical requirements but fall short of user expectations [2]. Much of the existing research primarily focuses on physical accessibility, with less attention given to the emotional and psychological needs of disabled users [3,4]. Consequently, bridging emotional and physical considerations through systematic, data-driven methods has become an urgent priority in inclusive design research.

From a symmetry-oriented design viewpoint, an inclusive restroom must sustain a dual equilibrium—spatial geometry and user experience—by choreographing circulation routes, sanitary fixtures, and assistive elements into a coherent layout that meets code requirements and nurtures users' affective comfort. Yet prevailing practice remains anchored in technical compliance, leaving the symmetrical coupling among emotional response, functional performance, and ergonomic support [5] insufficiently explored—a gap this study tackles through its KE–RST–SVM workflow. Recently, Kansei Engineering (KE) has emerged as a crucial framework for exploring users' emotional responses, thereby informing more user-centered, symmetrically balanced solutions [6]. Yet, when numerous and multifaceted design features must be evaluated, computational complexity increases, and existing research may fail to capture the symmetrical relationships that underlie user satisfaction. These challenges can be mitigated through Rough Set Theory (RST), a mathematical tool adept at processing ambiguous and partial data by trimming superfluous variables [7]. Parallel developments in machine learning, such as Backpropagation Neural Networks (BPNN) [8], Interactive Genetic Algorithms (IGA) [9], Convolutional Neural Networks (CNN) [10], and Generative Adversarial Networks (GAN) [11], have further facilitated the simulation of intricate, nonlinear dynamics linking user emotions to design parameters [12]. Among these, Support Vector Machine (SVM) stands out as a robust model well-suited for predicting and optimizing complex systems [13,14]. Despite the recent infusion of machine learning techniques into built environment scholarship, three critical gaps remain unresolved. First, KE investigations are still largely confined to building appearance or removable fittings rather than full-scale spatial configurations. For example, Montañana et al. [15] translated perceived "prestige" merely into real estate advertising imagery, without addressing the architectural context itself. Second, most SVM applications in architecture pursue performance optimization alone. Cheng et al. [16] applied machine learning to thermo-environmental audits, while Han et al. [17] employed an SVR–PSO hybrid model to predict ventilation efficiency. Although these studies demonstrate the technical viability of KE and SVM, they largely overlook the coupling of affective data. Third, existing KE–SVM workflows stop at the component level and lack a replicable path for mapping user emotions onto multi-attribute spatial layouts. Shao et al. [18], for instance, focused on AI-generated interior styles without addressing inclusive usability. Consequently, a data-driven paradigm capable of harmonizing emotional and functional objectives in accessible restroom design is still absent.

To bridge the identified gaps, this study introduces an integrated KE–RST–SVM paradigm and validates it through design-simulation experiments in inclusive restroom scenarios. The contributions are threefold. (1) Methodological—By systematically coupling affective factor extraction via KE with attribute reduction through RST and performing subsequent non-linear mapping using SVM, we establish a transferable emotion–function unified modeling framework. (2) Technical—By embedding a feature-importance-based interpretability mechanism within the SVM, the approach delivers quantitative decision criteria that balance affective and functional requirements. (3) Practical—By operationalizing this quantifiable workflow, the study yields actionable guidelines for accessible restroom design, equipping designers and policy-makers with implementable insights.

2. Literature Review

2.1. *Symmetry in Accessible Restrooms: From Regulation to Socio-Emotional Equity*

Striking a balance between functional requirements and emotional needs in public spaces remains a fundamental pursuit in modern inclusive design. The concept of accessible design can be traced back to legislative milestones such as the Architectural Barriers Act (ABA) of 1968 in the United States, which was among the earliest laws mandating accessibility features in public facilities [19]. Subsequently, the Americans with Disabilities Act (ADA) of 1990 expanded the scope to include employment, transportation, and other facets of public services, underscoring the principle of equal participation [20]. While these regulations primarily focus on physical accessibility, the broader idea of symmetry in design implies balancing functional requirements with the psychosocial well-being of all users, including those with varied cultural or social backgrounds [21].

Universal Design and Inclusive Design have further refined this concept by advocating anticipatory barrier removal rather than retro-fit solutions [22,23]. This approach aligns closely with the notion that environments should offer equitable experiences for individuals across diverse abilities, rather than relying on subsequent modifications. However, accessible restroom facilities pivotal for privacy, safety, and autonomy still frequently demonstrate mismatches between official standards and actual user satisfaction [24,25]. Although significant progress has been made, research shows that purely technical adherence may overlook cultural nuances and emotional or psychological comfort [26,27]. Consequently, genuine symmetry in accessible restrooms demands design protocols that meet ergonomic benchmarks and actively respond to disabled users' subjective and emotional expectations.

2.2. Empirical Progress in KE and Machine Learning for Emotion Design

Achieving genuine symmetry requires an evidence-based linkage between affective wellbeing metrics and verifiable usability performance within a unified design framework. KE, introduced in the 1970s, systematically incorporates users' subjective impressions—"Kansei"—into design processes [6,28]. After demonstrating success in automotive applications, the approach was extended to multiple fields such as consumer products [29], furniture [30], electronic interfaces, military systems [31], and service design [32]. Central to KE is the principle of quantifying emotional responses and translating them into actionable design guidelines [33]. Eliciting how "safe" or "comfortable" a user feels can inform precise modifications of physical features, contributing to an overall sense of symmetry between user expectation and product attributes.

Although KE has demonstrated wide-ranging applicability, its presence in accessible design has been comparatively limited. Existing studies highlight its promise in assistive devices [34], wheelchair design [35], specialized wudhu facilities [36], and modular furniture solutions for disabled users [37]. Table 1 shows a comparative overview of the algorithms applied to the study of architectural space, through which the advantages and adaptability of the workflow of this study are fully realized. By integrating KE frameworks in restroom design, researchers can more systematically measure and address the emotional perceptions of disabled users, thus creating facilities that better align functional practicality with human-centered symmetry. This research aims to bridge that gap by applying KE, in conjunction with RST and SVM, to decipher the emotional and functional attributes that lead to an optimally symmetrical accessible restroom experience.

Table 1. Overview of the algorithms applied to the study of architectural space.

Algorithm	Spatial Application	Fuzzy Expression	Reasoning Complexity	Loss Rate	Data Representation
PSO-SVR [38]	Outdoor micro-space form-finding	Medium	Medium	Low	Medium
RF [39]	Office thermal comfort diagnostics	Low	Medium	Medium	Medium
CNN [40]	Complex indoor way-finding meshes	Low	High	Medium	High
LSTM [41]	Smart-building load forecasting	Low	High	Medium	Medium
KNN [42]	Indoor location	Medium	Low	High	Medium
SVM [43]	Urban Streetscape Visual	Medium	Medium	Low	Medium

3. Methodological Framework for Applying KE to Accessible Restrooms

By integrating artificial intelligence with KE, this research develops an innovative approach to accessible restroom design that prioritizes disability inclusion and sustainability. An integrated, mixed-methods approach combining qualitative and quantitative techniques underpins this investigation. Through questionnaire surveys and interviews with disabled users, nine emotional words were initially collected and extracted, and three representative emotional words were screened through factor analysis (FA) and clustering methods. Next, the study identified eight design features of accessible restrooms through morphological analysis, simplified these features using RST, and finally identified six key design factors with the most significant impact on users' emotional needs. The research employed SVM technology to establish a precise correlation model between emotional needs and design parameters. Generated design solutions were visually rendered, then evaluated by 50 participants with diverse disabilities to validate the approach's practical efficacy.

This specific research steps are as follows: First, the emotional responses of the disabled group to the use of accessible restroom were extracted through FA, and representative emotionalized words were screened from them. Accessible restroom design elements were analyzed through morphological decomposition. RST simplified these features to determine the core design features that ultimately significantly impact users' emotional needs. Finally, based on the analysis of the questionnaire data, the relationship matrix between emotional evaluation and design features was constructed using the Analytic Hierarchy Process (AHP). The SVM algorithm successfully generated an optimized emotion-to-design mapping system, with Figure 1 illustrating the complete research framework and implementation workflow.

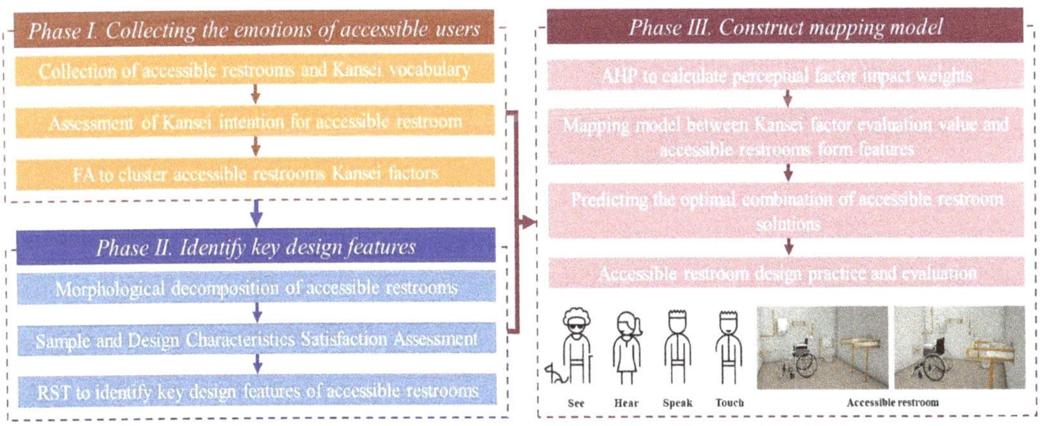

Figure 1. Proposed research structure for product biomimetic design.

3.1. Introduction to RST Methodology and Steps

Developed by Polish researcher Pawlak in 1982, RST represents a novel mathematical approach for processing ambiguous, contradictory, and partial datasets to determine approximate attribute sets in decision systems [44,45]. This methodology demonstrates particular efficacy in analyzing vague emotional preference data, enabling researchers to examine relationships between users' affective requirements and design characteristics while extracting implicit patterns and decision rules from complex datasets [46,47]. Within the attribute approximation framework, conditional attributes assigned zero weighting coefficients are identified as irrelevant to decision attributes, and can consequently be removed during the reduction procedure [48]. Therefore, this study adopts the RST attribute approximation algorithm to extract the critical design features of an accessible restroom. The following briefly describes the calculation process.

Step 1: An information system is formally defined as the quadruple $S = (U, A, V, f)$, where U represents the universe of discourse containing all evaluation instances; $A = C \cup D$ denotes the attribute set with $C \cap D \neq \emptyset$; $C \in A$ corresponds to conditional attributes (design parameters); $D \in A$ represents decision attributes (emotional assessments); V denotes the range of values of the attributes. The function f represents the mapping relationship between attributes and evaluation records, which can also be understood as the set of connections between evaluation records and attributes. This function describes the dependency between attribute values and evaluation records, providing a foundation for the subsequent attribute reduction process. f represents the information function that establishes correspondence between the universe U and attribute set A.

Step 2: The equivalence relation of R to U (as specified in Equation (1)) generates classification classes for design feature attributes, i.e., $U/IND(C)$, and the equivalence class of decision-making attributes (emotional evaluations), i.e., $U/IND(D)$. The equivalence class division based on the original set of features without considering specific conditional attributes is as $U/IND(D)$,

$$IND(R) = \{(x, y) \in U \times U | \forall a \in A, f(x, a) = f(y, a)\} \tag{1}$$

As formalized in Equation (1), the information function f assigns to each object–attribute pair $(x, a) \in U \times A$ its corresponding attribute value $f(x, a)$, completely characterizing the system's information structure. Here, A includes both conditional attributes and decision attributes. For example, in a dataset containing product features, the conditional attributes may include characteristics such as color and size, while the decision attribute may represent user satisfaction.

Step 3: The dependency measures are defined as follows. $r_c(D)$ quantifies how strongly decision attribute D relies on conditional attribute set C; $r_{c-|c_e|}(D)$ measures D's dependency on the reduced attribute set $C - \{c_e\}$; $\sigma(c_e)$, computed via Equation (2), represents the significance of attribute c_e for D.

$$\sigma(c_e) = r_c(D) - r_{c-|c_e|}(D) \tag{2}$$

Equation (2) formally defines the dependency measure $r_c(D)$ as the predictive capacity of conditional attribute set C for determining decision attribute D. This metric is computed through object similarity analysis based on attribute C. For example, if for a set of objects, those with identical values across all conditional attributes also share the same value for the decision attribute, we can consider the degree of dependence to be high.

Step 4: W_e denotes the condition attribute indicator weights—see Equation (3).

$$W_e = \frac{\sigma(c_e)}{\sum_{\lambda=1}^{n} \sigma(c_\lambda)} \tag{3}$$

3.2. Introduction to SVM Methodology and Steps

SVM constitutes a supervised learning algorithm primarily employed for binary classification and regression tasks [49]. Its fundamental operation involves identifying an optimal separating hyperplane within the feature space to maximize the margin between classes [50]. When the training samples are linearly indistinguishable, a nonlinear support vector machine is learned through kernel tricks and software interval maximization [51]. This study involves user emotions, which are uncertain, nonlinear, and obey a non-normal distribution [52]. Therefore, a nonlinear support vector machine is selected to model the Kansei mapping of the accessible restroom in this study, and the computational process is briefly described below.

Step 1: Let $\Phi(x)$ denote the eigenvector after mapping x. The model corresponding to dividing the hyperplane in the eigenspace can be represented as

$$f(x) = w \cdot \Phi(x) + b \tag{4}$$

Step 2: Based on the above equation, the original problem is transformed into

$$min\frac{1}{2} \parallel w \parallel^2 + C\sum_{i=1}^{n}\left[\xi_i + \xi_i^*\right] \tag{5}$$

$$s.t.\begin{cases} f(x_i) - y_i \le \varepsilon + \xi_i \\ y_i - f(x_i) \le \varepsilon + \xi_i \\ \xi_i \ge 0 \\ \xi_i^* \ge 0 \end{cases}, i = 1, 2, \cdots n \tag{6}$$

Step 3: The constrained optimization formulation is converted to a dual problem through Lagrangian relaxation, as formally expressed in Equation (7); the kernel function is applied to avoid encountering high-dimensional or infinite-dimensional problems (see Equation (8)), and Equation (9) is obtained after the solution.

$$max\left[-\frac{1}{2}\sum_{i=1}^{n}\sum_{j=1}^{n}\left(\alpha_i - \alpha_i^*\right)\left(\alpha_j - \alpha_j^*\right)K(x_i, x_j) - \varepsilon\sum_{i=1}^{n}\left(\alpha_i + \alpha_i^*\right) + \sum_{i=1}^{n} y_i\left(\alpha_i - \alpha_i^*\right)\right] \tag{7}$$

$$K(x_i, x) = \exp\left(-\gamma \cdot \parallel x_i - x \parallel^2\right) \tag{8}$$

$$f(x) = \sum_{i=1}^{n}\left(\alpha_i - \alpha_i^*\right)K(x_i, x) + b \tag{9}$$

4. Accessible Restrooms Design Study

4.1. Morphological Feature Samples and Emotional Word Collection

This study begins from the perspective of an accessible restroom and selects a sample for in-depth analysis. A wide range of 80 samples of accessible restrooms eligible for the study were collected through a specially designed research website. The samples used in this study encompass a variety of accessible restroom types, including differences in color, material, fixture shapes, and lighting intensity, with the goal of ensuring representativeness and diversity. The impacts of image quality on Kansei data validity were mitigated through expert review by five doctoral researchers in accessible environment design disciplines. Images that were blurry, obstructed, or excessively recognizable were excluded to reduce visual biases and ensure the integrity of participants' emotional responses.

The selection process prioritized both data sample diversity for machine learning training and computational efficiency considerations. It was essential to ensure that the sample size was neither too small nor too large, in order to avoid a decrease in the predictive performance or overfitting of the model. The final dataset comprised 60 representative accessible restroom images (6 × 10 matrix), forming the basis for emotional analysis and design optimization (Figure 2).

Figure 2. A database of 60 samples of accessible restrooms.

In the first stage of emotion word collection, 45 emotion words related to accessible restrooms were collected through user interviews, related literature, and design websites. In the second stage, focus groups were organized to pre-cull similar, functional, and repetitive words, and then a questionnaire was created to designate accessible groups to fill in, and nine groups of words with a voting rate of 60% or more were selected. Table 2 presents the randomized ordering and numbering of emotional vocabulary.

Table 2. Emotional words related to accessible restrooms.

Comfort	Spacious	Clean	Bright	Organized	Warmth	Durable	Safe	Convenient

4.2. Downscaling the Emotional Word of Accessible Restroom with FA

Focusing on the sample of accessible restrooms, 50 disabled users were selected for this study. To improve the experiment's efficiency and shorten the time of user participation, the research group suggested a three-point Likert scale to assess the emotional intention test used for the 60 samples. Through this test, a matrix of affective evaluations for accessible restrooms was collected and constructed (see Table 3). FA using IBM SPSS Statistics 27 assessed affective word consistency, with the results confirming data suitability—KMO = 0.857, and Bartlett's test of sphericity value = 271.684 (df = 36, $p < 0.01$) (see Table 4).

Table 3. Kansei evaluation of accessible restrooms.

No	Comfort	Spacious	Clean	Bright	Organized	Warmth	Durable	Safe	Convenient
1	2.80	1.97	2.50	2.37	2.33	2.43	2.27	2.40	2.40
2	2.43	2.53	2.60	2.47	2.50	2.07	2.43	2.33	2.30
3	2.70	2.80	2.67	2.50	2.43	2.53	2.50	2.60	2.47
...	...	...	...	...	...	...	...	...	...
58	2.40	2.23	2.47	2.43	2.17	2.27	2.27	2.17	2.30
59	2.47	2.10	2.27	2.43	2.27	2.40	2.20	2.17	2.43
60	2.50	2.77	2.63	2.47	2.33	2.50	2.40	2.40	2.47

Table 4. Factor analysis suitability tests.

KMO Sampling Adequacy Measure.		0.857
Bartlett's sphericity test	Appro. Chi-Square	274.684
	df	36
	Sig.	<0.01

Principal component analysis yielded three factors explaining 75.8% of total variance (exceeding the 75% criterion), as documented in Table 5. In the FA rotated component matrix (Table 6), things such as spacious, clean, and bright were classified as the "tidiness factor"; durable, safe, organized, and convenient were classified as the "utility factor"; and warmth and comfort were classified as the "care factor". After extracting the core user emotional words, the focus group discussed the rationality of the three refined and summarized emotion words, and finally identified the emotional words as tidiness factor, utility factor, and care factor, which were used for the subsequent construction of the emotion data set of accessible restroom users.

Table 5. FA total variance explained.

IE	Sum of Squared Loadings						Sum of Squared Rotated Loadings		
	Sum	**Var./%**	**Cum./%**	**Sum**	**Var./%**	**Cum./%**	**Sum**	**Var./%**	**Cum./%**
1	4.721	52.452	52.452	4.721	52.452	52.452	2.702	30.02	30.02
2	1.314	14.601	67.054	1.314	14.601	67.054	2.135	23.728	53.748
3	0.786	8.732	75.785	0.786	8.732	75.785	1.983	22.037	75.785
4	0.639	7.096	82.882						
5	0.459	5.097	87.979						
6	0.354	3.933	91.912						
7	0.29	3.225	95.137						
8	0.251	2.785	97.922						
9	0.187	2.078	100						

Table 6. FA rotated component.

Kansei Word	Component		
	1	**2**	**3**
1	0.038	0.201	0.845
2	0.867	0.209	0.079
3	0.829	0.177	0.202
4	0.733	0.243	0.180
5	0.505	0.632	0.367
6	0.240	0.119	0.862
7	0.528	0.616	−0.058
8	0.342	0.666	0.516
9	0.127	0.851	0.205

4.3. RST Defines Important Design Features

To ensure a rigorous evaluation of the restroom morphology database, we assembled a panel of five design experts with complementary expertise—one PhD candidate in architectural design (specializing in universal design), two professors of environmental design (with extensive experience in public space planning), and two Master's researchers in human-centered design (focusing on ergonomic evaluation). Experts were selected based on (1) academic qualifications in relevant design disciplines, (2) a minimum of five years of research or professional experience in spatial design, and (3) prior publications or projects related to accessibility or user-centered design. After initial restroom morphology organization, the expert group objectively evaluated the 60-sample database for core feature identification. Through expert deliberations, the samples were decomposed into eight design features: overall color, safety handrail, handrail material, floor material, room brightness, spatial layout, toilet elevation, and wash basin top side. The analysis yielded 41 distinct design variants across all features (Figure 3).

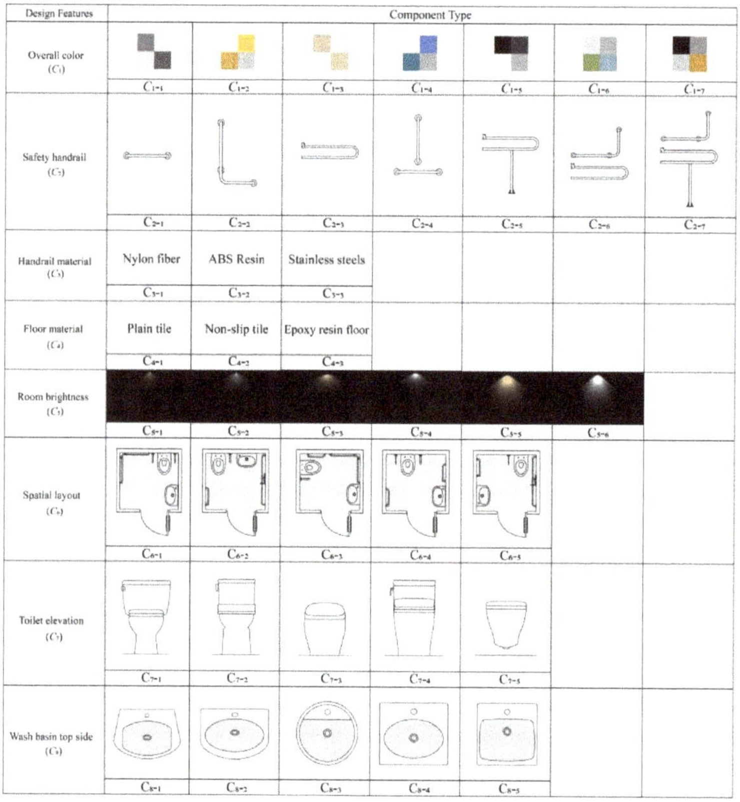

Figure 3. Morphological deconstruction of an accessible restroom.

Given the varying degrees of importance of different design features to user needs, lower-importance features may affect the prediction accuracy of subsequent models. Therefore, this study used the RST algorithm to approximate the attributes of eight design features to improve the recognition accuracy. Sixty restroom samples were systematically classified before conducting user evaluations. A demographically diverse panel of 50 disabled users (gender-balanced) assessed all samples and 41 derived design types through standardized five-point Likert measurements, with one indicating "not at all enhancing the willingness to use", three indicating "moderate", and five "significantly increased". Since the original emotional evaluation data are more continuous, it is necessary to discretize the constant data. To address potential concerns regarding information loss during discretization, we conducted a correlation analysis between the original continuous evaluation scores and their discretized counterparts. The Spearman's rank-order correlation test revealed a statistically significant strong positive relationship ($\rho = 0.89$, $p < 0.001$, 95% CI [0.85, 0.92]), indicating that the discretization process preserved the ordinal structure and discriminative power of the original data. This high correlation coefficient (exceeding the conventional 0.7 threshold for strong associations) empirically validates that the rough set reduction based on discretized data maintains the essential information integrity required for accurate feature selection. Therefore, based on the characteristics of the distribution of the evaluation data, the discretization method proceeds as follows: the evaluation level of interval [1, 2.08] is 1; the evaluation level of interval [2.08, 2.68] is 2; the evaluation level of interval [2.68, 3.28] is 3; the evaluation level of interval [3.28, 3.88] is 4; interval [3.88, 5] has an evaluation rating of 5. The RST conditional attribute, known as C, comprises eight design features; the decision attribute regarding user satisfaction with the accessible restroom is denoted as D. The transformed discrete values can be found in Table 7.

Table 7. Discrete decision table for accessible restrooms.

No.	Overall Color	Safety Handrail	Handrail Material	Floor Material	Room Brightness	Spatial Layout	Toilet Elevation	Wash Basin Top Side	D
1	3	4	2	2	3	3	4	5	2
2	4	5	5	5	3	4	5	2	3
3	4	2	4	4	3	2	4	3	4
...	...	...	...	...	...	...	...	...	...
58	3	5	4	4	3	5	4	3	3
59	5	3	5	2	1	4	4	3	3
60	3	3	2	4	4	5	4	4	2

The design feature weight coefficients are computed using the RST attribute approximation algorithm implemented in MATLAB R2024b¯$w_1 = 0.0.142857$, $w_2 = 0$, $w_3 = 0$, $w_4 = 0.142857$, $w_5 = 0.285714$, $w_6 = 0.142857$, $w_7 = 0.142857$, $w_8 = 0.142857$. To summarize, the expert panel carefully evaluated the design characteristics, excluding those with zero weighting values during approximation, and ultimately identified six essential features for accessible restroom design—overall color, floor material, room brightness, spatial layout, toilet elevation, and wash basin top side.

4.4. SVM Algorithm to Build Mapping Models

The study initially employed the three clustered emotional words and 60 samples to develop a five-point Likert scale. Fifty participants were then recruited to evaluate the perceived intention of each sample, followed by calculating the mean emotional assessment score. Secondly, this study used AHP hierarchical analysis [53,54] to calculate the three emotional word weight indicators after clustering, and the specific process is described below. Four members of the expert group (one PhD, two professors and one researcher) were again invited to participate in the hierarchical analysis of hierarchy (AHP) weighting assessment, and another postgraduate student acted as an independent supervisor for the process compliance review. The experts compared the feature importance two by two through two rounds of anonymous scoring (five-level scale method); the consistency ratios (CR) were controlled at <0.1, and the final weighting results were aggregated through a group decision matrix. Supervisors verified the data entry and calculation process to ensure that there was no conflict of interest and that the Delphi specification was met. The assessment details and final indicator weights of the 4 experts are shown in Table 8.

Table 8. AHP expert assessment rules and consistency tests.

Expert	Kansei Factor	Utility Factor	Tidiness Factor	Care Factor	Weight (W)	W	λ_{max}	C_I	C_R
1	Utility Tidiness Care	1 1/4 1/5	4 1 1/3	5 3 1	0.67 0.23 0.10	Utility Tidiness Care	3.08	0.04	0.08
2	Utility Tidiness Care	1 1/2 1/5	2 1 1/4	5 4 1	0.57 0.33 0.10	Utility Tidiness Care	3.02	0.01	0.02
3	Utility Tidiness Care	1 1/3 4	3 1 5	1/4 1/5 1	0.23 0.10 0.67	Utility Tidiness Care	3.08	0.04	0.08
4	Utility Tidiness Care	1 1 1/2	1 1 1/2	2 2 1	0.25 0.25 0.50	Utility Tidiness Care	3	0	0

According to the results, it can be seen that the test coefficients C_R of the four experts are less than 0.1, which satisfies the consistency test. In summary, in order to make the results of the weighted values more objective, it is necessary to summarize the weighted values of the four experts and find their mean values, concerning utility factor (0.43), tidiness factor (0.2275), and care factor (0.3425). The weights were multiplied with the corresponding emotional assessment means and then summed to obtain the emotional composite assessment value. A matrix (Table 9) was constructed between the emotional composite assessment values and the critical design features of accessible restrooms.

Table 9. Mapping model between restroom design features and Kansei factor evaluation value.

No.	Overall Color	Floor Material	Room Brightness	Spatial Layout	Toilet Elevation	Wash Basin Top Side	Evaluation Value
1	1	1	4	1	4	1	3.0337
2	4	2	4	2	3	5	2.9866
3	2	3	4	3	1	4	2.6042
...	...	...	...	...	...	...	...
58	1	3	4	4	1	2	3.2265
59	3	1	2	2	4	4	2.9953
60	1	3	3	4	4	3	2.9385

This study employs SVM for predicting user emotions. Although SVM is typically used for classification tasks, its extended version, Support Vector Regression (SVR), effectively handles regression tasks. In this study, user emotions are treated as continuous values, and therefore, SVR is selected for prediction. Depending on the research requirements, one or more Kansei mapping models can be constructed to achieve the final prediction. In the previous phase, to better align the results with users' diverse emotional needs, the AHP was used to calculate the weights of various emotional needs, ultimately yielding the final average emotional value. As a result, only one Kansei mapping model needs to be constructed for this study. The six key design features, after encoding, are used as input layer data, and the average emotional value, after encoding, serves as the output layer data corresponding to the key emotional needs.

To mitigate potential overfitting and bias due to limited dataset sharing, we implemented rigorous data preprocessing and validation techniques. The quantified emotional demands were analyzed using SVM in MATLAB R2024b, with the dataset first normalized (z-score) to minimize feature scaling effects. To ensure robustness, the data were partitioned into training (90%) and holdout test sets (10%), followed by stratified five-fold cross-validation on the training set to optimize hyperparameters (penalty coefficient C and RBF kernel width gamma). This approach prevents data leakage and provides a more reliable estimate of model generalizability. After iterative optimization, the best-performing parameters (C = 1, gamma = 5.6569) were selected based on cross-validated performance metrics. The test set converges to the better data parameter, the R^2, RMSE, MAE, and MBE parameters (0.99768, 0.087851, 0.0068429, 0.00048986). The model achieved strong training stability and low error variance, indicating minimal overfitting risk. Meanwhile, the remaining 10% samples of the mapping model were used as a test set to test the model performance with the test set's R^2, RMSE, MAE, and MBE parameters (0.93068, 0.08462, 0.08462, 0.08462)—these give SVM parameter results, as shown in Table 10. The close alignment between the cross-validation and the test results ($\Delta R^2 < 0.07$) suggests the model captures underlying patterns rather than noise, despite dataset size constraints. The mapping models a true value and actual value fitting plot, as shown in Figure 4.

Table 10. SVM parameter results.

Parameter	R^2	RMSE	MAE	MBE
Training set	0.99768	0.087851	0.0068429	0.00048986
Test set	0.93068	0.08462	0.08462	0.08462

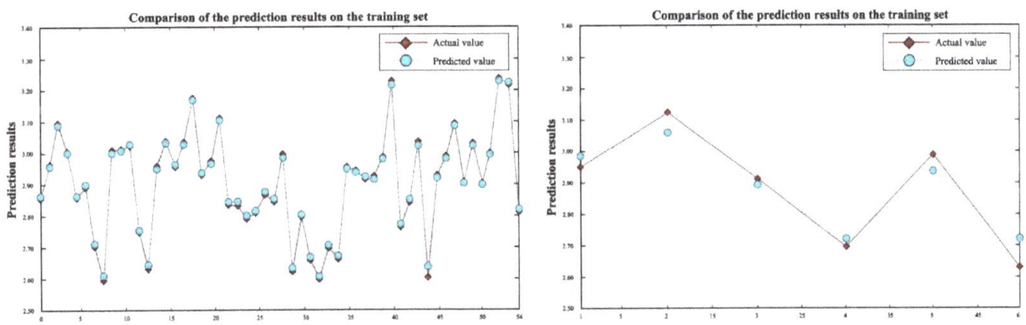

Figure 4. SVM mapping model's predictive set and test values set fitting plot.

The results demonstrate that the SVM-based emotional mapping model developed in this study exhibits strong predictive performance, with high accuracy, excellent fit, and minimal deviation between predicted and actual values. This model effectively establishes the relationship between special populations' needs and accessible restroom design features. Consequently, it identifies the optimal design combination that maximizes satisfaction among special populations—specifically, the configuration yielding the highest emotional evaluation score. There are $7 \times 3 \times 6 \times 5 \times 5 = 15{,}750$ combination scenarios for the five key design features obtained by using RST approximation. All the combination schemes are coded and brought into the trained SVM model to predict each combination scheme's corresponding emotional evaluation value, as shown in Table 11. Meanwhile, Table 11 highlights the top three coding schemes in terms of average value for designers' reference. The analysis reveals that Design Combination #4135 yields the highest predicted emotional value (3.187211) according to the model's evaluation, corresponding to the design features of overall color ($C_{1\text{-}2}$), floor material ($C_{4\text{-}3}$), room brightness ($C_{5\text{-}4}$), spatial layout ($C_{6\text{-}1}$), toilet elevation ($C_{7\text{-}2}$), and wash basin top side ($C_{8\text{-}5}$).

Table 11. SVM mapping model for accessible restroom Kansei predictive value.

No.	Overall Color	Floor Material	Room Brightness	Spatial Layout	Toilet Elevation	Wash Basin Top Side	Predictive Value
1	1	1	1	1	1	1	2.893248
2	1	1	1	1	1	2	2.861935
3	1	1	1	1	1	3	2.861935
...	...	...	...	...	...	...	...
1885	1	3	4	1	2	5	3.169948
...	...	...	...	...	...	...	...
4135	2	3	4	1	2	5	3.187211
...	...	...	...	...	...	...	...
4260	2	3	5	1	2	5	3.147104
...	...	...	...	...	...	...	...
15,748	7	3	6	5	5	3	2.877173
15,749	7	3	6	5	5	4	2.871681
15,750	7	3	6	5	5	5	2.877758

5. Results and Proposed Design Implementation

The algorithmic framework developed in this study underwent comprehensive evaluation through detailed analysis of the proposed design solutions. During morphological analysis, a comprehensive optimization was performed based on six critical design attributes: overall color ($C_{1\text{-}2}$), floor material ($C_{4\text{-}3}$), room brightness ($C_{5\text{-}4}$), spatial layout ($C_{6\text{-}1}$), toilet elevation ($C_{7\text{-}2}$), and washbasin top side ($C_{8\text{-}5}$). Beyond these algorithmically derived morphological features, three key emotional dimensions, namely, neatness, practicality, and caring, were also thoroughly integrated to ensure a balanced synergy between functional usability and users' psychological comfort. The conceptual scheme was developed in Cinema 4D R24, utilizing parametric modeling tools to construct the 3D structure from 2D CAD plans. Photorealistic rendering was achieved through Octane Render 2024, employing PBR materials with measured reflectance values and hybrid HDRi lighting. The final 4K output (Figure 5) was generated using path tracing with AI denoising, combining accurate material representation and optimized lighting simulation for design validation. The final result is the conceptual scheme shown in Figure 5. This conceptual scheme serves as a holistic reference for accessible restroom projects, yet it may require context-specific adjustments to accommodate distinct spatial and cultural factors.

A stratified purposive sample of 50 adults with disabilities from East China was recruited; 30 wheelchair users with mobility limitations, 8 individuals with visual impairments (3 totally blind, 5 low-vision), 6 with hearing impairments, and 6 with multiple disabilities. The gender ratio was 60% male to 40% female; ages ranged from 18 to 65 years (M = 37.4, SD = 11.2). Prior to rating, participants reviewed digital renderings of the proposed restroom; blind participants received step-by-step, concrete verbal descriptions to ensure information equivalence. They then rated overall satisfaction and three secondary attributes—neatness, practicality, and caring—on a seven-point Likert scale (1 = "not at all satisfied", 7 = "very satisfied").

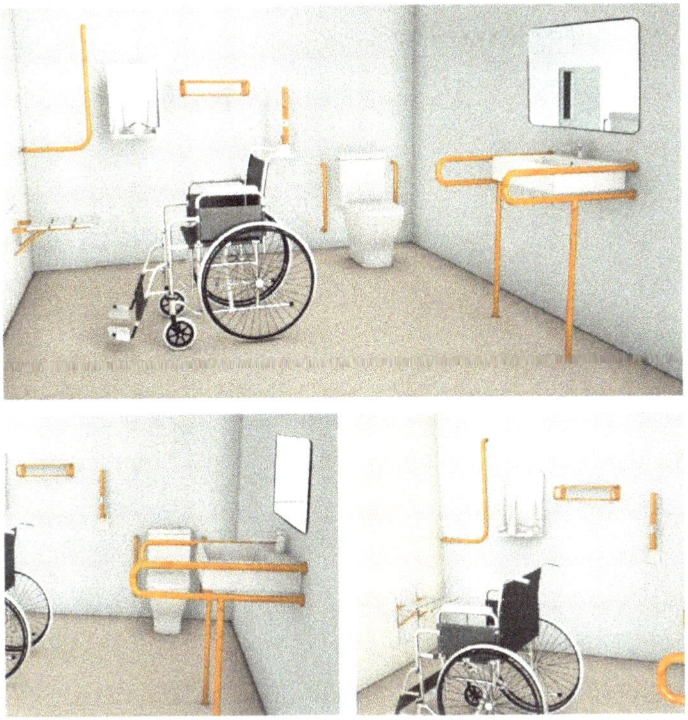

Figure 5. Optimized design solutions for accessible restroom.

Mean overall satisfaction reached 5.10 (SD = 0.93), exceeding the neutral midpoint of 4. Mean scores for neatness, practicality, and caring were 5.34, 5.06, and 5.02, respectively, indicating a broadly positive reception. Because the sample was limited to the East China region and was modest in size, and descriptive statistics alone were reported, no significance testing was conducted to compare disability groups or cultural backgrounds; the single-region sample may also reflect culture-specific expectations. Future studies should broaden geographical coverage through multi-site sampling and apply non-parametric or mixed-effects analyses to examine how disability type and regional culture modulate satisfaction, thereby validating the design for a wider disabled population.

6. Discussion

6.1. RST–SVM as a Science-Based Pathway to Emotion–Function Symmetry

The value of this study is not to merely add another machine learning layer, but to demonstrate how RST-guided dimensionality reduction and SVM's non-linear learning can jointly implement the principle of emotion–function symmetry, thereby providing a scientific design solution that satisfies both affective resonance and functional adequacy. In this research process, RST acts as a symmetry-preserving filter and SVM operates as a joint optimizer, together unifying emotional needs with technical performance. First, RST compresses the original attributes into the smallest subset that still contains key functional indicators and affective surrogates mapped from the Kansei word tidiness factor, utility factor and care factor, thus locking both dimensions at the feature level. The resulting vector is then fed into an RBF-kernel SVM, whose hyper-parameters are tuned via ten-fold cross-validation to prevent the model from biasing toward any single dimension. Empirical validation with fifty participants confirms that the top-ranked configuration attains a satisfaction score well above the neutral midpoint while meeting all accessibility standards, demonstrating that the emotion–function symmetry established in the workflow successfully translates into a tangible design solution.

6.2. Comparison of the Efficiency of Models for Extracting Key Elements

The RST attribute reduction algorithm adopted in this study demonstrates significant theoretical innovation and practical value in the multi-criteria decision-making process for accessible restroom design. Its core advantage lies in the formalized processing mechanism of knowledge granularity: through the mathematical definitions of lower and upper approximation sets, it systematically identifies redundancy and complementarity among design elements. This characteristic enables outstanding performance in scenarios with incomplete engineering data. For instance, in the face of missing user research data, RST can directly perform feature simplification based on knowledge granules without traditional interpolation. Experimental data confirm that the algorithm intelligently reduces eight-dimensional design features to six core attributes (information retention rate $\geq$ 98%) in just 15.7 milliseconds, improving efficiency by 63.7% compared to the analytic hierarchy process (AHP, 43.2 ms) and by 77.1% compared to the entropy weight method (68.5 ms). More importantly, unlike AHP—which relies on manual judgment matrices (requiring consistency tests with CR < 0.1)—and the entropy weight method—which is based on probabilistic assumptions—RST's data-driven nature not only avoids subjective bias, but also overcomes the technical limitations of traditional methods in handling incomplete information. This approach provides a more robust and automated solution for design feature optimization in complex, real-world engineering environments.

6.3. Comprehensive Emotional Evaluation and Model Efficiency Comparison

This research employed the SVM model to predict a unified score reflecting users' composite emotional evaluations. Consolidating multiple emotional factors into a single metric simplifies the mapping between restroom design features and user satisfaction, improving computational efficiency. A potential alternative would be to develop multiple SVM-based models, each targeting a distinct emotional dimension, to provide more granular insights into the influence of each design element. However, such a strategy could substantially raise computational complexity and complicate subsequent design recommendations. This study compared the mean error of constructing three affective models (R^2 = 0.89524, RMSE = 0.19472) versus a single affective model (R^2 = 0.96418, RMSE = 0.0862355) prior to developing the SVM model. The results demonstrate that this accessible restroom scenario is more suitable for a unified perceptual mapping model. Adopting a single SVM model in this study was a deliberate decision made in order to achieve comprehensive emotional predictions while keeping computational demands manageable. Although multiple models might offer finer distinctions between emotional attributes, our unified approach adequately encapsulates the primary emotional components, thereby lowering overfitting risks and streamlining implementation.

In this study, the superior performance of SVM in emotion–function mapping modeling essentially reflects the principle of structural risk minimization in statistical learning theory. Meanwhile, the comparative experiments revealed that the SVM model with the RBF kernel achieved a prediction accuracy of 0.96418, significantly outperforming the CNN (R^2 = 0.9380), BPNN (R^2 = 0.9439), and F-QFD method. Table 12 provides a detailed comparison of the performance of each model on the training and test sets.

Table 12. Comparison of mean values of parameter results for Kansei mapping models.

Model	R^2	RMSE	MAE	MBE
SVM	0.9642	0.0862	0.0457	0.0426
BPNN	0.9439	0.1545	0.1285	0.1495
CNN	0.9380	0.1600	0.1340	0.0250

In-depth analysis reveals the three-fold nature of its advantages, as follows: (1) Computational efficiency—SVM only takes 2.7 min to complete the training, which is 12.9% faster than CNN (3.1 min) and 22.9% faster than BPNN (3.5 min). (2) Model complexity level—SVM is known for efficient optimization (especially with the RBF kernel on moderate-sized datasets). BPNN and CNN require more epochs and hyperparameter tuning (e.g., learning rate, batch size), increasing training time. MBE shows that SVM (0.0426) has lower systematic bias than BPNN (0.1495) and CNN (0.0250), suggesting more stable predictions. (3) The generalization performance exhibits the smallest generalization gap (training set–test set R^2 difference 0.067), outperforming BPNN (0.0792) and CNN (0.0823). It is particularly noteworthy that although BPNN and CNN outperform the training set, their test set performance drops sharply, exposing typical overfitting characteristics. SVM, on the other hand, achieves optimal generalization stability while maintaining prediction accuracy through a structural risk minimization mechanism. These findings provide a methodological paradigm with both theoretical rigor and engineering practicality for use in intelligent decision-making systems in buildings.

6.4. *Pathways Toward Explainability and Cross-Context Generalization*

Although the existing KE–RST–SVM workflow exhibits high predictive validity, stakeholders cannot fully trust the model because of its black-box nature. As a next step, explainable-AI (XAI) techniques—specifically SHAP value decomposition and counterfactual sensitivity analysis—will be introduced to reveal the marginal contribution of each spatial attribute to the affective prediction scores. This transparency is expected to support evidence-based decision-making by designers and to empower users with disabilities to engage in critique and co-creation, thereby advancing participatory accessibility research. A second limitation is that the perceptual lexicon was derived from a Mandarin-speaking East China cohort, and its semantic loadings may vary across languages and cultures. To improve semantic robustness, a multilingual affective lexicon will be constructed, and multi-group confirmatory factor analysis will be employed to test metric equivalence across at least three geo-cultural contexts (North China, South China, and Southwest China). With respect to scenario scalability, accessible restrooms share structural homology with other high-dependency public facilities such as outpatient clinics, special-education classrooms, and inter-modal transit hubs. We therefore plan to adapt the trained SVM model to these settings via a transfer-learning protocol that (i) freezes lower-level feature weights, (ii) fine-tunes higher-level layers on a small, locale-specific dataset, and (iii) validates performance through a mixed-method user study. This strategy will both test the robustness of the proposed workflow and broaden its social impact. Finally, to control for geographic and demographic heterogeneity, a longitudinal multi-center trial has been scheduled in Shanghai, Guangzhou, and Chengdu. Each site will recruit at least 80 participants, stratified by disability type, gender, and age, and a stratified mixed-effects model will be used to distinguish design fixed effects from site-specific random effects, thereby providing generalizable design guidelines for local policymaking.

7. Conclusions

7.1. Key Findings and Contributions

Through surveys and expert interviews, nine core emotional descriptors regarding accessible restroom use were identified, from which factor analysis and clustering extracted three representative emotional dimensions: tidiness, utility, and care. Morphological analysis followed by RST simplification distilled eight design features into six pivotal elements. An SVM-based mapping model connecting these features with user emotional responses demonstrated satisfactory predictive accuracy, as confirmed by experimental validation.

By integrating KE with data-driven intelligent algorithms, this research makes several contributions to the field of accessible design, as follows:

(a) Methodological framework. The proposed system bridges the gap between user emotional requirements and design practice, thereby enhancing precision and providing scientific underpinnings for barrier-free facility design;
(b) By integrating cutting-edge AI methods, the decision support system enhances design choices, bringing advanced computer science and data-driven insights into accessible design practices;
(c) Novel RST–SVM integration. By fusing RST-based attribute reduction with SVM modeling, we offer a robust barrier-free design process that addresses diverse emotional needs, thus fostering inclusiveness and user satisfaction.

7.2. Broader Implications and Limitations

Beyond facilitating individual well-being for disabled users, this method significantly contributes to broader social inclusion and emphasizes design's pivotal role in promoting equity and dignity. In addition, it aligns with sustainability principles, minimizing excessive trial and error and resource consumption through data-driven methodologies.

Nevertheless, several limitations merit attention. (1) The study's participant pool was limited in size, particularly with only 50 participants included in the analysis, and lacked sufficient demographic diversity such as variations in age and disability type, potentially affecting the generalizability of findings. Future research should expand recruitment to ensure broader representation. (2) User preferences and emotional responses were captured through subjective scales or questionnaires, which may lack temporal precision or fail to account for subconscious bias. Integrating physiological instruments such as EEG, eye-tracking, or galvanic skin response for continuous emotion monitoring could offer finer-resolution insights into stress triggers or comfort levels. (3) Although SVR demonstrated efficacy, its performance may plateau with highly nonlinear or high-dimensional data. Hybrid models combining SVR with deep learning for feature extraction or emerging techniques like federated learning to address data privacy in multi-site studies warrant further exploration. (4) The current design framework assumes static user needs, yet disabilities often involve progressive conditions or fluctuating symptoms. Longitudinal studies tracking user interactions with dynamic adjustable systems such as IoT-enabled fixtures would strengthen practical relevance. By addressing these areas, subsequent investigations can refine the proposed approach and continue to elevate the standard of accessible restroom design in the pursuit of symmetrical, user-centered solutions.

Author Contributions: Conceptualization, Z.C., J.T., H.Z. and D.W.; formal analysis, Z.C.; funding acquisition, H.Z.; investigation, J.T.; methodology, J.T. and D.W.; project administration, H.Z.; resources, H.Z.; software, Z.C.; supervision, H.Z. and D.W.; validation, Z.C.; visualization, J.T.; writing—original draft, Z.C. and J.T.; writing—review and editing, D.W. All authors have read and agreed to the published version of the manuscript.

Funding: This research was funded by the Tracking Program for Specially Appointed Professors (Oriental Scholars) in Shanghai Universities, grant number 1400152019.

Data Availability Statement: The original contributions presented in the study are included in the article; further inquiries can be directed to the corresponding authors.

Acknowledgments: All authors would like to thank the anonymous referees for their constructive comments.

Conflicts of Interest: The authors declare no conflicts of interest.

References

1. Iwarsson, S.; Ståhl, A. Accessibility, usability and universal design—Positioning and definition of concepts describing person-environment relationships. *Disabil. Rehabil.* **2003**, *25*, 57–66. [CrossRef]
2. Shinohara, K.; Wobbrock, J.O.; Pratt, W. Incorporating social factors in accessible design. In Proceedings of the 20th International ACM SIGACCESS Conference on Computers and Accessibility, Galway, Ireland, 22–24 October 2018; pp. 149–160.
3. Quintero, C. A review: Accessible technology through participatory design. *Disabil. Rehabil. Assist. Technol.* **2022**, *17*, 369–375. [CrossRef] [PubMed]
4. Sarmiento-Pelayo, M.P. Co-design: A central approach to the inclusion of people with disabilities. *Rev. Fac. Med.* **2015**, *63*, 149–154. [CrossRef]
5. Lid, I.M. Universal design and disability: An interdisciplinary perspective. *Disabil. Rehabil.* **2014**, *36*, 1344–1349. [CrossRef] [PubMed]
6. Nagamachi, M. Kansei engineering: A new ergonomic consumer-oriented technology for product development. *Int. J. Ind. Ergon.* **1995**, *15*, 3–11. [CrossRef]
7. Pawlak, Z.; Skowron, A. Rudiments of rough sets. *Inf. Sci.* **2007**, *177*, 3–27. [CrossRef]
8. Liu, Z.; Wu, J.; Chen, Q.; Hu, T. An improved Kansei engineering method based on the mining of online product reviews. *Alex. Eng. J.* **2023**, *65*, 797–808. [CrossRef]
9. Kang, X.; Nagasawa, S.y. Integrating kansei engineering and interactive genetic algorithm in jiangxi red cultural and creative product design. *J. Intell. Fuzzy Syst.* **2023**, *44*, 647–660. [CrossRef]
10. Su, Z.; Yu, S.; Chu, J.; Zhai, Q.; Gong, J.; Fan, H. A novel architecture: Using convolutional neural networks for Kansei attributes automatic evaluation and labeling. *Adv. Eng. Inform.* **2020**, *44*, 101055. [CrossRef]
11. Gan, Y.; Ji, Y.; Jiang, S.; Liu, X.; Feng, Z.; Li, Y.; Liu, Y. Integrating aesthetic and emotional preferences in social robot design: An affective design approach with Kansei Engineering and Deep Convolutional Generative Adversarial Network. *Int. J. Ind. Ergon.* **2021**, *83*, 103128. [CrossRef]
12. Li, T. Construction and Optimization of an Artificial Intelligence-Assisted Kansei Engineering Product Design Mapping Model. In Proceedings of the 2024 5th International Conference on Education, Knowledge and Information Management (ICEKIM 2024), Chengdu, China, 19–21 April 2024; pp. 920–927.
13. Cherkassky, V.; Ma, Y. Practical selection of SVM parameters and noise estimation for SVM regression. *Neural Netw.* **2004**, *17*, 113–126. [CrossRef] [PubMed]
14. Kang, X. Combining rough set theory and support vector regression to the sustainable form design of hybrid electric vehicle. *J. Clean. Prod.* **2021**, *304*, 127137. [CrossRef]
15. Montañana, A.; Nolé, M.L.; Llinares, C. Strategic design approaches for eliciting the perception of 'prestige' in housing consumers. *Buildings* **2024**, *14*, 853. [CrossRef]
16. Cheng, Q.; Wang, H.; Xu, X.; He, T.; Chen, Z. Indoor Thermal Comfort Sector: A Review of Detection and Control Methods for Thermal Environment in Livestock Buildings. *Sustainability* **2024**, *16*, 1662. [CrossRef]
17. Han, M.; Zhang, C.; Yin, S.; Jia, J.; Kim, C. Optimization Design of Indoor Environmental Ventilation in Buildings Based on Improved SVR-PSO Model. *Sustainability* **2024**, *16*, 5256. [CrossRef]
18. Shao, Z.; Chen, J.; Zeng, H.; Hu, W.; Xu, Q.; Zhang, Y. A new approach to interior design: Generating creative interior design videos of various design styles from indoor texture-free 3D models. *Buildings* **2024**, *14*, 1528. [CrossRef]
19. Goldman, C.D. Architectural barriers: A perspective on progress. *W. N. Eng. L. Rev.* **1982**, *5*, 465.
20. Hymas, D.R.; Parkinson, B.R. Architectural barriers under the ADA: An answer to the judiciary's struggle with technical non-compliance. *Cal. WL Rev.* **2002**, *39*, 349.
21. Petersen, J.C.; Piletic, C.K. Facility accessibility: Opening the doors to all. *J. Phys. Educ. Recreat. Danc.* **2006**, *77*, 38–44. [CrossRef]
22. Persson, H.; Åhman, H.; Yngling, A.A.; Gulliksen, J. Universal design, inclusive design, accessible design, design for all: Different concepts—One goal? On the concept of accessibility—Historical, methodological and philosophical aspects. *Univers. Access Inf. Soc.* **2015**, *14*, 505–526. [CrossRef]

23. Story, M.F. Maximizing usability: The principles of universal design. *Assist. Technol.* **1998**, *10*, 4–12. [CrossRef] [PubMed]
24. Yuan, H.; Shen, X. Optimizing floor plans of accessible restrooms in elderly long-term care facilities: A path planning approach. *Archit. Eng. Des. Manag.* **2024**, *20*, 891–910. [CrossRef]
25. Swayne, M.R.; Calzo, J.P.; Felner, J.K.; Welsh Carroll, M. Developing evidence for building sanitation justice: A multi methods approach to understanding public restroom quantity, quality, accessibility, and user experiences. *PLoS ONE* **2023**, *18*, e0288525. [CrossRef] [PubMed]
26. Zallio, M.; Clarkson, P.J. Inclusion, diversity, equity and accessibility in the built environment: A study of architectural design practice. *Build. Environ.* **2021**, *206*, 108352. [CrossRef]
27. Arora, S.; Deshpande, A. Inclusive design—Designing barrier-free public spaces. In *Design for Tomorrow—Volume 1: Proceedings of ICoRD 2021*; Springer: Singapore, 2021; pp. 133–146.
28. Nagamachi, M. Kansei engineering as a powerful consumer-oriented technology for product development. *Appl. Ergon.* **2002**, *33*, 289–294. [CrossRef]
29. Yang, C.; Liu, F.; Ye, J. A product form design method integrating Kansei engineering and diffusion model. *Adv. Eng. Inform.* **2023**, *57*, 102058. [CrossRef]
30. Fu, L.; Lei, Y.; Zhu, L.; Lv, J. An evaluation and design method for Ming-style furniture integrating Kansei engineering with particle swarm optimization-support vector regression. *Adv. Eng. Inform.* **2024**, *62*, 102822. [CrossRef]
31. Ge, Y.; Wang, S.; Han, R.; Peng, J.; Zhang, Z.; Hong, Y.; Yang, Y. Application of Kansei Engineering in aircraft design. *Ind. Textila* **2023**, *74*, 534–541. [CrossRef]
32. Wu, L.; Gao, H.; Wang, K.-C.; Yang, C.-H. A Green-IKE Inference System Based on Grey Neural Network Model for Humanized Sustainable Feeling Assessment about Products. *Math. Probl. Eng.* **2020**, *2020*, 6391463. [CrossRef]
33. Schütte, S.T.; Eklund, J.; Axelsson, J.R.; Nagamachi, M. Concepts, methods and tools in Kansei engineering. *Theor. Issues Ergon. Sci.* **2004**, *5*, 214–231. [CrossRef]
34. Ismail, M.N.; Lokman, A.M.; Abdullah, N.A.S. Formulating Kansei concept of assistive device for people with physical disabilities. In Proceedings of the 2014 3rd International Conference on User Science and Engineering (i-USEr), Shah Alam, Malaysia, 2–5 September 2014; pp. 30–35.
35. Lokman, A.M.; Ismail, M.N.; Abdullah, N.A.S.; Omar, A.R. Kansei wheelchair design based on KJ method. *Adv. Sci. Lett.* **2017**, *23*, 4349–4353. [CrossRef]
36. Prabaswari, A.; Fernanda, M.; Jati, N.; Indrawati, S.; Aldhiza, R. Redesign of wudhu facility for disabilities using integrated kansei method. *IOP Conf. Ser. Mater. Sci. Eng.* **2020**, *722*, 012016. [CrossRef]
37. Bonenberg, A. Aiding Self-reliance of the elderly and the disabled-modular cupboard with mobile internal units. In *Universal Access in Human-Computer Interaction. Access to the Human Environment and Culture, Proceedings of the 9th International Conference, UAHCI 2015, Held as Part of HCI International 2015, Los Angeles, CA, USA, 2–7 August 2015*; Proceedings, Part IV 9; Springer: Cham, Switzerland, 2015; pp. 403–412.
38. Tian, J.; Chen, Z.; Yuan, L.; Zhou, H. Optimizing Outdoor Micro-Space Design for Prolonged Activity Duration: A Study Integrating Rough Set Theory and the PSO-SVR Algorithm. *Buildings* **2024**, *14*, 3950. [CrossRef]
39. Aparicio-Ruiz, P.; Barbadilla-Martín, E.; Guadix, J.; Nevado, J. Analysis of Variables Affecting Indoor Thermal Comfort in Mediterranean Climates Using Machine Learning. *Buildings* **2023**, *13*, 2215. [CrossRef]
40. Yang, Y.; Yang, H.; Meng, F. A Bluetooth Indoor Positioning System Based on Deep Learning with RSSI and AoA. *Sensors* **2025**, *25*, 2834. [CrossRef]
41. Anan, M.; Kanaan, K.; Benhaddou, D.; Nasser, N.; Qolomany, B.; Talei, H.; Sawalmeh, A. Occupant-Aware Energy Consumption Prediction in Smart Buildings Using a LSTM Model and Time Series Data. *Energies* **2024**, *17*, 6451. [CrossRef]
42. Xiang, L.; Xu, Y.; Cui, J.; Liu, Y.; Wang, R.; Li, G. GM(1,1)-Based Weighted K-Nearest Neighbor Algorithm for Indoor Localization. *Remote Sens.* **2023**, *15*, 3706. [CrossRef]
43. Zhao, J.; Suo, W. Research on the Construction and Application of a SVM-Based Quantification Model for Streetscape Visual Complexity. *Land* **2024**, *13*, 1953. [CrossRef]
44. Pawlak, Z. Rough sets. *Int. J. Comput. Inf. Sci.* **1982**, *11*, 341–356. [CrossRef]
45. Pawlak, Z. Rough set theory and its applications. *J. Telecommun. Inf. Technol.* **2002**, *3*, 7–10. [CrossRef]
46. Shieh, M.-D.; Yeh, Y.-E.; Huang, C.-L. Eliciting design knowledge from affective responses using rough sets and Kansei engineering system. *J. Ambient. Intell. Humaniz. Comput.* **2016**, *7*, 107–120. [CrossRef]
47. Chen, Z. An Elderly-Oriented Form Design of Low-Speed New Energy Vehicles Based on Rough Set Theory and Support Vector Regression. *World Electr. Veh. J.* **2024**, *15*, 389. [CrossRef]
48. Chen, Z. An Elderly-Oriented Design of HMI in Autonomous Driving Cars Based on Rough Set Theory and Backpropagation Neural Network. *IEEE Access* **2024**, *12*, 26800–26818. [CrossRef]
49. Chaabane, S.B.; Hijji, M.; Harrabi, R.; Seddik, H. Face recognition based on statistical features and SVM classifier. *Multimed. Tools Appl.* **2022**, *81*, 8767–8784. [CrossRef]

50. Yue, S.; Li, P.; Hao, P. SVM classification: Its contents and challenges. *Appl. Math.-A J. Chin. Univ.* **2003**, *18*, 332–342. [CrossRef]
51. Shuo, W.; Ming, M. Exploring online intelligent teaching method with machine learning and SVM algorithm. *Neural Comput. Appl.* **2022**, *34*, 2583–2596. [CrossRef]
52. Yang, C.-C. A classification-based Kansei engineering system for modeling consumers' affective responses and analyzing product form features. *Expert Syst. Appl.* **2011**, *38*, 11382–11393. [CrossRef]
53. Zuo, Y.; Wang, Z. Subjective product evaluation system based on Kansei Engineering and analytic hierarchy process. *Symmetry* **2020**, *12*, 1340. [CrossRef]
54. Du, Y.; Zhang, M.; Cai, M.; Park, K. Construction of Product Appearance Kansei Evaluation Model Based on Online Reviews and FAHP: A Case Study of Household Portable Air Conditioners. *Sustainability* **2024**, *16*, 3132. [CrossRef]

MDPI AG
Grosspeteranlage 5
4052 Basel
Switzerland
Tel.: +41 61 683 77 34

Buildings Editorial Office
E-mail: buildings@mdpi.com
www.mdpi.com/journal/buildings

www.ingramcontent.com/pod-product-compliance
Lightning Source LLC
Chambersburg PA
CBHW061608220926
51850CB00005B/456
* 9 7 8 3 7 2 5 8 5 9 4 5 0 *